BASIN ANALYSIS IN PETROLEUM EXPLORATION

BASIN ANALYSIS IN PETROLEUM EXPLORATION

A case study from the Békés basin, Hungary

edited by

PAUL G. TELEKI
U.S. Geological Survey, Reston, Virginia, U.S.A.

ROBERT E. MATTICK
U.S. Geological Survey, Reston, Virginia, U.S.A.

and

JÁNOS KÓKAI
Mining Bureau of Hungary, Budapest
formerly Hungarian Oil and Gas Trust, Budapest, Hungary

KLUWER ACADEMIC PUBLISHERS
DORDRECHT / BOSTON / LONDON

Library of Congress Cataloging-in-Publication Data

Basin analysis in petroleum exploration : a case study from the Békés basin, Hungary / edited by Paul G. Teleki, Robert E. Mattick, and János Kókai.
p. cm.
Includes index.
ISBN 0-7923-3014-5 (acid-free paper)
1. Petroleum--Geology--Bekes Basin (Hungary and Romania)
2. Petroleum--Prospecting--Bekes Basin (Hungary and Romania)
I. Teleki, P. G. (Paul G.), 1930- . II. Mattick, Robert E.
III. Kókai, János.
TN271.P4B37 1994
553.2'8'09439--dc20 94-22321

ISBN 0-7923-3014-5

Published by Kluwer Academic Publishers,
P.O. Box 17, 3300 AA Dordrecht, The Netherlands.

Kluwer Academic Publishers incorporates
the publishing programmes of
D. Reidel, Martinus Nijhoff, Dr W. Junk and MTP Press.

Sold and distributed in the U.S.A. and Canada
by Kluwer Academic Publishers,
101 Philip Drive, Norwell, MA 02061, U.S.A.

In all other countries, sold and distributed
by Kluwer Academic Publishers Group,
P.O. Box 322, 3300 AH Dordrecht, The Netherlands.

Printed on acid-free paper

Printed in the Netherlands

TABLE OF CONTENTS

FOREWORD

Basin analysis is a comprehensive approach to decipher the geological evolution of a basin by constructing the stratigraphic framework and facies relationships, by mapping and understanding the geological structures in light of their historical development, and by examining the physical and chemical properties of rocks and their sedimentological and petroleum attributes. The purpose of the analysis is to identify potential source rocks and reservoir rocks, and to evaluate the maturation, generation, migration, and entrapment of hydrocarbons in a sedimentary basin.

This book is a compendium of chapters reporting on the results of a basin analysis of the Békés basin, a subbasin of the Pannonian Basin system in Hungary. The study, carried out in 1986-87, was conducted at the request of the Hungarian Oil and Gas Company (then called OKGT, and now MOL Ltd.) and the World Bank as part of a broader program designed to improve oil and gas exploration in Hungary. The work was accomplished by a joint effort of the Hungarian Oil and Gas Company and the U.S. Geological Survey.

The original evaluation of the Békés basin contained information on possible structural and stratigraphic targets, but for reasons of confidentiality, this information is omitted in this volume.

The Békés basin covers an area of about 4000 km^2 in southeastern Hungary and continues eastward into Romania (see Fig. 2, Kovács and Teleki, this volume). In Hungary, the basin is bounded on the north, west and south by buried basement highs composed of igneous and metamorphic rocks that range in age from Precambrian to Mesozoic. Seventeen oil and natural gas fields are associated with these peripheral basement highs. The thickness of Neogene and younger sedimentary deposits in the basin varies from 1000-2600 m along its periphery to as thick as 6500 m in central parts of the basin. The area has an average surface elevation of 85-90 m above sea level and is part of the Great Hungarian Plain. As this name implies, topographic relief is minor.

The Békés basin is one of many small subbasins in the larger Pannonian Basin (Royden and Horváth, 1988), most of which are bounded in the subsurface by basement blocks. The Pannonian Basin system, composed of these subbasins, is a Mediterranean back-arc basin that formed behind the Carpathian mountain chain (Horváth, 1988). The subbasins developed as a result of relatively rapid differential subsidence of thinned crust, beginning in mid-Miocene time and continuing to the present (Stegena and others, 1975; Horváth and Royden, 1981; Royden, 1988). The relatively undeformed cover of mid-Miocene to Recent sedimentary rocks, as thick as 7 km, hides a basement complex that was extensively deformed by Mesozoic thrusting and subsequently by a complex system of normal and wrench faults (Channel and Horváth, 1976; Horváth and Royden, 1981; Royden and others, 1982; Balla, 1985, 1987; Royden, 1988). The structural complexity of the region is the result of Cretaceous to Miocene convergence and collision of the European plate with several small continental fragments to the south (Royden and Báldi, 1988). These tectonic processes have strongly affected the occurrence of mineral resources, including oil and natural gas (Pogácsás, Mattick, Tari, and others, this volume).

Worldwide interest in the Pannonian Basin by geoscientists increased during the past two decades. The reasons are four-fold: (1) the desire to increase the exchange of information between scientists of the West and those of Central and Eastern Europe; (2) the interest in a unique geological basin used as a model in the classification of basins (Bally and Snelson, 1987); (3) the increasing interest in testing and probing concepts of the earth's evolution on regional, as well as global scales; and (4) the economic need in Central and Eastern Europe to develop mineral resources requiring the use of modern methods available through using Western sources. An example of (1) can be found in Killényi and Teleki (1985), (2) in Földváry (1988), (3) in Royden and Horváth (1988), and (4) in this volume.

In this volume, we summarize the petroleum geology of the Békés basin with respect to its geological setting in the Pannonian Basin. In mid-1991, petroleum production in Hungary amounted to 1.89×10^3 metric tons/day of oil and 5.2×10^6 m^3/day of natural gas, with the Békés basin contributing about 3% of the total oil production and 11.5% of the natural gas production. Although Hungary's production of oil and gas is small by world standards, the country is able to satisfy approximately 25% of its oil and 50% of its natural gas demand - significant percentages for a small country in the midst of economic change.

The Pannonian Basin is highly explored by world standards. Exploration began in 1915, when the Eötvös torsion balance was used to measure the earth's gravitational field (Kovács and Teleki, this volume). Since that time, more than 11,000 km of drilling in wildcat, stepout, and production wells have been accomplished, and over 40,000 km of multifold seismic-reflection profiles have been collected in Hungary. Several areas, however, remain relatively unexplored, such as the Mesozoic section throughout Hungary, and the Paleogene basin in north Hungary (Pogácsás, Mattick, Tari, and others, this volume).

Much of the present hydrocarbon production of Hungary is from middle Miocene (Badenian and Sarmatian), synrift sedimentary rocks and middle Miocene-Pliocene (Pannonian *sensu lato*), postrift sedimentary rocks. The synrift sedimentary rocks represent two cycles of sedimentation separated by an unconformity at the margins of the basin (Szentgyörgyi and Teleki, this volume). Early Badenian time is characterized by a continuous transgression of shallow, marine waters, and late Badenian time by a regression. From the beginning of Sarmatian time, the Pannonian Basin was isolated from the Eastern Paratethys and the salinity of the basin's waters decreased (Szentgyörgyi and Teleki, this volume). Following a regression of the sea in late Sarmatian-early Pannonian (*sensu lato*) time, subsidence accelerated in the area of the Békés basin and postrift sediments were deposited in a deep, lacustrine environment (Mattick and others, this volume). The postrift sedimentary rocks of Hungary are probably developed best in the Békés basin, where synrift sediments are relatively thin (Grow and others, this volume; Mattick and others, this volume).

Interpretation of cores, well logs and seismic-reflection profiles from the 6500-m-thick postrift section indicates that the Neogene postrift history of the Békés basin was one of the last episodes of filling of the Pannonian basin. Mapping of seismic sequences shows that, during this time, the river systems had advanced to the area of the Békés basin, and then contributed to a rapid deltaic filling of an initially wide and relatively deep lake. With time, sediment deposition became more and more localized as the areal extent of the progressively shallowing basin became restricted (Mattick and others, this volume). The sedimentary rocks and sedimentary processes are related to 3 major facies of lacustrine sedimentation (Phillips and others, this volume; Molenaar and others, this volume): (1) basal calcareous to silty organic-rich marl, (2) prodelta turbidites, and (3) delta-slope deposits. The relative chronostratigraphy of late Miocene and Pliocene deposits in the Békés basin can be correlated with deposition in the larger Pannonian Basin by the correlation of high-resolution magnetic polarity records from 4 continuously-cored drill holes (Elston and others, this volume; Pogácsás, Mattick, Elston, and others, this volume). This work, however, suggests that the chronostratigraphic framework and the model for timing of the accumulation of late Miocene and Pliocene deposits in the Pannonian Basin needs revision.

Geochemical analyses indicate that synrift (Badenian and Sarmatian) and postrift (lower Pannonian) shales and marls in the deeper parts of the Békés basin and adjacent basins are the probable source of the natural gas and oil produced in the Békés basin (Szalay, 1988; Clayton and Koncz, this volume; Clayton and others, this volume). These rocks contain predominantly type III (gas-prone) kerogen and it is not surprising, therefore, that chiefly natural gas is produced in the Békés basin with only minimal amounts of oil. The source rocks are inferred to have passed into the oil generation window about 8 million years ago when, in the deep troughs, fluid migration was downward and/or lateral due to overpressuring (Spencer and others, this volume). Upward migration of oil and gas appears to have occurred mainly in areas of maximum basement relief on the periphery of the Békés basin where the oil and natural gas fields have been discovered.

Reservoir rocks range in age from Precambrian to late Pliocene in the Békés basin. In the deepest reservoirs, fractured igneous and metamorphic rocks of the basement complex and fractured and brecciated synrift sedimentary rocks form a common oil and gas reservoir (Spencer and others, this volume; Kovács and Teleki, this volume). Above this zone, oil and natural gas are produced from reservoirs composed of lower Pannonian conglomerate, calcareous marl, and sandstone. Small amounts of natural gas are also produced from upper Pannonian sandstones; part of this gas is of biogenic origin (Clayton and Koncz, this volume).

Production of oil and natural gas is from relatively simple structures, namely compaction anticlines cored by basement rocks, in the Békés basin. In other areas of Hungary, however, production of oil and natural gas is from a wide variety of structural, stratigraphic, and combination traps (Pogácsás, Mattick, Tari, and others, this volume).

Deep exploration for oil and gas in central parts of the Békés basin has been relatively unsuccessful to date (Kovács and Teleki, this volume; Spencer and others, this volume). The deepest well, Békés-2, was drilled to a total depth of 5500 m (Grow and others, this volume; Lee and others, this volume). This well penetrated 2161 m of Mesozoic rocks but no hydrocarbon shows were reported. For the most part, only methane gas shows, mixed with CO_2 and commonly accompanied by salt water, have been encountered at depths greater than 3000 m.

In spite of the fact that the Pannonian Basin in Hungary is a highly explored region, relatively unexplored plays remain. In the Békés basin, the best targets for future exploration are stratigraphic traps on the downdip flanks of currently producing structures between depths of 3000-4000 m (Clayton and Koncz, this volume; Spencer and others, this volume). Although the limited-scope exploration of the Mesozoic section in the Békés basin has been disappointing, it must be restated, that this part of the stratigraphic section here, and in other parts of Hungary as well, remains relatively unexplored. At the Nagylengyel field, in Western Hungary, oil and natural gas are produced from Cretaceous rocks that are believed to have been charged from underlying, organic-rich Triassic source rocks (Clayton and others, this volume). In the Vienna basin of Austria, natural gas is produced from tightly folded and thrusted Mesozoic rocks (Wessely, 1988). Thrust-related structures are probably ubiquitous in the Mesozoic section of the Pannonian Basin (Royden, 1988; Grow and others, this volume), yet, these structures are difficult to recognize on reflection seismic profiles. A few high-amplitude reflections from the Mesozoic section appear to occur below and parallel to the major thrust faults and probably represent fracture zones associated with complex Mesozoic decollement folds as inferred from the vertical seismic profile (VSP) studies of Lee and others (this volume) and Miller and Véges (this volume).

An assessment of the recoverable, undiscovered resources of the Békés basin suggests that 5.22×10^6 metric tons of oil and 18.05×10^9 m^3 of natural gas remain to be discovered in 12 plays in the Békés basin (Charpentier and others, this volume). These estimations were made with the help of a package of computer programs (FASPUM) based on analytical probabilistic methodology and developed by Crovelli and Balay (this volume).

As noted by Royden and Horváth (1988), confusion has been created by the usage of the term "Pannonian" both for a late Miocene biostratigraphic stage defined by Papp (1953) and, throughout Hungary, for a stratigraphic sequence that begins at the base of Papp's Pannonian biostratigraphic stage and includes rocks up to latest Pliocene age. In this volume, we follow the terminology used by Royden and Horváth (1988) that distinguishes between the two usages by referring to the former as "Pannonian *sensu stricto*" (Pannonian s.s.) and the latter as "Pannonian *sensu lato* (Pannonian s.l.)". This classification has been further compounded in Hungary by dividing the Pannonian (s.l.) into two parts (Lower Pannonian and Upper Pannonian) based on rock lithology. This lithologic boundary has been shown to be strongly time-transgressive (Mattick and others, 1985). To minimize confusion in stratigraphic nomenclature, we use the terms "lower Pannonian" and "upper Pannonian", when the distinction between the units is based on lithology.

Commonly, the term "Miocene rocks" in Hungary is applied to rocks of Miocene age that are older than "Pannonian s.l.". Again, in order to avoid confusion, we refer to Miocene age rocks that are older than "Pannonian s.l." either as pre-Pannonian Miocene rocks, or by the appropriate stage names (typically, Badenian or Sarmatian in the Békés basin).

On behalf of all the authors, we wish to thank MOL Ltd. for allowing publication of their data. We also wish to express our appreciation to Dr. György Szabó, Vice-President of the former OKGT (now MOL Ltd.), Dr. Béla Bardócz, Chief Geologist of MOL Ltd., Dr. Károly Molnár, General Manager of the Geophysical Exploration Company (GKV), and Mr. Niels Fostvedt and Dr. Anton Smit of the World Bank for their patient and generous support of this study. The cooperation of many other colleagues from the GKV, MOL and the USGS is also appreciated , and in this regard special thanks are due to János Rumpler of GKV, who was instrumental in making seismic data and their interpretations available, to Éles Zsolt of MOL Ltd., who assisted in a major way in compiling this book, and to the drafting staff of the GKV for the many figures in this volume.

We hope that the results will serve well the interests of those involved in the study and exploration of Pannonian-type basins everywhere in the world.

Paul G. Teleki   Robert E. Mattick   János Kókai

REFERENCES

Balla, Z., 1985, The Carpathian loop and the Pannonian Basin: a kinematic analysis; Eötvös Lóránd Geophys. Inst., Geophysical Transactions, 30 (4), 313-354.

Balla, Z., 1987, Tertiary paleomagnetic data for the Carpatho- Pannonian region in the light of Miocene rotation kinematics; Tectonophysics, 139, 67-98.

Bally, A.W. and Snelson, S., 1987, Realms of subsidence; *in* Foster, N.H. and Beaumont, E.A. (eds.), Geologic basins I, classification, modeling, and predictive stratigraphy; Amer. Assoc. Petrol. Geol. Treatise of Petroleum Geol. Reprint Series, No. 1, p. 1-86.

Channel, J.E.T. and Horváth, F., 1976, The African/Adriatic promontory as a paleogeographic premise for alpine orogeny and plate movements in the Carpatho-Balkan region; Tectonophysics 35, 71-101.

Földváry, G.Z., 1988, Geology of the Carpathian region, World Scientific Pub. Co., Singapore and Teaneck, N.J., 571 p.

Horváth, F., 1988, Neotectonic behavior of the Alpine-Mediterranean region; *in* Royden, L.H. and Horváth, F. (eds.), The Pannonian Basin a study in basin evolution; Assoc. Petrol. Geol. Mem. 45, 27-48.

Horváth, F. and Royden., L., 1981, Mechanism for the formation of the intra-Carpathian basins: A Review; Earth Evol. Sci. 1, 307-316.

Kilényi, É. and Teleki, P.G. (eds.), 1985, Cooperative research in the geosciences between the U.S. Geological Survey (USGS) and the Central Office of Geology (KFH) during 1978-84; Proc. Joint Conf., Eötvös Lóránd Geophys. Inst., Geophysical Transactions, 31, (also: U.S. Geol. Surv. Open-File Rept. 85-291), 335 p.

Mattick, R.E., Rumpler, J. and Phillips, R.L., 1985, Seismic stratigraphy of the Pannonian basin in southeastern Hungary; *in* Killényi, É. and Teleki, P.G. (eds.), 1985, Cooperative research in the geosciences between the U.S. Geological Survey (USGS) and the Central Office of Geology (KFH) during 1978-84; Proc. Joint Conf., Eötvös Lóránd Geophys. Inst., Geophysical Transactions, 31, (also: U.S. Geol. Survey Open-File Rept. 85-291), 13-54.

Papp, A., 1953, Die Molluskenfauna des Pannon in Wiener Becken; Mitteilungen der geologischen Gesellschaft in Wien, 44, 86-222.

Royden, L.H., 1988, Late Cenozoic tectonics of the Pannonian basin system; *in* Royden, L.H. and Horváth, F. (eds.), The Pannonian Basin a study in basin evolution; Amer. Assoc. Petrol. Geol. Mem. 45, 27-48.

Royden, L.H. and Báldi, T., 1988, Early Cenozoic tectonics and paleogeography of the Pannonian and surrounding regions; *in* Royden, L.H. and Horváth, F. (eds.), The Pannonian Basin a study in basin evolution; Amer Assoc. Petrol. Geol. Mem. 45, 27-48.

Royden, L.H. and Horváth, F., 1988, The Pannonian basin a study in basin evolution; Amer. Assoc. Petrol. Geol., Mem. 45, 27-48.

Royden, L.H., Horváth, F. and Burchfiel, B.C., 1982, Transform faulting, extension and subduction in the Carpathian Pannonian region; Geol. Soc. Amer. Bull. 93, 717-725.

Stegena, L., Géczy, B., and Horváth, F., 1975, Late Cainozoic evolution of the Pannonian Basin; Tectonophysics, 26, 71-90.

Szalay, Á., 1988, Maturation and migration of hydrocarbons in the southeastern Pannonian Basin; *in* Royden, L.H. and Horváth, F. (eds.), The Pannonian Basin a study in basin evolution; Amer. Assoc. Petrol. Geol. Mem. 45, 27-48.

Wessely, G., 1988, Structure and development of the Vienna Basin; *in* Royden, L.H. and Horváth, F., (eds.), The Pannonian basin a study in basin evolution; Amer. Assoc. Petrol. Geol. Mem. 45, 333-346.

1 Structure of the Békés Basin Inferred from Seismic Reflection, Well and Gravity Data

John A. Grow[1], Robert E. Mattick[2], Anikó Bérczi-Makk[3], Csaba Péró[4], Dénes Hajdú[5], György Pogácsás[6], Péter Várnai[6] and Ede Varga[6]

ABSTRACT

The Békés basin (areal extent 3900 km^2) is a northwest-trending, Neogene basin located in southeast Hungary. The basin contains over 6500 m of synrift and postrift sedimentary fill. Middle Miocene synrift deposits are relatively thin and no Paleogene rocks have been reported to be present. The prerift section (basement) is composed of Mesozoic carbonate and clastic rocks and Paleozoic and older volcanic, igneous and metamorphic rocks. The Mesozoic rocks represent dominantly shallow-water environments and are up to 5000 m thick in the Békés-Doboz Mesozoic trough and 2000 m thick in the Battonya-Pusztaföldvár Mesozoic trough.

In the basement numerous examples of a repeated section occur, as reported in exploration well reports, and these are inferred to have resulted from overthrusting. At the Tótkomlós-I well, Triassic rocks overlie Jurassic rocks, and at the Békés-2 well, Triassic and Jurassic rocks overlie Cretaceous rocks. Nappes composed of Paleozoic and older rocks overlying Mesozoic rocks have not been reported; the interpretation of seismic data, however, indicates that such nappes could be present.

Mesozoic nappes in the basement occur primarily in two parallel, northeast-southwest-trending belts. Rock units of the southern belt (Battonya-Pusztaföldvár trough) can be correlated to lithologic successions in the allochthonous Codru nappes of the Apuseni Mountains of western Romania. Rock units in the northern belt (Békés-Doboz trough) can be correlated to lithologic successions in the Bihor autochthon of the northern Apuseni Mountains.

A gravity model constructed across the eastern part of the Békés basin near the Hungarian-Romanian border, indicates that the crust near the axis of the Békés basin is composed of either dense (mafic?) rocks and/or that the mantle lies at relatively shallow depth. The dense rocks may represent fragments of a Mesozoic oceanic crust, and, as such, may be similar in composition to the ophiolites in the Apuseni Mountains. An alternate possibility is that rocks of high density were intruded into the crust during an extensional phase in Neogene time.

[1]U.S. Geological Survey, Denver, Colorado 80225, USA
[2]U.S. Geological Survey, Reston, Virginia 22092, USA
[3]MOL Rt. (Hungarian Oil and Gas Co., Ltd.), H-1311, Budapest, Hungary
[4]Eötvös Lóránd University, Budapest, Hungary
[5]MOL Rt. (Hungarian Oil and Gas Co., Ltd.), H-5001, Szolnok, Hungary
[6]MOL Rt. (Hungarian Oil and Gas Co., Ltd.), H-1068, Budapest, Hungary

P. G. Teleki et al. (eds.), Basin Analysis in Petroleum Exploration, 1–38.

Along the northwest boundary of the Békés basin, several northeast-southwest-trending, buried, basement ridges and small troughs occur. These appear to be associated with a middle-Miocene, left-lateral, strike-slip fault zone. Quantitative estimates of the amount of strike-slip movement cannot be made using the existing seismic reflection and well data. Above several of the basement ridges, listric normal faults are present. These faults are inferred to have resulted from differential compaction of sediments, and not from reactivation of strike-slip faults.

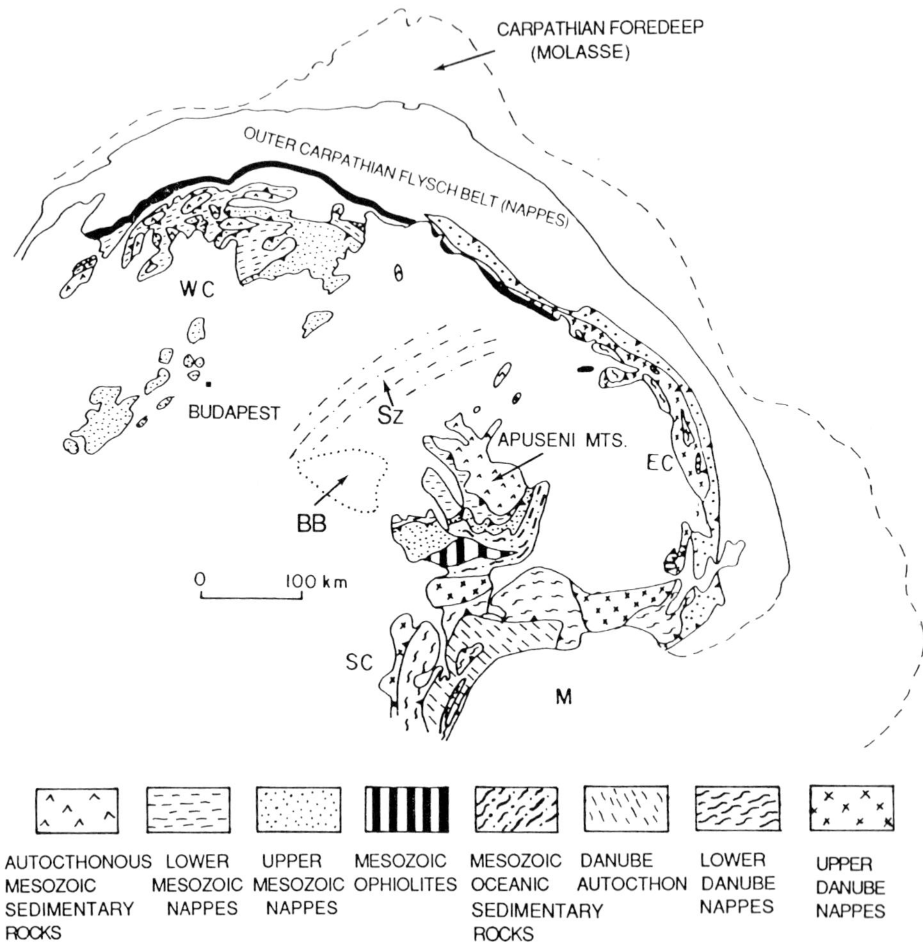

Figure 1. Major tectonic units of the inner Carpathians showing Békés basin (BB) and tentative correlations between units in west Carpathians (WC) and Apuseni Mountains (A) and between south (SC) and east Carpathians (EC) (modified from Fig. 3 of Burchfield and Royden, 1982, and Sandulescu, 1975, 1980, 1988). Correlations cannot be made from the Apuseni Mountains to the south or east Carpathians. Unit shown in black is Pieniny klippen zone and vertical lines indicate ophiolites in the southern Apuseni Mountains. M: Moesian platform; Sz: Szolnok trough (subsurface).

INTRODUCTION

The Békés basin is a northwest-southeast-trending trough in southeast Hungary and has an areal extent of 3900 km^2. The basin is more than 6500 m deep and is filled with sedimentary rocks of Neogene age (Figs. 1 and 2). The basin is one of several subbasins in the Pannonian Basin. The geographical location of the Békés basin is the Great Hungarian Plain which has an average elevation of about 100 m above sea level and nominal topographic relief.

Prior to the formation of the Békés basin, Paleozoic and Mesozoic sedimentary rocks, which now form the base-

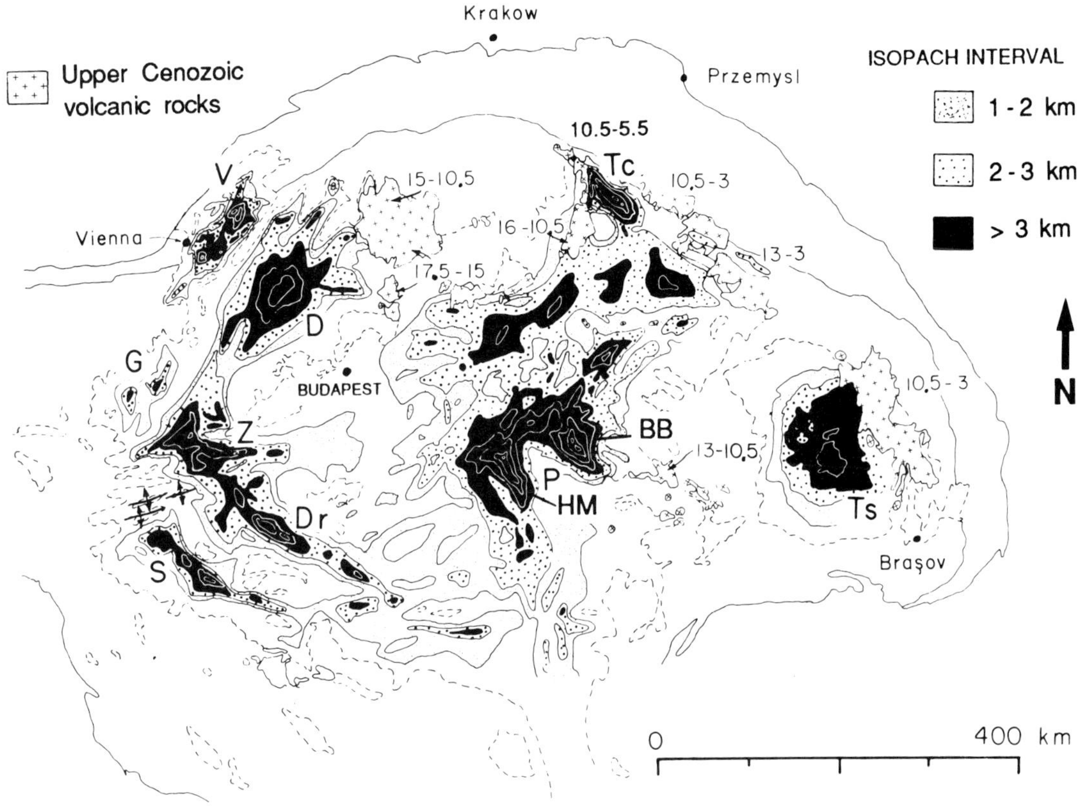

Figure 2. Map showing sediment thickness map above the base of the Miocene in the Carpathian region, and locations of Békés basin (BB) and Hódmezővásárhely-Makó trough (HM) (modified from Fig. 2 of Royden and others, 1983). Contour interval = 1 km. Other basins: S: Sava; Dr: Drava; Z: Zala; G: Graz; D: Danube; V: Vienna; P: Pannonian; Tc: Transcarpathian; Ts: Transylvanian; Sz: Szolnok trough. Dashed lines show outcrops of pre-Neogene rocks. The numerical values (e.g., 13-3) are age ranges (x10^6 years) of Neogene volcanic centers from Póka (1982). The term "Pannonian basin" sometimes includes parts of the Danube and Drava basins, and areas in between, but in this paper we use it to designate only the subsided area to the east of the Danube; i.e., the Békés basin and Hódmezővásárhely-Makó trough and the smaller basins to the north.

ment, were thrust northward in the area of the present western Carpathians and Apuseni Mountains of western Romania (Bleahu and others, 1981). In western Romania, numerous thrust sheets or nappes developed. These nappes consist of a lithologically heterogeneous assemblage of shallow-water carbonates, crystalline gneisses, deep-water clastics, and ophiolites (Fig. 1). Late Cretaceous(?) and Paleogene rocks are absent in the Békés basin, but were penetrated by wells in the Szolnok trough to the north (Fig. 1) and occur elsewhere in the Carpathian Mountains (Burchfiel and Royden, 1982).

In middle-to-late Miocene time the Pannonian basin began to subside rapidly, probably as a result of lithospheric extension that was coupled to outward thrusting of the Carpathian arc. This orogenic phase was followed by cooling of the lithosphere and its loading by sediments (Sclater and others, 1980). Most of the deformation appears to be characterized by strike-slip faulting, whereby oblique extensional and compressional zones alternate locally between the strike-slip zones (Royden and others, 1982; Royden and others, 1983; Royden, 1988). During this phase of extension and subsidence, numerous depocenters began to develop. For the most part, these depocenters contain less than 3000 m of

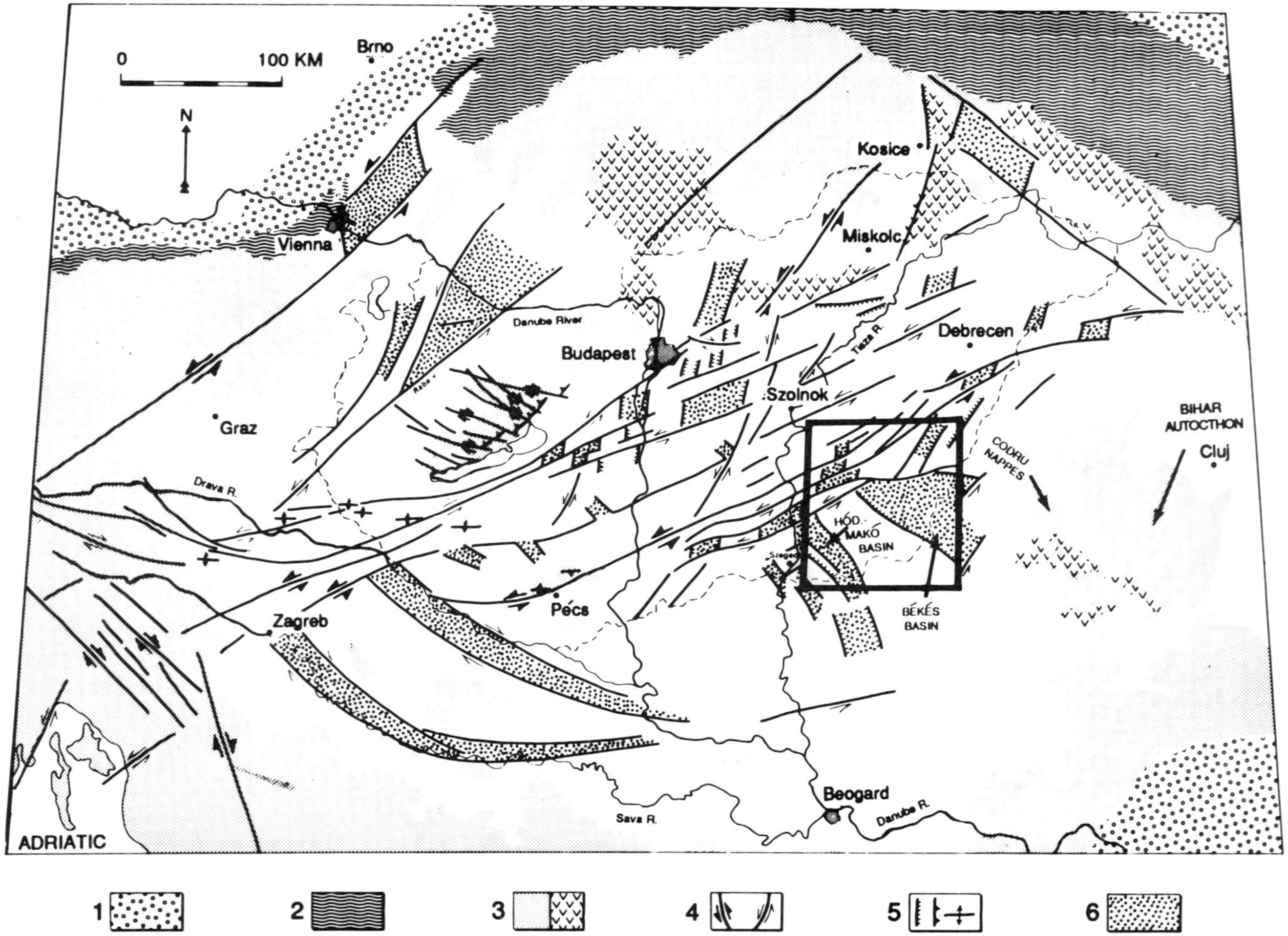

Figure 3. Tectonic map of Pannonian basin and surrounding regions showing the main faults and folds of Neogene age (modified from Rumpler and Horváth, 1988). Box shows location of study area. Tectonic activity culminated during middle Miocene time and has been greatly reduced since that time. Explanation: (1) molasse foredeep; (2) Alpine-Carpathian flysch belt; (3a) Inner Alpine-Carpathian Mountain belt and the Dinarides; (3b) outcrops of Neogene calcalkaline volcanic rocks; (4) strike-slip faults; the sense (and usually the amount) of displacement is well constrained (thick arrows) or unconstrained (thin arrows); (5) normal fault, thrust fault, and fold; and (6) areas of major crustal extension and subsidence.

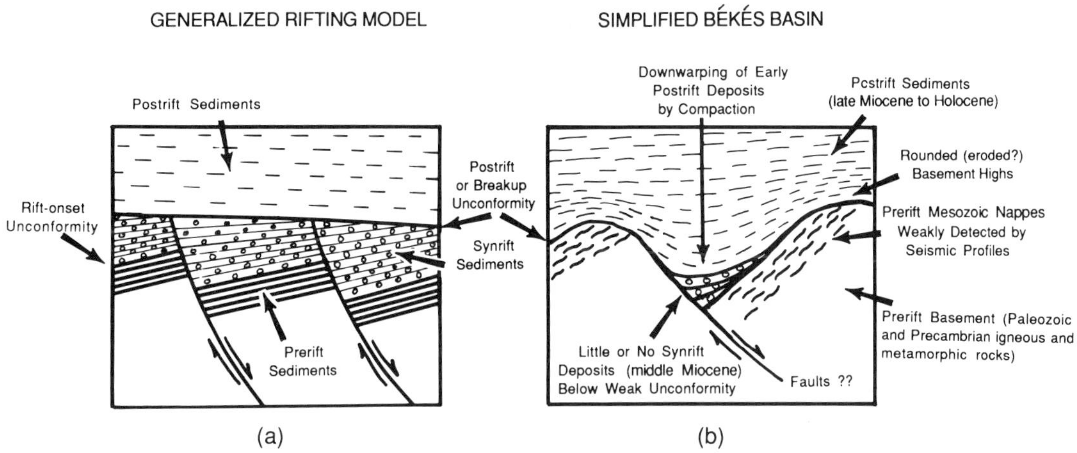

Figure 4. (a) Simplified extensional graben containing prerift, synrift and postrift sediments. (b) Simplified model of the Békés basin with relatively thin synrift deposits and rounded (planed) basement (prerift rocks) surface.

sediment, but local sediment accumulations exceed 6000 m (Fig. 2).

Recent tectonic interpretations of the Pannonian Basin, such as that by Rumpler and Horváth (1988), indicate that a complex system of east-northeast-trending, left-lateral, strike-slip faults occur in the area northwest and outside of the Békés basin (Fig. 3). According to Balla (1985, 1990), crustal blocks were rotated within the Carpathian orogenic loop and the Pannonian Basin during Neogene time, and these may be analogous to rotations within the San Andreas strike-slip system of California where blocks have experienced a rotation of 30-90^0 during the Neogene (Hornafius and others, 1986). The Békés basin is most likely part of this complex zone of Neogene strike-slip faults resulting from oblique extension, and has probably also undergone oblique compression and rotation. Although the magnitude of strike-slip displacement cannot be estimated in the Békés basin, it is possible that the scale of displacement is similar to the 80-km long, left-slip displacement along northeast-southwest-trending faults in northwestern Hungary estimated by Royden (1988). Generally, the Neogene-age faults in the Pannonian Basin are either low-angle listric or detachment faults (Royden and others, 1983; Pogácsás, 1985; Rumpler and Horváth, 1984, 1988) that may, in some cases, represent partial reactivation of early Late Cretaceous-age thrust faults.

Extensional terranes and Atlantic-type continental margins commonly display three kinds of sedimentary sequences (Fig. 4a): (1) sedimentary rocks deposited before the onset of rifting referred to as "prerift sediments", (2) sedimentary rocks deposited during active rifting known as "synrift sediments", and (3) sedimentary rocks deposited after the cessation of rifting that are called" postrift sediments". Each type of sedimentary sequence may vary in thickness or be absent altogether, depending on several factors, such as existing geologic structures, rate and duration of crustal extension, quantity of sediment supply, and variations in climate. Where both prerift and synrift sediments are present, they are separated commonly by a "rift onset unconformity" (Falvey, 1974). In typical extensional grabens, the synrift deposits dip into listric normal faults in a fan-shaped pattern, signifying that deposition took place contemporaneously with rifting and faulting (Bally and others, 1981). When extensional rifting terminates abruptly, postrift sediments are deposited unconformably on the synrift sediments (Fig. 4a). Such an unconformity, referred to as the "postrift unconformity" (Grow and Sheridan, 1988; Klitgord and others, 1988), between the synrift and postrift sedimentary sequences is usually well developed. Others authors, such as Falvey (1974), prefer the term "breakup unconformity".

The Mesozoic and older rock sections in the Békés basin comprise the prerift sequence as evidenced by a well-developed erosional unconformity between the

Mesozoic-Paleozoic section and the superimposed basin fill of middle Miocene to Holocene age (Phillips and others, this volume; Szentgyörgyi and Teleki, this volume), and the fact that the Mesozoic and older basement units have been subjected to compression and subsequently to extension prior to the deposition of middle Miocene-Holocene sediments.

The boundary between synrift and postrift deposits, although difficult to deduce from seismic data, is placed at the Sarmatian-Pannonian boundary (postrift unconformity of Fig. 4b). Badenian and Sarmatian rocks (16.5-12 Ma) are inferred to comprise the synrift sequence, and overlying Pannonian and younger rocks the postrift sequence. In the area of the Pannonian basin system, the period of active rifting and extension lasted from early Miocene (22 Ma) to Plio-Pleistocene, the active zone of which shifted eastward (Royden, 1988). In the area of the Békés basin, active rifting is thought to have taken place in middle Miocene time (Badenian and Sarmatian stages, 16.5-12 Ma). During this period, relatively thin synrift sediments (0-275 m where penetrated by wells on the periphery of the basin and, perhaps, 1000 m in the deepest parts of the basin as estimated from seismic data) were deposited in shallow marine and brackish-water environments. An unconformity or disconformity may separate the synrift and postrift sequences. According to Szentgyörgyi and Teleki (this volume), the absence of Sarmatian sediments in some areas indicates that erosion took place in pre-Pannonian time. In the deep, central parts of the basin, the Sarmatian-Pannonian boundary may be marked by a period of non-deposition. Although Mattick and others (this volume) report little or no unconformable relation at the base of the Pannonian section in the deepest, central parts of the Békés basin based on the interpretation of seismic data, Pogácsás and others (this volume) report that little or no sedimentation occurred in the time period 12-9 Ma. All of these authors agree, however, that the Sarmatian-Pannonian boundary represents a change from shallow-water to deep-water deposition - from water depths of less than 200 m to as deep as 1000 m (Mattick and others, this volume) - and that no other unconformities exist in the Pannonian section in central parts of the Békés basin.

The Békés basin is an atypical extensional basin in that it contains relatively thin synrift deposits (in general less than 1000 m) in comparison to thick postrift deposits (6000 m), and the surface of the prerift rocks has a smooth (planed), rounded character (Fig. 4b). In typical extensional basins, the surface of the prerift section is usually offset by numerous high-angle, normal faults, and the synrift deposits reach great thicknesses in local, fault-bounded grabens (Fig. 4a). The synrift deposits are usually thick because local relief is normally greatest during the stage. These aspects of the Békés basin are developed in subsequent sections of this chapter.

The purpose of this investigation was to: (1) identify the primary Neogene fault systems, (2) map the distribution of prerift Mesozoic sedimentary rocks beneath Neogene and postrift sedimentary cover, (3) construct a model of the crustal structure beneath the Békés basin from gravity data, and (4) where applicable, relate the above to occurrences of oil and gas in the Békés basin. This report focuses primarily on the analysis of seismic reflection profiles; however, well data is presented in order to document the presence of overthrusts and to compare rock assemblages of Mesozoic age in the Békés basin with Mesozoic-age rock assemblages in the Apuseni Mountains.

PREVIOUS SEISMIC REFLECTION STUDIES FROM THE PANNONIAN BASIN

Although voluminous amounts of multi-channel seismic data have been collected by the Hungarian Oil and Gas Co. (MOL) and other institutions in Hungary, only a small number of profiles showing detailed analysis of structure and tectonic style have been published. The majority of publications report on extensional grabens, strike-slip faults, and growth faults from widely-separated regions of Hungary (for example, Varga and Pogácsás, 1981; Horváth and Rumpler, 1984; Pogácsás, 1985; Rumpler and Horváth, 1984, 1988). Other articles report examples of strike-slip faults which appear to cut the entire postrift sequence as well as the underlying prerift sequence (Horváth and Rumpler, 1984, 1988). This report differs from these earlier articles by focusing entirely on the tectonic regime of a single basin, that is, the Békés basin.

MORPHOLOGY OF THE BÉKÉS BASIN

The dominant basement feature in the Békés basin is the Battonya-Pusztaföldvár high which bounds the Békés basin on the southwest as shown on the structure contour map drawn on the surface of pre-Neogene rocks (Fig. 5). From this high, covered by less than 1000 m of Neogene sediments (at the Romanian border), the basement surface plunges northeastward and southwestward to depths of more than 6500 m in the Békés and Hódmezövásárhely (Hód)-Makó

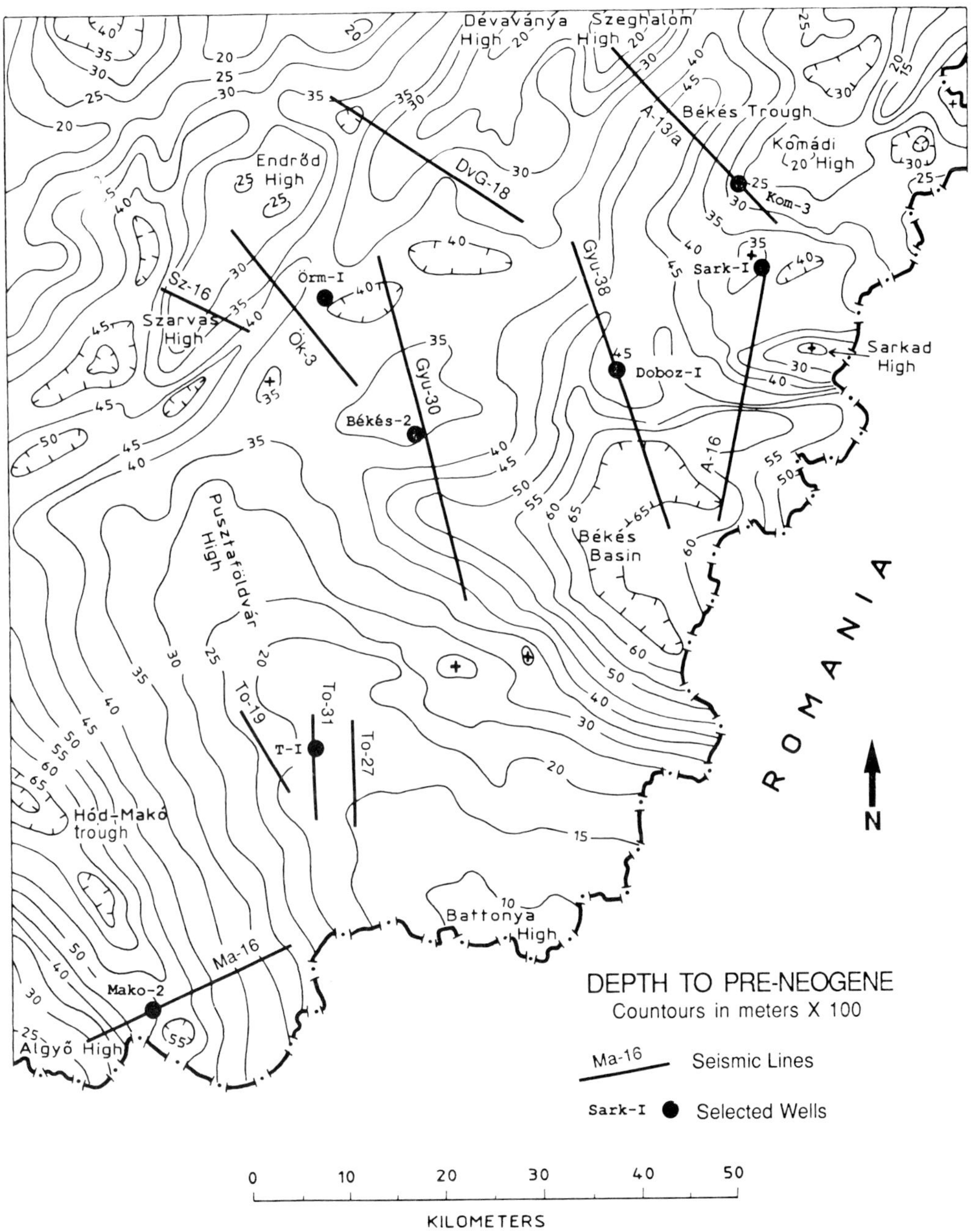

Figure 5. Map showing depth to pre-Neogene rocks in the vicinity of the Békés basin and Hódmezövásárhely-Makó (Hód-Makó) trough. The location of seismic profiles discussed in text and selected wells also are shown. Contours modified from Kilényi and Rumpler, 1984. Örm: Örménykút; Sark: Sarkadkeresztúr; Makó: Hódmezövásárhely-Makó.

basins, respectively. On the northwest, the Békés basin is bounded by the Szarvas and Endröd highs, on the north by the Dévaványa and Szeghalom highs, and on the northeast by the Komádi and Sarkadkeresztúr highs. The Algyö high bounds the Hód-Makó basin on the southwest. The Békés basin extends into Romania as well. Lack of access to data for the Romanian part of the basin prevented study of the basin in its entirety.

EXTENSIONAL AND OBLIQUE EXTENSIONAL FEATURES

The best example of extensional tectonics in the Békés basin is seen on seismic line Ök-3 (Fig. 6) which extends southeastward from the Endröd high (Fig. 5). Prerift Jurassic and Triassic rocks, encountered in the Örménykút (Örm)-I well, appear to be offset by a series of southeast-dipping, normal faults. Miocene deposits appear to have been deposited in the resulting grabens. The best developed graben occurs 4 km northwest of the Örm-I well where middle Miocene synrift deposits appear to reach a maximum thickness of several hundred meters. Further northwestward, the sequence thins against a southeast-dipping, normal fault. Smaller grabens occur southeast of the Örm-1 well. Other faults may be present in the prerift section, but these cannot be resolved from the available seismic data. Six kilometers southeast of the southeast end of the profile, at the Békés-2 well, the synrift deposits remain relatively thin (168 m) with Sarmatian sediments absent as a result of non-deposition or erosion.

A larger extensional graben exists in the vicinity of seismic profile A-13/a (Fig. 7). This graben, named the Békés trough, is located in the northeast corner of the Békés basin (Fig. 5). Inferred synrift sediments dip gently southeastward and terminate (or become thin) against a northwest-dipping normal fault. At the Komádi (Kom)-3 well, Badenian deposits are 178 m thick and Sarmatian sediments are reported to be absent, although they may occur in the graben. The synrift (Badenian and, if present, Sarmatian) sequence is 1000-1500 m thick (assuming a velocity of 4000-5000 m/sec) in the axis of the Békés trough (Fig. 7). Within the prerift basement complex, at the northwest end of the profile, a south-dipping, normal fault with a left-lateral, strike-slip component is inferred to be present based on the interpretation of nearby seismic lines. The existence of strike-slip faults cannot be proven, however, because horizontal offsets of unique marker structures, or piercing points, cannot be demonstrated. The postrift Pannonian section appears to be cut by at least two faults at the Kom-3 well site (Fig. 7), and these faults are interpreted to be local growth faults that do not penetrate the basement rocks. Such growth faults are commonly the result of differential sediment compaction.

Structurally, the Hódmezövásárhely-Makó (Hód-Makó) basin (Fig. 8, seismic profile Ma-16) appears to be less complex than the Békés basin. This seismic profile displays a slightly asymmetric extensional graben with the thickest sediment accumulation in the southwestern part of the trough. At the Makó-2 well, Badenian-age, synrift sedimentary rocks (Sarmatian is absent) are 272 m thick and appear to thicken to 800-1000 m in the central part of the trough. The basement surfaces on the southwest and northeast flanks of the Hód-Makó basin are relatively smooth (planed) and postrift sediments onlap these surfaces.

None of the seismic profiles (Figs. 6, 7 and 8) allow one to discriminate clearly between thin synrift sediments and the basal part of the postrift deposits in the bottom of the troughs. Although there appears to be a slight angular unconformity between the and postrift sequences in Figure 6, clearly the sequences are conformable in Figures 7 and 8. As mentioned earlier, the two sequences may be separated by a 3-million-year, non-depositional hiatus (Pogácsás and others, this volume), but this is uncertain. It is also possible that an erosional unconformity, that cannot be resolved by the seismic data, exists between the synrift and postrift sequences, as suggested by the absence of Sarmatian deposits in many wells.

The most remarkable aspect of these seismic profiles from the Békés basin and Hód-Makó basin is that the synrift section appears to be extremely thin in comparison to the postrift section. In addition, the fact that the basement surface is relatively smooth (planed) suggests that a significant amount of erosion occurred. This raises an obvious contradiction - high rates of erosion coupled with low rates of deposition? We offer two explanations or, perhaps, a combination of these explanations: (1) if erosion rates were high during the period, then the bulk of the sediments must have bypassed the Békés basin; or (2) rifting, abruptly started and abruptly terminated, occurred during a relatively short period of time. Most of the seismic profiles from the Békés basin area show little or no indication of renewed faulting during the postrift period. Numerous authors infer that sediments bypassed the area (Spencer and others, this volume) or were trapped in extensional grabens closer to the major source terrains, probably to the west and north (Mattick and others, this volume). Royden (1988) concludes that thick

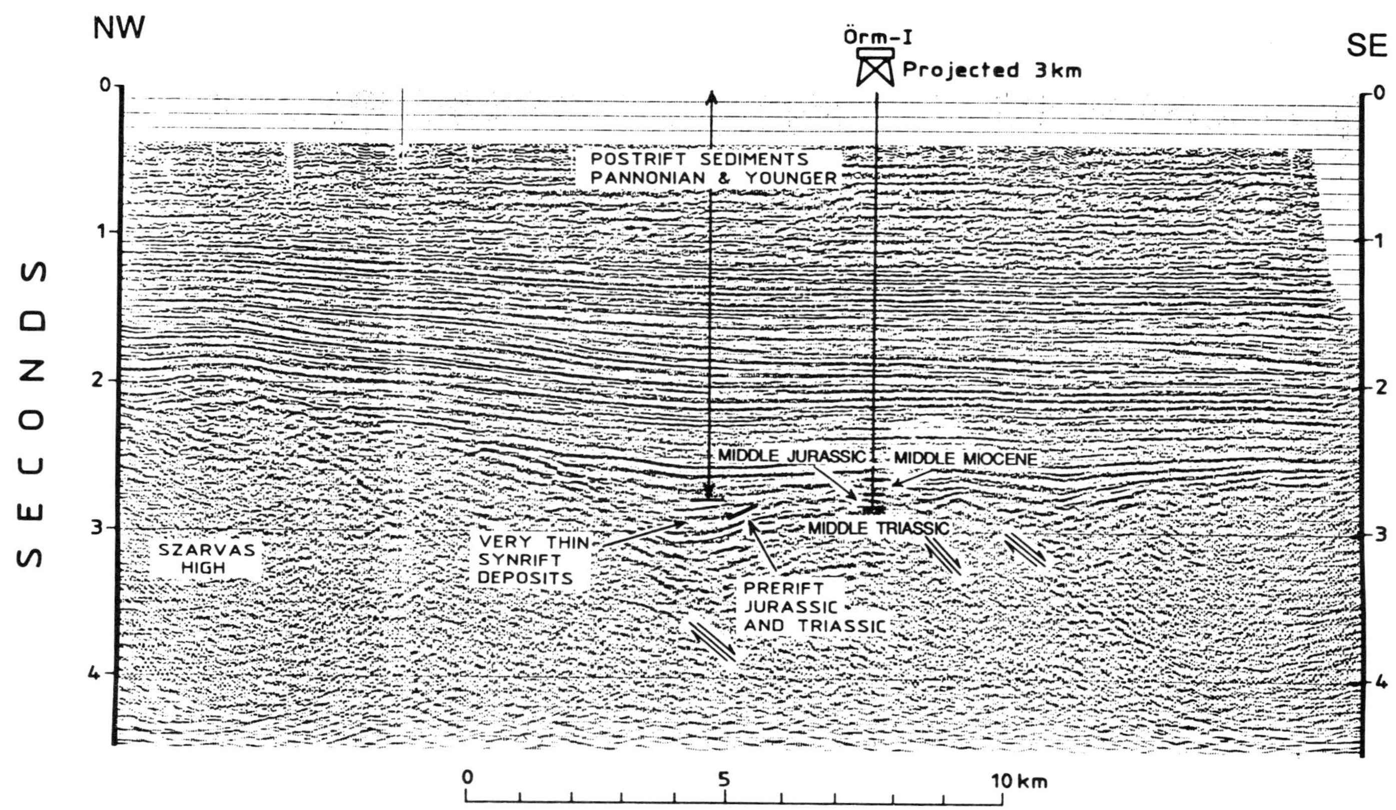

Figure 6. *Seismic line Ök-3 showing oblique-extensional half-graben containing very thin synrift deposits. The location of profile is shown in Figure 5. Örm: Örménykút.*

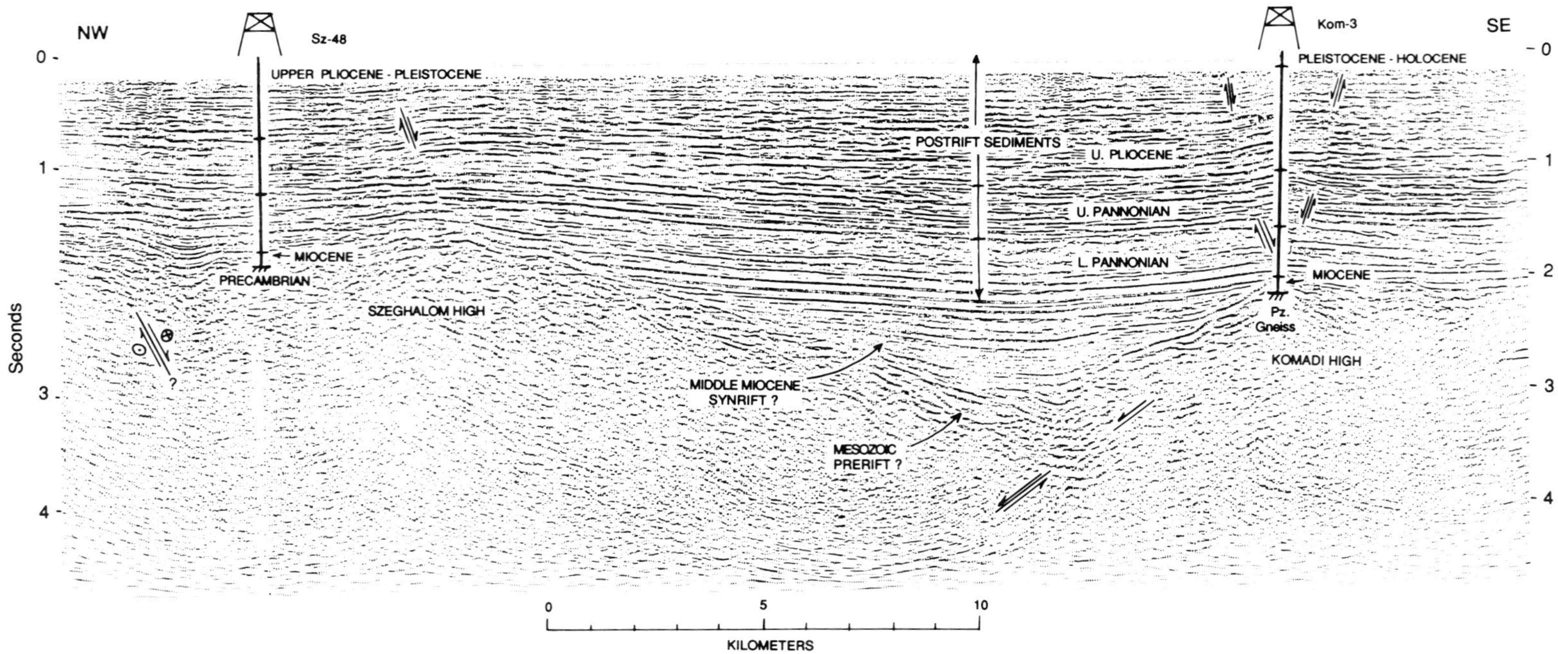

Figure 7. Seismic line A-13/a over Békés trough showing extensional half-graben with possible oblique-extensional, left-lateral, strike-slip fault northwest of the Szeghalom high (based on regional morphology of the pre-Neogene surface, Figs. 5 and 23). Sz: Szeghalom; Kom: Komádi; Pz: Paleozoic; Mio: Miocene.

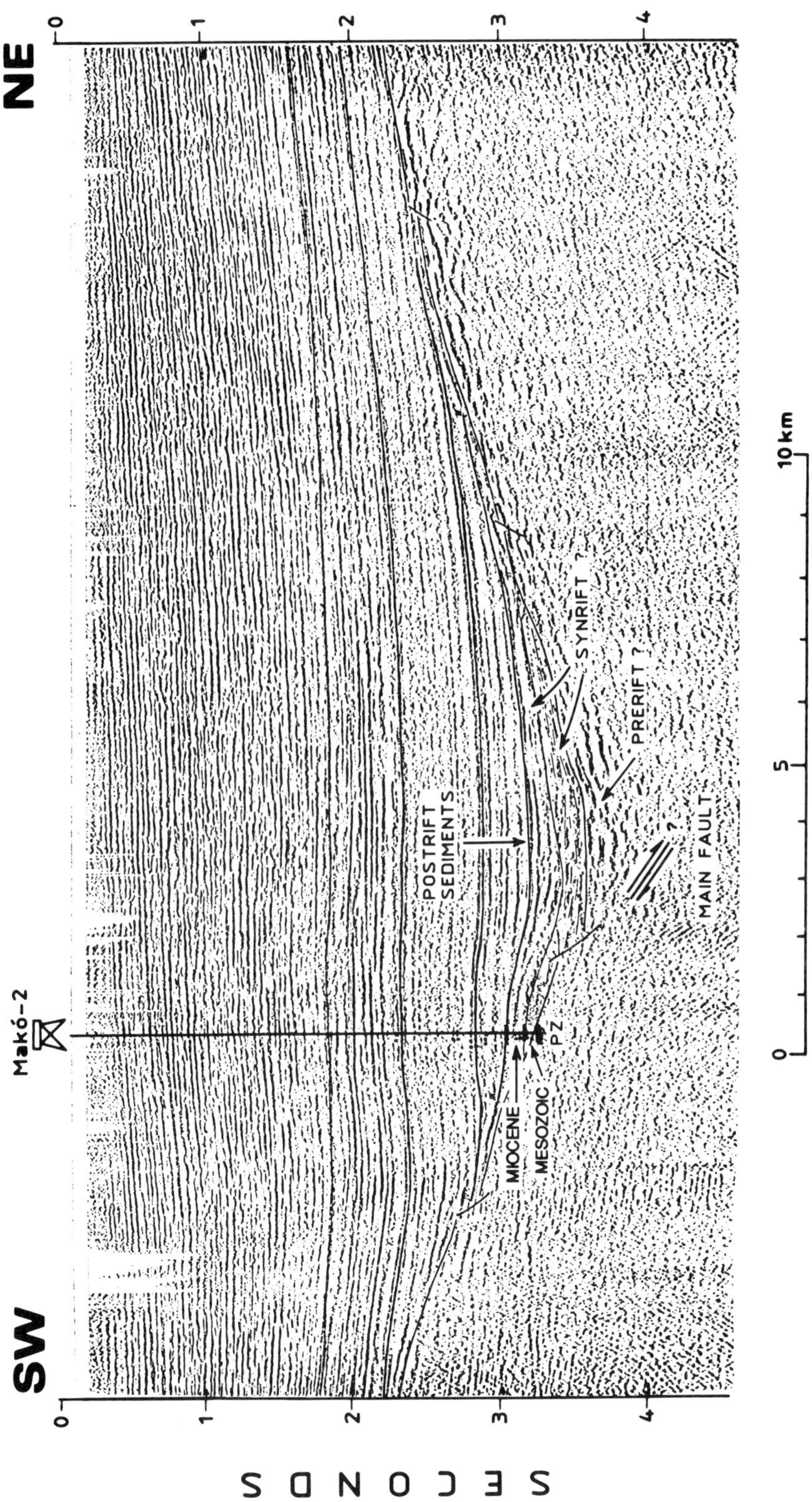

Figure 8. Seismic line Ma-16 over the Hódmezővásárhely-Makó (Hód-Makó) basin showing large extensional half-graben with northeast-dipping, low-angle, normal fault. Mio: Miocene; Mez: Mesozoic. Location of profile is shown in Figure 5.

synrift sedimentary deposits occur near the Carpathian Mountains and that only minor amounts of sediments reached the deep basins beneath the Great Hungarian Plain during the synrift stage.

MESOZOIC BELTS

Mesozoic sedimentary rocks, which comprise the basement complex in parts of the Békés basin, occur mainly in two northeast-southwest-trending belts (Fig. 9). The southern belt, referred to as the Battonya-Pusztaföldvár Mesozoic trough, is 10-15 km wide and extends across the Battonya high. Mesozoic rocks are about 500 m thick in the southern part of the trough and thicken northward to about 2000 m (Fig. 9). The second belt, referred to as the Békés-Doboz Mesozoic trough, underlies the northern part of the Békés basin. It is 20-50 km wide and contains up to 5000 m of Mesozoic rocks based on interpretation of seismic data.

The buried Mesozoic rocks in the Békés basin consist chiefly of limestone, dolomite, calcareous marl, oolitic limestone and detrital clastic deposits of Triassic-Cretaceous age. These rocks represent deposition in mainly shallow-marine environments.

To date, no significant discoveries of oil or gas have been made in the Mesozoic section in the deep, central parts of the Békés basin (Kovács and Teleki, this volume), although oil and gas is produced from the upper few meters of fractured Mesozoic and older rocks on basement highs and on the flanks of basement highs along the periphery of the Békés basin. This is attributed to: (1) the source of the hydrocarbons is in the Neogene section, (2) good quality source rocks may not be present in the Mesozoic section, and (3) the Mesozoic section in the Békés basin area is thermally overmature (Clayton and Koncz, this volume).

Battonya-Pusztaföldvár Mesozoic Trough

Past interpretations of geological data have concluded that the Mesozoic rocks in the Battonya-Pusztaföldvár trough were deposited in an extensional graben or trough bounded by normal faults (unpublished reports of the Hungarian Oil and Gas Co. (MOL)). There is evidence, however, that suggests that much of the Mesozoic section is allochthonous and is composed of several overthrust sheets (nappes) associated with Late Cretaceous compression, as explained below.

The seismic profiles across the Battonya-Pusztaföldvár Mesozoic trough (Figures 10, 11 and 12; profiles To-37, To-31 and To-19, respectively) exhibit discontinuous (possibly a result of faulting), northwest-dipping reflections below relatively flat-lying reflections from the postrift sedimentary sequence. The profiles show generally northward thickening Mesozoic strata that dip to the north above a north-dipping Paleozoic or older basement surface. Internal reflections from within the Mesozoic and older sections are generally poor, probably because there is little density or velocity contrast within these rock units.

The Tótkomlós (T)-I and Pusztaszöllös (Psz)-1 wells, located near seismic line To-31 (Fig. 11), were extensively cored. The results of paleontologic analyses document the presence of age inversions or reversals (older rocks overlying younger rocks) in the Mesozoic section (Figs. 13 and 14). In the T-I well, Middle Triassic dolomite overlies Lower Jurassic limestone (contact at 2693 m, Fig. 13). In the Psz-1 well, Upper Triassic dolomite overlies Lower Cretaceous marl at a depth of 1818.5 m (Fig. 14). These age inversions indicate the presence of one or more thrust faults. The unconformities identified at 2905 m and 3237 m in the T-I well may mark other thrust fault surfaces.

Although these well data only document inverted sequences in the Mesozoic section, it is conceivable that thrusting involved Paleozoic and older rocks as well. On seismic profile To-19 (Fig. 12), Paleozoic units are shown to overlie Mesozoic units. This interpretation resulted from projections along adjacent seismic lines. Accordingly, Paleozoic rocks are inferred to underlie postrift and sediments below about 1.8 seconds on the northeast side of the profile, and Mesozoic rocks are inferred to underlie basin fill on the southwest side of the profile. The strong change in reflection character that occurs below 1.8 seconds about 2 km southwest of the northwest end of the profile is interpreted to represent a northeast-dipping, thrust surface with Paleozoic rocks thrust over Mesozoic rocks.

Mesozoic units which comprise the nappes in the Battonya-Pusztaföldvár trough appear to correlate with rocks units of the Codru nappe system of the Apuseni Mountains in nearby Romania (Fig. 1). A description of the lithostratigraphic units penetrated by wells in the Battonya-Pusztaföldvár trough is shown in Figure 15, and tentative correlations between rock units of the Battonya-Pusztaföldvár trough and the Codru nappe system are shown in Figure 16. According to Bérczi-Makk (1986), almost all of the rock units exposed in the Codru

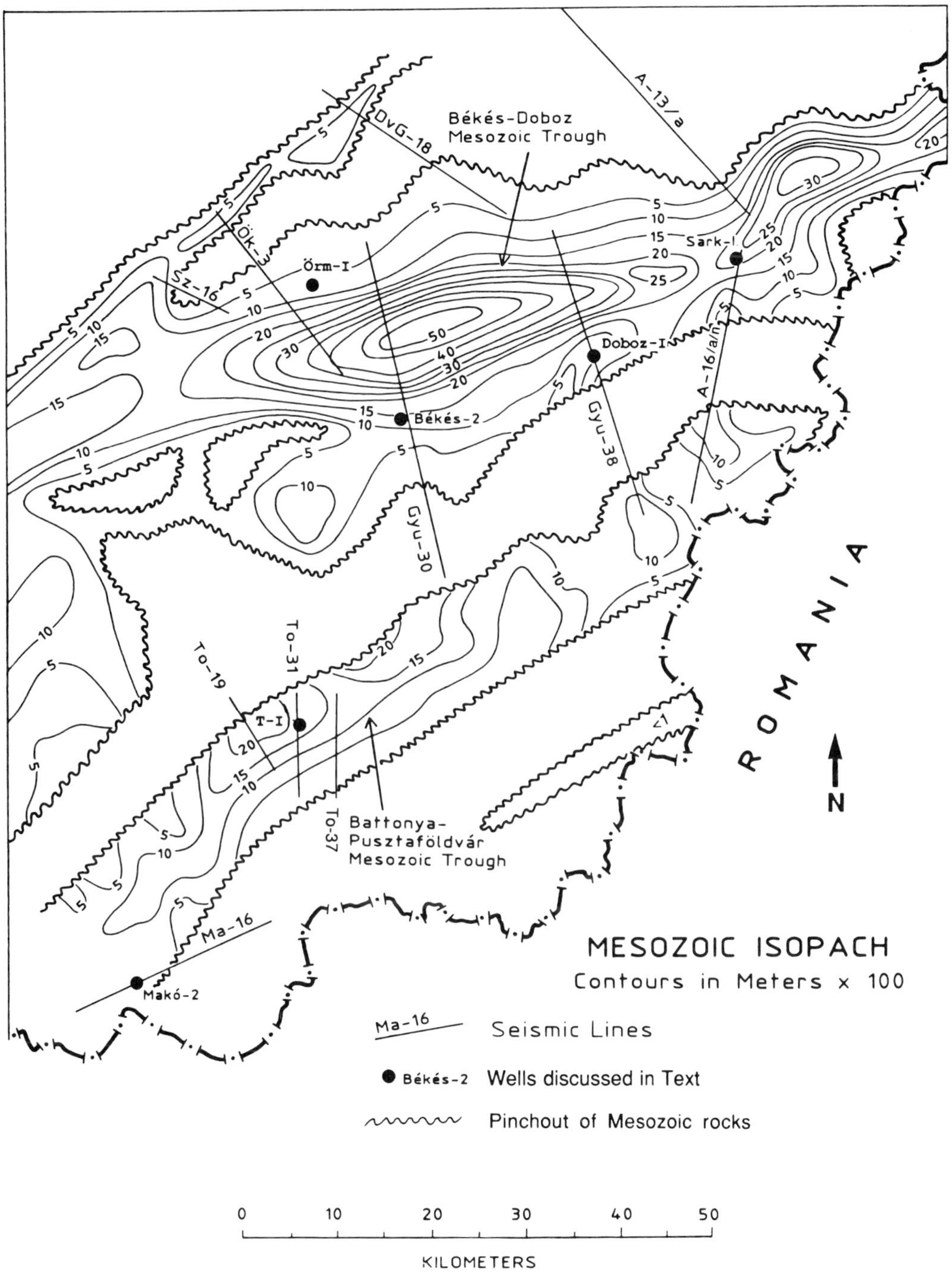

Figure 9. Isopach map showing inferred thickness of Mesozoic sedimentary rocks based on well data and interpretation of seismic reflection profiles. The two major belts of Mesozoic rocks are referred to as the Battonya-Pusztaföldvár and Békés-Doboz Mesozoic troughs. Blank zones represent areas where Neogene rocks were deposited directly on Paleozoic or older rocks. Hód-Makó: Hódmezövásárhely-Makó; Örm: Örménykút; Sark: Sarkadkeresztúr.

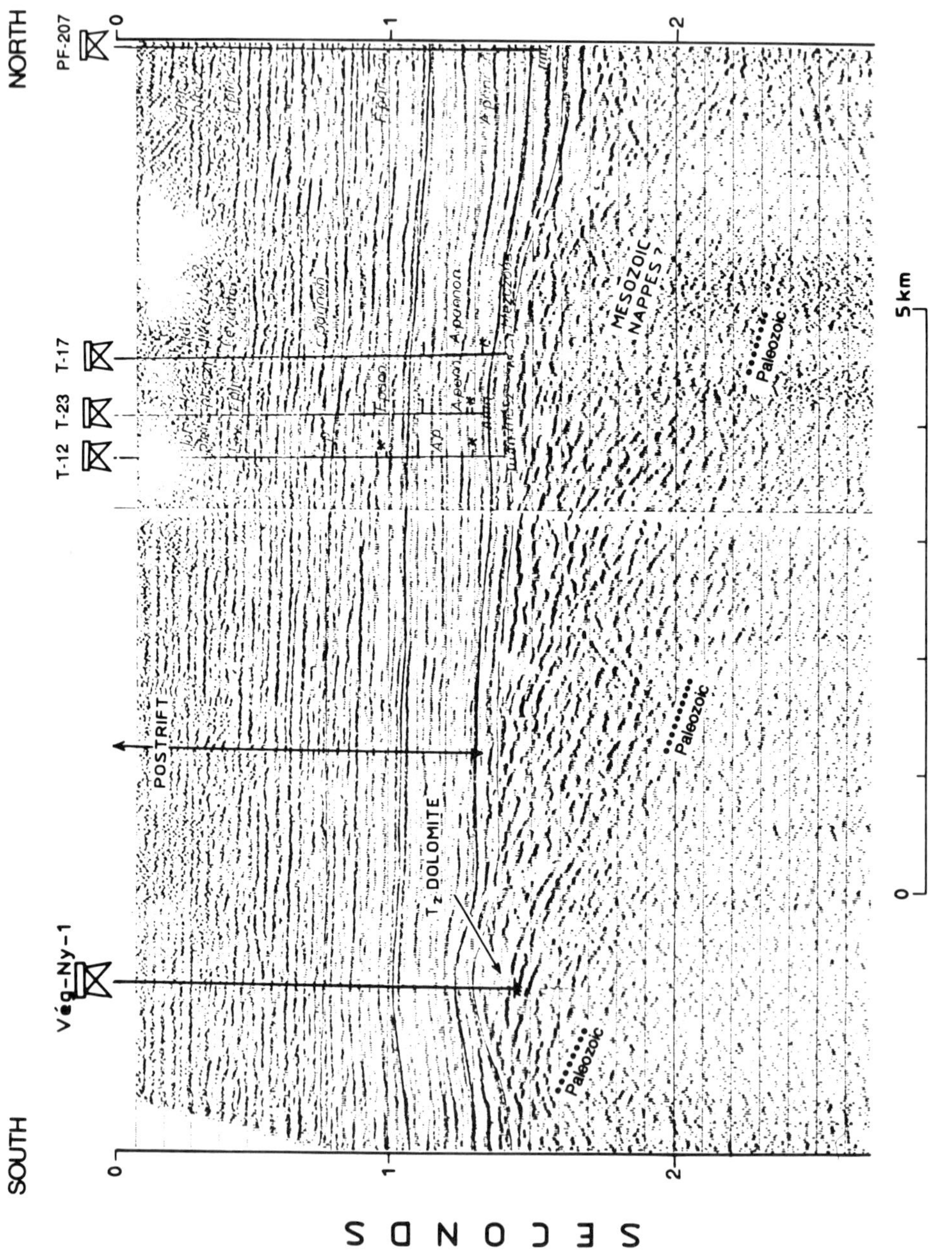

Figure 10. Seismic line To-37 showing the pinchout of deformed Mesozoic sedimentary rocks at the southern edge of the Battonya-Pusztaföldvár Mesozoic trough. The location of line is shown in Figure 9. T_2: Middle Triassic; Vég: Végegyháza; T: Tótkomlós; Pf: Pusztaföldvár.

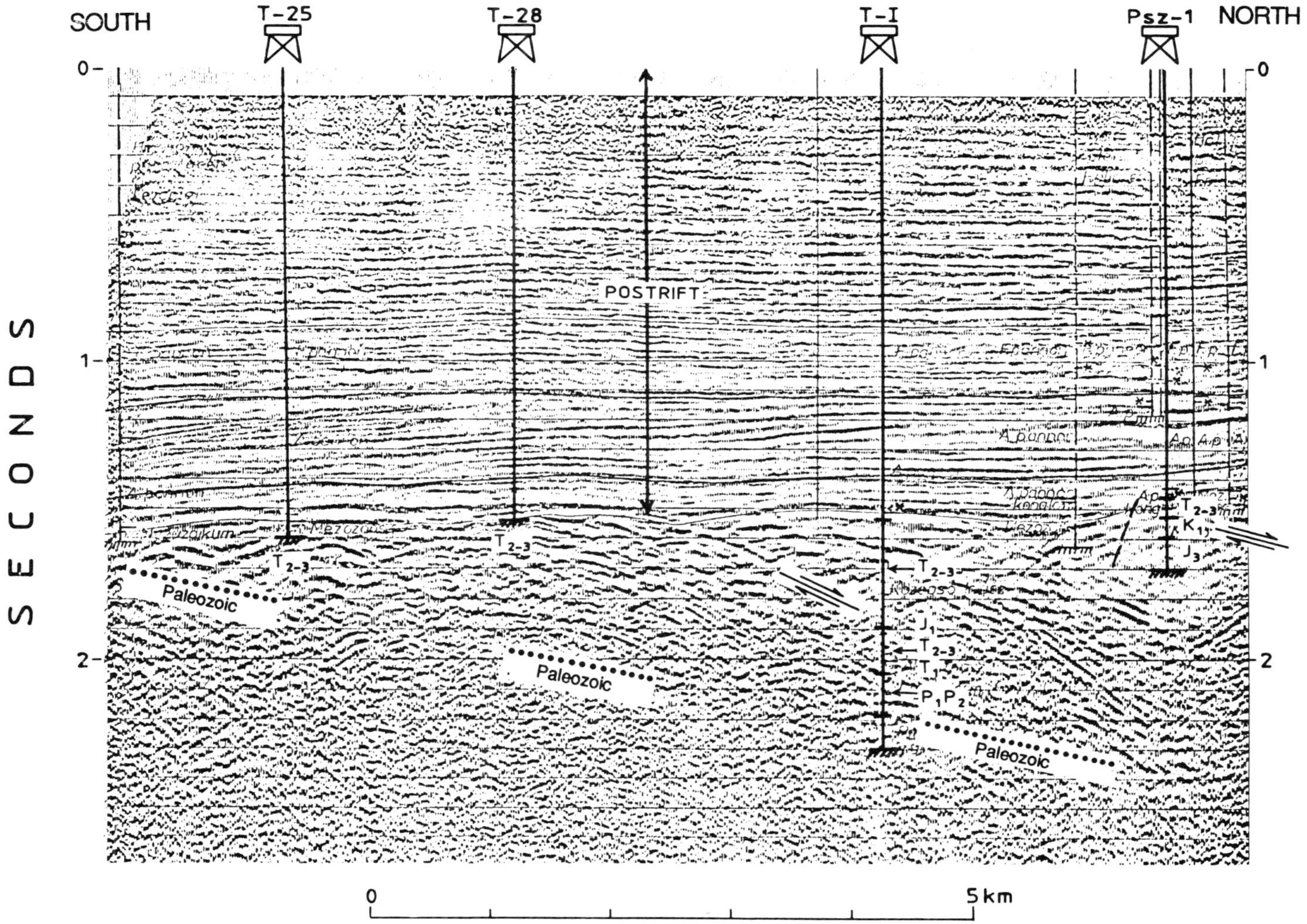

Figure 11. Seismic line To-31 across the center of the Battonya-Pusztaföldvár Mesozoic trough. Age inversions reported from the Tótkomlós (T)-I and other nearby wells indicate the presence of thrust faults. These faults are believed to have formed initially in Late Cretaceous time when nappes were displaced northward, similar to the direction of displacement of nappes in the Apuseni Mountains. Faults may have been reactivated in the Miocene as extensional faults. Location of line is shown in Figure 9. Psz: Pusztaszölös; P_1: lower Permian; P_2: upper Permian; T_2: Middle Triassic; T_3: Upper Triassic; J_1: Lower Jurassic; J_3: Upper Jurassic; K_1: Lower Cretaceous.

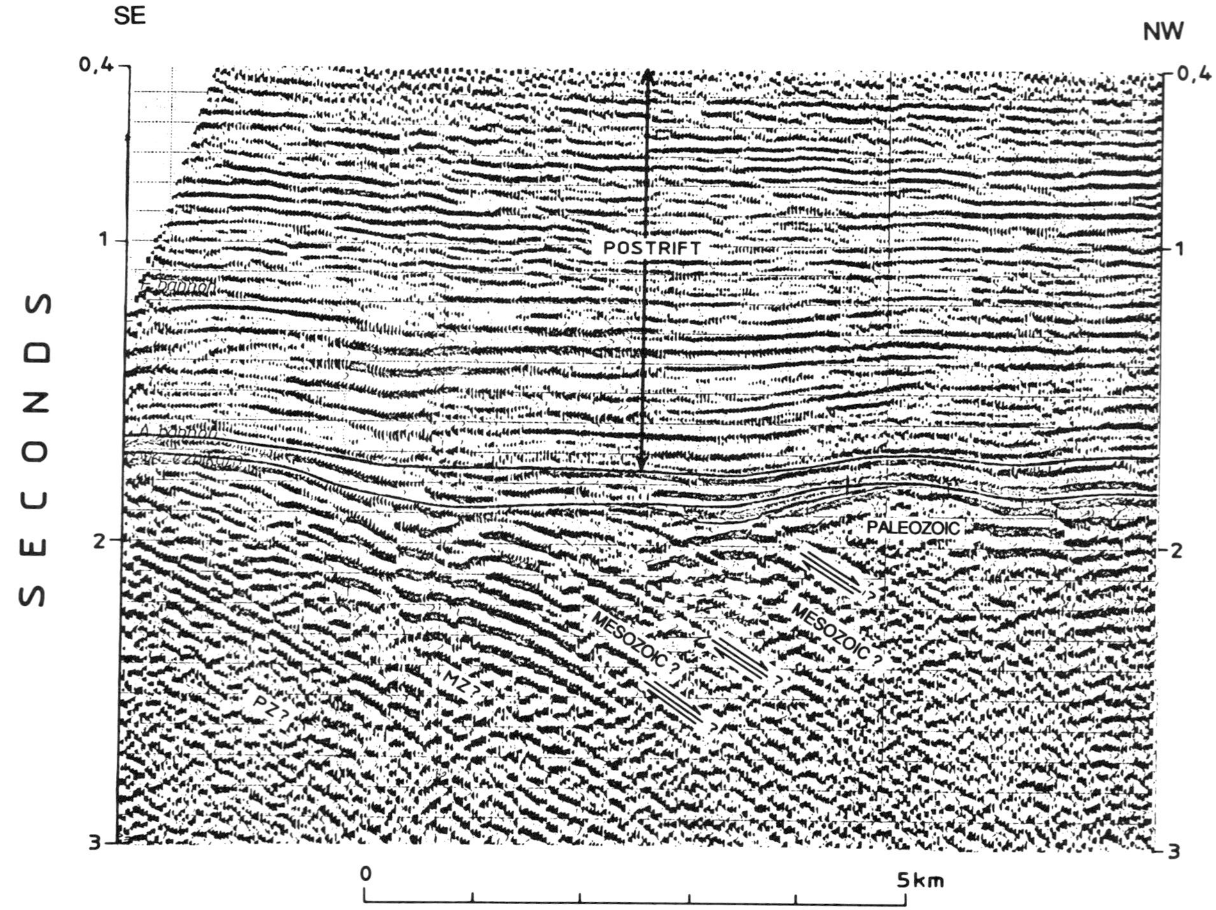

Figure 12. Seismic line To-19 across the north flank of the Battonya-Pusztaföldvár Mesozoic trough showing interpreted north-dipping thrust faults of early Late Cretaceous age. Note that on the northwest side of the profile, Paleozoic (Pz) rocks are inferred to have been thrust over Mesozoic (Mz) rocks. Location of profile is shown in Figure 9.

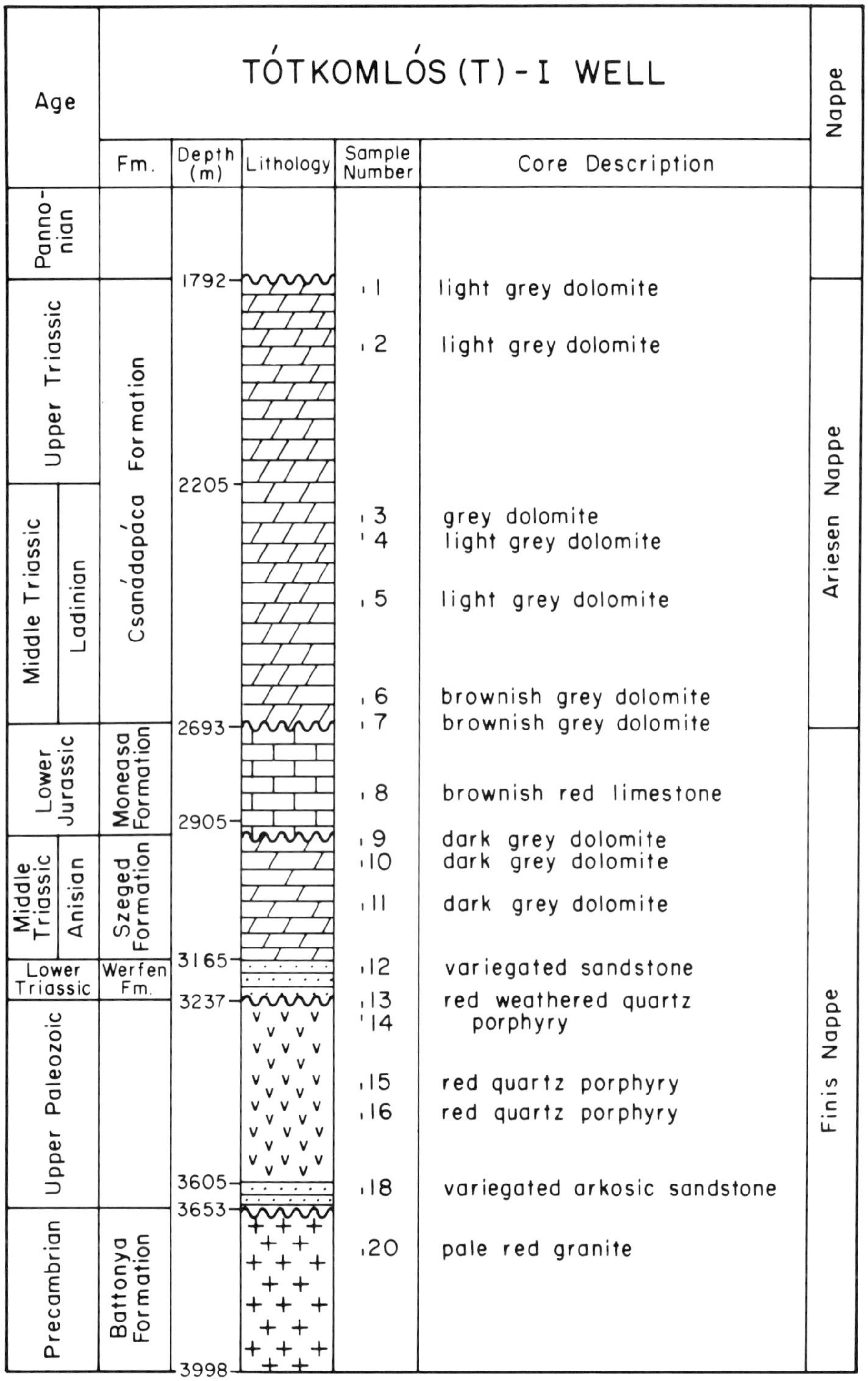

Figure 13. Summary of ages and lithologies of Mesozoic rock units penetrated in the Tótkomlós (T)-I well. Numbers show locations of cores. Note age inversion at a depth of 2693 m.

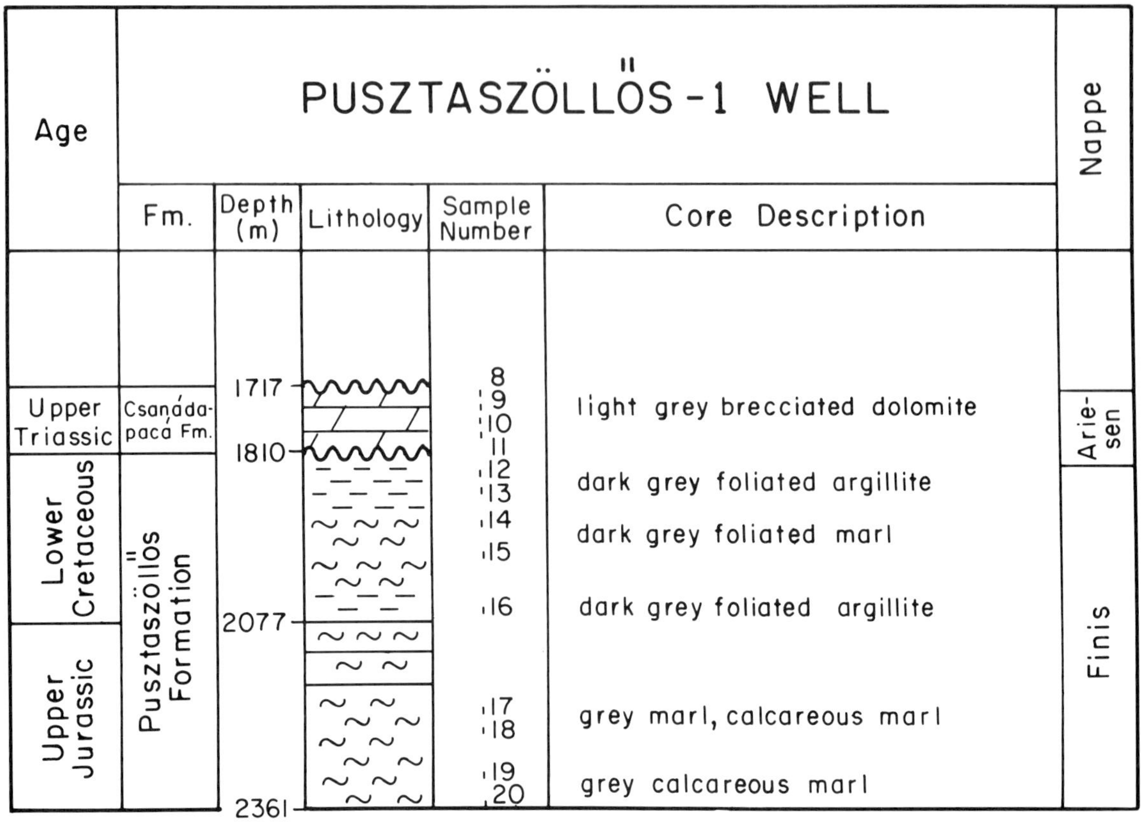

Figure 14. Summary of ages and lithologies of Mesozoic rock units penetrated by the Pusztaszöllös (Psz)-1 well. Numbers 8-20 show locations of cores. Note age inversion at a depth of 1810.5 m.

nappe system can be correlated to Mesozoic rocks penetrated by wells in the Csanádapáca, Pusztaszöllös and Tótkomlós areas (Fig. 15) of the Békés basin.

The best correlations between rock units of the Battonya-Pusztaföldvár area and the Codru nappe system occur in the Triassic section. In the Battonya-Pusztaföldvár trough, gray to purple sandstones occur at the base of the Lower Triassic section (Domb DNy-1 well) and grade upward to variegated shales and red sandstones (locally containing thin layers of anhydrite) of the Werfen Formation that represents deposition in a shallow-water environment (Fig. 15). The Werfen Formation is recognized by Bleahu and others (1981) in the Finis, Dieva and Arieseni nappes of the Codru nappe system in Romania (Fig. 16). At the base of the Middle Triassic section in the Battonya-Pusztaföldvár trough, a thick sequence of dark gray dolomite of the Szeged Formation overlies the Werfen Formation (Csa-2, -4, -5, -6, Oros-2, T-2, -6, -28, -29, -30, -31, Tk-1, and Vég Ny-1 wells). This sequence of dark gray dolomite is lithologically similar to a sequence of dark gray dolomite of Early-Middle Triassic age that overlies the Werfen Formation in the Codru nappe system (Fig. 16). In the Battonya-Pusztaföldvár trough, the Csanádapáca Formation of Middle-Late Triassic age is ubiquitous and is characterized by shallow-water carbonate rocks and lagoonal facies. The basal part of the Csanádapáca Formation consists of dark gray, marly limestone and shaley marl (Fig. 15). These basal deposits are overlain by a brown-gray, breccia dolomite that contains a characteristic assemblage of algae. The brown-gray dolomite grades upward to a light gray dolomite that is ubiquitous in the Battonya-Pusztaföldvár trough area. In the Codru nappe system, exactly the same sequence of Middle-Upper Triassic rocks can be found with the exception of white dolomite (Patrulius and others, 1979).

Correlations between Jurassic rock units of the Battonya-Pusztaföldvár trough and Codru nappe system are not as conclusive as were correlations of the Triassic section. The Jurassic section of the Battonya-Pusztaföld-

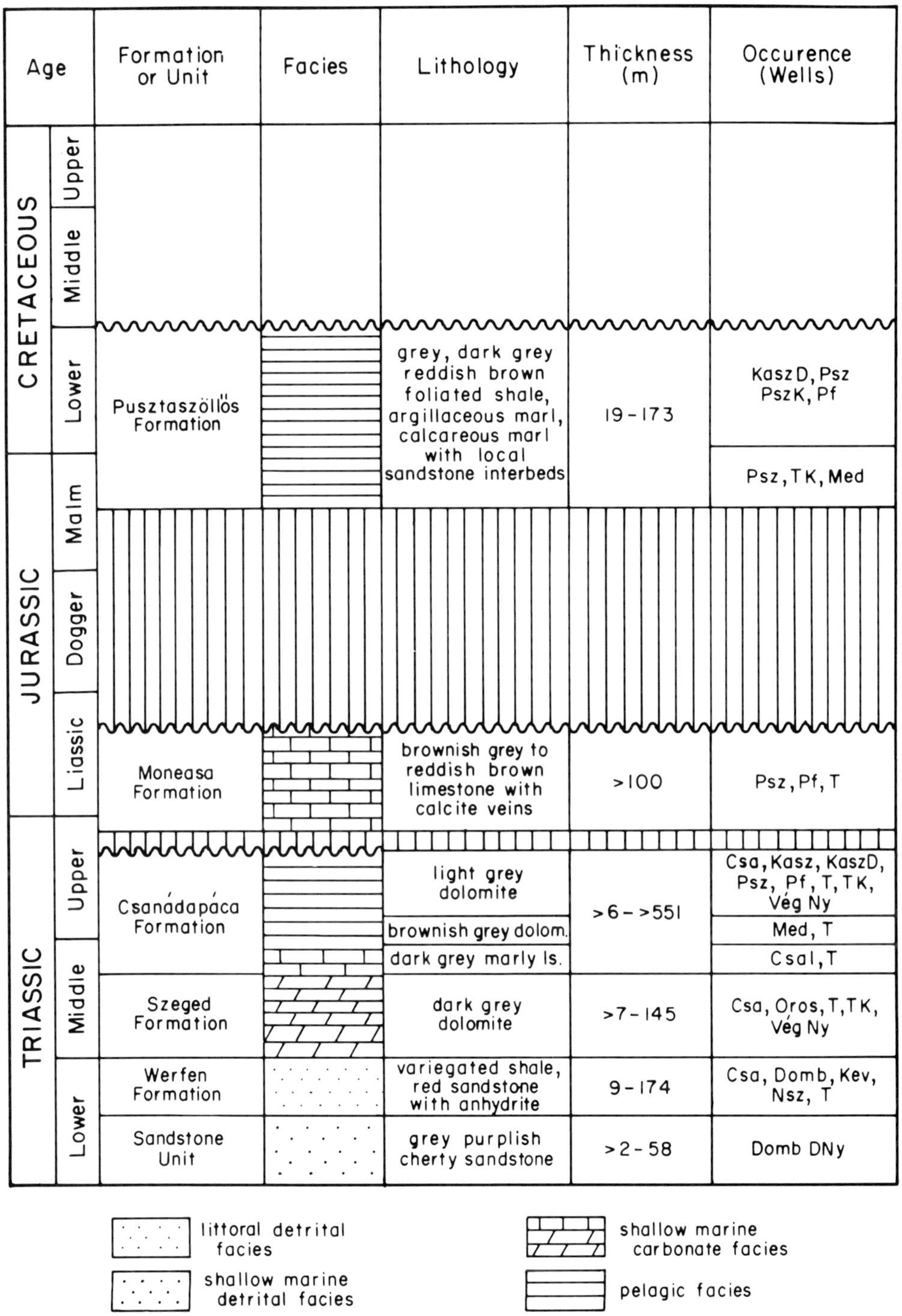

Figure 15. Summary of ages, facies and lithologies of Mesozoic rock units in the Battonya-Pusztaföldvár Mesozoic trough. Csa: Csanádapáca; Oros: Orosháza; T: Tótkomlós; Pf: Pusztaföldvár; Med: Medgyesbodzás; Domb: Magyardombegyháza; Psz: Pusztaszöllös; Nsz: Nagyszénás; Kasz: Kaszaper; Kev: Kevermes; Vég: Végegyháza; Csal: Csanádalberti.

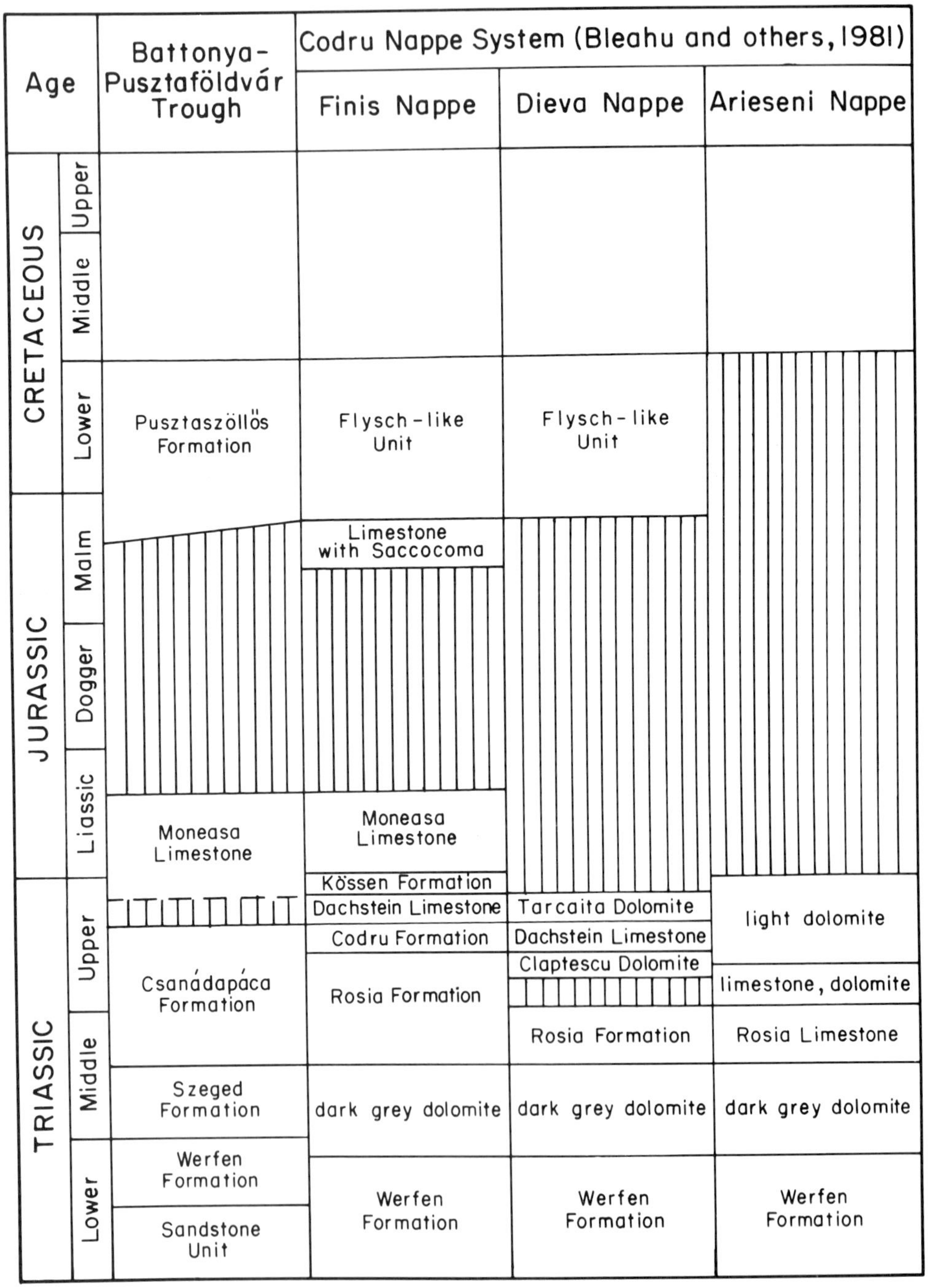

Figure 16. Inferred correlation of the Mesozoic rock units in the Battonya-Pusztaföldvár Mesozoic trough with rock units of the Codru nappe system in the Apuseni Mountains (Bérczi-Makk, 1986; Bleahu and others, 1981).

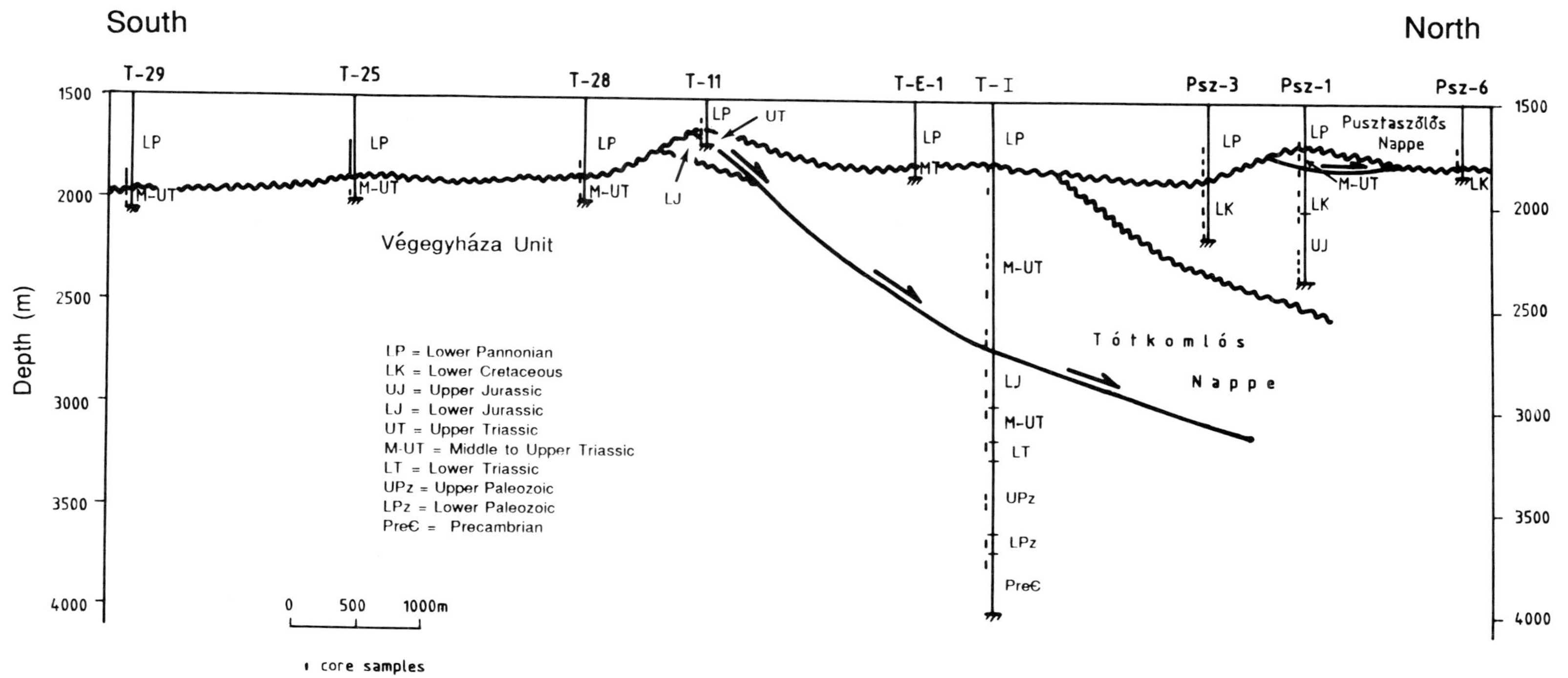

Figure 17. North-south section across the Battonya-Pusztaföldvár Mesozoic trough. At least two thrust faults are inferred to exist. Wavy lines indicate erosional unconformities. T: Tótkomlós; Psz: Pusztaszöllös; LP: lower Pannonian; LK: Lower Cretaceous; UJ: Upper Jurassic; LJ: Lower Jurassic; UT: Upper Triassic; MT: Middle Triassic; LT: Lower Triassic; UP: upper Permian; LP: lower Permian; PZ: Paleozoic.

vár trough is incomplete compared to the same age section in the Codru nappe system. The upper Liassic system in the Battonya-Pusztaföldvár trough is known from only a few wells (Psz-12, -13, Pf-128, -129, T-I and T-11) and is characterized by shallow-water limestones and a red-brown, clayey, crinoid sandstone of the Moneasa Limestone Formation (Fig. 15). The Moneasa Formation can be correlated with a similar assemblage of rocks found in the Codru nappe system (Bleahu and others, 1981).

Whether Cretaceous rock units of the Battonya-Pusztaföldvár Mesozoic trough can be correlated with rock units of similar age in the Codru nappe system is inconclusive. The Lower Cretaceous part of the Pusztaszöllös Formation (Fig. 15) in the Battonya-Pusztaföldvár trough consists of pelitic and tectonized gray to brownish-red, shaley and clayey marl, marl, carbonate marl and limestone with several interbeds of sandstone. Fossils, which could be used to determine age, have not been recovered from the Pusztaszöllös Formation. Although the rocks of the Pusztaszöllös Formation are tentatively correlated with the "flysch-like" deposits of the Finis and Dieva nappes of the Codru nappe system (Fig. 16), the only certain similarity between the units is that they both represent deposition in relatively deep water.

The Finis, Dieva, and Arieseni nappes of the Apuseni Mountains are inferred to represent allochthonous units thrust from south to north (Bleahu and others, 1981; Bérczi-Makk, 1986). By analogy, nappes in the Battonya-Pusztaföldvár trough also are inferred to represent allochthonous units thrust from south to north.

A diagrammatic, north-south section across the Battonya-Pusztaföldvár trough summarizes the main thrust surfaces and nappes that are inferred to occur in the area (Fig. 17). The Pusztaszöllös nappe, composed of Middle-Upper Triassic rocks, overlies the Tótkomlós nappe composed of Middle Triassic-Lower Cretaceous rocks. In the Tótkomlós nappe, Upper Jurassic rocks unconformably overlie Upper Triassic rocks - the Lower Jurassic section is absent because of non-deposition or erosion. The basal Végegyháza unit is composed of Middle Triassic-Lower Jurassic rocks with an erosional unconformity between the Triassic and Jurassic sections. Seismic reflection data (Figs. 10 and 11) suggest that Mesozoic bedding planes dip north. Thrust sheets mapped in the Apuseni Mountains (Fig. 1) are sub-parallel to bedding, and are interpreted to have been thrust in a south-to-north direction (Bleahu and others, 1981). Therefore, by analogy with nappes in the Apuseni Mountains, thrust sheets in the Battonya-Pusztaföldvár Mesozoic trough are inferred to have been displaced northward. Thrusting probably occurred during Late Cretaceous time as the youngest rocks involved in the thrusting are of Early Cretaceous age. The original thrust planes may have been low-angle, south-dipping surfaces; however, later uplift may have reversed the of dip of the faults during latest Cretaceous folding or, possibly, during Cenozoic extensional deformation.

Békés-Doboz Mesozoic Trough

Three seismic profiles across parts of the Békés-Doboz Mesozoic trough are shown in Figures 18-20 (profiles A-16/a/m, Gyu-38 and Gyu-30, respectively). Figure 20, and to a lesser degree Figures 18 and 19, show low-frequency reflections which dip northwestward beneath a unit which is interpreted to consist of deformed Mesozoic sedimentary rocks. Past interpretations of these low-frequency reflections suggest that they mark a south-verging thrust fault of Cretaceous age or the contact between Mesozoic and Paleozoic rocks (unpublished reports of the Hungarian Oil and Gas Co. (MOL Ltd.)). The maximum thickness of Mesozoic rocks in the Békés-Doboz trough is difficult to discern on these seismic profiles; however, they are estimated to be 3750-5000 m thick based on a two-way travel time of 1.5-2.0 seconds (Fig. 20) using an average velocity of 5000 m/sec.

The Békés-2 well (Fig. 21) penetrated a thrust fault in the Mesozoic section at a depth of 4145 m (about 2.73 seconds in Fig. 21). The evidence for this thrust is that Middle Triassic rocks overlie Upper Cretaceous rocks at this depth (Upper Cretaceous-Middle Triassic rocks occur in the 3339-4145-m interval and Upper Cretaceous(?)-Upper Triassic in the 4145-5500-m interval) (Fig. 22). This thrust surface appears to be marked on the northwest and southeast sides of the Békés-2 well by weak reflections (Fig. 21).

Strong, low-frequency, northwest-dipping reflections on the northwest side of Békés-2 well appear to occur within the lower Mesozoic unit (Fig. 21). Vertical seismic experiments in the Békés-2 well (Lee and Göncz, this volume) and a model constructed from the seismic data of profile Gyu-30 (Miller and Véges, this volume) indicate that the strong, low frequency reflections represent discontinuous, low-impedance zones in the Mesozoic section. The fact that these zones are not mappable on seismic profiles for distances of more than a few kilometers suggests that strata were strongly folded and beds offset by faults. In addition, core descriptions from the

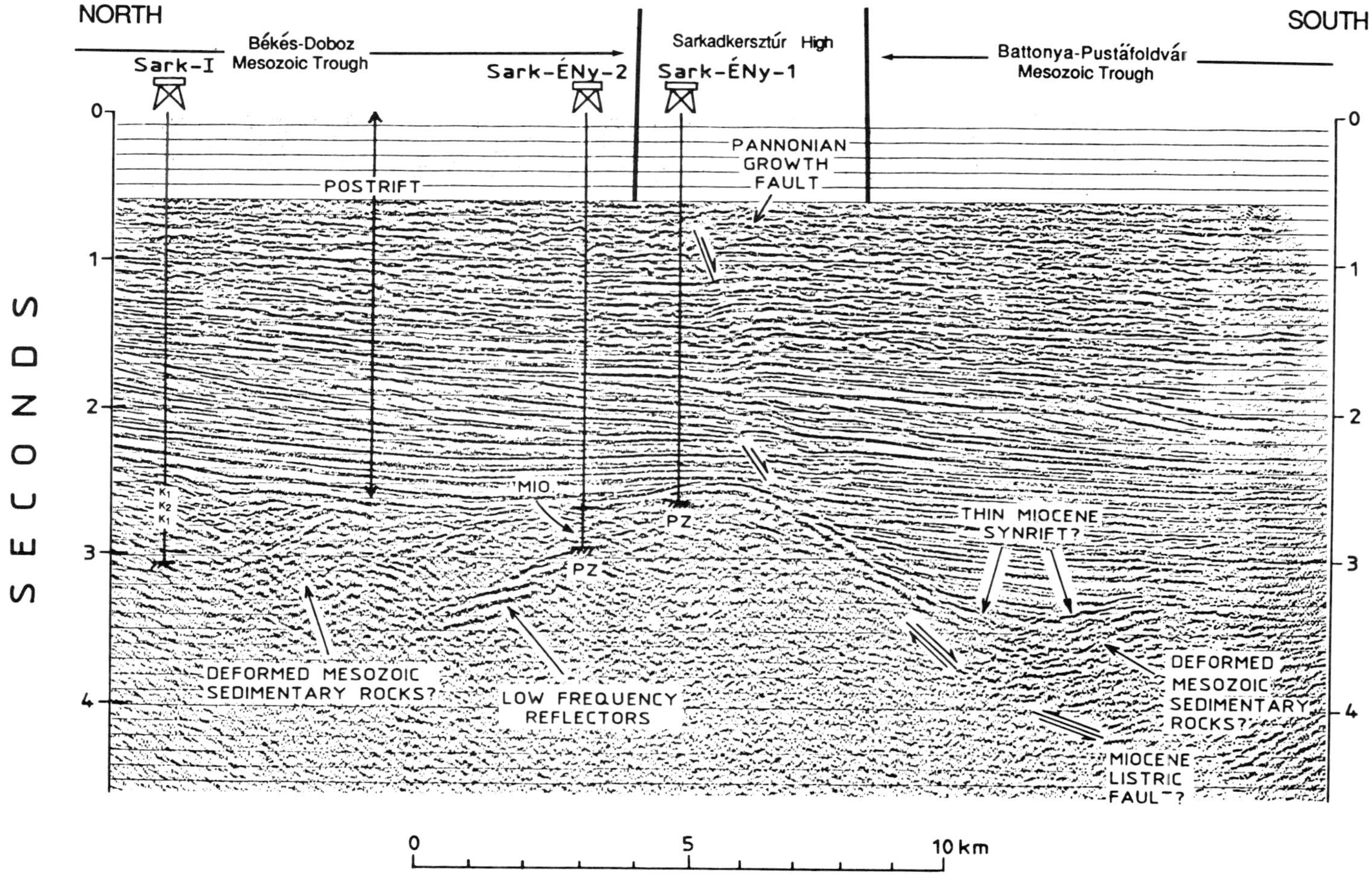

Figure 18. Seismic line A-16/a/n over the Sarkadkeresztúr high. In the Sarkadkeresztúr (Sark)-I well, Lower Cretaceous (K1) rocks overlie Upper Cretaceous (K2) rocks, as a result of thrust faulting. In this area, low frequency, north-dipping reflections appear to correlate with the Paleozoic (Pz) basement rocks penetrated in the Sark-ENy-1 well. Fractured Miocene (Mio) rocks overlie the Paleozoic basement in the Sark-ENy-2 well (Phillips and others, this volume). Deformed Mesozoic rocks occur at the south end of the line in the Békés-Doboz trough. These Mesozoic rocks are similar in age, lithology and depositional environment to rocks of the Bihor autochthon in the Apuseni Mountains (Fig. 24). Location of profile is shown in Figure 9.

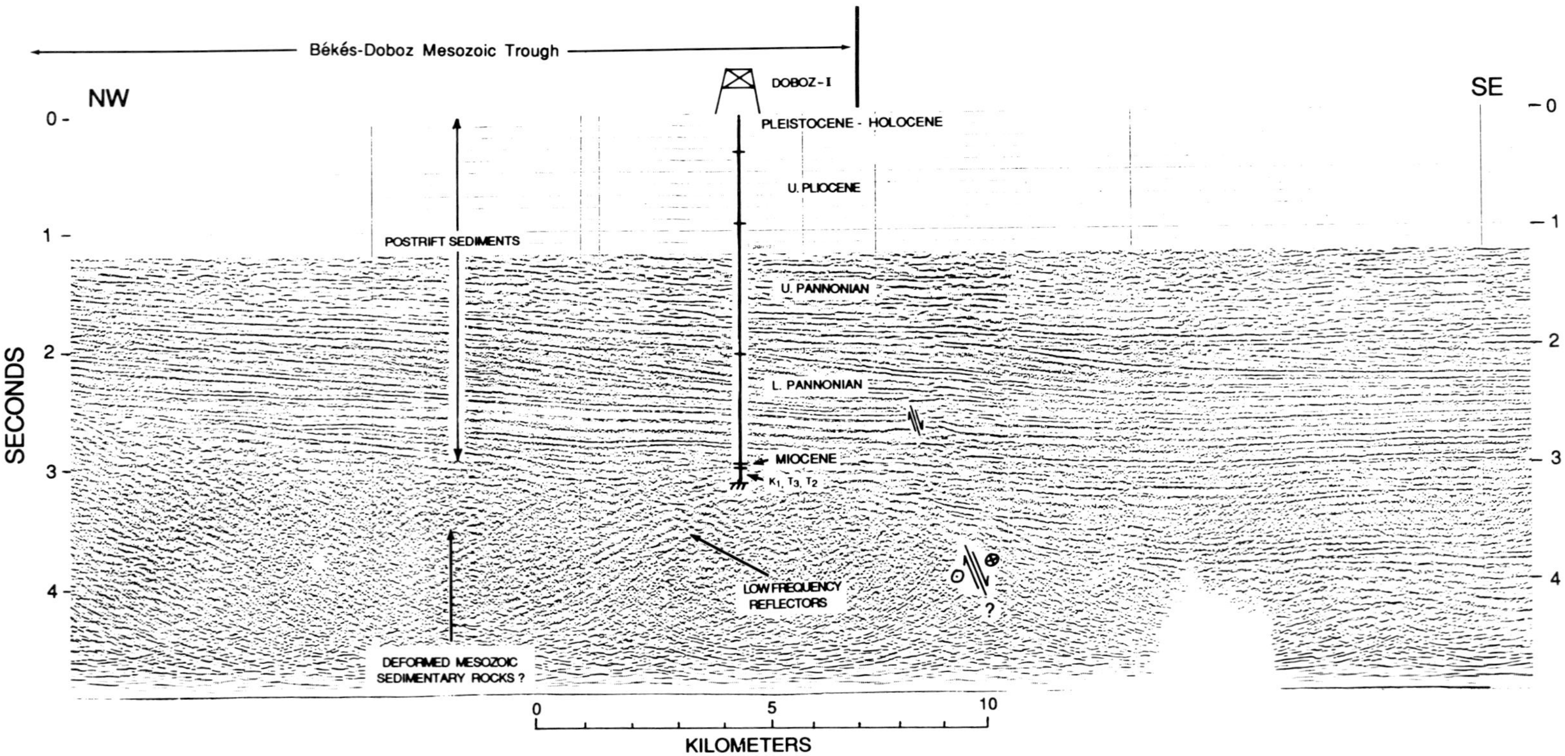

Figure 19. Seismic line Gyu-38 across the Békés-Doboz Mesozoic trough. The Doboz-I well encountered an unreversed sequence of Lower Cretaceous (K_1), Middle Triassic (T_2), and Upper Triassic (T_3) rocks which can be correlated to the rocks of the Bihor autochthon (Fig. 24). Location of seismic profile is shown in Figure 9.

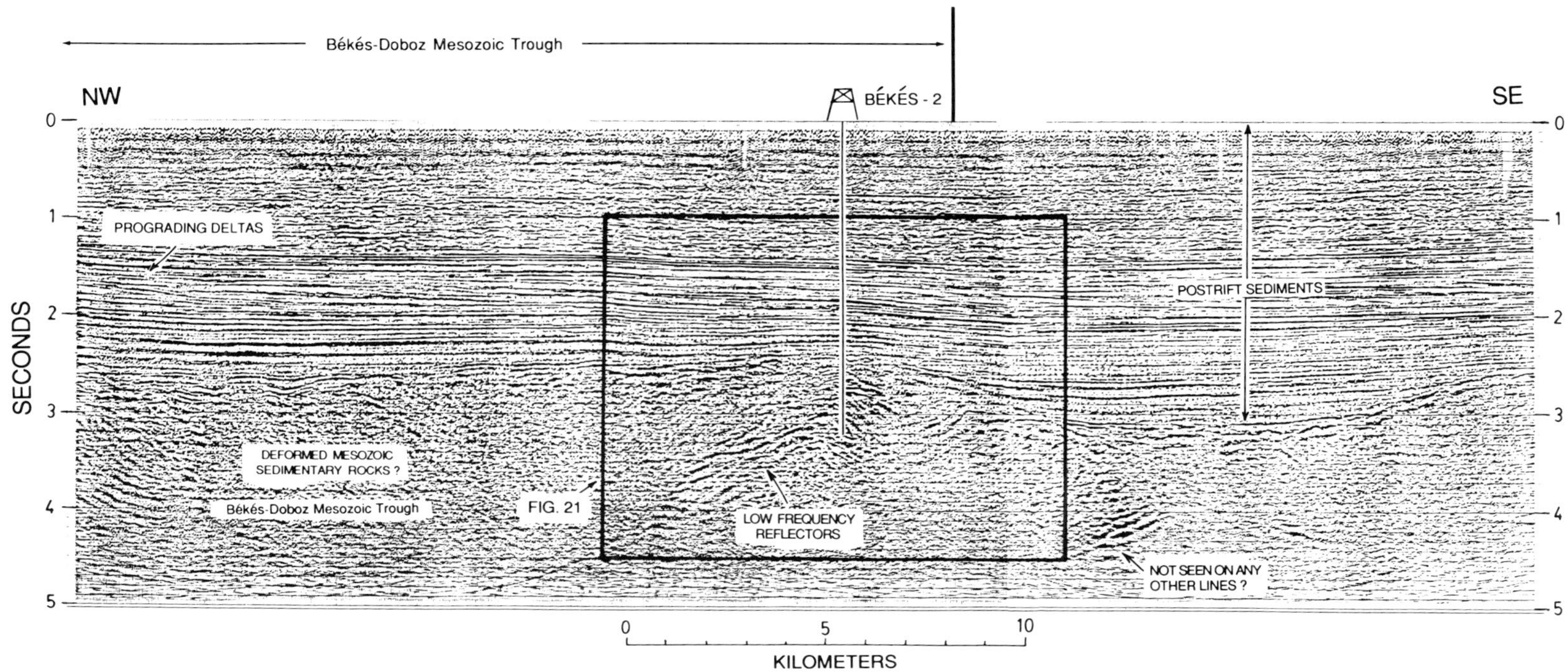

Figure 20. Seismic line Gyu-30 over the Békés-Doboz Mesozoic trough. The Békés-2 well was projected 0.6 km from the west. Location of profile is shown in Figure 9. Interpretation of that part of the profile enclosed in box is shown in the enlargement of Figure 21.

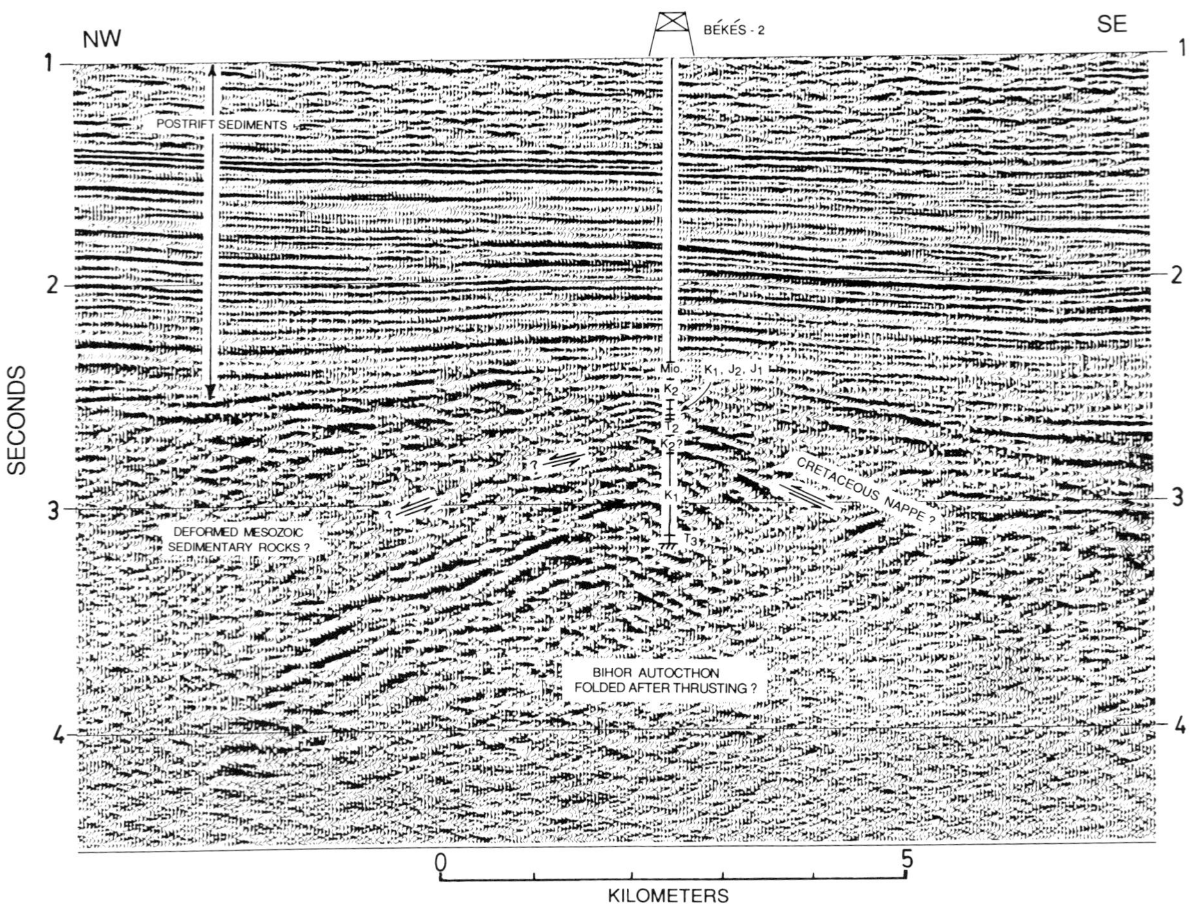

Figure 21. Enlargement of part of seismic profile Gyu-30 (Fig. 20) near the Békés-2 well showing age inversion in the Mesozoic section. The southeast dipping reflections on the right-hand side of the figure (between 2.5 and 2.7 seconds) correlate with the base of the overthrust section. The deeper Mesozoic section (Lower Cretaceous) between 2.7 and 3.1 seconds on the northwest side of the well appears to correlate with northwest-dipping, low-frequency reflectors (see also Fig. 20). These Mesozoic rock units can be correlated with rock units in the Bihor autochthon. K1: Lower Cretaceous; K2: Upper Cretaceous; J1: Lower Jurassic; J2: Middle Jurassic; T2: Middle Triassic; T3: Upper Triassic; Mio: Miocene.

Sarkadkeresztúr (Sark)-I and Békés-2 (Fig. 22) wells indicate that Mesozoic rocks are highly fractured (Phillips and others, this volume). All of these observations indicate that the Mesozoic section in the Békés-Doboz trough was highly deformed as a result of intense compressional thrusting and folding. It is difficult to establish the age of compression and thrusting.

In Figure 18, the Miocene deposits appear to be highly deformed, although to a lesser extent than the Mesozoic section. Clearly, except for growth faults that resulted from differential sediment compaction, the Pannonian postrift section is undeformed. Most likely, the main phase of compression and the associated thrusting occurred during Late Cretaceous time. Deformation of the Miocene deposits are believed to be the result of Miocene extensional tectonism.

The Mesozoic rock units in the Békés-Doboz trough appear to correlate to similar age rock units of the Bihor autochthon in the Apuseni Mountains of Romania rather than to rock units of the Codru nappe system. Rock units of the Bihor autochthon differ from rock units of the Codru nappe system in 3 main aspects: (1) Lower Cretaceous rock units (flysch) in the Codru nappe system represent deeper water deposition compared to rock units of the same age in the Bihor autochthon; (2) in the Codru nappe system, the Jurassic section is relatively thin to absent in comparison with the Bihor autochthon; and (3) the Upper and Middle Triassic sections in the Codru nappe system contain some basinal facies, whereas these sections represent mainly shallow-marine, carbonate facies in the Bihor autochthon.

A summary of the ages, formations, lithologies and depositional environments of Mesozoic units penetrated by wells in the Békés-Doboz trough is given in Figure 23. Although good correlation between the Békés-Doboz trough and the Bihor autochthon (Fig. 24) can be made by comparing the Cretaceous sections of the two areas, correlation of Triassic rock units from the two areas is less conclusive. Comparisons between the Jurassic sections of the two areas can not be made because the Jurassic section in the Békés-Doboz trough is incomplete (Fig. 23); however, future drilling may reveal a more complete Jurassic section.

The Lower Triassic section in the areas of the Békés-Doboz trough, the Bihor autochthon, the Battonya-Pusztaföldvár trough and the Codru nappe system are similar. In these four areas, the Lower Triassic section consists of a basal sandstone unit overlain by rocks of the Werfen Formation, which was discussed in the previous section of this chapter (Figs. 16 and 24).

Middle and Upper Triassic rocks units penetrated by wells in the Békés-Doboz trough appear to differ from similar age rock units in the Battonya-Pusztaföldvár and Codru nappe systems, but are inferred to correlate to Middle and Upper Triassic rocks units in the Bihor autochthon as shown in Figure 24. Rocks of the Anisian-age (Middle Triassic), Siklós Formation in the Békés-Doboz trough consist of thick, brown-gray, light to gray dolomites deposited in shallow water in a lagoonal environment (Bihu-I, Doboz-I, Fáb-4, Kom-4, Köt-I and Sár-I wells). According to Bérczi Makk (1985), these rocks of Anisian age contain a fossil assemblage (*Glomospira sp., Glomospira sinensis Ho, Glomospira cf. tanuifistula Ho, Glomospirella sp.*) that can be correlated to the Crisul Repede (Sebes Körös) Dolomite Formation of the Bihor autochthon. Both the Doboz Limestone Formation (Békés-Doboz trough) and the Wetterstein Limestone Formation (Bihor autochthon) of Early-Middle Triassic age (Fig. 24) represent shallow-water, carbonate platform deposition. The Doboz Formation, however, consists of dark gray to black limestone, whereas, the Wetterstein Formation consists chiefly of white to gray limestone. Although the Upper Triassic Kárpáti (Carpathian) Keuper Formation in the Békés-Doboz trough is tentatively correlated with the Scarita Formation of the Apuseni Mountains (Fig. 24), the Kárpáti Keuper Formation consists of carbonate rocks grading upward to clastic rocks, the Scarita Formation consists only of clastic rocks.

Good correlation exists between Lower and Middle Cretaceous rock units of the Békés-Doboz trough and rock units of the Bihor autochthon (Fig. 24). The base of the Cretaceous section in the Békés-Doboz trough is marked by a strong unconformity which is correlated to the occurrence of bauxite in the Bihor autochthon. In both the Békés-Doboz trough and the Bihor autochthon, the basal part of the Cretaceous section contains coarse-grained sandstones which are inferred to mark the initial Cretaceous transgression. The Biharugra Formation of the Békés-Doboz trough consists mainly of dark gray limestone and carbonate marl, and oolitic limestone occurs at the base and the top of the formation (Békés-2, Bihu-I and Doboz-I wells). According to Dragastan and others (1986), the Biharugra Formation is lithologically similar to the Hodobana Formation of the Bihor autochthon. However, it is also possible that the Biharugra Formation is of Late Jurassic age and correlative with the Albioara Formation (Fig. 24). Middle-Early Cretaceous deposition (Nagyharsány Formation, Fig. 24)

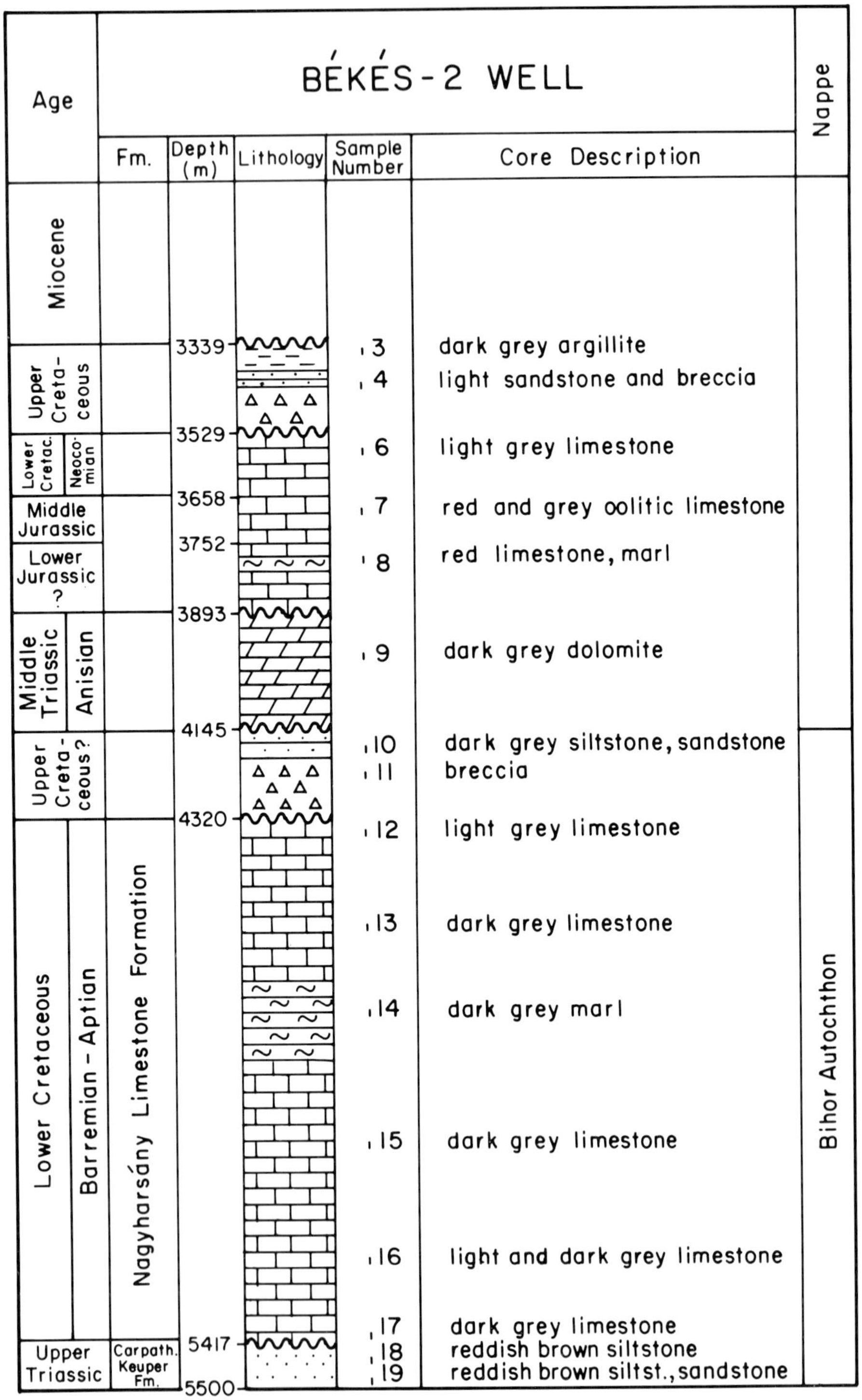

Figure 22. Summary of ages and lithologies of rocks penetrated in the Békés-2 well. Numbers show locations of cores. Note age inversion at a depth of 4145 m.

Age		Formation or Unit	Facies	Lithology	Thickness (m)	Occurence (Wells)
CRETACEOUS	Upper					
CRETACEOUS	Middle					
CRETACEOUS	Lower	Nagyharsány Formation		light brownish grey argillaceous limestone with calcite veins	>77 - >1081	Békés, Sark, Bihu
CRETACEOUS	Lower	Biharugra Formation		dark grey marl, calcareous marl, oolitic limestone	20-200	Békés, Doboz, Bihu
CRETACEOUS	Lower			red siltstone & congl.	51-110	Bihu, Doboz
JURASSIC	Malm					
JURASSIC	Dogger	?		red & grey oolitic lms.	94-135	Bihu, Békés
JURASSIC	Liassic					
TRIASSIC	Upper	Carpathian Keuper Fm.		red siltstone, sandstone with grey limestone	13-305	Bihu, Doboz, Békés
TRIASSIC	Middle	Doboz Formation		dark grey argillaceous limestone calcareous marl	>160-197	Bihu, Doboz
TRIASSIC	Middle	Siklós Formation		brownish grey, compact dolomite with light grey limestone	>16-295	Bihu, Doboz, Fáb, Kom, Köt, Örm, Sár
TRIASSIC	Lower	Werfen Formation		variegated shale, grey sandstone	>49-243	Fáb, Gyoma, Köt
TRIASSIC	Lower	Sandstone Unit		variegated massive sandstone, red shale	77	Bihu

littoral detrital facies — shallow marine carbonate facies

shallow marine detrital facies — neritic-pelitic facies

Figure 23. Summary of ages and lithologies of Mesozoic rock units in the Békés-Doboz Mesozoic trough. Sark: Sarkadkeresztúr; Kom: Komádi; Fáb: Fábiánsebestyén; Köt: Köröstarcsa; Örm: Örménykút; Sár: Sáránd; Bihu: Biharugra.

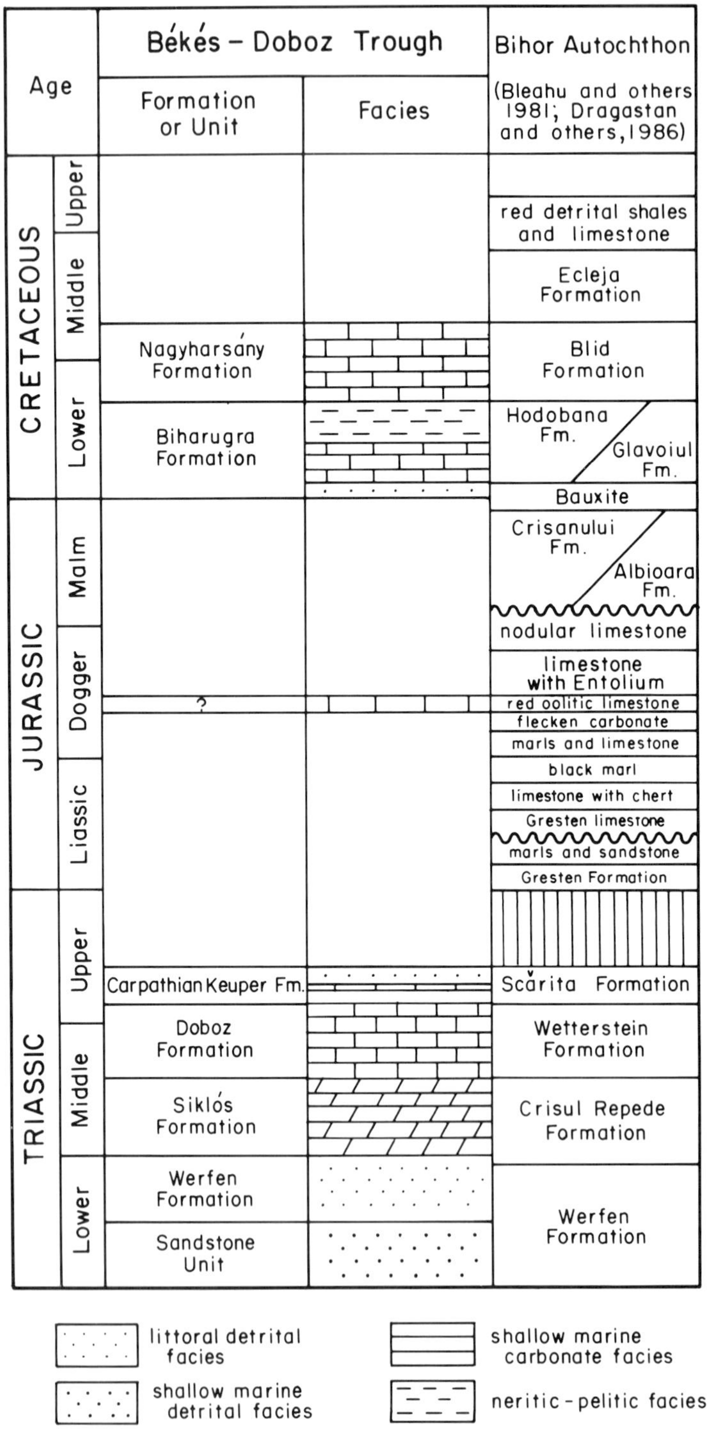

Figure 24. Correlation of Mesozoic formations and rock units in the Békés-Doboz Mesozoic trough with those of the Bihor autochthon in the Apuseni Mountains.

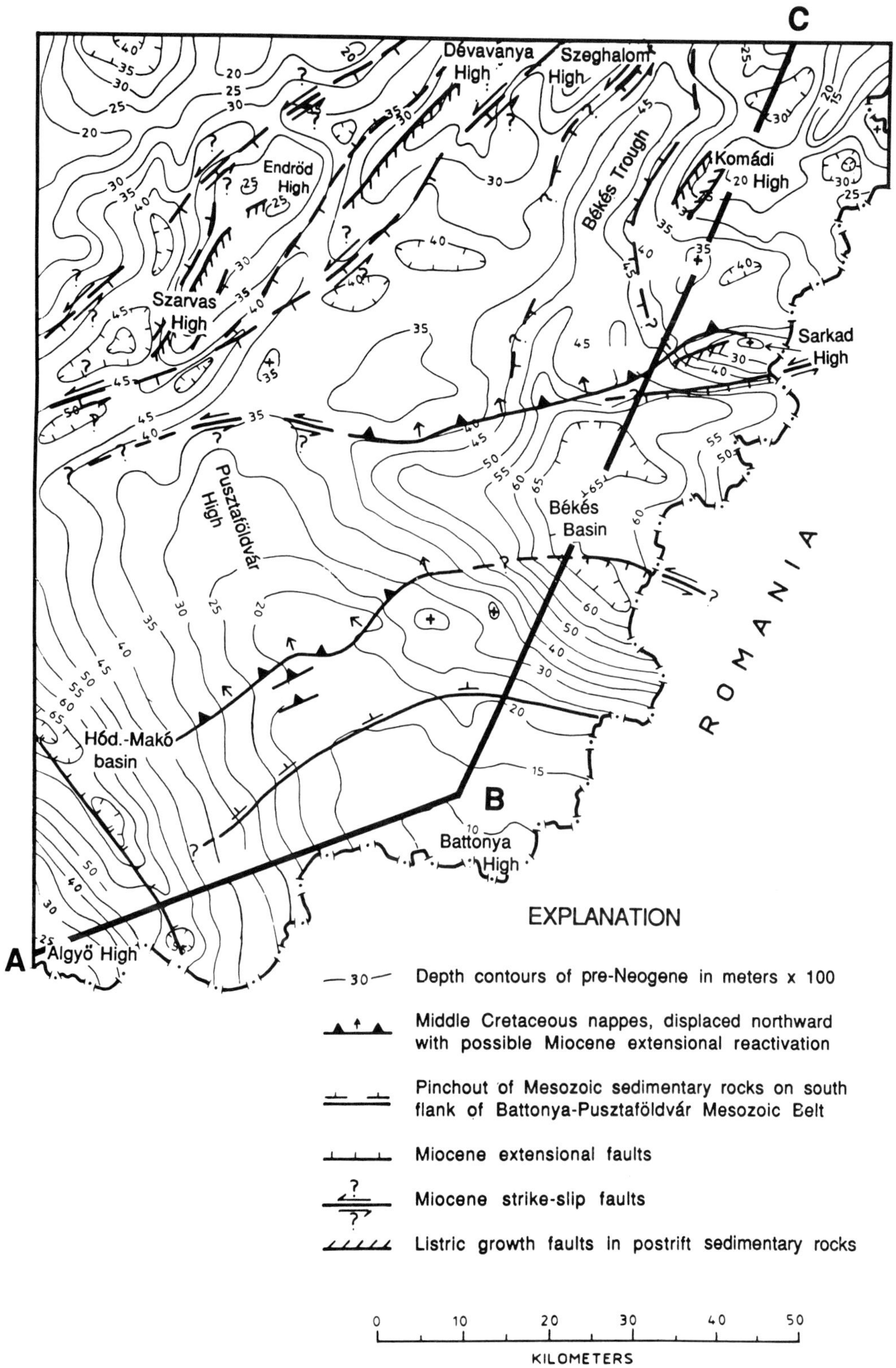

Figure 25. Tectonic map of the Békés basin and surrounding region. ABC shows location of gravity and magnetic profiles of Figure 28. Hód- Makó: Hódmezövásárhely-Makó; Sark: Sarkadkeresztúr.

in the Békés-Doboz trough is represented by a thick (1081 m), brown to gray, clayey, calcitic *Orbitolina* limestone that represents deposition in warm, shallow water (Bihu-I, Bihu-1, Bihu Ny-2 and Sark-I wells). The *Orbitolina* limestone of the Nagyharsány Formation is similar to the limestones of the Blid Formation in the Bihor autochthon.

By analogy with the Bihor autochthon in the Apuseni Mountains, we assume that the nappes of the Békés-Doboz trough, were displaced northward only a short distance. The present northward dip of the low-frequency reflections in the Békés-Doboz Mesozoic trough (Figs. 18-20) may be due to late-stage folding of the Mesozoic section during Late Cretaceous time. Alternatively, extensional rotations during Miocene time may have caused this northward dip.

LISTRIC NORMAL FAULTS IN THE NEOGENE SECTION

Throughout most of the Békés basin, the postrift Neogene sedimentary section, which mainly consists of lacustrine deltaic and prodeltaic sandstones, shales and marls, is relatively undisturbed by faulting with the exception of listric normal faults. The strike of these faults is parallel to the northwestern and northeastern bounda-

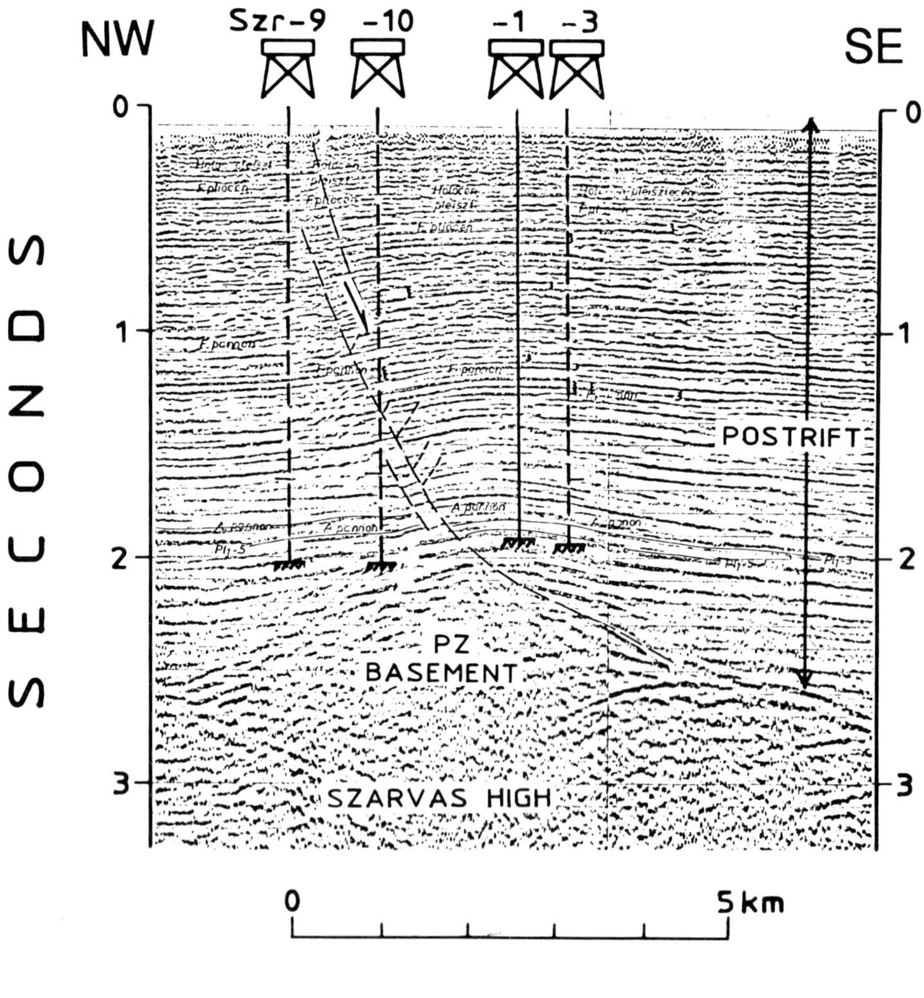

Figure 26. Seismic line Szr-16 over the Szarvas (Szr) high showing growth faults in the postrift sedimentary section. Numerous growth faults similar to these occur in the Békés basin along the crest of basement highs. Note that the faults do not appear to offset basement, but flatten and parallel the basement surface at depth. The faults are inferred to be the result of differential compaction of sediments. Location of profile is shown in Figure 5. Pz: Paleozoic.

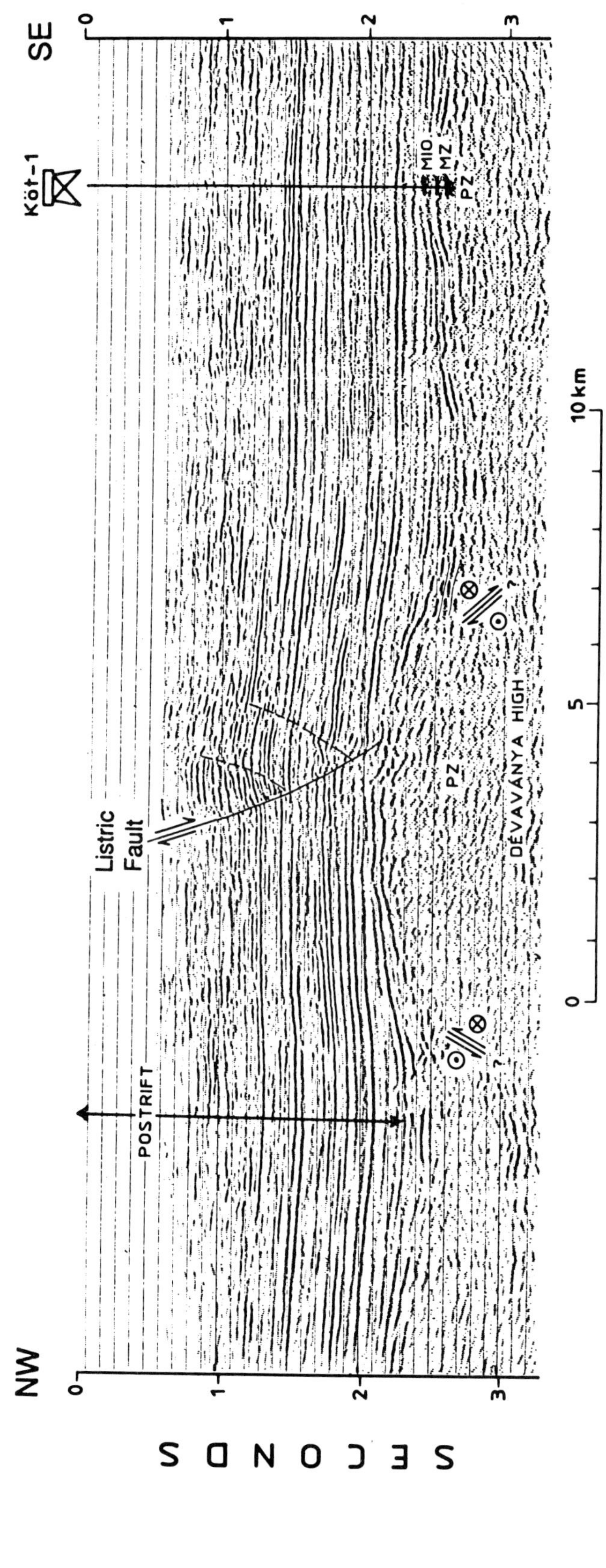

Figure 27. Seismic line DvG-18 over the Dévaványa high showing another example of growth faults in the postrift sediments. Oblique extensional faults of middle Miocene age with left-slip offsets are interpreted on the southeast and northwest flanks of the Dévaványa high, but the younger growth faults are probably due primarily to differential compaction of sediments. Location of profile is shown in Figure 5. Pz: Paleozoic; Mz: Mesozoic; Mio: Miocene; Köt: Köröstarcsa.

ries of the Békés basin in the vicinities of the Szarvas, Endröd, Dévaványa, Komádi, and Sarkadkeresztúr basement highs (Fig. 25). Seismic profiles A-16/a/n, Szr-16, and DvG-18 (Figs. 18, 26 and 27, respectively) illustrate several of these faults. In the profiles across the Szarvas and Dévaványa basement highs (Figs. 26 and 27, respectively), the listric normal faults appear as curvilinear features with dips decreasing with depth. The faults are inferred to cut only the Neogene section and to parallel the basement surface at depth. The fact that the faults cut across relatively recent sedimentary sequences, almost reaching the surface, suggests that at least part of the movement on these faults has occurred quite recently.

Other authors (unpublished reports of the Hungarian Oil and Gas Co. (MOL)) have suggested that some of these faults represent strike-slip or oblique extensional motion that started in Middle Miocene time during the rifting stage and continued into Plio-Pleistocene time. However, interpretations involving strike-slip motion are incompatible with the indication by the seismic data that the faults do not cut basement rocks. Alternatively, the data suggest that these listric features may be growth faults that resulted primarily from differential compaction of postrift sediments. Evidence which supports this alternative interpretation is that the faults most commonly occur above the crests of basement highs along the northwestern and northeastern boundaries of the Békés basin. According to Mattick and others (1985) and Pogácsás and others (this volume), the basement highs on the northwestern and northern sides of the basin were oriented approximately perpendicular to the direction of deltaic progradation in Pannonian time and acted as dams to southeastward and southward sedimentation. The damming of sediments resulted in the concentration of coarse-grained sediments (delta plain and delta front) on the northwestern and northern flanks of the basement highs. Finer grained (delta slope and prodelta) sediments were deposited on the southeastern and southern flanks. Later, as a result of differential compaction during sediment dewatering, the Pannonian sedimentary sections on the southeastern and southern flanks of the basement highs compacted more than the sedimentary sections on the crests of the highs. This mechanism of differential compaction promoted the development of southeastern dipping listric normal growth faults above the crests of the basement highs. Vertical displacements on the faults are relatively small, on the order of tens of meters or less.

If, as postulated, the faults in the synrift section involved growth through Pannonian-Pleistocene time, then the throw along the faults has to increase with depth, and sedimentary beds are thicker on the downthrown sides of the faults than on the upthrown sides. Because of the small amount of vertical displacement along the faults, any large amount of throw or thickening of beds with depth is difficult to document on the seismic records. On the downthrown sides of the faults, significant rollover structures appear on the seismic records (Figs. 26 and 27), and some of these have been shown to be associated with hydrocarbon traps (Pogácsás and others, this volume).

GRAVITY AND MAGNETIC INTERPRETATION

Gravity and magnetic profiles along the southeastern border of Hungary reveal a Bouguer gravity high of about 21 mGal and a total magnetic field high of about 100 nT (peak-to-trough amplitude in nannoTeslas; 1 nT = 1 gamma) over the axis of the Békés basin where Miocene and younger sedimentary fill is about 6700 m thick (Fig. 28). Assuming that the sedimentary fill usually has a lower average density (2.3-2.55 g/cm^3) than the underlying basement rocks (2.6-2.7 g/cm^3), we would expect a gravity low to occur over the axis of the basin. The presence of gravity and magnetic highs suggests that either crust of very high density and/or a thin crust must underlie the central part of the Békés basin.

A two-dimensional gravity model for the Békés basin (Fig. 28) shows that if the average densities of the upper and lower crust are assumed to be 2.7 and 3.0 g/cm^3 respectively, then a large mass of high density material (2.9 g/cm^3 in the upper crust and 3.2 g/cm^3 in the lower crust) must exist within the crust beneath the axis of Békés basin. The high density body is calculated to be 15 km wide between depths of 8 and 23 km. The model also includes an upwarp of the mantle from normal depths of 26-29 km (based on the regional seismic studies of Posgay and others (1986)), to 23 km beneath the Békés basin and 25 km beneath the Hód-Makó trough. It should be noted that the deep reflection lines of Posgay and others (1986) were not located over the deepest parts of the Békés basin or Hód-Makó trough.

The high density body beneath the axis of the Békés basin probably represents mafic igneous rock. An assemblage of mafic rock could represent fragments of oceanic crust (ophiolites) that were incorporated between pieces of continental crust during episodes of Mesozoic and/or younger continental collisions. Alternatively, the high density material may represent mafic rocks intruded into the crust in Miocene time during an episode of extension. Episodes of collision were responsible for the

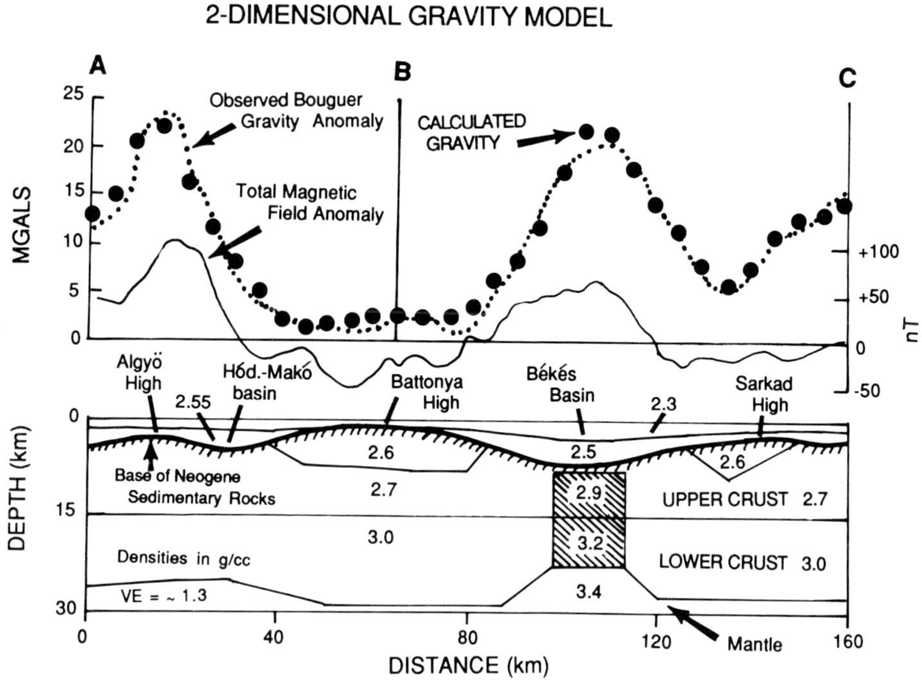

Figure 28. Gravity and magnetic profiles across the Hódmezövásárhely-Makó (Hód-Makó) trough and Békés basin, and a crustal model calculated from the gravity data. A high density crust and/or shallow mantle is required to explain the gravity high that occurs over the central part of the Békés basin. See text for detailed discussion. Mgals: milligals; nt: nannoTeslas (1 nannoTesla = 1 gamma). Densities are expressed in g/cm^3 and the vertical exageration is 1.3. Location of profile is shown in Figure 25. Sarkad: Sarkadkeresztúr.

formation of nappes in the Apuseni Mountains of northwestern Romania (Bleahu and others, 1981) and extensive outcrops of high density ophiolites(?) have been reported from the southern Apuseni Mountains (Fig. 1). Whether these postulated ophiolites accreted during the Cretaceous (Bleahu and others, 1981) or during the Cenozoic (Hamilton, 1990) is still not known well.

The gravity lows over the Battonya-Pusztaföldvár and Sarkadkeresztúr basement highs are also difficult to explain. In the results of Figure 28, the gravity lows were modeled assuming that bodies composed of low density rock (2.6 g/cm^3) are present in the upper part of the basement. Reports from wells drilled on the Battonya and Sarkadkeresztúr do not indicate that anomalously low density basement rocks were penetrated; therefore, the data appear to be in conflict with the gravity model. However, if nappes composed of low density rocks are present at shallow depths beneath the basement surface, the gravity model could be plausible. More detailed geophysical studies, including deep-crustal reflection and refraction seismic, and magnetotelluric measurements will be required to fully explain the measured gravity anomalies in the vicinity of the Békés basin.

TECTONIC INTERPRETATION

The major tectonic elements observed in the seismic reflection and well data from the Békés basin are summarized in Figure 25. Northeast-trending, extensional and oblique extensional faults, assumed to be of middle Miocene age, bound the basin to the northwest in the vicinity of the Szarvas, Endröd, Dévaványa and Szeghalom highs. Northward-verging, Cretaceous thrust faults are inferred to exist in the Battonya-Pusztaföldvár and Békés-Doboz Mesozoic troughs. These thrusts may have been reactivated during Miocene time as extensional faults.

Poor seismic resolution in the Mesozoic section makes identification of the major faults difficult; however, the overall pattern is compatible with a system dominated during middle-to-late Miocene time by northeast-trending, strike-slip faults and zones of local extension, oblique extension and oblique compression (Rumpler and Horváth, 1988). No definitive data are available from the Békés basin to either prove or disprove the existence of strike-slip faults with large offsets. faults inherited from earlier deformational episodes probably have been partially reactivated as oblique compressional, oblique extensional, or strike-slip faults. The tectonic map is intended to illustrate possible fault patterns in a region of multiple tectonic episodes and, because of the poor seismic resolution below the Neogene section, many of the faults are speculative, especially the strike-slip faults.

Importantly, the existence of Cretaceous thrusts in the Békés-Doboz and Battonya-Pusztaföldvár Mesozoic troughs can be documented. Examination of the rock units penetrated by wells in the Battonya-Pusztaföldvár Mesozoic trough brings forth strong evidence that the southern part of the Békés basin is underlain by stacked, allochthonous nappes, composed of rocks of Cretaceous age, with rock affinities to the Codru nappes of the Apuseni Mountains of Romania. The nappes could have been moved from the south significant distances. These nappes were probably later deformed in Late Cretaceous time and then again in the Miocene by extensional and strike-slip movements. Many of the low-angle thrust surfaces inherited from the Cretaceous were probably reactivated in extensional and strike-slip modes during the Miocene. The Mesozoic section in the Békés-Doboz Mesozoic trough is also inferred to have been overthrust and probably intensively folded. The rocks in the overthrust units are similar in lithology to rocks of the Bihor autochthon in the Apuseni Mountains; therefore, it appears that the nappes in the Békés-Doboz trough were not displaced long distances during Late Cretaceous time.

Hamilton (1990) has recently proposed an alternative interpretation of the intra-Carpathian region. In his view, the "region was produced by the squashing together of a number of small continental and island arc fragments during Tertiary time, a view opposite to the commonly accepted concept that its Tertiary history records primarily the moderate disruption of a previously coherent mass." Hamilton's concept downplays the significance of major strike-slip faults in the Pannonian Basin. The apparent correlation between nappes of the Battonya-Pusztaföldvár and Békés-Doboz Mesozoic troughs with those in the Apuseni Mountains would seem, at first glance, to favor the "coherent mass" model, rather than Hamilton's "small continental and island arc fragments" model. Nevertheless, more detailed studies of well data from the pre-Neogene section and deep crustal geophysical studies are needed in the Békés basin and surrounding regions before Hamilton's model can be rejected.

Although a few Neogene growth faults are observed in the Békés basin, these likely represent differential compaction of sediments rather than zones of significant horizontal displacement (strike-slip faults) during the postrift period.

In general, Neogene strike-slip and extensional deformation in the Békés basin appears to have occurred rapidly during the middle-to-late Miocene and then terminate abruptly. The basin was virtually starved of sediments during the synrift period, and was filled subsequently by massive volumes of deltaic and lacustrine sediments during the postrift period.

CONCLUSIONS

First, the Békés basin and Hódmezövásárhely-Makó trough contain extensional or oblique-extensional asymmetric grabens which suggests the presence of east- and northeast-trending, strike-slip faults as inferred by Rumpler and Horváth (1988). However, the amount of strike-slip movement in the study area cannot be determined from available data.

Secondly, good examples on seismic records of listric normal faults associated with Miocene extension and clear reflections from their faults were not observed in the study area; therefore, it is difficult to determine the amount of extension that occurred in Miocene time. Miocene sediments are relatively thin in comparison to typical extensional basins, and occur in poorly developed grabens. This suggests that the Békés basin was starved of sediments during the rifting phase, likely because of ponding of the sediments upstream or because of downstream bypass into the Caspian or Mediterranean seas.

Thirdly, poor seismic resolution in the Mesozoic section makes it difficult to map deep faults and internal structures in the Mesozoic troughs; however, well and seismic data indicate that overthrusting occurred within the Békés-Doboz Mesozoic trough during Late Cretaceous time. The age and lithologies of the Mesozoic rock units in the Békés-Doboz Mesozoic trough (both hanging-wall and foot-wall blocks) are similar to those of the Bihor autochthon in the Apuseni Mountains of Romania and probably were not subjected to significant horizontal displacement. Analyses of well and seismic data from the Battonya-Pusztaföldvár Mesozoic trough suggest that two or more north-dipping nappe structures are present. The rocks of these Cretaceous nappes are similar with respect to lithology, age and depositional environment to rocks of the allochthonous Codru nappes in the Apuseni Mountains. Seismic and core data indicate that complex structures exist within the Mesozoic section and that large velocity inversions (10-20%) occur at depth. These low velocity zones may represent zones of fractured Mesozoic rock, perhaps with cavernous porosity. However, limited exploration for oil and gas in the Mesozoic section in deep (central) parts of the Békés basin has not been successful (Kovács and Teleki, this volume). This is probably due to the fact that good quality source rocks do not exist in the Mesozoic section in the area of the Békés basin and the fact that the Mesozoic section in the area is thermally overmature (Clayton and others, this volume).

A fourth result is that the two-dimensional gravity model constructed across the Békés basin indicates either high density crustal material existing beneath the basin and/or that depths to the mantle are relatively shallow (approximately 23 km). The body of high density, mafic(?) crustal rock could represent fragments of oceanic crust incorporated into the crust during Late Cretaceous compression or mafic rocks intruded into the crust during Miocene extension. Deep crustal reflection, refraction, and magnetotelluric soundings are needed to further delineate this anomalous crust.

Although existing 24- and 48-channel seismic profiles from the Békés basin provide excellent penetration and resolution in the Neogene postrift section, seismic resolution in the complexly folded and thrusted prerift Mesozoic and older sections is poor. Higher-fold seismic profiles or 3-dimensional seismic grids might significantly improve the quality of seismic data from the Mesozoic section, but the high cost of obtaining such data may not be justified unless (or until) better prospects for commercial quantities of oil and/or natural gas in the Mesozoic section can be demonstrated.

REFERENCES

Balla, Z., 1985, The Carpathian Loop and the Pannonian Basin: A kinematic analysis; Geophysical Transactions 30, (4), 313-353.

Balla, Z., 1990, The Pannonian basin - A study of basin evolution: discussion; Amer. Assoc. Petrol. Geol. Bull., 74, (8), 1273-1280.

Bally, A.W., Bernoulli, D., Davis, G.A., and Montadert, L., 1981, Listric normal faults; Proceedings of 26th International Geological Congress, Geology of Continental Margins Symposium, Paris, July 7-17, 1980, Oceanologica Acta, 1-10.

Bérczi-Makk, Anikó, 1985, Triász mikrofauna Kelet-ma-gyarországi szénhidrogénkutató mélyfúrásokból (Triassic microfauna from deep hydrocarbon exploratory wells in eastern Hungary): Földtani Közlöny, Budapest, 115, (3), 303-313.

Bérczi-Makk, A., 1986, Mesozoic formation types of the Great Hungarian Plain: Acta Geologica Hungarica, 29, 261-282.

Bleahu, M., Lupu, M., Patrulius, D., Borde, S., Stefan, A., and Panin, S., 1981, The structure of the Apuseni Mountains: Proceedings Carpatho-Balkan Geol. Assoc. Conference 12, Guide to Excursion B3, Bucharest, 106 p.

Burchfiel, B.C. and Royden, L., 1982, Carpathian Foreland Fold and Thrust Belt and its relation to Pannonian and other basins: Amer. Assoc. Petrol. Geol. Bull., 66, (9), 1179-1195.

Clayton, J.L., King, J.D., and Tatár, E., Organic geochemistry of crude oil and source rocks, Békés basin, this volume, 161-186.

Dragastan, O., Purecel, R., and Brustus, T., 1986, Mesozoic formation types of the Great Hungarian Plain: Acta Geologica Hungarica, 29, 261-282.

Falvey, David A., 1974, The development of continental margins in plate tectonic theory: APEA Journal, 95-106.

Grow, J.A. and Sheridan, R.E., 1988, U.S. Atlantic continental margin: A typical Atlantic-type or passive continental margin: *in* Sheridan, R.E. and Grow, J.A. (eds.), The Atlantic continental margin, U.S.: Geol. Soc. Amer., 1-2, 1-7.

Hamilton, W.B., 1990, On terrane analysis: Phil. Trans. R. Soc. London, A331, 511-522.

Hornafius, J.S., Luyendyk, B.P., Terres, R.R., and Kamerling, M.J., 1986, Timing and extent of Neogene tectonic rotation in the western Transverse Ranges, California: Geology, 97, 1476-1487.

Horváth, F. and Rumpler, J., 1984, The Pannonian basement: extension and subsidence of an alpine orogen: Acta Geologica Hungarica, 27, 229-235.

Kilényi, É. and Rumpler, J., 1984, Pre-Tertiary basement relief map of Hungary: Geophysical Transactions, 30, (4), 425-428.

Klitgord, K.D., Hutchinson, D.R., and Schouten, H., 1988, U.S. Atlantic continental margin: Structure and tectonic framework, *in* Sheridan, R.E. and Grow, J.A. (eds.), The Atlantic continental margin, U.S.: Geol. Soc. Amer., 1-2, 19-55.

Kovács, A. and Teleki, P.G., History of oil and natural gas production in the Békés basin; this volume, 237-258.

Lee, M.W. and Göncz, G., Vertical seismic profile experiments at the Békés-2 well, Békés basin; this volume, 259-298.

Mattick, R.E., Rumpler, J., and Phillips, R.L., 1985, Seismic stratigraphy of the Pannonian Basin in southeastern Hungary; Eötvös Lóránd Geophysical Institute, Geophysical Transactions, 31, (1-3), 13-54.

Mattick, R.E., Rumpler, J., Ujfalusy, A., Szanyi, B., and Nagy, I., Sequence stratigraphy of the Békés basin; this volume, 39-66.

Miller, J.J. and Véges, I., Modeling of seismic reflection data in the vicinity of the Békés-2 well; this volume, 279-296.

Patrulius, D., Bleahu, E., Antonescu, E., Baltres, A., Bordea, S., Bordea, J., Gheorghian, D., Tordan, M., Mirauta, E., Panin, S., Popa, E., and Tomescu, C., 1979, The Triassic formation of the Bihor autochthon and Codru nappe system (Apuseni Mountains); Carpato-Balkan Geol. Assoc. Bucharest, 3rd Triassic Guide Book to Field Trips, 21 p.

Phillips, R.L., Révész, I., and Bérczi, I., Lower Pannonian deltaic-lacustrine processes and sedimentation, Békés basin; this volume, 67-82.

Pogácsás, Gy., 1985, Seismic stratigraphic features of Neogene sediments in the Pannonian Basin; Geophysical Transactions, 30, (4), 373-410.

Pogácsás, Gy., Mattick, R.E., Elston, D.P., Hámor, T., Jámbor, Á., Lakatos, L., Simon, E., Vakarcs, G., Várkonyi, L., and Várnai, P., Correlation of seismo- and magnetostratigraphy in southeastern Hungary; this volume, 143-160.

Póka, T., 1982, Chemical evolution of the Inner-Carpathian Neogene and Quaternary magmatism and the structural formation of the Carpathian basins; Proc., Workshop/Discussion Meeting on evolution of extensional basins within regions of compression, with emphasis on the Intra-Carpathians, Hung. Central Geol. Office, Hung. Oil and Gas Trust, Hung. Geol. Inst., Hung. Acad. Sci., and U.S. National Sci. Foundation, Veszprém, Hungary, June 20-26, 1982.

Posgay, K., Albu, A., Ráner, G., and Varga, G., 1986, Characteristics of the reflecting layers of the earth's crust and upper mantle in Hungary; *in* Barazangi, M. and Brown, L. (eds.), Reflection Seismology: A Global Perspective; Amer. Geophysical Union Geodynamics Series, 13, 55-65.

Royden, L.H., 1988, Late Cenozoic Tectonics of the Pannonian Basin; *in* Royden, L.H. and Horváth, F. (eds.), The Pannonian Basin - A study in Basin Evolution; Amer. Assoc. Petro. Geol. Mem. 45, 27-48.

Royden, L.H., Horváth, F., and Burchfiel, B.C., 1982, Transform faulting, extension and subduction in the Carpathian Pannonian region; Geol. Soc. Amer. Bull. 93, 717-725.

Royden, L. H., Horváth, F., and Rumpler, J., 1983, Evolution of the Pannonian Basin System, 1. Tectonics; Tectonics, 2, (1), 63-90.

Rumpler, J., and Horváth, F., 1984, Extenziós tektonika szeizmikus szelvényeken és ennek köolajkutatási jelentösége a Pannon medencében (Extensional tectonics of the Pannonian Basin based on seismic and exploration well data); Földtani Kutatás, 27, 49-61.

Rumpler, J. and Horváth, F., 1988, Some representative seismic reflection lines from the Pannonian Basin and their structural interpretation; *in* Royden, L.H. and Horváth, F. (eds.), The Pannonian Basin - A study in Basin Evolution; Amer. Assoc. Petrol. Geol. Mem. 45, 153-169.

Sandulescu, M., 1975, Essai de synthèse structurale des Carpathes; Bull. Geol. Soc. France, 17, 299-358.

Sandulescu, M., 1980, Analyse géotectonique des chaines alpines situées au tour de la Mer Noire occidentale; Annuarul Institutului Geologie si Geofizika., Bucharest, 56, 5-54.

Sandulescu, M., 1988, Cenozoic tectonic history of the Carpathians; *in* Royden, L.H. and Horváth, F. (eds.), The Pannonian Basin - A study in Basin Evolution; Amer. Assoc. Petrol. Geol. Mem., 45, 17-25.

Sclater, J.G., Royden, L., Horváth, F., Burchfiel, B.C., Semken, S., and Stegena, L., 1980, The formation of the intra-Carpathian basins as determined from subsidence data; Earth and Planet. Sci. Letters, 51, 139-162.

Szentgyörgyi, K. and Teleki, P.G., Facies and depositional environments of Miocene sedimentary rocks, this volume, 83-98.

Varga, I. and Pogácsás, Gy., 1981, Reflection seismic investigations in the Hungarian part of the Pannonian Basin; Earth Evol. Sci., 1, (3), 232-239.

2 Sequence Stratigraphy of the Békés Basin

Robert E. Mattick[1], *János Rumpler*[2], *Antal Ujfalusy*[2], *Béla Szanyi*[2], *and Irén Nagy*[2]

ABSTRACT

The 6500-m-thick section of Neogene and Quaternary fill in the Békés basin (3900 km^2) reflects a normal cycle of rift-related sedimentation that began in late Badenian time (16.5 Ma) when marine waters were shallow (<200 m). By late Sarmatian or early Pannonian time (12-10 Ma), lacustrine conditions prevailed and water depths in the basin increased as a result of basin subsidence rates that greatly exceeded sedimentation rates. The latest history of the basin, from approximately 5.78 Ma, reflects continuously shallowing waters, which resulted from sedimentation rates that were generally higher than basin subsidence rates. By late Pannonian time (4.25 Ma), water depths in the lake were 200-400 m, and continued shoaling culminated in the eventual disappearance of the lake in Pliocene time.

Detailed mapping of sequences, the boundaries of which are attributed to the cyclic shifting of delta lobes, indicates that the Neogene post-rift history of the Békés basin reflects one of the last episodes of filling of the much larger Pannonian Basin. It was not until other subbasins to the northwest, north, and northeast became filled by southward-prograding sediment wedges that the major river systems were able to advance to the area of the Békés basin. Thereafter, rapid deltaic filling of the basin proceeded as major river systems converged and lacustrine sedimentation became confined to a progressively shoaling basin that was rapidly decreasing in areal extent.

During much of the early Pannonian history of the Békés basin, large volumes of clastic sediments transported by rivers were channeled along structural depressions; the depressions or troughs were derived from prerift or early rift topography of the basin. In some of these troughs, large subaqueous fans, consisting chiefly of very fine grained material, developed near the base of the slope. In later stages of basin filling, sand was derived primarily from prograding delta fronts and moved downslope by slumping, sliding, mass flow, and turbidity currents. In this later stage, subaqueous fan systems and channels on the slope were poorly developed.

INTRODUCTION

According to Lerner (1981), the general geographic name for the region inside the arc formed by the Eastern Alps, the Carpathian mountain system, and the Dinaric Alps is the Carpathian basin or the intra-Carpathian region (Fig. 1). This intramontane region, however, is not topographically uniform; emergent ranges divide it

[1] U.S. Geological Survey, Reston, Virginia 22092, USA
[2] MOL Rt. (Hungarian Oil and Gas Co., Ltd.), H-1062, Budapest, Hungary

P. G. Teleki et al. (eds.), Basin Analysis in Petroleum Exploration, 39–65.

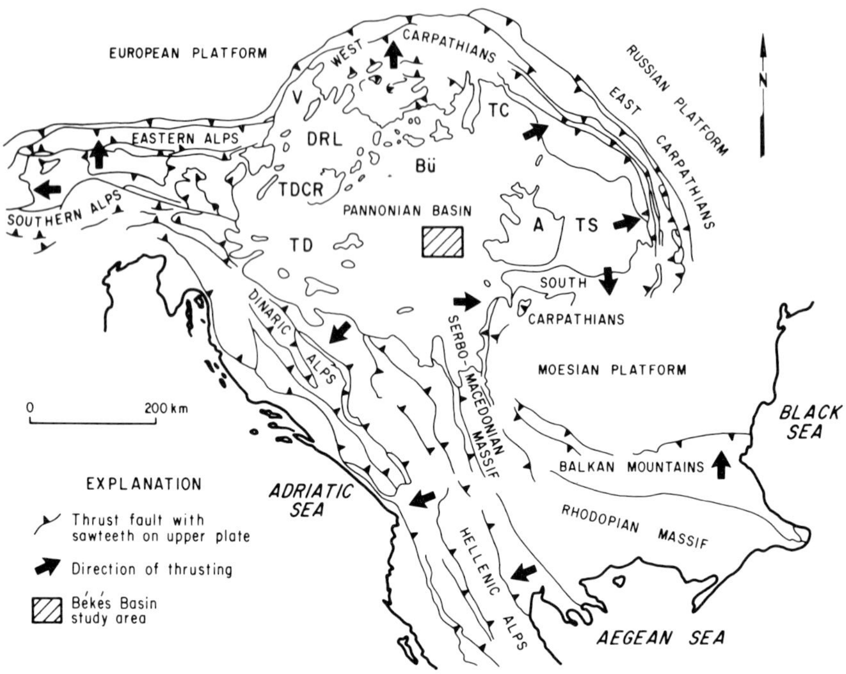

Figure 1. Location of the Pannonian Basin and the study area within the intra-Carpathian region. The Carpathian Basin consists of the Pannonian, Vienna (V), Southwest Transdanubian (TD), Transcarpathian (TC), and Transylvanian (TS) Basins. Danube-Rába Lowland: DRL; Transdanubian Central Range: TDCR; Bükk Mountains: Bü; Apuseni Mountains: A. Figure modified from Burchfiel and Royden (1982).

into several subbasins, the largest of which is the Pannonian basin. The Pannonian Basin is superimposed on structural elements of the Carpathians and represents a post-tectonic basin, partly created by extension in middle and late Tertiary time (Burchfiel, 1976). According to Royden (1988) and Grow and others (this volume), basins of the intra-Carpathian region are underlain by nappes of Mesozoic age, and initial subsidence of the basins was synchronous with Miocene thrusting of the outer Carpathians toward the European foreland. Thrust belts in the area are controlled by overall convergence of Europe and Africa and by the arrangement of small continental fragments between the converging continents (Horváth, 1988).

The part of the Pannonian Basin within Hungary is divided, mainly on the basis of buried basement topography, into subbasins, one of which is called the Békés basin (Fig. 1). The Békés basin is in southeastern Hungary, covers an area of about 3900 km^2, and is roughly bounded by buried basement highs to the southwest, northwest, and northeast (Fig. 2). The Békés basin extends southeastward into Romania, where it is referred to simply as the Pannonian Basin (Burchfiel, 1976).

Because the purpose of this study was to map Neogene and Quaternary stratigraphy, older rocks are referred to as basement; these rocks are chiefly Paleozoic and Mesozoic igneous, metamorphic, and deformed sedimentary rocks.

Neogene subsidence in the Békés basin probably began during the late Styrian orogenic cycle in early Badenian time (Szentgyörgyi and Teleki, this volume). The Neogene transgression of the sea is represented by Miocene sedimentary rocks of Badenian and early Sarmatian age, which were deposited in relatively shallow-water, near-shore marine to brackish-water environments during subsidence. The combined Badenian (16.5-13 Ma) and Sarmatian (13-11.5 Ma) stages represent deposition from 16.5 to 11.5 Ma (Steininger and others, 1985; Hámor and others, 1987). As indicated by well and seismic data, the combined thickness of the Badenian and Sarmatian sections in the Békés basin ranges from 0 to about 275 m where penetrated by wells. These rocks are absent on some basement highs as a result of non-deposition (Szentgyörgyi and Teleki, this volume).

The lower part of the Badenian section contains coarse-grained conglomerates and/or breccia that grade upward

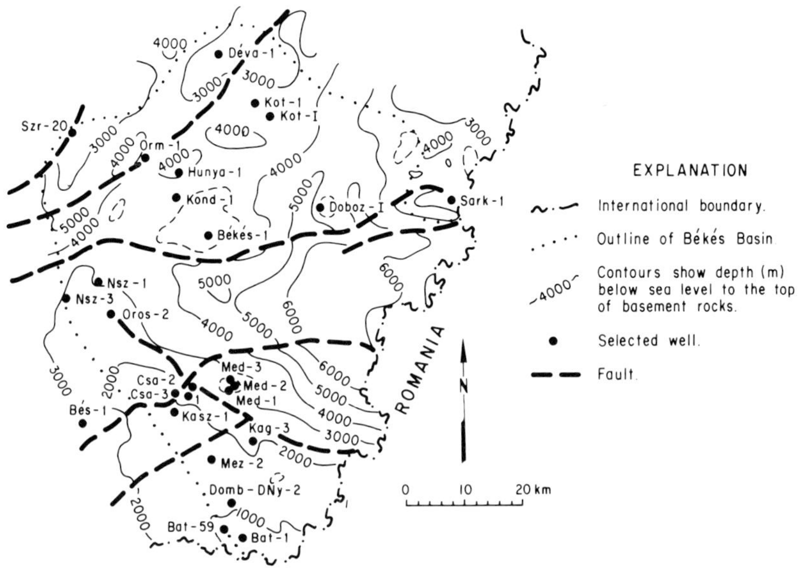

Figure 2. Structure map on pre-Cenozoic rocks in the area of the Békés basin, southeast Hungary. Figure modified from Kilényi and Rumpler (1984).

to sandstone and marl and biogenic limestone. According to Szentgyörgyi and Teleki (this volume), the Badenian section represents deposition in water depths of less than 200 m in a marine environment of normal salinity. During Sarmatian time the Pannonian basin became isolated from the eastern Paratethys and waters in the Békés basin became shallower and less saline in comparison to Badenian time.

In the central parts of the Békés basin, the Sarmatian shallow-water deposits are conformably(?) overlain by a section of lacustrine clayey marls, marls, and turbidites which, according to Lukács-Miksa and others (1983) and Bérczi and others (1985), were deposited in part in deep water, perhaps as deep as 800-900 m (Mattick and others, 1985), during late Sarmatian(?) and early Pannonian time. This part of the sedimentary rock section reaches thicknesses of about 2500 m.

Studies by Késmárky and others (1981), Pogácsás (1984), Marton (1985), and Mattick and others (1985) indicate that the clayey marls, marls, and turbidites are conformably overlain by a series of prograding sediment wedges composed of sediments deposited in fluvial, delta, prodelta, and basin environments. The sediment wedges in the Pannonian Basin are related to the progradation of deltas from the northwest, north, and northeast during Pannonian through Pliocene time when water depths in the basin were rapidly shallowing from about 900 to 0 m. These youngest sedimentary rocks (2500 m thick) are inferred to represent a final stage of deposition, which culminated in the eventual disappearance of the lake in Pliocene time (Kázmér, 1990).

PRINCIPLES USED IN INTERPRETATION

The Békés basin contains a relatively conformable sequence of sedimentary rocks that represents a complete cycle of rift-related basin fill, starting with sedimentary rocks that represent basin subsidence developed during initial rifting and ending when the rate of basin subsidence slowed and the basin was filled by sediments.

In order to reconstruct the Neogene depositional history of the basin, we divide the sedimentary rock section into three principal units: (1) a basinally restricted unit, (2) a retrogradational unit, and (3) a progradational unit. The progradational unit, in turn, is subdivided into seismic sequences based on lithologic and facies associations and objective criteria observable on seismic records. Approximately 2000 km of seismic-reflection data were

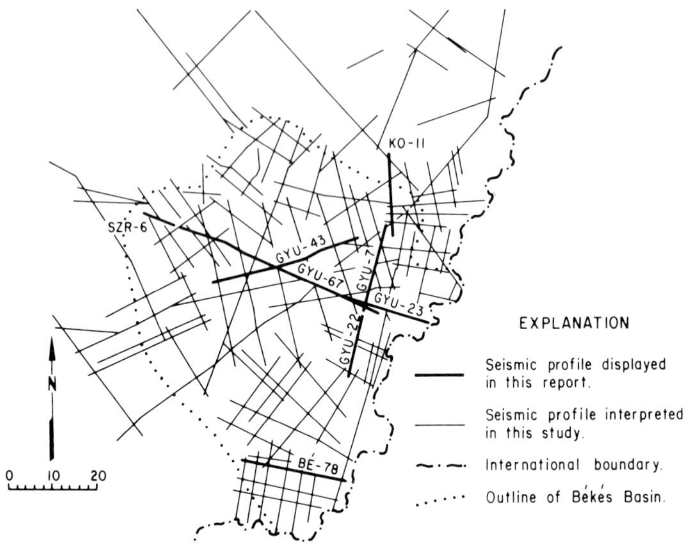

Figure 3. Location of seismic profiles interpreted in this study. Heavy lines (numbered) mark profiles displayed in other figures of this report. The numbering system of the seismic profiles follows that used by the Oil and Gas Trust of Hungary (OKGT, presently MOL Ltd.).

analyzed (Fig. 3). These data were correlated with the paleomagnetic work of Elston and others (this volume) in order to approximate the ages of the various units and sequences.

Basinally-Restricted Unit (BRU)

The basinally restricted unit consists of Miocene (Badenian and Sarmatian) sedimentary rocks deposited in relatively shallow-water, near-shore marine to brackish-water environments during the initial subsidence of the basin. As previously stated, this section includes conglomerates and/or breccia that grade upward to sandstone and marl and biogenic limestone (Szentgyörgyi and Teleki, this volume) and represents a time period of about 3.5 million years. Because this unit is relatively thin, it is not distinguishable on most seismic records.

Retrogradational Unit (RU)

On seismic records from central parts of the Békés basin, the retrogradational unit is represented by strong continuous, horizontal reflections that exhibit an onlap pattern at the base of the section (Fig. 4a). Laterally, beyond the northern and western margins of the basin, basal reflections form a downlap pattern. Well data (Phillips and others, this volume) indicate that much of this unit consists chiefly of marls and very fine grained sandstone turbidites which represent basinal facies deposited in a relatively deep-water lacustrine environment. The strong onlap pattern of reflections within the unit suggests that the sediments were deposited mostly during a period of rising lake level. The unit averages about 2000-2500 m in thickness and represents a time period of about 5.72 million years.

Progradational Unit (PU)

The progradational unit is divided into an early progradational unit (PU_1) and a late progradational unit (PU_2). In general, PU_1 is represented seismically by sigmoid and oblique-sigmoid reflections that are indicative of basinward progradation (Fig. 4a). Further basinward, reflections within the unit form a downlap pattern on the top of the retrogradational unit. As will be discussed, PU_1 is inferred to represent an episode of progradational delta construction within the Békés basin during which time water depths in the lake reached a maximum of about 1000 m and then rapidly shallowed to about 200

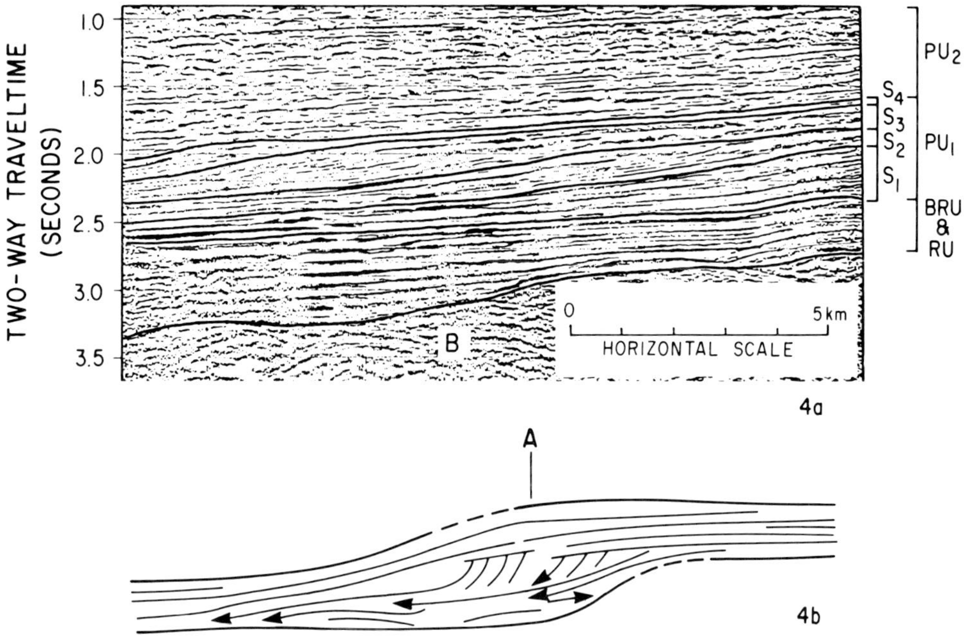

Figure 4. (a) Seismic record (GYU-7) from the Békés basin showing division of the sedimentary rock section into a basinally restricted unit (BRU), retrogradational unit (RU), early progradational unit (PU_1), and late progradational unit (PU_2) above basement (B). BRU is less than 100 m thick and, therefore, cannot be distinguished on the seismic record. The early progradational unit is subdivided into sequences (S) by surfaces of periodic nondeposition, which are believed to represent periods of delta-lobe switching. (b) A schematic diagram of a sequence. The upper and lower boundaries of the sequence are usually of high amplitude except in the uppper clinoform zone, where the reflections are weak to absent. The basal-most reflections above the bottom boundary show an onlap pattern at the base of the clinoform zone, followed by sigmoid-shaped reflections, followed by oblique reflections, and followed again by sigmoid-shaped reflections. The inflection point (A) represents the shelf break or shelf-slope boundary.

m. Well data indicate that PU_1 contains delta-plain, prodelta, and basin-floor facies including marls and turbidites. PU_1 averages about 1700 m in thickness and represents a time period of about 1.53 million years.

The late progradational unit (PU_2) is characterized seismically by moderately strong, wavy reflections (Fig. 4a). Well data indicate that this unit consists chiefly of shallow-water to delta-plain facies deposited in shallow-lake, fluvial, and marsh to terrestrial environments. PU_2 represents the last stage of basin filling, averages about 2000-2500 m in thickness, and represents a time period of about 4.25 million years.

Sequences

Local hiatuses separate progradational units in PU_1 (Fig. 4). These local hiatuses are probably best referred to as sequence boundaries (Mitchum and others, 1977; Vail and others, 1977). They appear, throughout much of the basin, to be related to short periods of non-deposition, which probably resulted from continuous delta-lobe switching and, as such, are inferred to represent 4th order stratigraphic cycles.

A schematic diagram of a sequence is shown in Figure 4b. The basic pattern (from bottom to top) of bottom boundary, basal onlap, sigmoid-shaped reflections, oblique reflections, repeated sigmoid-shaped reflections, and upper boundary is interpreted as representing a local cycle of sedimentation along a delta front. The change in reflection character within the sequence is believed to represent changes in sediment supply and deposition rates - from a period of non-deposition (bottom boundary), to relatively low to intermediate sediment supply and deposition rates (sigmoid reflections), to relatively high sediment supply and deposition rates (oblique reflections), and ending with relatively low to intermediate sediment supply and deposition rates (sigmoid reflections). The top of the sequence is believed to mark a temporary cessation of sedimentation resulting from channel or delta-lobe switching. The sequence shown in Figure 4b is an idealized model; those observable on seismic records differ somewhat from this and exhibit a wide degree of lateral variability as might be expected along a delta front where numerous delta lobes were built out into a shoaling lacustrine basin by many migrating distributary channels.

Other authors (Pogácsás and others, this volume) have attempted to correlate some of the stronger hiatuses, or sequence boundaries, with basinwide changes in lake level, related, perhaps, to global sea-level changes. Detailed analysis of closely spaced seismic records, however, indicates that most of the hiatuses or sequence boundaries are local events that extend for distances of 30 km or less along strike, and that beyond these distances the boundaries become conformable relations. This suggests that the hiatuses are related to temporal lateral movement of delta lobes and not to basinwide lake-level changes, which should result in basinwide unconformable events. If some of the hiatuses are related to basinwide lake-level changes, the important issue then is the cause of these changes or cycles. Further research may show that proximity to marine waters or climatic (precipitation/evaporation) cycles may be linked to lacustrine cycles.

The major units (basinally restricted, retrogradational, and progradational) are inferred to have genetic similarity to systems tracts as defined by Posamentier and others (1988) and Van Wagoner and others (1988, 1990). However, because the term "systems tract" is usually used in association with a eustatic sea-level cycle, the authors have used descriptive terms that separate the units from their genetic history.

Similarly the term "sequence" is used rather than parasequence or subsequence. Although parasequence boundaries have been interpreted as resulting from the drowning of delta lobes (Van Wagoner and others, 1988), parasequences are typically several meters thick and usually not resolvable by seismic-reflection data. Seismically, the "sequences" identified herein resemble typical high-frequency water-level cycles superimposed on a longer basin subsidence curve associated with rifting of the basin. Although technically a seismic sequence must be bounded by an erosional unconformity and this feature appears to be lacking for the sequences identified in the Békés basin, the inability to identify erosion-related features may be due to problems associated with seismic resolution.

The inflection point on the upper sequence boundary (Fig. 4b) is interpreted as representing the shelf break or the shelf-slope boundary. In Figure 5, the loci of inflection points along the tops of the major, mappable prograding sequences are plotted. These are interpreted as representing shelf breaks associated with major lacustrine deltas or delta systems that filled the subsiding Békés basin.

There follows a discussion of sedimentation patterns inferred from structure contour maps constructed on RU and the major sequences, a discussion of water depths, ages, and depositional rates for the various units.

STRUCTURE MAP ON RU

The retrogradational unit (RU) is marked seismically by a strong onlap pattern (Fig. 6). Correlation with core data and well-log data (Phillips and others, this volume; Gy. Juhász and I. Révész, oral communication, 1989) indicates that this unit represents, for the most part, periodic influxes of turbidites into a deep basin where the precipitation of $CaCO_3$ and the deposition of mud from suspension occurred. According to Mattick and others (1985), the marls and turbidites represent basinal facies which were coeval with the development of turbidite-fronted deltas or delta systems located chiefly to the north, northwest, and west of the Békés basin. During the early stage of deposition, much of the sediment being transported basinward from these delta areas was captured by intervening subbasins. The clastic sediment that reached the Békés basin was transported mainly through subaqueous canyons or troughs and was distributed by bottom currents as evidenced by the strong onlap pattern of reflections.

A structure contour map constructed on the top of RU is shown in Figure 7. Comparison with the structure contours on the top of the basement complex (Fig. 2) indicates that most of the sediments represented by RU simply filled preexistent basement lows. As will be shown by seismic evidence, the large northeast-trending trough, herein referred to as the Békés trough and located at the northern boundary of the basin (Fig. 7), controlled sedimentation throughout much of the early history of the Békés basin.

To the south, RU pinches out against the north flank of the Battonya-Pusztaföldvár basement high (Fig. 7). The

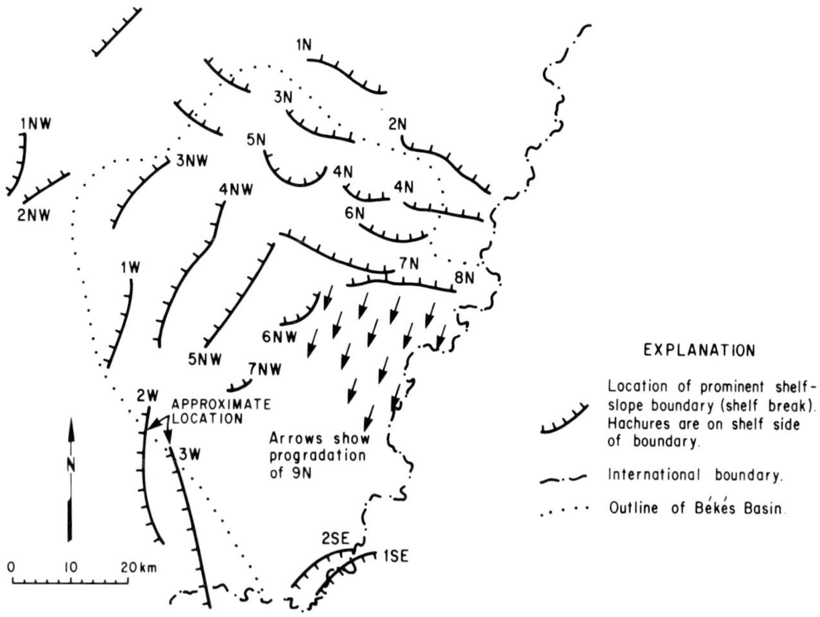

Figure 5. The loci of inflection points for major prograding sequences in the area of the Békés basin. These are interpreted as representing shelf breaks or shelf-slope boundaries. The sequences prograding from a particular direction (N, NW, W, and SE) are sequentially numbered from oldest to youngest. Relative ages of unnumbered sequences could not be determined.

configuration of contour lines in the vicinity of the pinchout suggests that differential subsidence of the basement high occurred after deposition of RU. Apparently, the basement surface to the southeast remained relatively stable while, to the northwest, it subsided.

The maximum depth (4100 m) to the top of RU and its maximum thickness (2700 m) occur in the central part of the basin - the area outlined by the 2.70-second contour line in Figure 7. As will be shown, this central area remained a major depocenter throughout much of the basin's subsequent history.

MAJOR SEQUENCES OF PU_1

As the advancing deltas prograded to the margins of the Békés basin, sediments began to spill directly into the basin. The start of this stage of deposition is marked on seismic records by sigmoid and oblique-sigmoid reflections of the early progradational unit (PU_1, Fig. 6). The early progradational unit (PU_1) was divided into 25 sequences (Fig. 5) and structure contour maps prepared on some of these units were used to reconstruct the spatial and temporal filling of the basin. Each of the sequences is inferred to represent a delta lobe containing sediments deposited in delta-plain, prodelta, and basin-floor environments and, as indicated by results from drilling, many are fronted by a thin turbidite apron. In the discussion below, the delta lobes are labeled by the direction from which they prograde (N, NW, W, and SE) and consecutively numbered from oldest to youngest.

Sequences 1N-6N

Structure contour maps on sequences 1N-6N show progressive southward deltaic encroachment into the basin (Figs. 8-12). The oldest of these sequences (1N and 2N, Fig. 8) can be traced about 20 km south of the basin's margin, where they terminate in a downlap pattern on the underlying marls and turbidites of RU.

On seismic records, sequences 1N and 2N are separated by an area of chaotic reflections (Fig. 8), which is interpreted as representing subaqueous mass movement

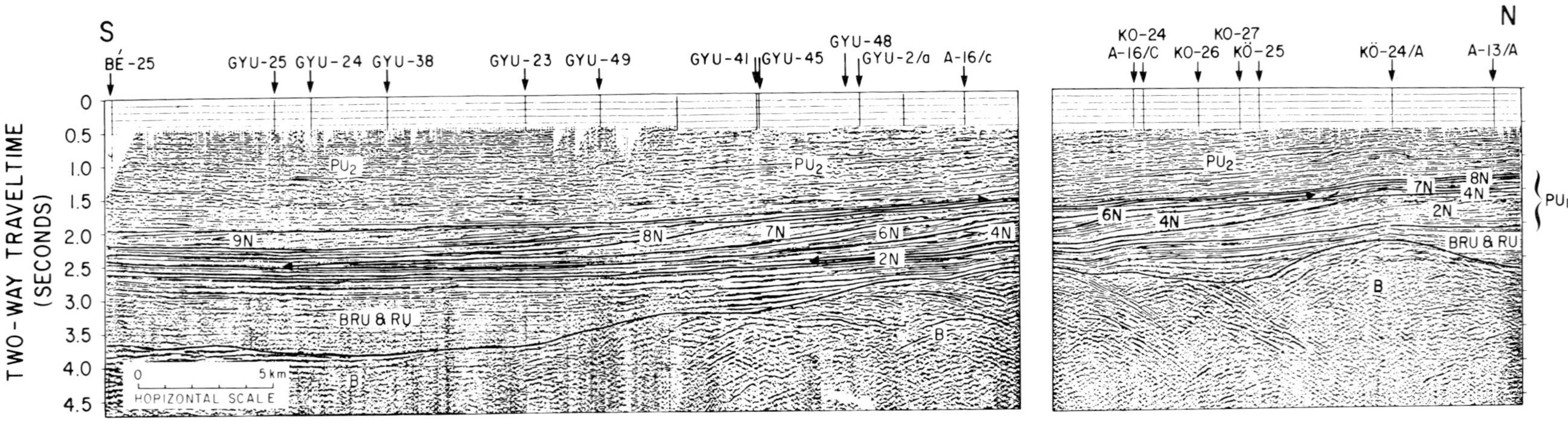

Figure 6. Composite of seismic profiles KO-11, GYU-7, and GYU-22 showing late progradational unit (PU2), early progradational unit (PU1) and combined basinally restricted and retrogradational units (BRU & RU). PU1 is subdivided into sequences 2N, 4N, 6N, 7N, 8N, and 9N. Basement complex, B. The locations of the profiles are shown in Figure 3.

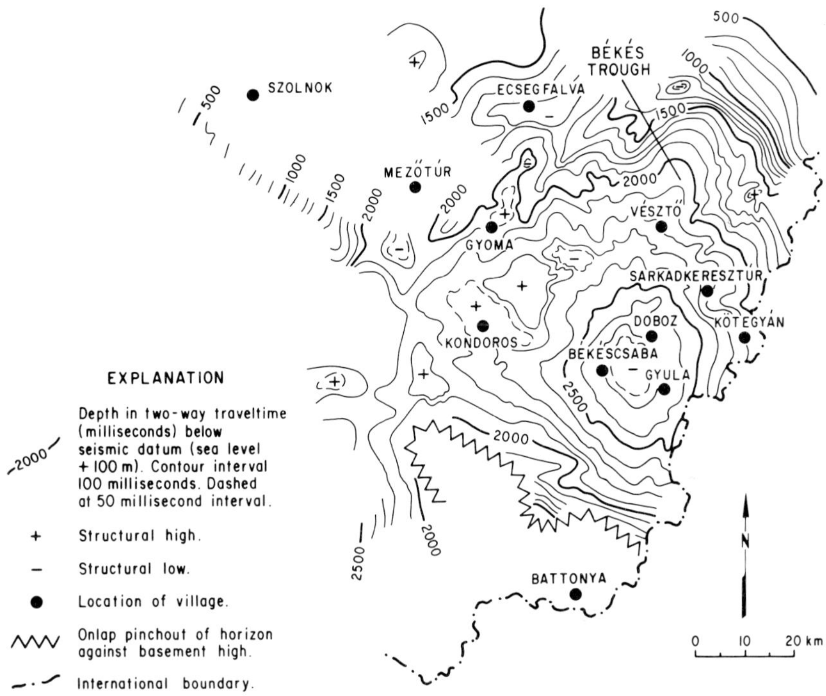

Figure 7. Structure contour map on the top of the retrogradational unit (RU).

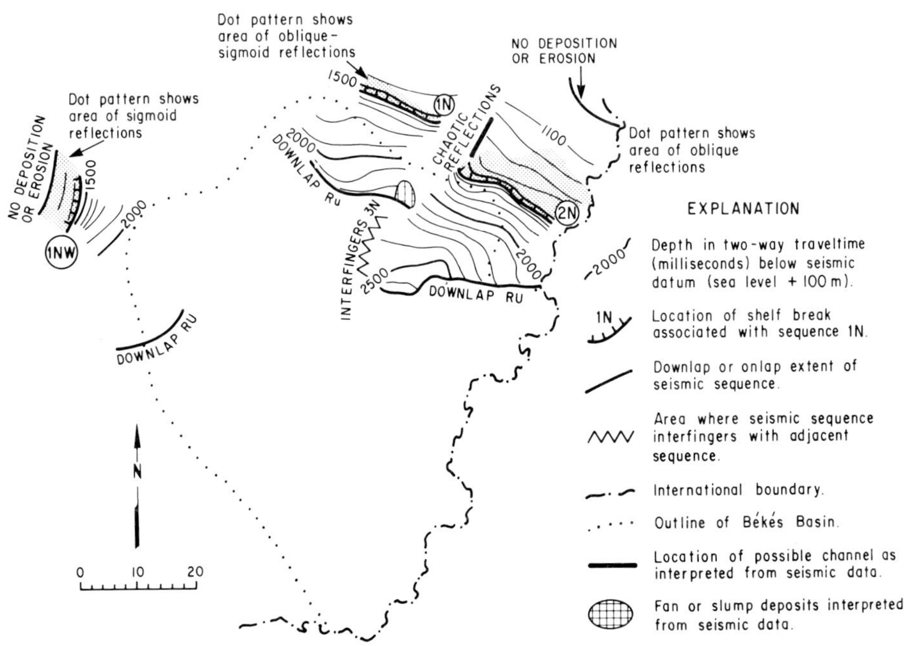

Figure 8. Structure contour maps on the tops of sequences 1N, 2N, and 1NW. Reflections in these sequences show a downlap pattern onto RU. Chaotic reflections occur between 1N and 2N in the area of the Békés trough, where sediments were funneled to central parts of the basin.

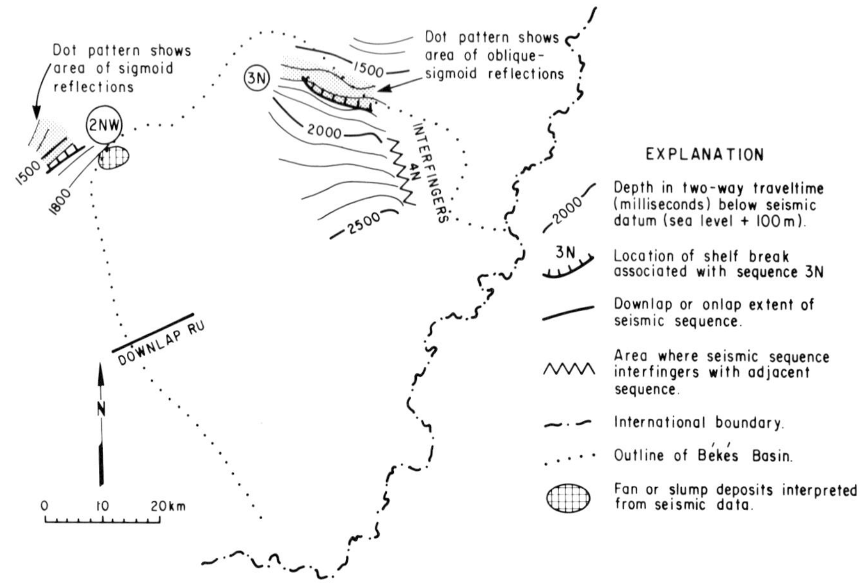

Figure 9. Structure contour maps on the tops of sequences 3N and 2NW. Deposition of 3N was coeval with the deposition of 4N (Fig. 10) because reflections from within the two units interfinger.

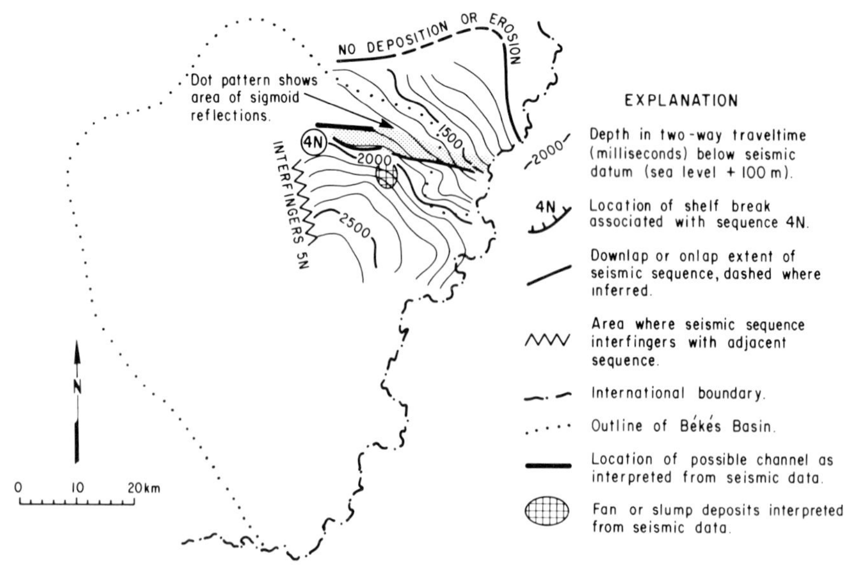

Figure 10. Structure contour map on the top of sequence 4N.

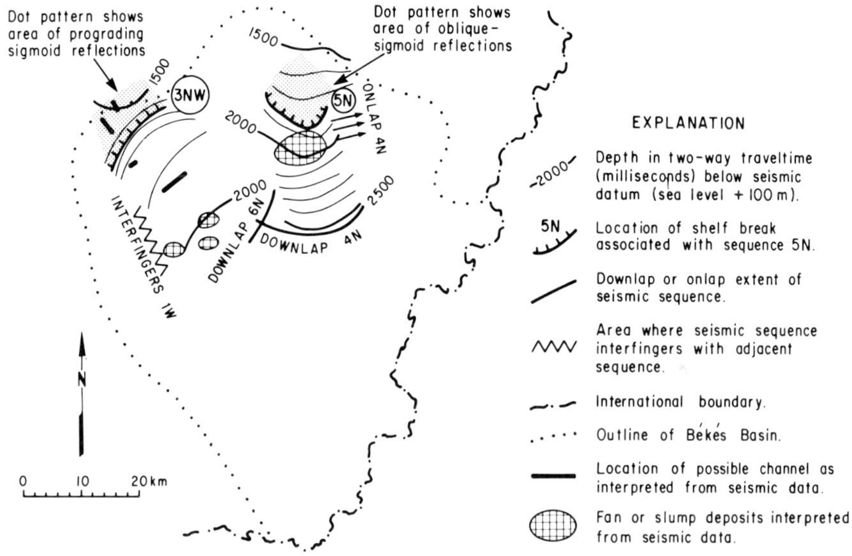

Figure 11. Structure contour maps on the tops of sequences 5N and 3NW. Onlap, downlap, and interfingering relations with adjacent and underlying units are shown.

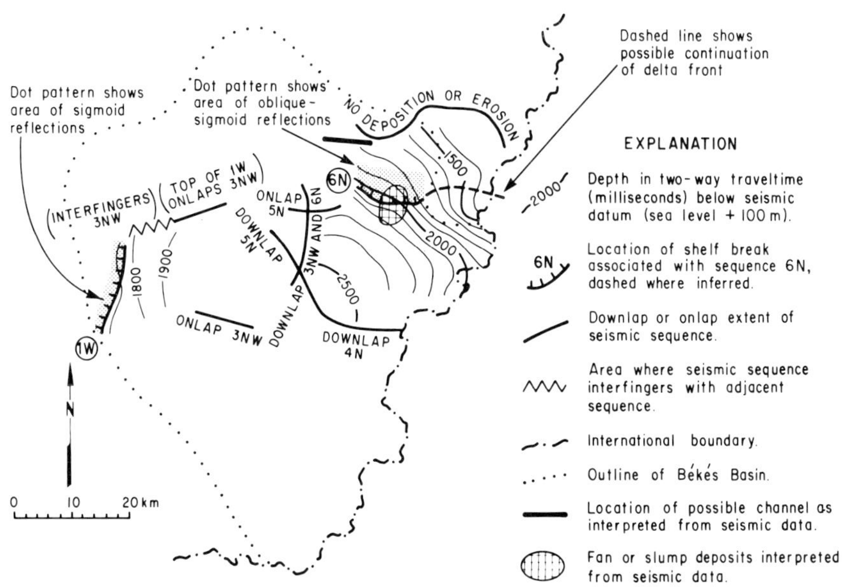

Figure 12. Structure contour maps on the tops of sequences 6N and 1W.

(slumps, slides, and irregular deposition) that occurred as sediments were funneled basinward, down the axis of the Békés trough. Reference to Figure 2 (basement map) indicates that the trough is a relatively old feature. Its trend, therefore, was probably determined by prerift or early rift topography and, apparently, it was a major conduit for moving sediments into the central parts of the Békés basin throughout much of its history, possibly until less than 5.78 Ma as indicated by results of magnetostratigraphic studies, which will be discussed in another section of this paper.

The vertical and lateral relations among sequences 1N-6N are summarized schematically in Figure 13. Sequences 1N and 2N, approximately coeval in the basal part of the section, terminate against the top of RU by downlap. The lower part of 3N is approximately time-equivalent to 2N, whereas, the upper part of 3N is approximately coeval with 4N. Similarly, the lower part of 5N is coeval with the upper part of 4N. Sequence 6N terminates by downlap on top of 4N and 5N and, therefore, is younger than both.

It is not clear whether this early progradational episode (deposition of 1N-6N) was linked to a single fluvial system that periodically shifted from the west side of the Békés trough to the east side of the trough, or whether two or more fluvial systems, one from the north and another from the northeast, transported the sediment load. It is clear, however, that the Békés trough, which controlled sedimentation during the deposition of RU, gradually became filled, and a nearly continuous delta front (6N, Fig. 12) formed south of the older progradational fronts.

Sequences 1NW-3NW

Deltaic encroachment from the north was concurrent with deltaic encroachment from the northwest (sequences 1NW, 2NW, and 3NW; Figs. 8, 9, and 11, respectively). The time-stratigraphic relations between

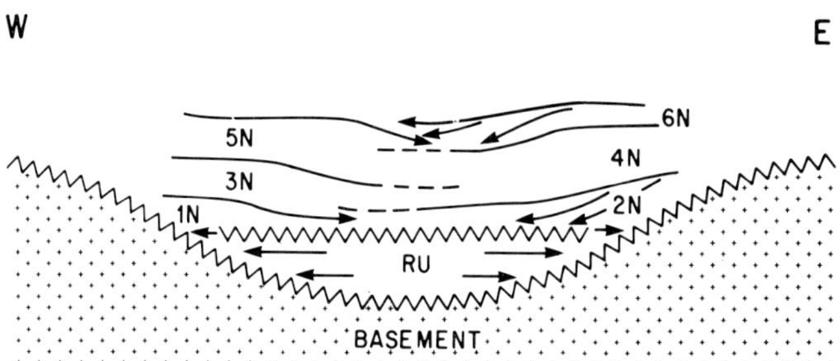

Figure 13. Schematic diagram showing the relations between RU, sequences 1N-6N, and basement in the northern part of the Békés basin. Arrows show direction of onlap or downlap. Dashed lines indicate interfingering of units. Wavy line represents an unconformity.

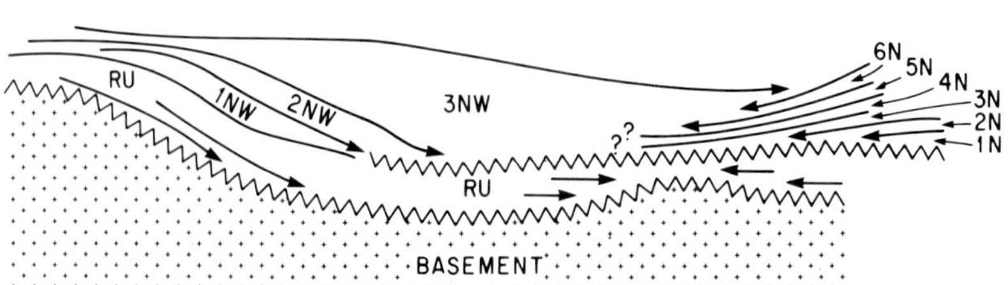

Figure 14. Schematic diagram showing the relations between early sequences prograding from the northwest (NW) and north (N) above RU. Arrows show onlap and downlap relations. The ages of sequences 1NW and 2NW relative to the ages of the sequences prograding from the north are not known. The top of 3NW, however, is younger than 6N.

RU, the sequences prograding from the northwest (1NW-3NW), and the sequences prograding from the north (1N-6N) are shown diagrammatically in Figure 14. Although the relative time-stratigraphic relation between the sequences mapped in the northwestern part of the basin is clear (based on the principal of superposition), their time-stratigraphic relation to the sequences mapped in the northern part of the basin is not. Clearly, however, 1N and 1NW are younger than RU, and the top of 3NW is younger than 6N (Fig. 14).

Sequence 1W

During deposition of the sedimentary rocks represented by 1NW-3NW, or possibly subsequent to the deposition of 2NW, sediments were entering the basin from the west (sequence 1W, Fig. 12). Sequence 1W extends from the western margin of the basin eastward to about the center of the basin. At its eastward extent, reflections in the upper part of 1W terminate by downlap against the tops of 3NW and 1N; whereas, laterally to the north, the sequence interfingers 3NW and to the south, it terminates in an onlap configuration against 3NW (Fig. 12).

The relation of sequence 1W to other sequences mapped nearby is shown schematically in Figure 15. This figure illustrates that the top of 1W is younger than the top of 6N and 3NW, and that the base of 1W is younger than 2NW. The time-stratigraphic relation between the lower part of 1W and 3NW (based on the principal of superposition), however, is not obvious on the seismic records. Because of this uncertainty, it can not be determined if progradation from the west was contemporaneous with progradation from the northwest during deposition of the sediments represented by 1W and 3NW or if progradation from the west postdated deposition of sediments represented by 3NW.

Sequences 7N-9N and 4NW-6NW

After deposition of sequences 1W, 3NW, and 6N, the lake in the Békés basin was reduced to about half its former extent. Much of the subsequent history of the basin is one of converging fluvial systems depositing their sediment loads in a lacustrine basin shrinking in size, while the main depocenter remained relatively stationary in the east-central part of the basin. Confined to a relatively small basin, many of the main prograding depositional sequences began to interfinger downdip and laterally. In this manner, the younger sequences became more widespread and, as such, represent more varied depositional environments.

Structure contour maps on the tops of sequences 4NW and 7N are shown in Figure 16. Although approximately coeval, the top of 7N extends slightly higher up into the sedimentary section than does the top of 4NW as indicated on seismic records from the center of the basin where the two sequences interfinger.

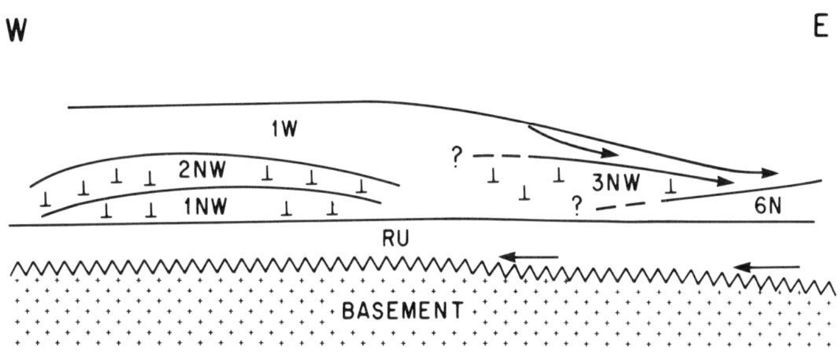

Figure 15. Schematic diagram showing the relation of sequence 1W to other mappped sequences. Symbol ⊥ indicates progradation perpendicular to direction of section. Arrows show onlap and downlap relations.

In the central part of the basin, 4NW and 7N are overlain by 5NW to the northwest (Fig. 17), 6NW to the north-northwest (Fig. 18), and 8N to the north-northeast (Fig. 17). Sequence 5NW terminates by onlap against 4NW to the west and by downlap against the same sequence in the deeper parts of the basin. South of the main depocenter, 8N interfingers with age-equivalent sequences that represent sediments brought in from the south; whereas, north of the depocenter, 5NW and 8N interfinger with 6NW (Fig. 17).

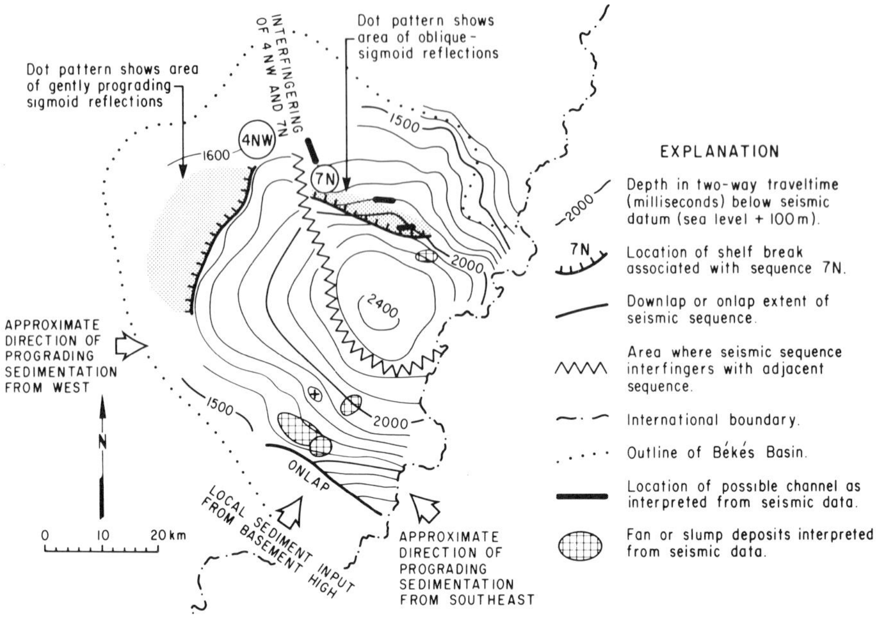

Figure 16. Structure contour maps on tops of coeval sequences 4NW and 7N. By this time, the areal extent of the lake had decreased by more than 50% in comparison to its size during deposition of 1N (compare with Figure 8). This decrease in size was a result primarily of rapid deltaic progradation from the north and northwest.

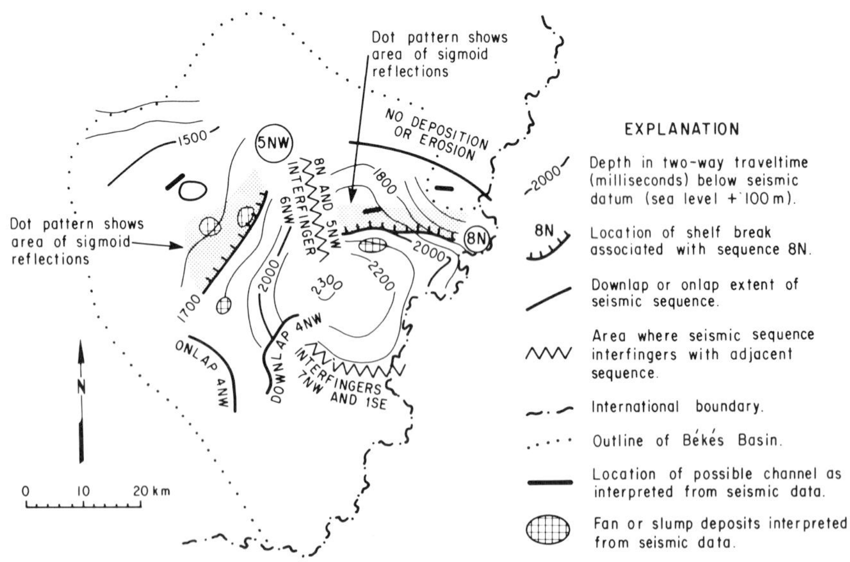

Figure 17. Structure contour maps on tops of sequences 5NW and 8N. The advance of delta fronts from the north and northwest became more rapid as the areal extent of the lake decreased (compare with Figures 8, 11, and 16).

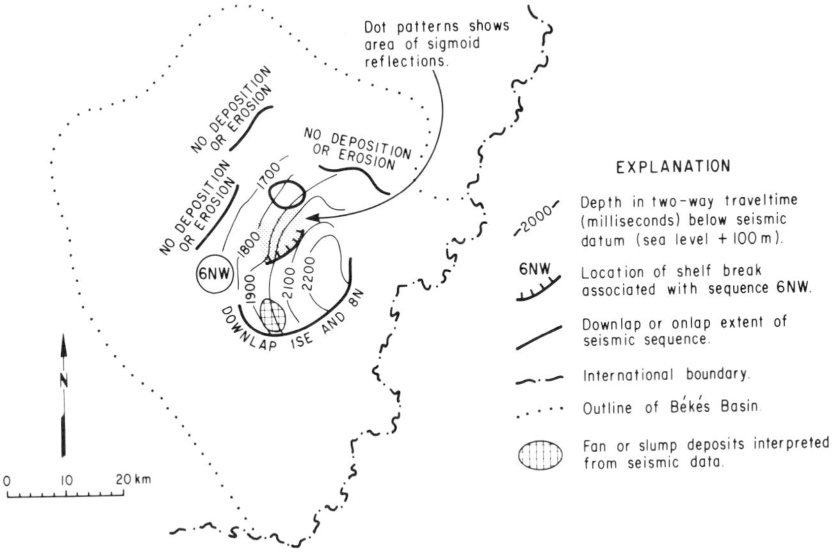

Figure 18. Structure contour map on top of sequence 6NW. Following the convergence of delta fronts that were prograding chiefly south and southeast (Fig. 17), progradation now was to the south-southeast.

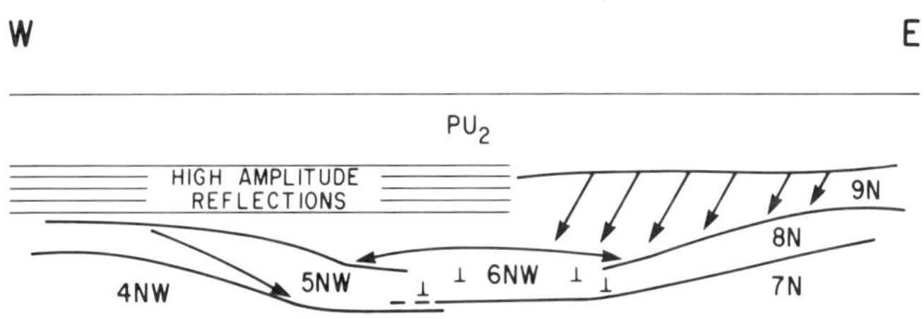

Figure 19. Schematic diagram showing the relations between sequences mapped in the central part of the Békés basin. Symbol ⊥ indicates progradation perpendicular to direction of section. Arrows show downlap relations. Area marked "high-amplitude reflections" is believed represent a shallow-lake or marsh environment that developed behind prograding delta fronts.

The lateral and vertical relations among sequences 5NW, 6NW, 8N, 9N, and the sequences underlying them are shown schematically in Figure 19. The somewhat complicated interfingering relation among 5NW, 6NW, and 8N is interpreted as resulting from the convergence of fluvial systems. Reference to Figure 5 shows that the composite delta lobes of 1NW-5NW formed a southeastward-prograding sediment wedge. Similarly, the composite delta lobes of 1N-8N formed a separate southwestward-prograding sediment wedge. The two rivers associated with these sediment wedges apparently converged about 10 to 15 km north-northwest of the basin depocenter, and a sediment wedge began to prograde south-southeast from the confluence of these rivers. The new southward-prograding sediment wedge is represented by sequence 6NW.

During and following deposition of the sediments represented by 6NW, progradation from the north continued in the northeastern part of the basin. This episode is represented by 9N (Figs. 5, 6, 19, 20, and 21). Seismically, 9N is characterized by strongly oblique reflections

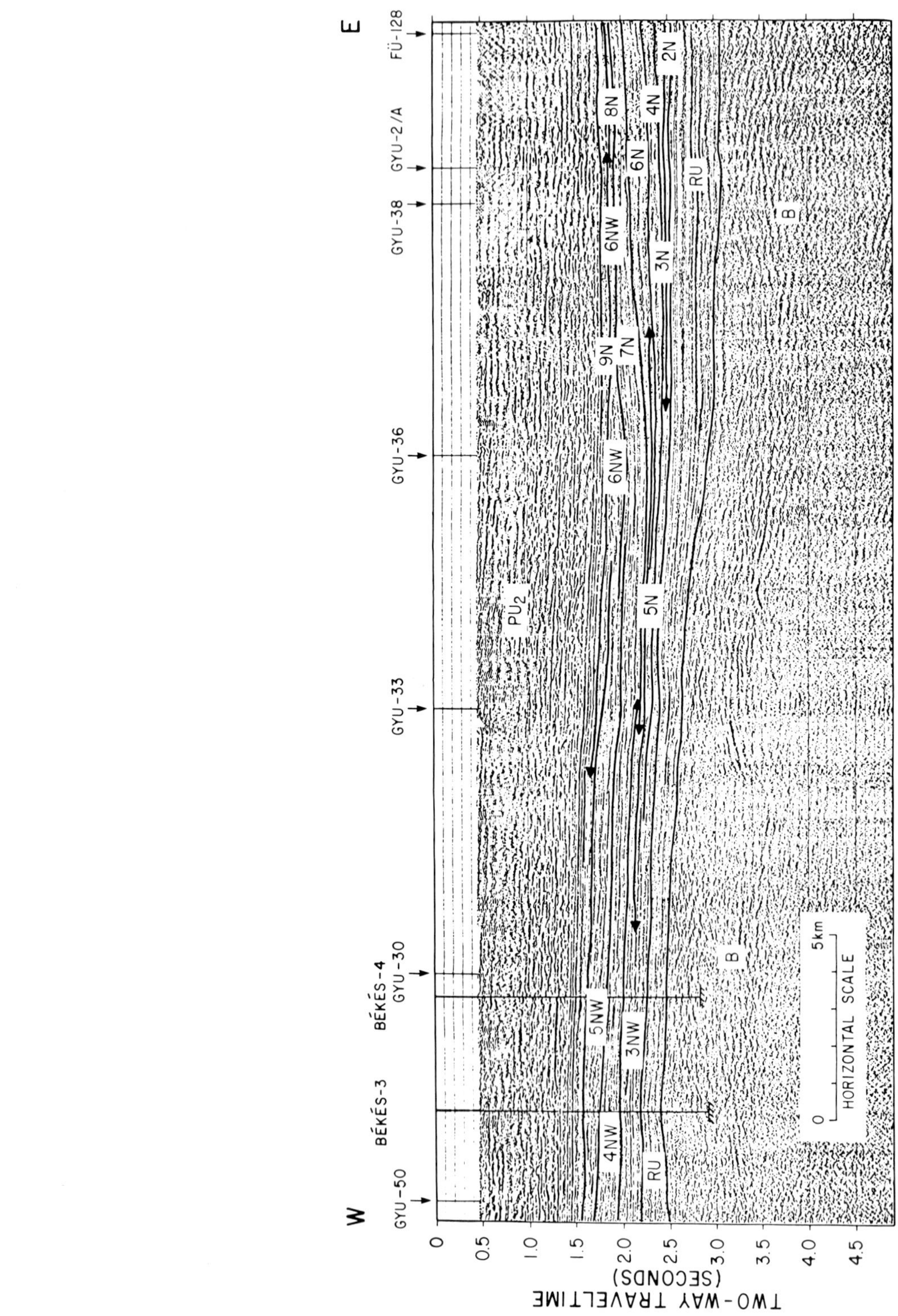

Figure 20. Seismic profile GYU-43 showing relations between many of the sedimentary units mapped in the Békés basin area. The strong reflection in the basement complex at about 3.2 sec near the west end of the profile is believed to be associated with low-angle thrust faulting of Mesozoic rocks and is discussed by Grow and others (this volume). The location of the profile is shown in Figure 3.

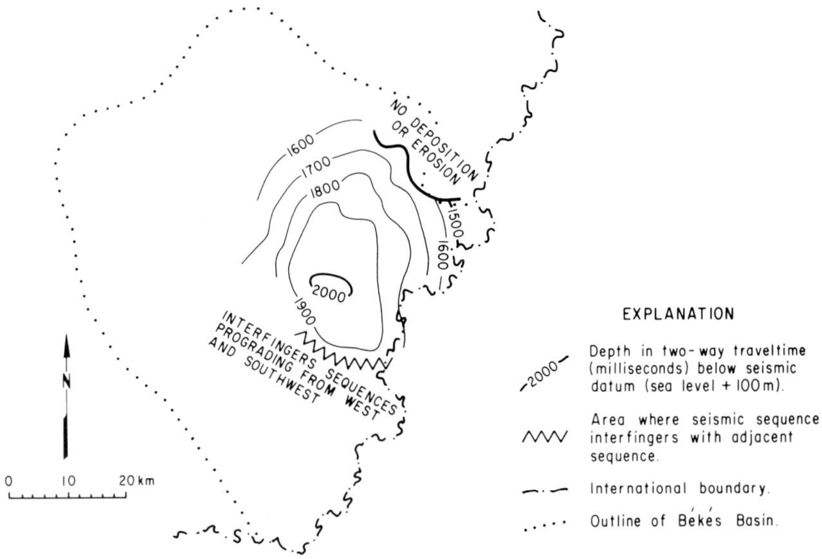

Figure 21. Structure contour map on top of sequence 9N. At this time, the main depocenter began to shift southward (compare with Figure 17).

(Fig. 6). These oblique reflections are interpreted as evidence of a rapid increase in the rate of delta progradation and basin filling (indicated by arrows in Figure 5).

The time-stratigraphic relations among sequences 9N, 8N, 6NW, and a series of high-amplitude, continuous reflections are illustrated schematically in Figure 19. Sequence 9N is clearly younger than 8N, as evidenced by the downlap termination of oblique reflections of 9N against the top of 8N. The time-stratigraphic relation between 9N and 6NW is less clear. The uppermost part of 6NW appears to interfinger with the youngest part of 9N; interfingering suggests that the two units are somewhat coeval, at least in the youngest parts of their section. To the west (Figs. 19 and 20), flat-lying, high-amplitude, continuous reflections are seen to be, in part, time equivalent to 9N. These reflections are interpreted to represent deposition in a low-energy (quiet, shallow-water) environment that developed on the platform formed by 5NW and 6NW.

During the last stage of deposition represented by 9N, the main depocenter of the Békés basin was gradually shifting southward at the same time that sediments were being brought into the basin from the southwest (Fig. 21).

Sequences in the Battonya-Pusztaföldvár Area

Although the precise time is not clear, a depocenter began to form in the southern part of the Békés basin when the area of the Battonya-Pusztaföldvár basement high (Fig. 7) began to sink below lake level or became submerged as the lake level rose, as evidenced by reflection patterns that indicate aggradational stacking. By this time, much of the Hódmezövásárhely trough, located to the east-southeast of the Békés basin, had been filled by prograding sediment wedges. This filling allowed fluvial systems to advance over the trough and discharge their sediment loads into the Békés basin. This advance was from the west (Fig. 5, sequences 2W and 3W). At the same time, other delta fronts began to encroach from the south-southeast (Fig. 5), likely from highland areas (Apuseni Mountains and Eastern Carpathians(?)) in what is now Transylvania in Romania.

Submergence of the Battonya-Pusztaföldvár area is evidenced by a relatively thin layer (0-150 m) of basal conglomerate, including sandstones and shales, that unconformably overlies basement rocks and has been penetrated in numerous wells drilled on the Battonya-Pusztaföldvár basement high. The age of this basal conglomerate (Fig. 22) relative to the previously discussed depositional units is uncertain because, north of the Bat-

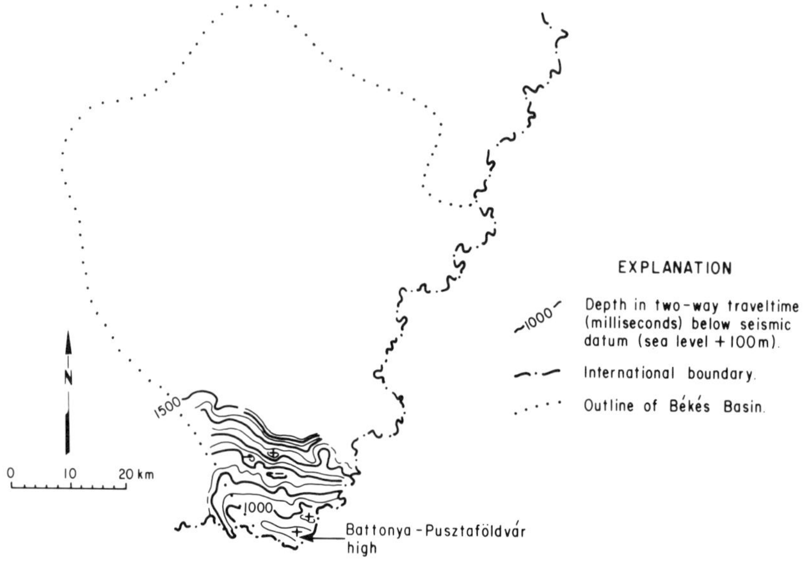

Figure 22. Structure contour map on top of basal conglomerate (BC), a diachronous unit which represents initial submergence of the Battonya-Pusztaföldvár high.

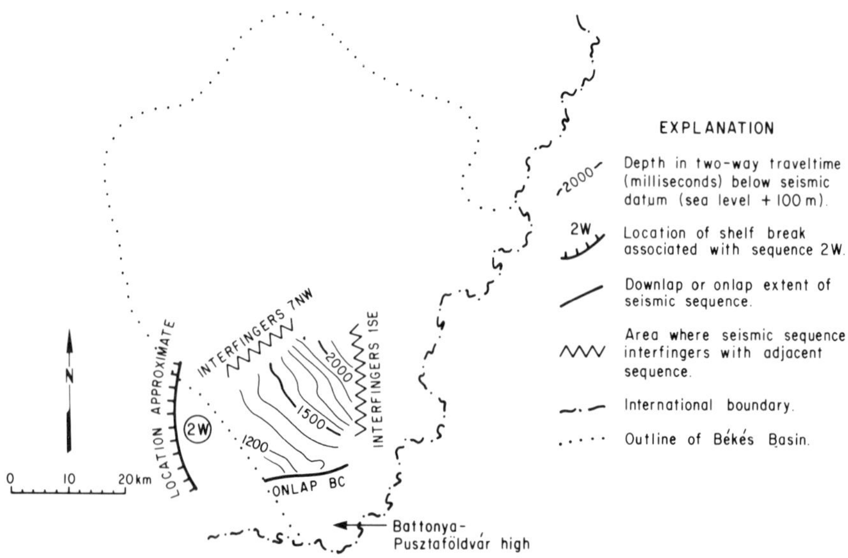

Figure 23. Structure contour map on the top of sequence 2W. Coeval sequences 7NW and 1SE (Fig. 24) mark the start of deltaic sedimentation in the southern part of the Békés basin as the Battonya-Pusztaföldvár area sank below lake level. BC: basal conglomerate.

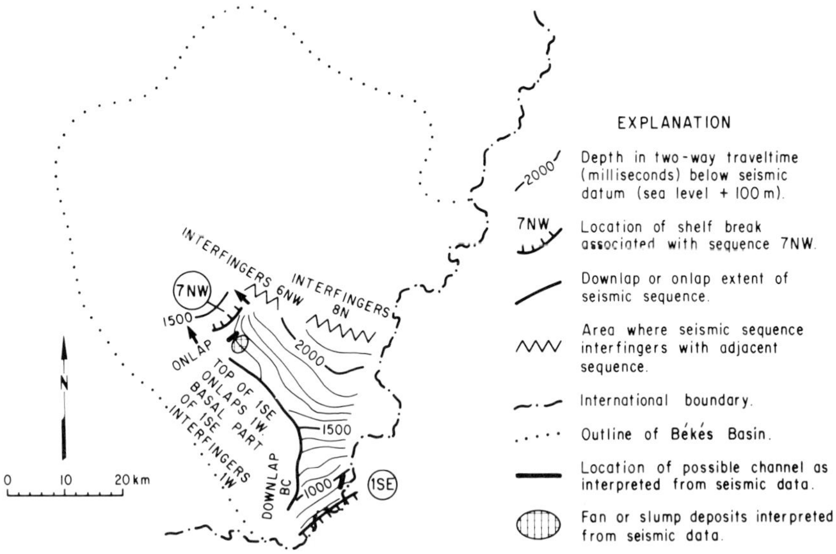

Figure 24. Structure contour map on the tops of sequences 1SE and 7NW. These sequences together with 2W (Fig.23) mark the start of deltaic sedimentation in the southern part of the Békés basin as the Battonya-Pusztaföldvár area (Fig. 23) sank below lake level. BC: basal conglomerate.

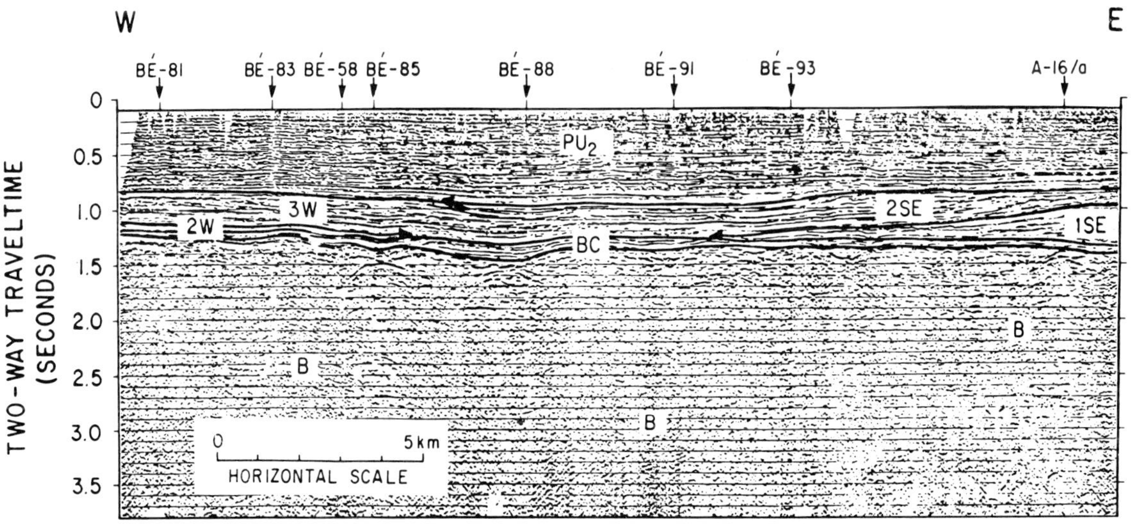

Figure 25. The relations between the basement complex, the basal conglomerate unit (BC), and sequences 2W, 3W, 2SE, and 1SE as they appear on seismic record Bé-78 (Fig. 3) recorded on the Battonya-Pusztaföldvár high in the southern part of the Békés basin. Sequences 2SE and 3W are approximately coeval. Arrows show direction of downlap.

Table 1. Approximate water depths during deposition of various units and sequences in the northern part of the Békés basin. These data indicate that waters of the lake in the Békés basin area were progres-

UNIT OR SEQUENCE	WATER DEPTH (m)
RU	1000 (?)
1N	1150
2N	1300
3N	900
4N	1000
5N	700
6N	850
7N	800
8N	650
Start of 9N	400
End of 9N	200
Pu_2	<200

tonya-Pusztaföldvár high, the conglomerate could not be traced seismically nor was it penetrated in wells. However, according to Molenaar and others (this volume), the unit represents deposition along the lake shoreline as the lake level rose, or the Battonya-Pusztaföldvár high subsided below lake level, and is time transgressive. From relations evident on seismic records, the basal conglomerate unit of the Battonya-Pusztaföldvár area is inferred to be younger than the previously discussed turbidites and marls of RU.

In the Battonya-Pusztaföldvár area, the oldest sequences above the basal conglomerate can be chronostratigraphically correlated with each other and with some of the previously discussed sequences (2W, Fig. 23; 1SE and 7NW, Fig. 24). Near the crest of the Battonya-Pusztaföldvár buried basement high, the chronostratigraphic relation between 2W and 1SE is not clear on seismic records (Fig. 25); but further north, the two units are inferred to be approximately coeval because of their interfingering relationship (Fig. 23). To the northwest, 2W interfingers with 7NW (Fig. 23); whereas, to the northeast, 1SE interfingers with 8N (Fig. 24). These time-stratigraphic relations are summarized schematically in Figure 26. On the basis of these relations, 8N, 7NW, 2W, and 1SE are inferred to represent sediments deposited as the main depocenter in the Békés basin began to shift south toward the Battonya-Pusztaföldvár area. At this time, fluvial systems were converging from the north, northwest, west, and southeast to a small area in the southeastern part of the Békés basin.

Following the deposition of sediments represented by 2W and 1SE, delta fronts continued to advance from the west, southwest, and north in the southern part of the Békés basin (3W, Fig. 27; 2SE, Fig. 28). At its northward extent, 3W interfingers with 9N (Fig. 27), and seismic data from the area of the interfingering indicate that the top of 3W is slightly younger than the top of 9N. In most areas, 3W and 2SE are approximately coeval, but locally, the upper part of 2SE is somewhat younger than the upper part of 3W, as evidenced by the fact that reflections in the upper part of 2SE show a downlap relation to the top of 3W (Fig. 29).

WATER DEPTHS

The inference that, on the average, the waters of the Békés basin were continuously shoaling after deposition of the retrogradational unit (RU) can be tested by estimating the approximate water depths of the lake during

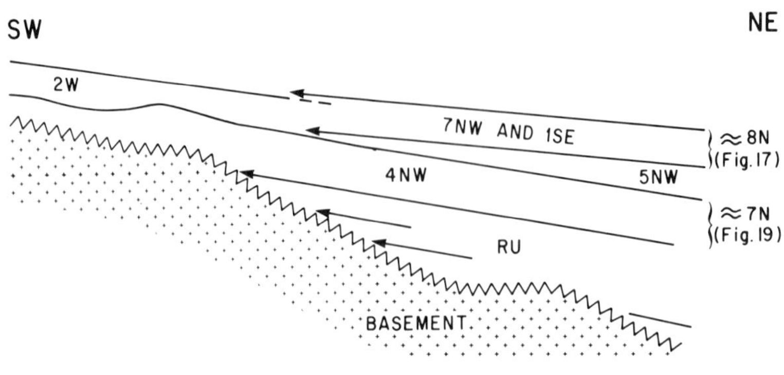

Figure 26. Schematic diagram showing relations between selected sequences in the southern part of the Békés basin. Sequences 7NW and 1SE were mapped as a single unit approximately coeval with 2W. Further north, 1SE appears to be coeval with 8N (Fig. 24). RU, retrogradational unit; B, basement.

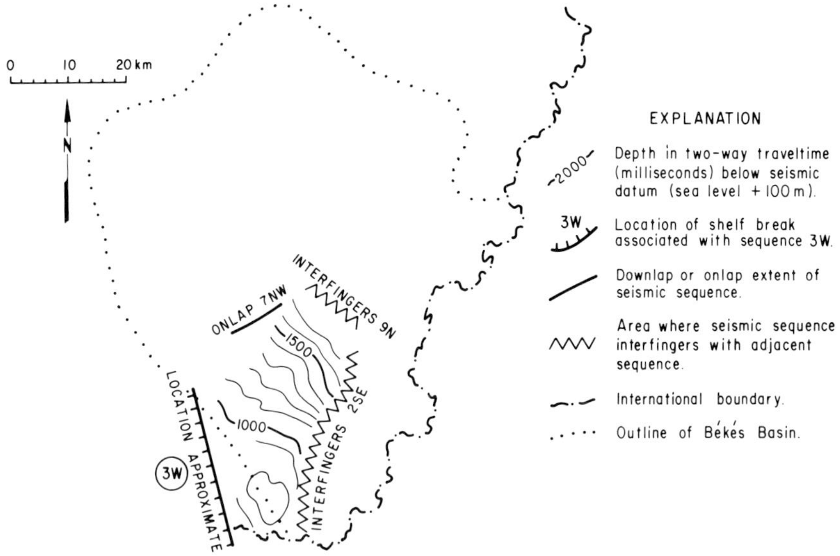

Figure 27. Structure contour map on the top of sequence 3W, which is approximately coeval with 2SE and 9N as inferred from the noted interfingering relations.

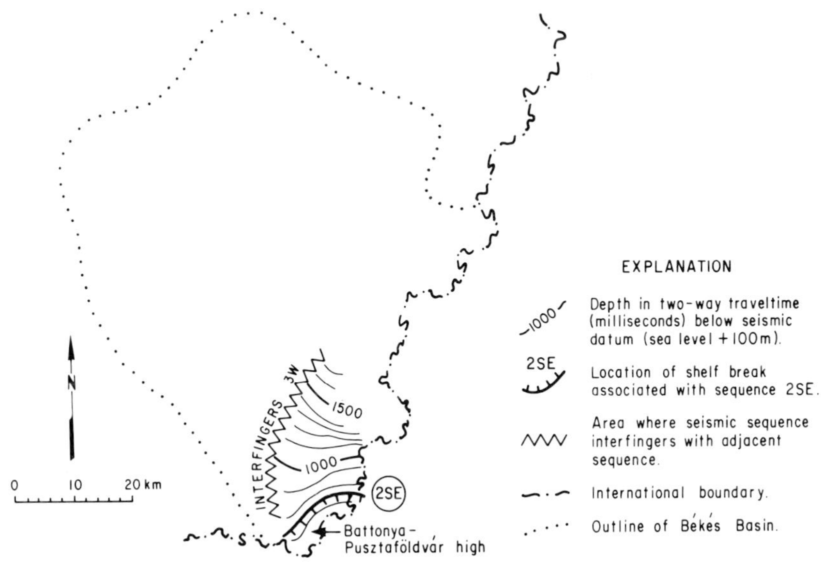

Figure 28. Structure contour map on the top of sequence 2SE, which is approximately coeval with 3W as inferred from noted interfingering relations.

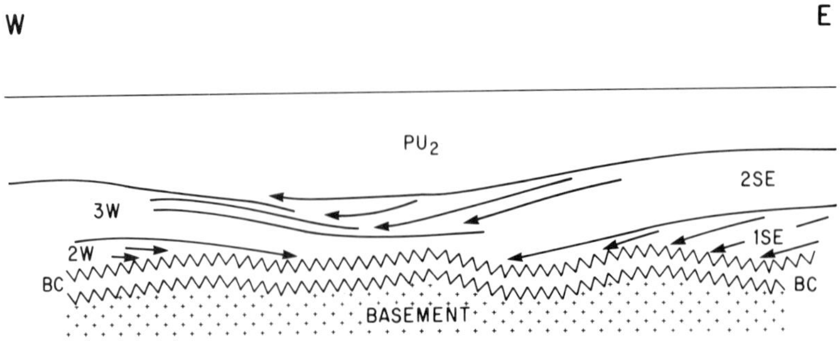

Figure 29. Schematic diagram showing relations between basement complex, the basal conglomerate unit (BC), sequences 2W, 3W, 1SE, and 2SE, and PU_2 in the Battonya-Pusztaföldvár area (Fig. 28). Wavy lines indicate unconformities and arrows show directions of downlap and onlap.

Table 2. Summary of thicknesses (T), ages, and average sedimentation rates for units in the Békés basin. With the exception of BRU, thicknesses of the units were calculated from Figure 30.

Sequence	T (m)	Age (Ma)	Duration (10^6)	Sedimentation ($m/10^6$ yrs)
BRU	275	15(?)-11.5	3.5 (?)	79
RU	2225	11.5-5.78	5.72	390
PU_1	1700	5.78-4.25	1.53	1100
PU_2	2500	4.25-0	4.25	590

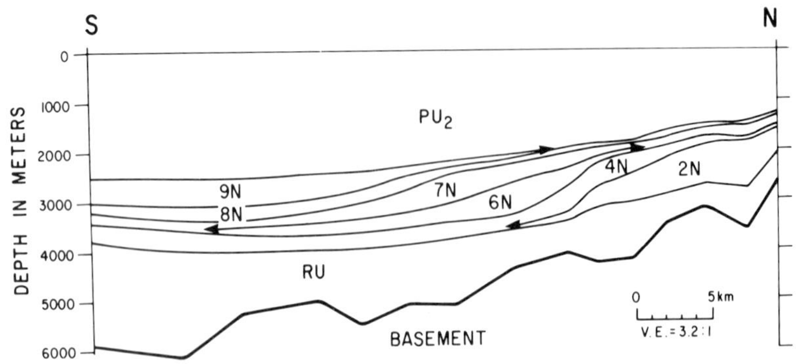

Figure 30. Interpretation of composited seismic profiles KO-11, GYU-7, and GYU-22 across the north and central parts of the Békés basin. The seismic profile is shown in Figure 6 and its location in Figure 3. Arrows show directions of downlap and onlap. Symbols designating the various units are explained in the text. Note that the vertical scale is in depth rather than time.

deposition of various units and sequences (Table 1). Approximate water depths were calculated by measuring the vertical distance from bottomset beds (undaform reflections) to topset beds (fondaform reflections) along continuous reflections on seismic records from the northern part of the Békés basin where the sigmoid and oblique-sigmoid patterns are best developed. Although the calculated depths are only approximate because many of the prograding units prograde from atop basement highs to basin deeps and corrections for differential compaction and subsidence were not applied, the general indications are that the water depths continuously decreased from 1000 m to <200 m as the basin filled during Pannonian and later time.

The water depths are estimated to have been about 1000 m during the last phase of deposition of the retrogradational unit (RU). The water depths increased somewhat during the initial phases of deposition of the early progradational unit (sequences 1N and 2N), and decreased to about 200 m during the latest phase (sequence 9N). Water depths associated with PU_2 were probably much less than 200 m as inferred from the absence of internal clinoform reflections within the unit and the description by Phillips and others (this volume) of the inferred depositional environments - perhaps tens of meters.

It should be noted that the water depths shown in Table 1 were estimated in the northern part of the study area where paleo-prodelta slopes are characterized by relatively steep gradients. In the southeastern part of the basin, water depths appear to have been shallower, and paleo-prodelta slopes were characterized by relatively gentle gradients.

AGES AND DEPOSITIONAL RATES

According to Mattick and others (1985), indirect evidence indicates that Neogene delta construction in the northernmost part of the Pannonian basin began sometime between Sarmatian and earliest Pannonian time, perhaps at 12 Ma. Magnetostratigraphic studies (Elston and others, this volume) indicate that distinct delta fronts were being constructed at about 7.41 Ma about 100 km north of the Békés basin. In addition, Pogácsás and others (this volume), using seismic stratigraphic methods, extend the work of Elston and others (this volume) to the northern margin of the Békés basin. The results of these authors indicate that the uppermost part of the retrogradational unit (RU) in the Békés basin was deposited about 6.68 Ma, sequence 1N was deposited about 5.78 Ma, and 9N was deposited earlier than 4.25 Ma. Accordingly, average sedimentation rates are calculated to be 79, 390, 1100, and 590 m/million years, respectively, for BRU (15(?)-11.5 Ma), RU (11.5-5.78 Ma), PU_1 (5.78-4.25 Ma), and PU_2 (4.25-0 Ma). These data are summarized in Table 2.

SUMMARY OF DEPOSITIONAL HISTORY

The Neogene and Quaternary sedimentary filling of the Békés basin can be divided into four depositional stages corresponding to the basinally restricted unit (BRU), the retrogradational unit (RU), the early progradational unit (PU_1), and the late progradational unit (PU_2).

BRU represents initial basin subsidence during Badenian and Sarmatian time (15(?)-11.5 Ma) and is marked by shallow-water, near-shore marine to brackish-water environments of deposition. Low sedimentation rates (79 m/million years), about equal to subsidence rates, prevailed throughout this period.

RU represents a period of increasing water depths in a lacustrine environment over a period of about 5.72 million years (Table 2). The early part of this stage is inferred to correspond to a period of rapidly increasing water depths (late Sarmatian or early Pannonian) during which time the Békés basin was largely starved of sediments because debris transported by rivers was being captured by subbasins along the northern and western margins of the Pannonian Basin. As a result of high subsidence rates coupled with low sedimentation rates, the water depths in the lake rapidly increased from 200 to about 1000 m. Support for this inference comes from Grow and others (this volume), who, as a result of their tectonic investigations, point out that synrift sediments are either relatively thin or absent in the basin. Pogácsás and others (this volume) conclude that the Békés basin was relatively starved of sediments between 12 and 9 Ma.

During the early part of this stage, delta fronts were located about 100 km north and west of the basin's margins. clastic sediments (turbidites) were periodically transported into the basin. marls (interbedded with the turbidites) resulted from the precipitation of $CaCO_3$ and mud from suspension.

During later phases, turbidite-fronted deltas approached the margins of the basin, and sediments spilled directly into the basin. Depositional patterns during this later phase of deposition of RU are illustrated in Figure 31. By this time, delta fronts had advanced to within about 35 km of the basin's

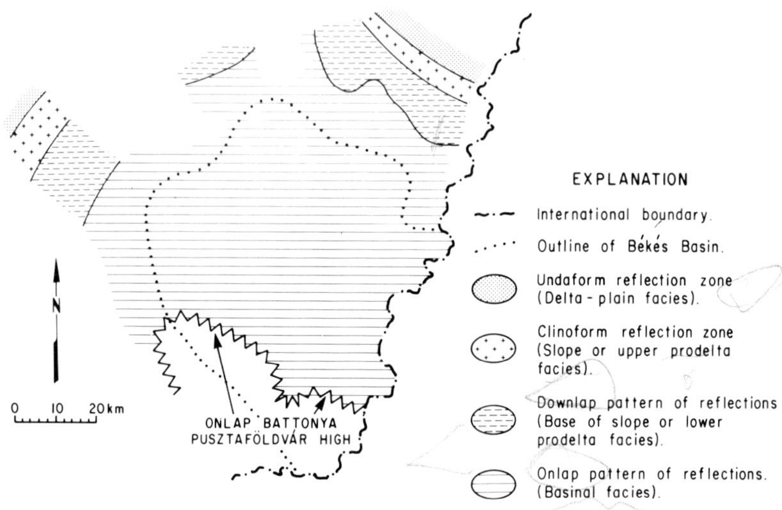

Figure 31. Map showing facies distribution the final phase of deposition of RU. By this time, delta fronts, which were prograding southward and southeastward, were about 35 km from the northern and northwestern margins of the Békés basin. Deposition within the Békés basin consisted chiefly of marls and turbidites representing distal basinal facies. The Battonya-Pusztaföldvár high was above lake level.

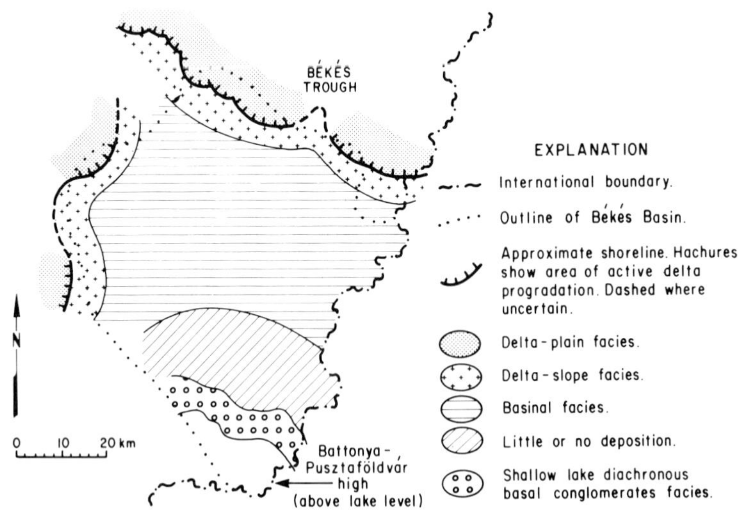

Figure 32. Map showing inferred facies distribution during the early phase of the deposition of PU_1. Delta fronts had reached the northern and western margins of the Békés basin and the Battonya-Pusztaföldvár high was starting to sink below lake level. The map was constructed using the following sequences: upper part of 3N (Fig. 9), middle of 4N (Fig. 10), lower part of 3NW (Fig. 11), middle of 1W (Fig. 12), and the lower part of the basal conglomerate unit (Fig. 22).

northern, northwestern, and western margins. Within the Békés basin, deposition was limited to basin facies - principally marls and turbidites. The Battonya-Pusztaföldvár basement high was above lake level during much of this time period.

About 6.25 Ma, delta fronts had advanced from the north and west to the margins of the Békés basin. This advance marks the beginning of the deposition of the early progradational unit (PU_1). Along the northern margin of the basin, a delta front was almost continuous. It was disrupted only by the Békés trough (Fig. 32) which, thereafter, played only a minor role in controlling sedimentation. These delta systems were composed of thinner sedimentary sequences than those formed during the earlier stage and lacked a thick turbidite apron. The dominant constructional mechanism was probably delta-lobe switching, which gave rise to a succession of prograding sequences that reflect multiple overlapping lobes. In central parts of the basin, deposition was represented by basin facies - chiefly turbidites and marls (Fig. 32). To the south, on the north flank of the Battonya-Pusztaföldvár high, locally derived conglomerates, sandstone, and shale were being deposited (Fig. 32).

During later phases of PU_1, the basin had shrunk by about 50% in comparison to its size at 6.25 Ma (Fig. 33). Major river systems began to converge, and the main depocenter moved southward as sedimentation was gradually restricted to a basin decreasing in size. Water depths in the deepest parts of the basin were probably on the order of 650 m (Table 1, sequence 8N). Behind the delta fronts, marshes and/or very shallow, ephemeral lakes began to form (Fig. 33). To the southeast in the vicinity of the Battonya-Pusztaföldvár high, rivers that were previously southeast of the basin had advanced to the margins of the basin. The ridge of the submerged Battonya-Pusztaföldvár high practically escaped sedimentation because sediments brought in from the southeast were largely funneled north of the high and laid down on its still-steep north flank (Fig. 33).

During the last phases of deposition associated with PU_1, the main delta fronts advanced southward and

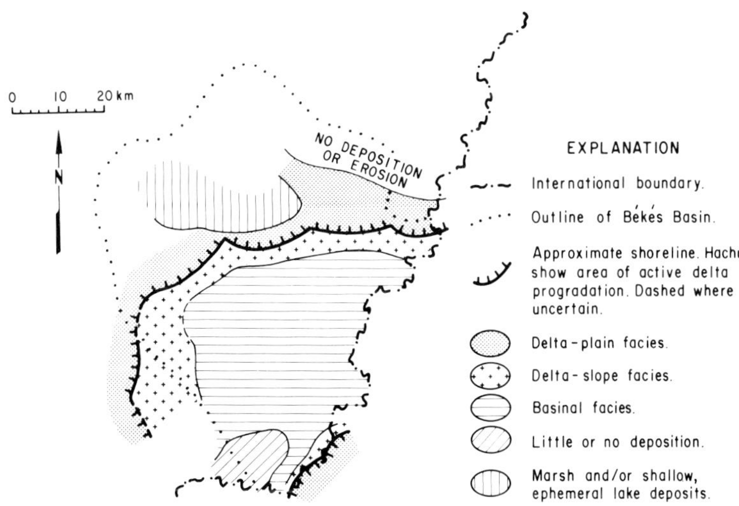

Figure 33. Map showing inferred facies distribution a later phase of the depositon of PU_1. As a result of continued delta prograding from the north and west, the areal extent of the Békés lake had decreased by more that 50% compared with its size at the start of deposition of PU_1 (Fig. 32). The Battonya-Pusztaföldvár high (Fig. 32) had sunk below lake level and delta fronts were encroaching from the southeast. Shallow lakes or marshes began to develop in the northern part of the basin behind the advancing delta fronts. The map was constructed using the following sequences: upper part of 8N (Fig. 17), upper part of 5NW (Fig. 17), lower part of 6NW (Fig. 18), upper part of 2NW (Fig. 23), and lower part of 1SE (Fig. 24).

southeastward, and the main depocenter migrated southeastward as well.

According to Collinson (1976), a progression from delta deposits associated with deep-water basins to delta deposits associated with shallow-water basins is part of the normal evolution of a basinal sequence of sedimentary fill. This evolution appears to characterize the Békés basin. Collinson (1976) further stated that deep-water turbidites that front deltaic sequences are the earliest in any basinal sequence and are generally overlain by younger, shallower water deposits composed of thinner sedimentary sequences that are characterized by delta-front sheet sandstones, which probably are related to lobate-type subdeltas. The dominant mechanism in the construction of the later delta deposits is often lobe switching, which results in a succession of stacked sequences (Coleman, 1976) as is interpreted for the Békés basin.

The average sedimentation rate during deposition of PU_1 was about 1100 m/million years (Table 2) - much greater than the average subsidence rate as evidenced by incessant shoaling waters, the depth of which decreased from about 1100 m to 200 m (Table 1). This stage was associated with rapidly prograding delta fronts. Prograding delta wedges in the northern part of the basin (sequences 1N-9N) are calculated to have advanced at an average rate of about 30 km/million years; the rate of advance was relatively slow during the early phases of deposition and increased during later phases.

The late progradational unit (PU_2) represents the youngest episode of deltaic deposition in the Békés basin. The sedimentary rocks associated with PU_2 represent shallow-water deposits consisting chiefly of delta-plain facies deposited in shallow lake, fluvial, and marsh to terrestrial environments. These deposits locally reach thicknesses of 2500 m, and, according to Bérczi and Phillips (1985) and Phillips and others (this volume), analyses of core data indicate that they represent shallower water environments than the underlying prograding sequences. Locally, the base of the deposits are marked by a strong onlap pattern which Mattick and others (1985) interpreted as evidence of a period of widespread shallow lakes and marshes. The absence of clinoform beds within the unit suggests that water depths in the Békés basin were very shallow at the time of deposition - probably much less than 200 m. This stage of basin filling represents a relatively long period of time (4.25-0 Ma) during which the average sedimentation rate was about 590 m/million years (Table 2) - about equal to or slightly higher than the average subsidence rate.

In his review of this paper, Molenaar (written communication, 1988) points out that in places the base of the late progradational unit (PU_2) is time transgressive. This relation would be expected if PU_2 was deposited on top of PU_1 as waters receded locally rather than basinwide.

ACKNOWLEDGEMENTS

The authors wish to thank the Hungarian Oil and Gas Trust (now MOL Ltd.) for their permission to publish the seismic data and interpretations contained in this paper. Also, we are deeply indebted to Györgyi Juhász and István Révész who provided well-log interpretations that were used for correlation of the seismic data. The authors appreciate the gracious help of Erzsébet Szabó and Zsuzsanna Orbán in preparing the many maps contained in this paper, of József Ádám in gathering the seismic records from company files, of László Wenner, Éva Jankura, Éva Fülöp, Zsuzsanna Halász and Éva Barabás in drafting the many illustrations, and of Mária Dominiák, who typed the first draft of this manuscript. Last, but most important, we wish to thank Kay Molenaar, John Schlee, Ferenc Horváth, and Mary Feeley, who reviewed the manuscript and provided many helpful suggestions.

REFERENCES

Bérczi, I., and Phillips, R.L., 1985, Processes and depositional environments within Neogene deltaic lacustrine sediments, Pannonian Basin, southeast Hungary; Eötvös Lóránd Geophysical Institute, Geophysical Transactions, 31, (1-3), 55-74.

Burchfiel, B.C., 1976, Geology of Romania; Geological Society of America Special Paper 158, 1-82.

Burchfiel, B.C. and Royden, L.H,, 1982, Carpathian foreland fold and thrust belt and its relation to Pannonian and other basins; Amer. Assoc. Petrol. Geol. Bull., 66, (9), 1179-1185.

Coleman, I.M., 1976, Processes of deposition and models for exploration: Continuing Education Publication Co., Champaign, Illinois, 102.

Collinson, J.D., 1976, Deltaic evolution during basin fill - Namurian of central Pennine Basin England (abs.); Amer. Assoc. Petrol. Geol. Bull., 60, (4), 1- 659.

Elston, D.P., Lantos, M. and Hámor, T., High resolution polarity records and the stratigraphic and magnetostratigraphic correlation of late Miocene and Pliocene (Pannonian, s.l.) deposits of Hungary; this volume, 111-142.

Grow, J.A., Mattick, R.E., Bérczi-Makk, A., Péró, C., Hajdú, D., Pogácsás, G., Várnai, P., and Varga, E., Structure of the Békés basin inferred from seismic reflection, well and gravity data; this volume, 1-38.

Hámor, G., Báldi, T., Bohn-Havas, M., Hably, L., Halmai, J., Hajós, M., Kókai, J., Kordos, L., Korecz-Laky, I., Nagy, B., Nagymarosy, A., and Völgyi, L., 1987, The bio-, litho-, and chronostratigraphy of the Hungarian Miocene; Proc. 8th Cong. Reg. Comm. Mediterranean Neogene Stratigraphy; Hung. Inst. Geol. Ann. Rept., 70, 351-353.

Horváth, F., 1988, Neotectonic behavior of the Alpine-Mediterranean region, in Royden, L.H. and Horváth, F., (eds.), The Pannonian Basin a study in basin evolution: Amer. Assoc. Petrol. Geol. Memoir 45, 49-55.

Kázmér, Miklós, 1990, Birth, life and death of the Pannonian Lake; Palaeogeography, Palaeoclimatology, Palaeoecology, 79, 171-188.

Késmárky, I., Pogácsás, Gy., and Szanyi, B., 1981, Seismic stratigraphic interpretation in Neogene-Quaternary depressions of eastern Hungary; Proc. 26th Geophys. Symp., Leipzig, GDR, 130-140.

Kilényi, É., and Rumpler, J., 1984, Pre-Tertiary basement relief map of Hungary; Geophys. Trans., 30, (4), 425-428.

Lerner, J., 1981, Satellite image map of the Carpatho-Pannonian region: Earth Evol. Sciences, 1, (3-4), 180-182.

Lukács-Miksa, M., Pogácsás, Gy., and Varga, F., 1983, Seismic facies analysis and stratigraphic interpretation of unconformably dipping Pliocene features in the Pannonian Basin (in Hungarian); Proc., Balatonszemes Symp., Paper 2874, Eötvös Geophys. Inst., 173-186.

Marton, Gy., 1985, Seismic stratigraphic analysis of the Derecske basin (in Hungarian, abs. in English); Magyar Geofizika (Hungarian Geophysics), 26, (5-6), 161-181.

Mattick, R.E., Rumpler, J., Phillips, R.L., 1985, Seismic stratigraphy of the Pannonian Basin in southeastern Hungary; Geophysical Transactions, 31, (1-3), 13-54.

Molenaar, C.M., Révész, I., Bérczi, I., Kovács, A., Juhász, Gy.K., Gajdos, I., and Szanyi, B., Stratigraphic framework and sandstone facies distribution of the Pannonian sequence in the Békés Basin; this volume, 99-110.

Phillips, R.L., Révész, I., and Bérczi, I., Lower Pannonian deltaic-lacustrine processes and sedimentation, Békés Basin; this volume, 67-82.

Pogácsás, Gy., Mattick, R.E., Elston, D.P., Hámor, T., Jámbor, Á., Lakatos, L., Lantos, M., Simon, E., Vakarcs, G., Várkonyi, L., and Várnai, P., Correlation of seismo- and magnetostratigraphy in southeastern Hungary; this volume, 143-160.

Pogácsás, Gy., 1984, Seismic stratigraphic features of Neogene sediments in the Pannonian Basin; Eötvös Lóránd Geophysical Institute, Geophysical Transactions, 30, (4), 373-410.

Posamentier, H.W., Jervey, M.T., and Vail, P.R., 1988, Eustatic controls on clastic deposition I - conceptual framework, in Wilgus, C.K., and others, (eds.), Sea-level changes: an integrated approach; Society Economic Paleontologists and Mineralogists Special Publ. 42, 109-124.

Royden, L.H., 1988, Late Cenozoic tectonics of the Pannonian Basin system, in Royden, L.H., and Horváth, F., (eds.), The Pannonian Basin a study in basin evolution; Amer. Assoc. Petrol. Geol. Memoir 45, 27-48.

Steininger, F.F., Senes, J., Kleemann, K., and Rögl, F., 1985, Neogene of the Mediterranean Tethys and Paratethys (stratigraphic correlation tables and sediment distribution maps); Inst. of Paleontology, Univ. Press of Vienna, 1, 189 p.; 2, 524 p.

Szentgyörgyi, K., and Teleki, P.G., Facies and depositional environments of Miocene sedimentary rocks; this volume, 83-96.

Vail, P.R., Mitchum, R.M., and Thompson, S., III, 1977, Seismic stratigraphy and global changes of sea level, part 3: relative changes of sea level from coastal onlap, in Payton, C.E., (ed.), Seismic stratigraphy - applications to hydrocarbon exploration; Amer. Assoc. Petrol. Geol. Memoir 26, 63-97.

Van Wagoner, J.C., Mitchum, R.M., Campion, K.M., and Rahmanian, V.D., 1990, Siliciclastic sequence stratigraphy in well logs, cores, and outcrops: concepts for high-resolution correlation of time and facies; Amer. Assoc. Petrol. Geol. Methods in Exploration Series, 7, 1-55.

Van Wagoner, J.C., Posamentier, H.W., Mitchum, R.M., Vail, P.R., Sarg, J.F., Loutit, T.S., and Hardenbol, J., 1988, An overview of sequence stratigraphy and key definitions, in Wilgus, C.K., and others, (eds.), Sea level changes: an integrated approach; Society of Economic Paleontologists and Mineralogists Special Publ. 42, 39-45.

3 Lower Pannonian Deltaic-Lacustrine Processes and Sedimentation, Békés Basin

R. Lawrence Phillips[1], *István Révész*[2] *and István Bérczi*[2]

ABSTRACT

Sedimentological investigations of the 6500-m-thick Neogene and younger section of the Békés basin reveal multiple sedimentary processes related to deltaic progradation in a deep lacustrine basin. Cores, well logs and seismic profiles were used to investigate the sedimentologic processes, the paleoenvironments of deposition, and the history of basin filling. A gradual termination of middle Miocene (Badenian-Sarmatian) shallow-marine bioclastic deposition marks the time of accelerated basin subsidence and the onset of lacustrine sedimentation(Pannonian *sensu-lato*). Three major stages of lacustrine sedimentation are identified. These stages (in ascending order) are represented by: (1) basal calcareous to silty organic-rich marl; (2) prodelta turbidites; and (3) delta-slope deposits.

The basal marl (248 m thick) reaches its maximum thickness in central parts of the basin, but also drapes most surrounding basement highs.

Abundant sand-rich turbidites derived from advancing deltas form a thick (963 m) prodelta sequence. Subsequences identified within the prodelta strata include, in ascending order, distal lobe, fan, rare channel, and slumped deposits. Successive fan systems, that originated from multiple deltas, formed solitary to interfingering sand bodies as the deltas advanced.

The slope deposits, with inclined strata (3-25° dips) can be divided into 3 main depositional sequences: (1) alternating marl, silt, and sand laminae; (2) amalgamated turbidite sands; and (3) deformed (slumped) beds. Density flows and slumping were the dominant depositional-erosional processes on the delta slope. During deposition, sand bypassed the slope and was carried through gullies and/or channels to the base of the slope. These gullies and/or channels later became filled with amalgamated turbidites or with large slumped sediment masses. The turbidites contain marl rip-up clasts interbedded with large- and small-scale cross-bedded sandstones. The thickness of the delta-slope deposits suggests that the lake was at least 600 m deep.

1 U.S. Geological Survey, Menlo Park, California 94025, USA
2 MOL Rt. (Hungarian Oil and Gas Co., Ltd.), H-2443, Százhalombatta, Hungary

P. G. Teleki et al. (eds.), Basin Analysis in Petroleum Exploration, 67–82.

INTRODUCTION

An analysis of cores from hydrocarbon exploration wells in a deep lacustrine basin, the Békés, indicates that its sedimentological history was dominated by sequential as well as coeval deltas that prograded from the north, northwest and west. This study attempts to document past depositional processes and reconstruct depositional environments for Pannonian (*sensu lato*) age rocks by determining the distribution and lithology of pre-Neogene and Neogene age strata.

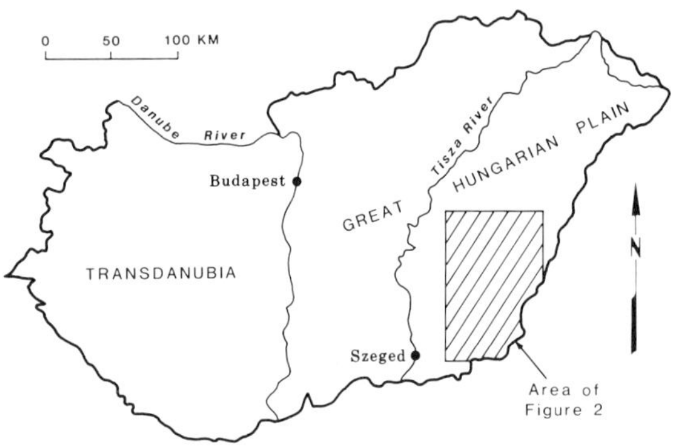

Figure 1. Index map of the study area in Hungary.

The area of investigation is located in southeastern Hungary in the the Great Hungarian Plain (Nagyalföld) (Fig. 1). The Békés basin contains more than 6500 m of Neogene and younger sedimentary rocks and is approximately 65 km in diameter. Beneath the post-Paleogene sedimentary fill, basement highs rise to within 1000-2000 m of the ground surface (the Battonya high to the southwest, the Szarvas-Endröd-Szeghalom highs to the north, and the Sarkadkeresztúr high to the northeast). These basement highs define the basin's boundaries in Hungary (Fig. 2).

Extensional tectonism, which began during the Miocene, resulted in subsidence and formation of the Békés basin and several other Neogene-age basins in Hungary (Horváth and Royden, 1981; Royden and others, 1982; Royden, 1988). Active Miocene basin rifting is recognized in the sedimentary column by the deposition of Badenian (middle Miocene) coarse, synrift basal conglomerates succeeded by shallow marine bioclastic sediments (Szentgyörgyi and Teleki, this volume).

The major rock stratigraphic units represented in core samples (in ascending order) are: (1) pre-Neogene rocks (basement), (2) middle Miocene shallow-marine bioclastics and clastics, and (3) Miocene and younger lacustrine sedimentary rocks. In this study the Miocene and younger lacustrine sedimentary rock sequence is divided into 5 units in ascending order: (1) basal conglomerates, (2) basal marls, (3) prodelta turbidites, (4) delta-slope deposits, and (5) delta-plain, fluvial-to-terrestrial deposits.

The goals of this study were: (1) define the processes and depositional environments within the deltaic-lacustrine Pannonian (*sensu lato*) age strata, (2) understand the stratigraphic history of basin filling, and (3) identify sand bodies that may be potential hydrocarbon reservoirs.

METHODS

Cores from 52 wells located both within as well as on the flanks of the Békés basin were examined (Fig. 2). Cores ranged in depth from 554 m (Szeghalom (Sz)-1 well) to 4525 m (Doboz-I well). Core diameters ranged from 6.5 to 8.5 cm. The standard core length was 1.5 m; but core recovery was highly dependent on sediment texture and degree of cementation. Sections as long as 37 m were recovered from the deepest part of the basin (Kondoros (Kond)-1 well). Continuous cores as long as 72 m (Pusztaföldvár (Pf)-190 well) were recovered on the Battonya high, and 34 m (Endröd (En)-23 well) on the Endröd high. Approximately 50 percent of the recovered cores from the Békés basin were investigated.

Where possible the cores were examined to identify vertical trends (beds thickening or thinning), sedimentary structures, texture, grading, evidence of bioturbation, and tectonic features (such as slickensides and fracturing), to as well as define the depositional environments.

SUMMARY OF THE STRATIGRAPHY

The Pannonian-age rock formations recognized in the Békés basin are shown in Figure 3. The Békés formation consists of basal conglomerates and is usually found on basement highs. The Tótkomlós, Vásárhely, Dorozsma, and Nagykörös Formations consist of basal calcareous to silty marls, which range from 50 to 90 percent $CaCO_3$. The Szolnok Formation consists of lacustrine prodelta turbidite deposits. The Algyö Formation represents delta-slope deposition. The Törtel and Zagyva Forma-

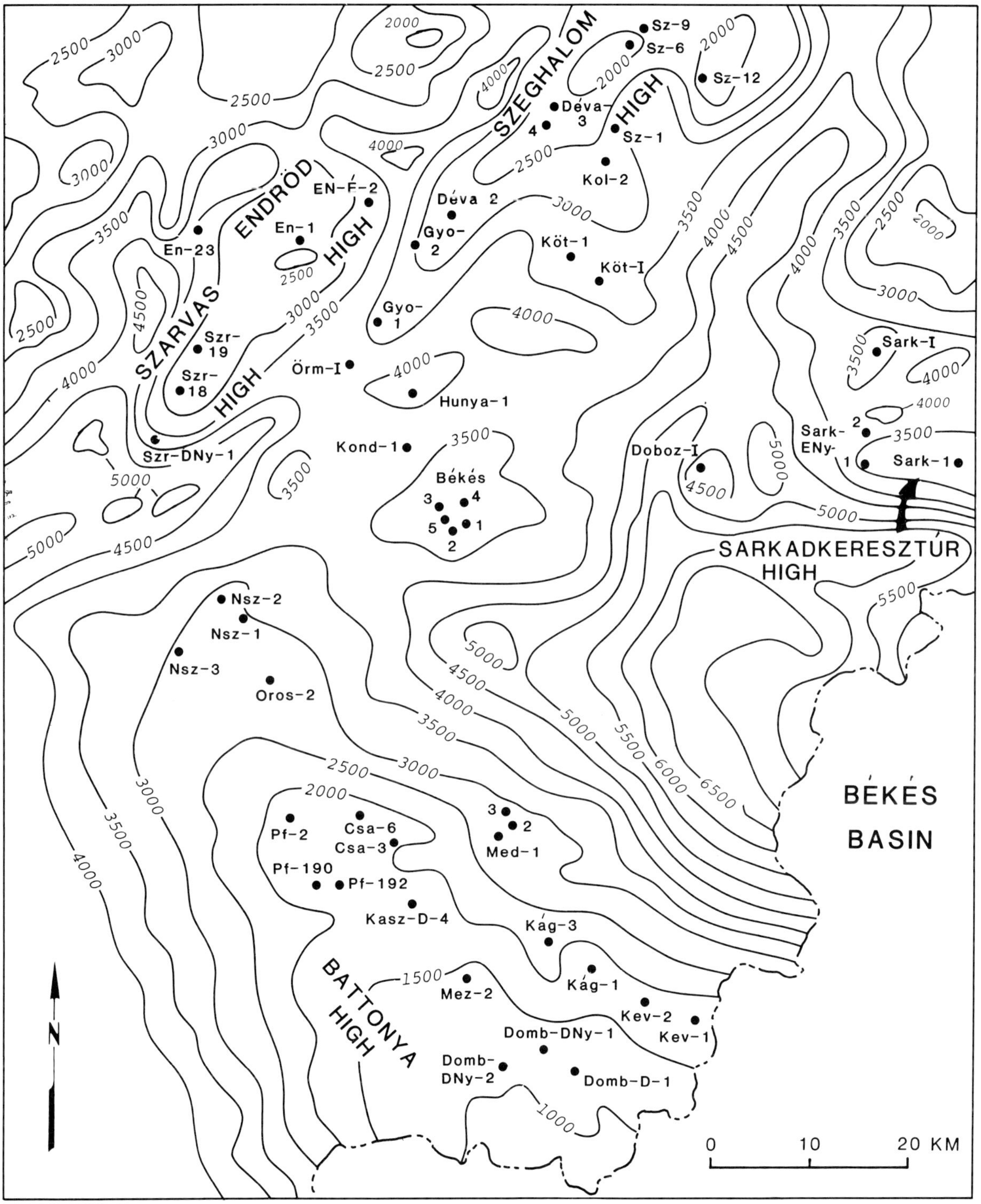

Figure 2. Isopach map of Neogene sedimentary fill in the Békés basin. Contours in meters. The wells whose cores were investigated are also shown.

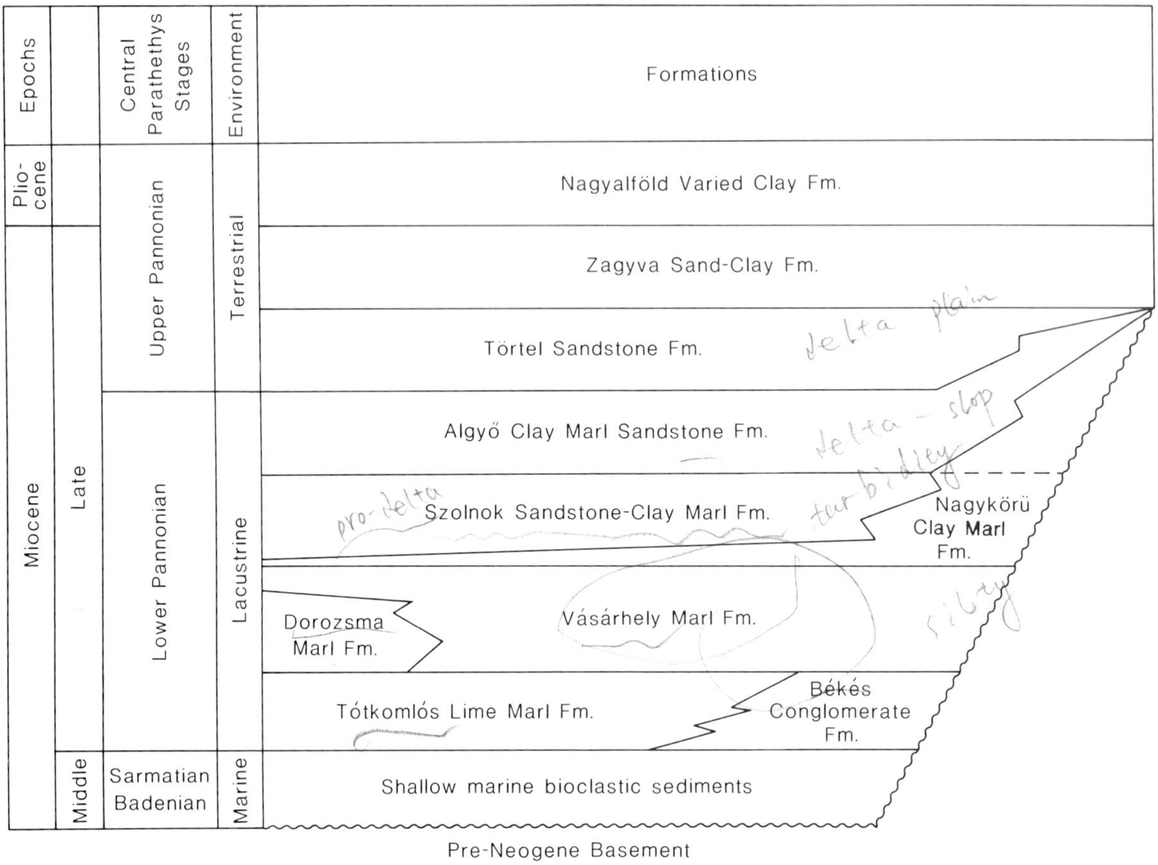

Figure 3. Rock-stratigraphic chart of formations, rock units, ages, and depositional environments within the Békés basin.

tions represent delta-plain deposition. In this study formation names were not used because many of the Hungarian formations are strongly time-transgressive.

PRE-NEOGENE (BASEMENT) ROCKS

The oldest (pre-Mesozoic) rocks of the basement complex consist chiefly of granite, schist and quartzite, and locally quartz porphyry. Mesozoic rocks consist of dolomite, conglomerates, sandstones and mudstones. Evidence of intense fracturing and shearing was observed in almost all rock cores from the basement complex; both open and filled fractures can be observed. Calcite fills many of the fractures (Fig. 4). Hydrocarbon traces (dead oil) were observed in open voids in fractured Mesozoic dolomite at depths of 3984-3997 m in the Örménykút (Örm)-I well. A discussion of the basement rocks and their tectonic history is given by Grow and others (this volume).

MIOCENE MARINE SEDIMENTARY ROCKS

Miocene (Badenian and Sarmatian stages) marine sedimentary rocks unconformably overlie pre-Neogene rocks (basement) over much of the Békés basin. The maximum penetrated thickness of this unit is about 275 m; however, it is inferred to be > 1000 m thick in central (deep) parts of the basin (Szentgyörgyi and Teleki, this volume). The unit is absent on some structurally high areas. The unit consists chiefly of shallow-marine bioclastic and clastic sedimentary rocks and contains conglomerate, sandstone, rhyodacite, limestone, and marl. Cores from this unit were recovered from 23 wells. In

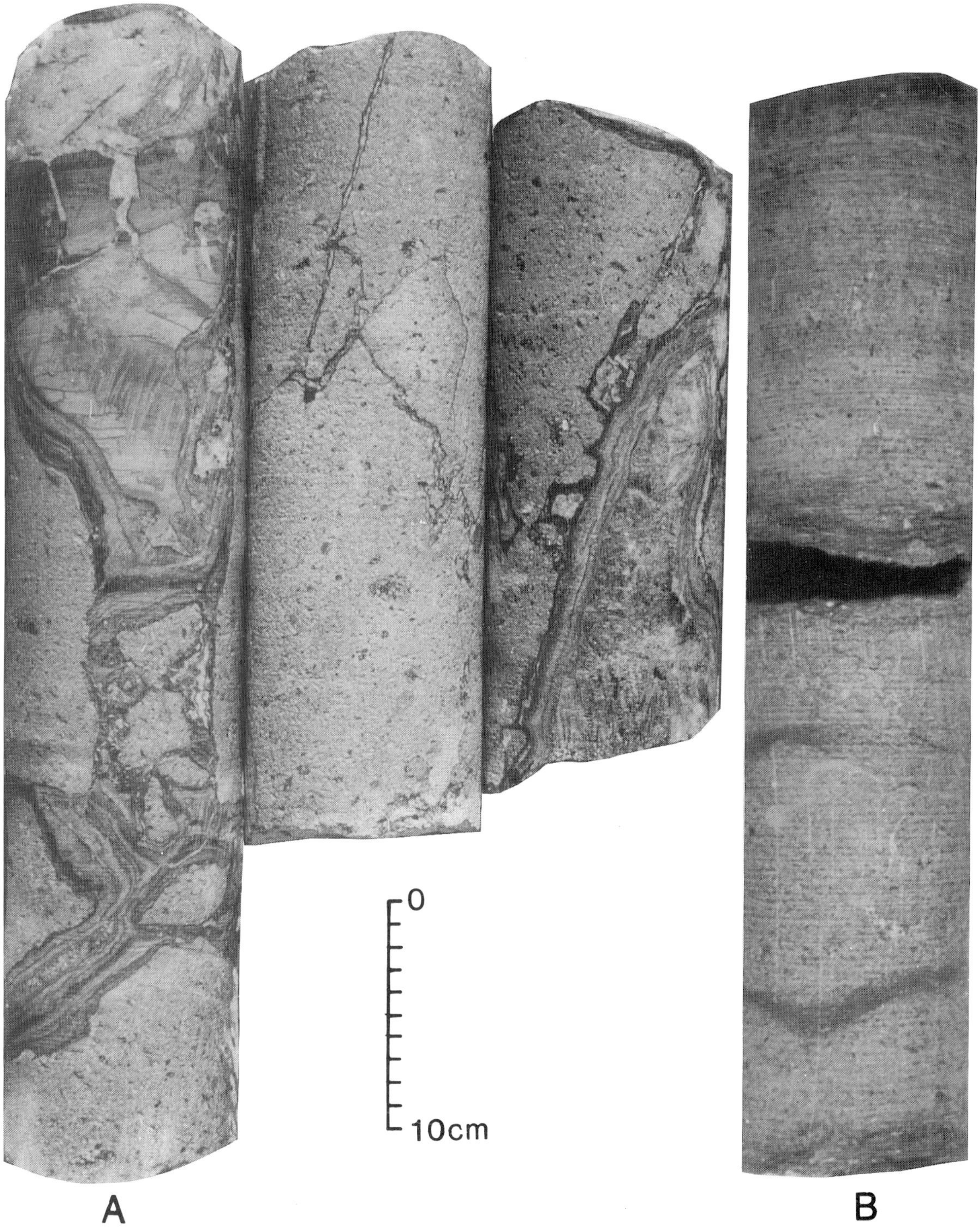

Figure 4. Photographs of cores of Miocene, shallow-marine sedimentary rocks: A) calcite-filled fractures in marine sandstone (Békés-2 well; depth: 3186-3194 m); B) bioclastic limestone consisting of coralline algal nodules and bioclastic debris (Nagyszénás (Nsz-2) well; depth: 3009-3018 m).

general the unit represents deposition in a high-energy, shallow-marine environment. The bioclastic limestone is composed of oolites, rhodoliths (algal nodules), coralline algae, and mega- and microfossils (Fig. 4). Black limestone reflecting deposition in a reducing environment is interbedded with the marine sediments. The presence of rounded to well-rounded monolithic mica schist conglomerate indicates marine reworking and sorting of basal Miocene conglomerates on the basement highs. A complete discussion of this marine sequence is found in Szentgyörgyi and Teleki (this volume). In 17 of the cores containing Miocene (Badenian/Sarmatian) marine rocks, there was evidence of intense fracturing and shearing as was observed in the underlying pre-Neogene rocks.

MIOCENE AND YOUNGER LACUSTRINE SEDIMENTARY ROCKS

Basal Conglomerate

Basal conglomerate of Pannonian (*sensu lato*) age was identified in 4 wells (Csanádapáca (Csa)-3, Csa-6, Pf-2, and Pf-190) on the Battonya high and in one well (Sz-9) on the Szeghalom high. The strata range in thickness from approximately 10 m (Pf-2 well) to 46 m (Csa-6 well). According to Molenaar and others (this volume), in the Battonya area, these conglomerates were deposited as the Battonya high subsided below lake level and are time-transgressive.

The sediment texture varies from coarse sandstone to conglomerate. The composition of the gravel indicates that it was derived from the underlying bedrock, which on the Battonya high consists of granite in the south and schist in the north. The texture of the basal conglomerate usually fines upwards.

Basal Marl

With the exception of the Battonya high and locally on the Szeghalom high, middle Miocene marine rocks are overlain by calcareous clay to silty marl. The contact between these units appears to be gradational, at least in central parts of basin. The basal marl was identified in 17 wells scattered throughout the basin. The marl grades upward from carbonate-rich clay to silty clay. The thickness of marl cored within the basin ranges from 173 m (Örm-I well) to 248 m (Hunya-1 well). In the Kunágota (Kág)-1 well on the flank of the Battonya high, at least 115 m of marl was penetrated; whereas, on the Battonya high, the Csa-3 well penetrated 92 m and the Magyardombegyháza (Domb)-DNy-1 well 86 m of marl. On the Szeghalom high the Dévaványa (Déva)-2 well penetrated at least 67 m of marl. These marls are of Pannonian age (Miocene) and represent deposition in a deep lacustrine environment (Bérczi and Phillips, 1985) following accelerated basin subsidence during extension. The lake system apparently covered much of the region as indicated by the fact the marl was penetrated on most basement highs. The southern part of the Battonya high apparently remained emergent and was only covered by the lake system during later stages of basin filling (Mattick and others, this volume).

In the deepest parts of the basin, the unit is represented by organic-rich, black, argillaceous-to-calcareous clay that was deposited in a reducing environment. In these areas sedimentary structures are rare and most of the marl appears to be without structures in the cores. Thin silt or very fine grained sandstone laminations increase upward in the marl section suggesting that clastic sediment influx into the basin increased as a delta system (or systems) advanced towards the margins of the basin from the north, northwest and west. bioturbation, observed in cores from depths of 3848 m in the Hunya-1 well, may have caused mixing of the sediment, resulting in the structure-free marls observed in the lower part of unit. However, in the Hunya-1 well, the evidence of bioturbation does not exist above approximately 3700 m and does not occur again until approximately 1971 m depth in sediments representing delta-slope deposits. This suggests that in most of the deeper parts of the basin, deposition took place in an euxinic environment. Water depths within the central part of the Békés basin may have been as deep as 900 m (Mattick and others, this volume) during deposition of the basal marls. Stratification of lake waters or poor circulation may have aided in the formation of a reducing environment in the deeper parts of the basin as has been observed for the basal marls deposited in the adjacent Hódmezövásárhely-Makó basin to the west (Phillips and Bérczi, 1985).

The marls in several wells are intensely sheared, as are the underlying pre-Neogene and Neogene marine strata. Within the deeper parts of the basin, the basal part of the marl sequence is fractured.

Prodelta Lacustrine Turbidites

The basal marls grades upward to lacustrine turbidites interbedded with marl. The prodelta lacustrine turbidites are interpreted as evidence of coarse clastic sediment derived from a delta system (or systems) that was advancing toward the Békés basin. Turbidites were identified in 11 wells within the deepest parts of the basin. Turbidite sandstones are found as deep as 4000 m (Doboz-I well) and as shallow as 2478 m (Köröstarcsa (Köt)-I well). The prodelta turbidite sandstones interbedded with marl range in thickness from 963 m (Hunya-1 well) to 737 m (Kond-1 well) and 700 m (Gyo-1 well). The turbidites rapidly thin from 700 m in the Gyo-1 well to 225 m in the Gyo-2 well and pinch out on the Szeghalom basement high north of these wells.

The turbidite sequence comprises a sand-rich sequence in which the sand content ranges from 68 to 90% in the basal beds and increases to 95% near the top of the sequence. Even in the deepest parts of the basin the sand content is high, the average sand content was 80% for cores from the Gyo-1 well and 92% for cores from the Kond-1 well. The sand-rich turbidites penetrated in some wells can be divided into depositional systems based on the presence of vertical sedimentary sequences. Depositional sequences identified within the turbidite unit include distal lobe deposits, amalgamated fans, and channel deposits.

After the deltas had advanced into the basin, coarse-grained sediment was transported through gullies and channels down the delta slope to the base of the slope by slumping and other density flow processes. Earlier, when the delta fronts were outside of the basin, sediment pathways were controlled by paleo-topography. For example, sediments were funneled into the basin through troughs between the Szeghalom and the Sarkadkeresztúr highs, between the Endröd and Szeghalom highs, and between the Szarvas and Battonya highs, as indicated by the presence of paleo-bathymetric troughs. The troughs eventually were filled as the deltas advanced over the highs. In the trough between the Szeghalom high and the Sarkadkeresztúr high, sediment was funneled from a delta system advancing from the north into the deepest part of the basin south of the Sarkadkeresztúr high. Turbidite sandstone beds, as much as 8.5 cm thick and containing Bouma Ta-b sequences, are found at depths of 4000 m. These are the deepest turbidites cored within the basin.

In the northern part of the basin (Hunya-1, Kond-1, and Gyo-1 wells), the earliest phase of distal turbidite deposition is interpreted to be represented by a lobe sequence of a lacustrine fan system. The turbidites may have been funneled between the Endröd and Szeghalom highs, or between the Szarvas and the Battonya highs. The lobe sequence, at depths of 3404-3419 m (Kond-1 well), consists of repetitive bundles of thickening upward turbidite sandstone beds that include thin laminations of marl. This sequence, which is 14.4 m thick, rests on the basal marl unit. Marl also overlies the sandstone beds. The maximum thickness of the turbidite beds is 164 cm; the average thickness is 23 cm. The sand content averages 87% and ranges from very fine-grained to fine-grained texture. Sedimentary structures in this sequence include graded beds with Bouma sequences Ta, Ta-e, Ta-b-e, and occasionally of Tc and Tc-e sequences. Load structures, internal deformation structures, dish structures, mud-draped scours, and thin sand laminae in marl are the remaining sedimentary structures (Fig. 5).

The early lobe sequence is overlain by stacked beds of interbedded turbidites and marl (Figs. 5 and 6). These turbidite beds comprise the major part of the prodelta deposits. Sandstone dominates in most of the beds; the sand content ranges from 75 to 95% in this sand-rich system. Depositional sequences are difficult to interpret owing to interfingering of multiple fan systems associated with the advancing multiple deltas. Repetitive upward-thickening sequences, which may represent interfingering lobe deposits are common. In addition, amalgamated turbidite beds, shallow meandering channel deposits, and possible deep channel deposits were also identified.

The maximum thickness of the prodelta turbidite beds is 211 cm (Békés-5 and Hunya-1 wells); the average turbidite bed thickness of each cored interval ranges from 10 to 48 cm. Average sandstone bed thickness for all prodelta cores ranges from 20 cm (Kond-1 well) to 31 cm (Hunya-1 well). Bouma sequences consist of Ta, Ta-e (both Ta and Ta-e sequences make up more than 60% of the turbidites), Ta-b, Ta-b-e, and Ta-b-c. Complete Bouma sequences, Ta to Te and Ta-b-c-d, are only found in the basal part of the turbidite section. Sedimentary structures within the prodelta deposits consist of graded and massive sandstone beds, load structures, flame structures, internal deformations, mud-draped scours, marl rip-up clasts, cross-beds, amalgamated sandstone beds, and laminated sandstone and marl.

Core data indicate that meandering channels were present on top of the fan-lobe systems. The migrating-channel, accretionary bank deposit shown in Figure 5c (Gyo-1 well) is 188 cm thick and consists of inclined beds (14° dip) of alternating marl and sandstone laminations; many of the sandstone beds are composed of small-scale cross-bedded strata. Such sequences of beds and their sedimentary structures are similar to those found in the modern channel-fan system of Lake Geneva as reported

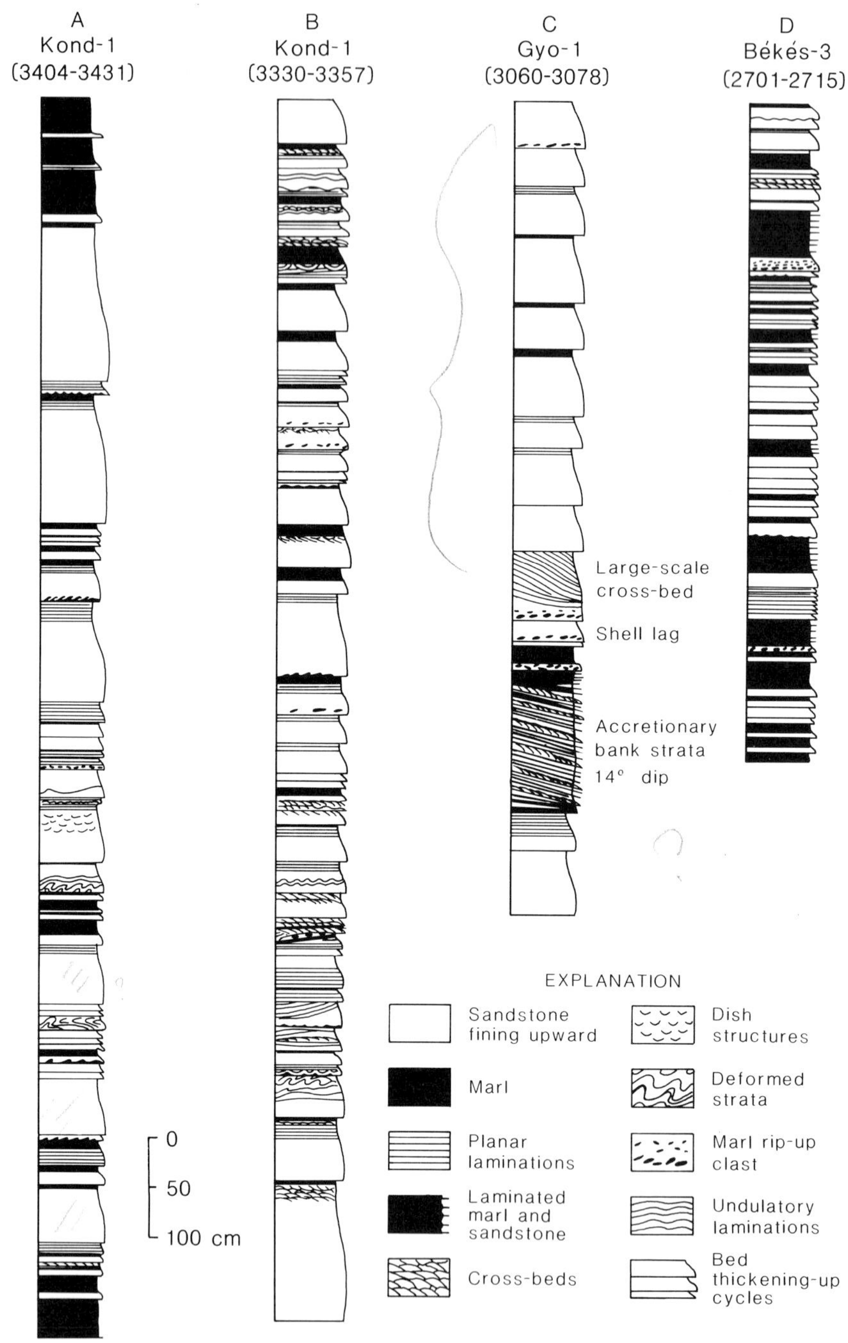

Figure 5. Measured stratigraphic sections within the prodelta depositional environment: A) lobe sequence (above basal marl) consisting of repetitive turbidite sandstone beds thickening upward (Ta, Ta-b-e sequences, Kondoros (Kond)-1 well; depth: 3404-3431 m); B) amalgamated turbidite sandstone beds within fan deposit (Kond-1 well; depth: 3330-3357 m); C) accretionary bank sequence dipping at 14° (in lower part of section) overlain by a shell lag deposit containing marl rip-up clasts, cross-beds, and amalgamated turbidite sandstone beds which may represent channel-fill (Gyoma (Gyo)-1 well; depth: 3060-3078 m); D) overbank channel deposits containing gently inclined strata consisting of amalgamated thin sandstone beds separated by laminated marl and sandstone (Békés-3 well; depth: 2701-2715 m).

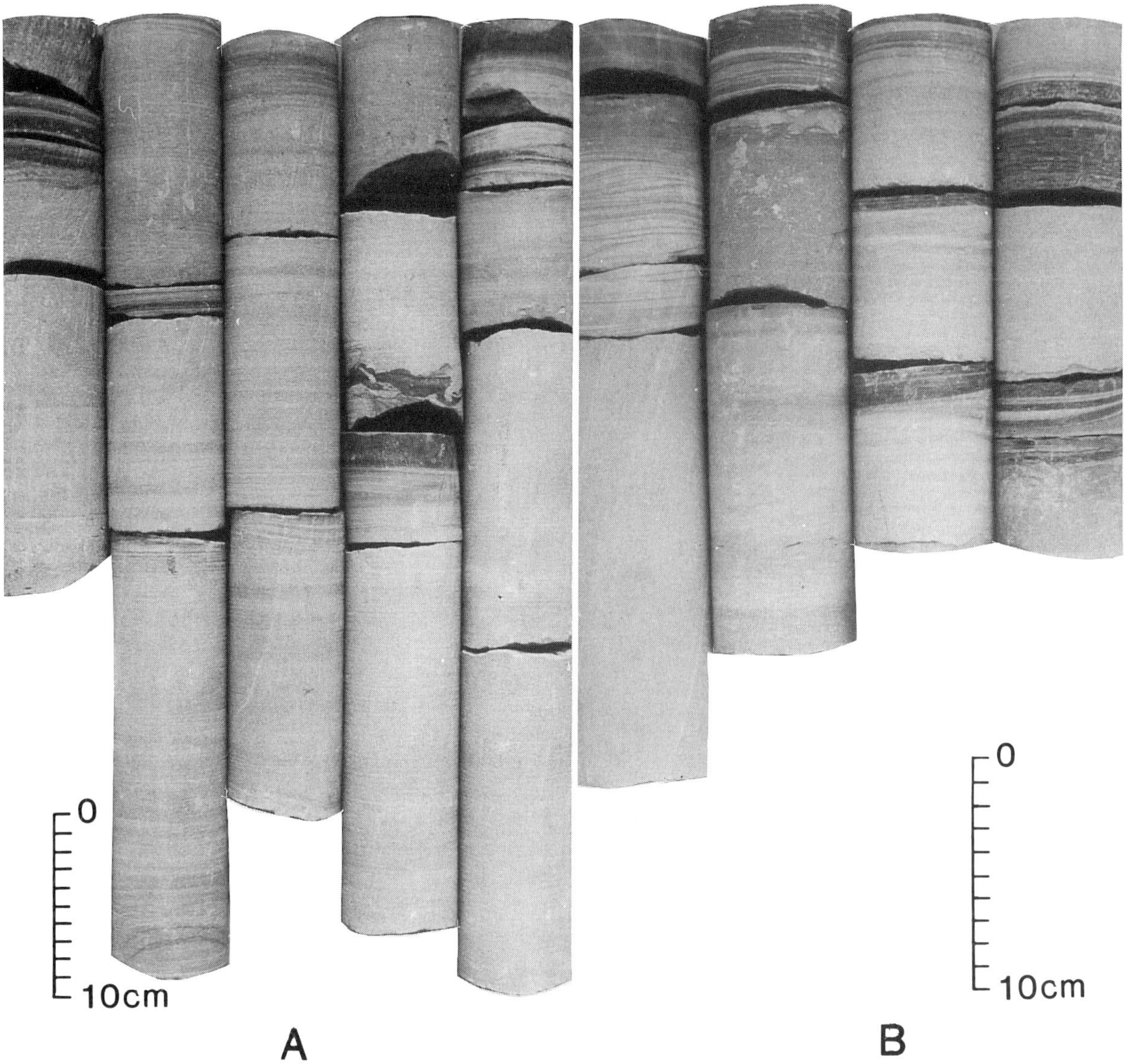

Figure 6. Photographs of cores of prodelta turbidite sandstones: A) amalgamated turbidite sandstone beds consisting mainly of Ta and Ta-b-e Bouma sequences (Kondoros (Kond)-1 well; depth: 3330-3357 m); B) turbidite sandstones containing Ta-b-e and Ta-b-c-e Bouma sequences (Gyoma (Gyo)-1 well; depth: 3200-3218 m).

by Houbolt and Jonker (1968) and to the ancient, meandering, deep-sea fan channels investigated by Mutti and Ricci Lucchi (1975), Mutti (1977), and Mutti and Normark (1987). The lower accretionary bank deposits are overlain by laminated marl and sandstone with marl rip-up clasts (containing abundant shells); the marl and sandstone are overlain by beds of large-scale cross-bedded sandstone and amalgamated turbidites. The cross-bedded sequence may represent the basal part of a channel system because the depositional sequence is similar in many respects to those reported from ancient deep-sea, channel-fan systems by Mutti and Normark (1987).

The existence of overbank channel deposits is also indicated by the presence of gently inclined strata consisting of Ta and Ta-e sequences (Fig. 5d). The sandstone beds in these sequence are thin (average 8 cm) and occur as amalgamated graded beds separated by thin sand-marl laminations. The bedded sequences, however, show no vertical trends (Fig. 5, Békés-3 well).

Figure 7. Photograph of cores of deformed turbidite sandstone containing overturned strata, load structures, and flame structures (Gyoma (Gyo-1) well; depth, 3060-3078 m).

Deformed turbidites, as well as deformed laminated sandstone and marl beds that represent slumped deposits, can be observed in cores at discontinuous intervals above depths of 2800 m in the Gyo-1, Hunya-1, and Kond-1 wells. The deformed strata are present in the upper part of the turbidite sequence but are rare below a depth of 2800 m. The slump deposits show evidence of overturned and vertically inclined beds that were faulted and deformed (Fig. 7). Horizontal-bedded strata occur above and below the deformed beds.

Turbidites in the upper part of the prodelta sequence, compared to those in the lower part of the sequence, contain more sand, have abundant amalgamated sandstone beds, and show evidence of increased frequency of tractive current flow as indicated by the presence of cross-beds. The sandstones in the upper part of the turbidite section usually have a coarser grain size (ranging from fine- to medium-grain sand) compared to the sandstones in the turbidites in the lower part of the section. In some cores from the upper part of the section, the sandstone is very friable. Sedimentary structures include massive to graded turbidites containing Ta, Ta-e, Ta-c and Tc-e Bouma sequences, marl rip-up clasts, large- and small-scale cross-beds, dish structures, load structures, and inclined or parallel-bedded thin sand and marl laminations. The sandstone beds, which average 18 to 23 cm in thickness, reach thicknesses of 64 cm.

The most striking character throughout the section of prodelta deposits in the Békés basin is the high sand content of the stacked turbidite beds. In the adjacent Hódmezövásárhely-Makó basin to the west, beds with similar sand content are present in deltaic facies (Phillips and Bérczi, 1985; Bérczi and Phillips, 1985).

The initial influx of coarse-grained sediment into the Békés basin is represented by the fan-lobe sequence. Progradation and aggregation of the repetitive fan-lobe sequences and development of multiple fan systems ahead of the advancing deltas produced the stacked sets of turbidites. Some parts of the section contain repetitive thickening-upward beds. In the northern part of the basin, these beds (between well depths of 3100-3460 m) represent either sheet-flow sands or fan-lobe turbidites deposited when delta fronts overtopped paleo-bathymetric highs that surrounded much of the Békés basin. Fan systems were constructed by broad (although apparently shallow) channels and narrow meandering channels. Modern and ancient turbidite fan systems that are sand-rich are characteristically channelized (Mutti and Normark, 1987). Successive fans originating from the multiple deltas rapidly filled the deep basin as the deltas advanced.

Delta-Slope Deposits

Delta-slope deposits were identified in cores from 22 wells. In the center of the basin, the top of the delta-slope deposits were penetrated at a depth of 1910 m in the Gyo-1 and at 2100 m in the Kond-1. The maximum penetrated thickness of these deposits is 600 m (Gyo-1 well). Based on interpretation of electric logs, delta-

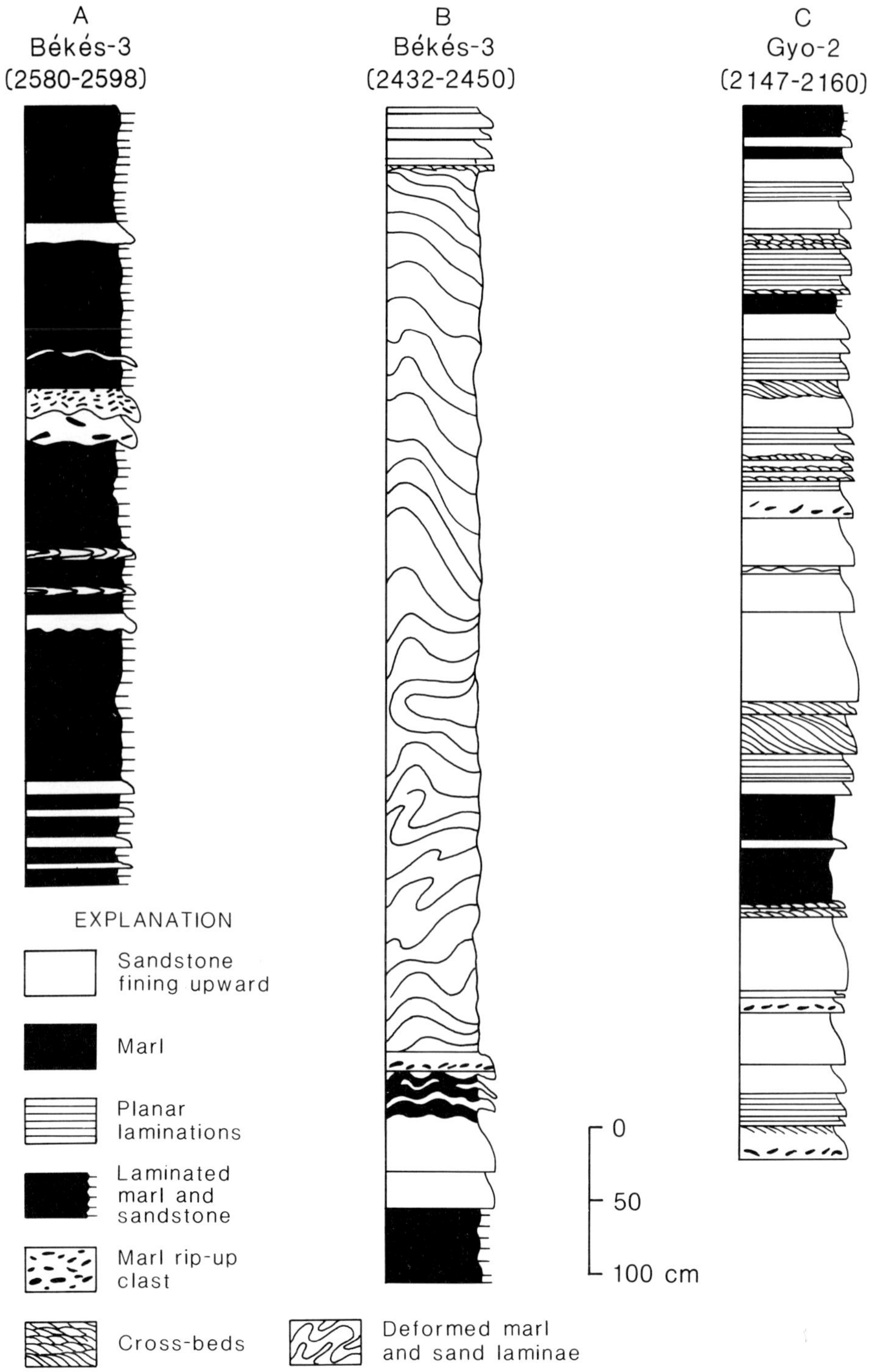

Figure 8. Measured stratigraphic sections of delta-slope deposits: A) laminated marl containing thin sand laminae and sandstone beds, some deformed, all dipping 3° (Békés-3 well; depth: 2580-2598 m); B) deformed (slumped) 5.9-m-thick bed of marl-sand laminae with dips ranging from 15 to 20° and overturned beds (Békés-3 well; depth: 2432-2450 m); C) amalgamated sandstone beds near base of slope. The beds contain marl rip-up clasts, abundant planar laminations, cross-beds and graded beds (Ta and Tb sequences) (Gyoma (Gyo)-2 well; depth: 2147-2160 m).

slope deposits vary in thickness from 475 to 600 m in the Békés basin (Molenaar and others, this volume).

The delta-slope deposits can be divided into three depositional sequences; in ascending order these are (Fig. 8): (1) alternating marl, silt, and sand laminae interbedded with thin sand beds; (2) amalgamated turbidite sands; and (3) deformed (slumped) beds. In comparison to the prodelta deposits, the sand content in the delta-slope deposits is noticeably less in most of the examined cores, ranging from 5% in the laminated marls and sandstones to 83% in the amalgamated turbidites. Sand texture ranges from fine- to medium-grained size. The sand beds usually contain abundant plant fragments. The average sandstone bed thickness is 15 cm; the maximum bed thickness (63 cm) occurs near the base of the delta-slope deposits.

The primary characteristic of the delta-slope deposits is gentle to steeply inclined strata; measured dips range from 3-25° but most of the laminae and beds dip between 3-8°. A few beds, however, appear to be horizontal or inclined at very low angles. Sedimentary structures include abundant silty marl laminae that alternate with laminated sandstone. In cores, small- to large-scale downslope oriented cross-beds (up to 26 cm thick), amalgamated thin to thick massive sand beds, and graded sandstone beds containing mainly Ta and Tb Bouma sequences could be recognized. A number of beds contain Ta-b-c, Tb-c, Tc, and Tc-d sequences. In additional, abundant marl rip-up clasts in sandstone beds, dewatering structures, abundant evidence of deformation (overturned beds, penecontemporaneous faulting, and folded and distorted strata), and sparse bioturbation represented by narrow vertical and horizontal burrows could be observed in the cores (Figs. 8 and 9).

Processes that were active on the delta slope can be inferred from the sedimentary structures. Alternating thin sand-marl laminations and associated small-scale cross-beds indicate that sediments were transported downslope by density currents. High and low density underflows that are common features of lake systems produce alternating "varved" light silt-rich and dark clay-rich marl laminations on delta slopes in modern lakes. Density underflows usually occur during river flooding when sediment discharge into the lake is high. Storms, however, will resuspend lake sediments and storm currents will also transport the sediment downslope as underflows, producing the characteristic "varved" laminae.

Alternating, inclined silt-sand laminae with sharp contacts that grade to silty marl are the most abundant

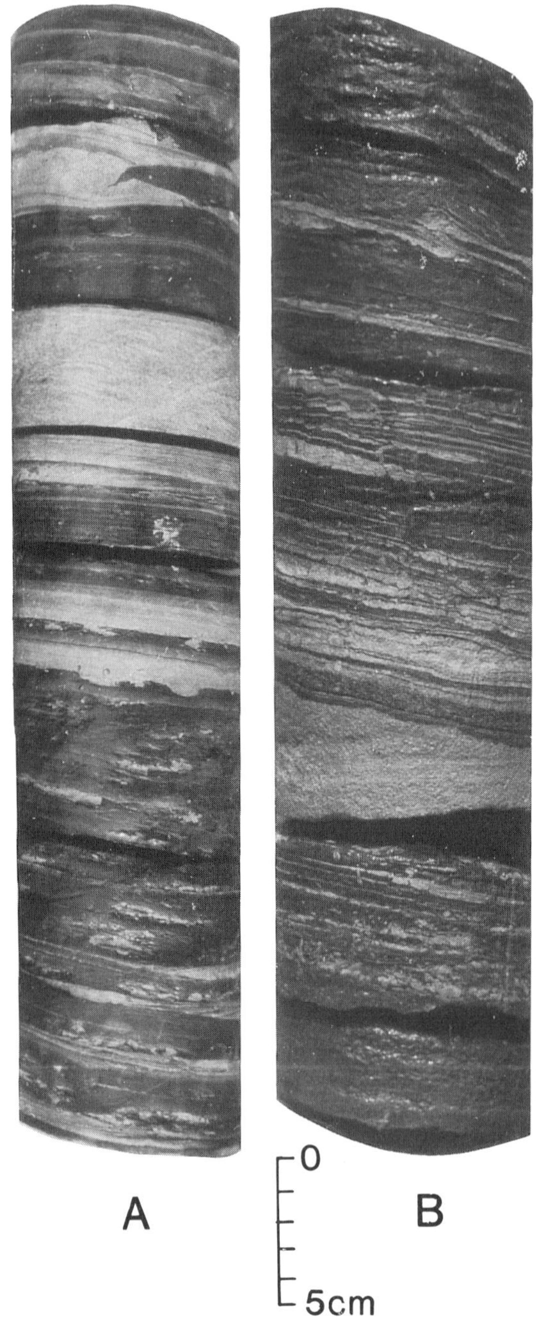

Figure 9. Photographs of cores of delta-slope deposits: A) gently dipping marl and sandstone laminae interbedded with sandstone beds that contain load structures, small-scale cross-beds, dewatered structures, and a deformed (slumped) sandstone bed (Sarkadkeresztúr (Sark)-1 well; depth: 2434-2439 m). B) inclined, alternating marl and sand laminae that may have been deposited by density underflow currents that flowed down the delta front (Szarvas (Szr)-19 well; depth: 2296-2299 m).

structures within the Békés delta-slope deposits (Fig. 9). These alternating laminae in the slope deposits are similar to laminae in cores from delta slopes of modern lakes (Houbolt and Jonker, 1968; Sturm and Matter, 1978; Pickrill and Irwin, 1983). Thick sand laminae and thin sandstone beds (1 to 5 cm thick) alternating with marl are common toward the base of the slope (Fig. 8). Many sandstone beds located near the base of the slope were observed to contain small-scale cross-beds. Traction currents, as well as downslope turbidite flows, are believe to be responsible for the formation of these sequences near the base of the slope.

As stated previously, the sand content of the delta-slope deposits is low in comparison to the sand content of the prodelta deposits in the Békés basin. Much of the sand apparently was transported in channels across the slope and deposited directly in the prodelta region. Deformed, slumped sandstone beds, which are interbedded with large- and small-scale cross-beds and form amalgamated sequences may represent channelized slope deposits (Fig. 8). The amalgamated sandstone beds suggest channelized flow that occurred within depressions formed initially from slumping or in broad low-relief channels. Some of the channels located on the slope may have had their origin (head) in the delta- plain-fluvial system as is the case with channels cut into the present-day delta slope in Lake Geneva (Houbolt and Jonker, 1968). Rapidly prograding, fluvial-deltaic systems apparently supplied abundant sediment to the delta front. According to Prior and Bornhold (1986), rapid sediment accumulation causes excess pore-fluid pressures to develop in unconsolidated sediments. This leads to unstable conditions which result in subaqueous landslides on the delta slope. Such slumps are common features on delta slopes both in lacustrine and marine environments (Prior and others, 1981, 1984; Prior and Bornhold, 1986; Prior and Coleman, 1982; Bornhold and others, 1986).

Evidence of significant slumping in a downslope direction can be observed in many of the cores from the delta-slope deposits. These slump deposits range from solitary deformed sandstone beds to slumped sequences as thick as 5.9 m composed of alternating sandstone and marl laminations and beds (Figs. 8 and 9). Much of the slumped sediment was transported from the delta slope by gravity flow mechanisms to the prodelta region in the form of turbidites

Delta-Plain Deposits

Delta-plain and overlying fluvial-terrestrial deposits were identified in cores from 12 wells. The depositional environments for these deposits were difficult to identify because these largely unconsolidated, fine-grained sediments are difficult to recover in cores. In general, delta-plain sediments are characterized by tan-to-brown or red silty claystone, rare medium-grained massive sand, deformed beds of sand and silt, sand and silt containing small-scale cross-beds, and organic-rich silty sediment.

The dominance of silty claystone suggests that deposition most likely occurred in the form of overbank flood deposits adjacent to river systems that crossed the delta plain. The apparent lack of abundant sandstone suggests that the river systems crossing the delta plain did not migrate laterally through time, because this would have resulted in the deposition of blanket channel sands. Instead, the rivers, through time, probably formed solitary, stacked channel systems. Some of these stacked channel systems can be recognized on seismic profiles. Sediments representing fluvial, shallow lake, bay, and marsh environments probably are also present but were not recognized in the cores because of poor core recovery.

SEDIMENTARY MODEL OF THE BÉKÉS BASIN

Sequential to coeval progradation of lacustrine deltas primarily from the north, northwest, and the west (Mattick and others, this volume) gradually filled the Békés basin during early Pannonian time. The lake was at least 600 m deep in early Pannonian time, as judged by the thickness of the delta-slope deposits. The active depositional processes and the history of basin fill can be determined from an interpretation of the distinctive depositional environments (Fig. 10).

Subsidence of the Békés basin in late Miocene time, accompanied by the closure of the connection to the sea resulted in a transition from a shallow-marine environment to a brackish-lacustrine environment (Paratethys). This transition is recognized in cores by a change in lithology from shallow-marine bioclastic sediments to bioturbated black calcareous marls, followed by non-bioturbated marls. Marl formed a blanket deposit in the basin and covered most of the surrounding bathymetric highs. Silt, settling from suspension, together with precipitation of carbonates was the dominant sedimentary process during the basin's early history. Later, as the prograding delta fronts approached the basin's margins, distal turbidite sands were deposited as thin laminae alternating with marl. During this time, on the marginal bathymetric highs, marl continued to be deposited while

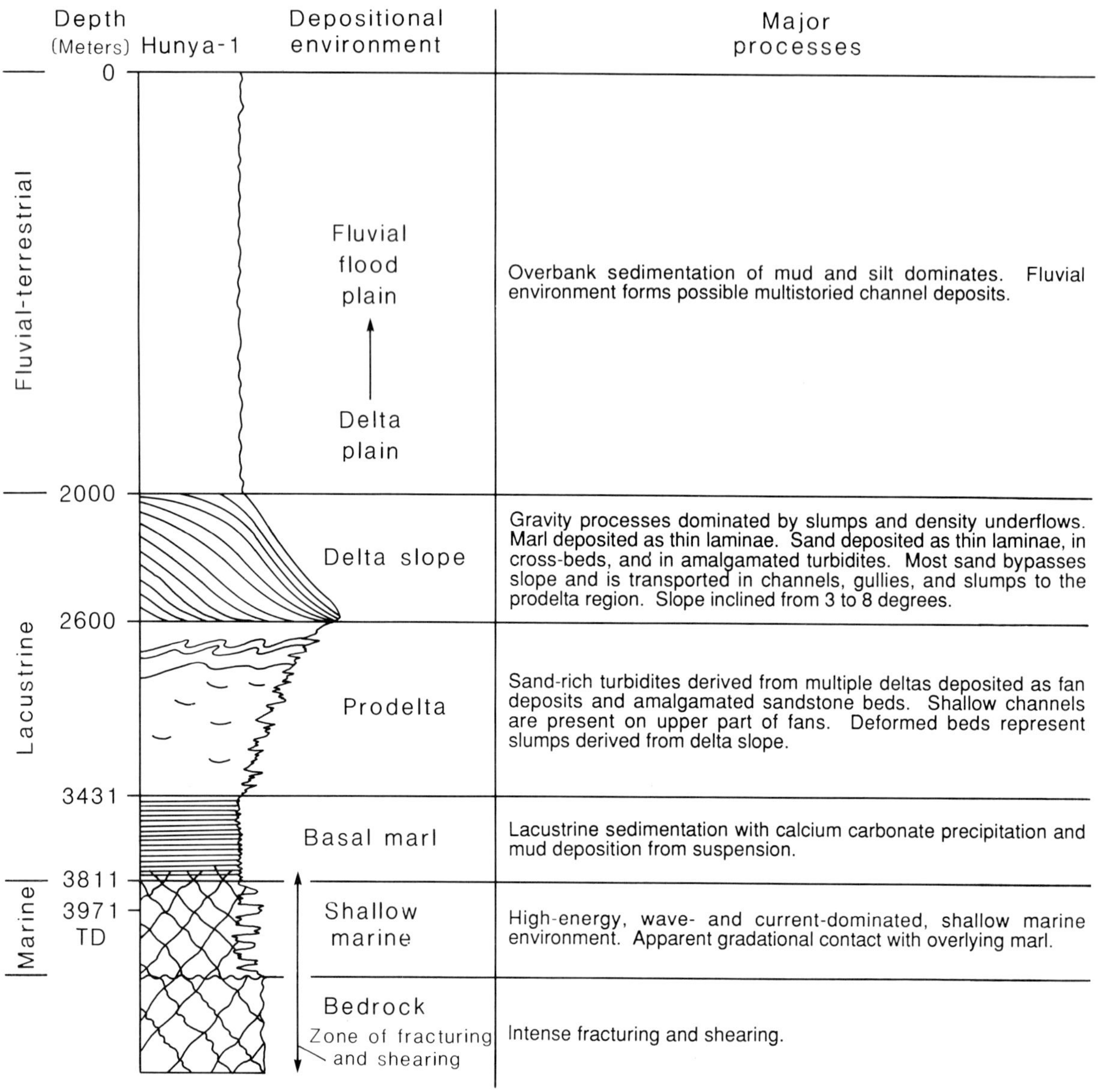

Figure 10. Stratigraphic section, depositional environments, and major depositional processes within the Békés basin. The depths of facies boundaries are from the Hunya-1 well. The thickness of the shallow-marine sedimentary rocks is shown to be 468 m, as interpreted from analysis of cores from the Hunya-1 well. This thickness conflicts with the interpretation of Szentgyörgyi and Teleki (this volume).

turbidite deposition accelerated in the deep parts of the basin.

Later, perhaps when the deltas reached the basin's margins, prodelta turbidite sands were deposited in deep parts of the basin. Topographic highs controlled initial turbidite deposition by funnelling sediments through troughs between highs to bathymetrically low areas. The distal prodelta deposits of the advancing deltas are identified as thin-lobe sequences; these sequences record the onset of sedimentation in prograding fan systems. Interfingering prodelta fan turbidites produced a thick, sand-rich sequence.

Sediments were distributed on early Pannonian lacustrine fans by shallow meandering channels. These chan-

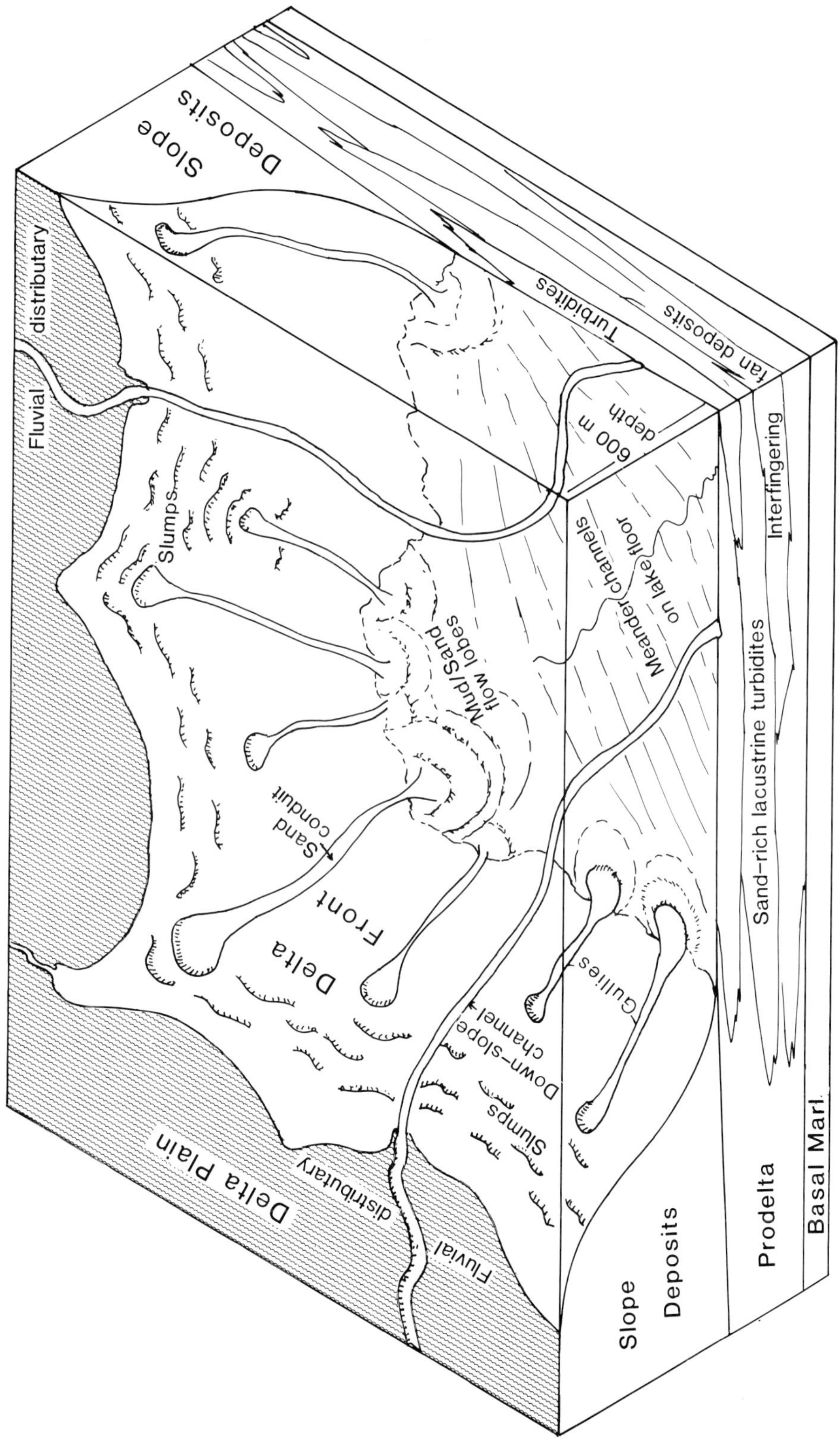

Figure 11. Schematic drawing of sedimentary model showing multiple prograding deltas in the Békés basin. Sand was transported by slumping and through channels downslope, and was deposited as amalgamated turbidites in interfingering fan systems.

nels were probably similar to the modern-day channels that feed the prodelta fan system in Lake Geneva (Houbolt and Jonker, 1968); they transport sediment to the outer fan region.

Above the turbidites and channel deposits, abundant deformed strata, interbedded with amalgamated turbidites, are present. The deformed strata indicate that slumping occurred on delta slopes during early Pannonian time.

Major sedimentation processes occurring on the advancing delta slopes included slumping, downslope channelized flow, and density current underflow. Density underflow, resulting in the deposition of alternating sand-silt-marl laminae, was probably the main depositional process acting on the delta slopes. Much of the sand bypassed the slopes and was transported, either in slump-carved depressions or gullies and channels, to turbidite fan systems in the prodelta region (Fig. 11).

Continued deltaic progradation into the basin eventually filled the lake and fluvial and flood-plain environments replaced the lacustrine environments.

REFERENCES

Bérczi, I., and Phillips, R. L., 1985, Processes and depositional environments within Neogene deltaic-lacustrine sediments, Pannonian Basin, southeast Hungary, *in* Kilényi, E., and Teleki, P. G., (eds.); Geophysical Transactions, Special Edition, Eötvös Lóránd Geophysical Institute of Hungary, 31, (1-3), 55-74 (also U.S. Geological Survey Open-File Report 85-291).

Bornhold, B. D., Yand, Z. S., Keller, G. H., Wiseman, W.J. Jr., Wang, Q., Wright, L. D., Xu, W. D. and Zhuang, Z., 1986, Sedimentary framework of the modern Huanghe (Yellow River) delta; Geo-Marine Letters, 6, 77-83.

Grow, J. A., Mattick, R.E., Bérczi-Makk, A., Péró, Cs., Hajdú, D., Pogácsás, Gy., Várnai, P., Varga, E., Structure of the Békés basin inferred from seismic reflection well and gravity data; this volume, 1-38.

Horváth, F., and Royden, L., 1981, Mechanism for the formation of the intra-Carpathian basins: a review; Earth Evolution Sciences, 1, 307-316.

Houbolt, J. J. H. G. and Jonker, J. M. B., 1968, Recent sediments in the eastern part of the lake of Geneva (Lac Leman); Geologie en Mijnbouw, 47, 131-148.

Mattick, R. E., Rumpler, J., Ujfalusy, A., Szanyi, B., and Nagy, I., Sequence stratigraphy of the Békés basin; this volume, 39-65.

Molenaar, G. M., Révész, I, Bérczi, I., Kovács, A., Juhász G., Gajdos, I. and Szanyi, B., Stratigraphic framework and sandstone facies distribution of the Pannonian sequence in the Békés Basin; this volume, 99-110.

Mutti, E., 1977, Distinctive thin-bedded turbidite facies and related depositional environments in the Eocene Hecho Group (south-central Pyrenees, Spain); Sedimentology, 24, 107-131.

Mutti, E., and Normark, W. R., 1987, Comparing examples of modern and ancient turbidite systems: problems and concepts; *in* J. K. Leggett, and G. G. Zuffa (eds.), Marine clastic sedimentology - concepts and case studies, Graham and Trotman, London, 1-38.

Mutti, E., and Ricci Lucchi, F., 1975, Turbidite facies and facies associations; *in* Mutti, E. (ed.) Examples of turbidite facies and associations from selected formations of the northern Apennines; Field Trip Guidebook A-11, 9th Cong., Int. Assoc. of Sedimentologists, Nice, France, 21-36.

Phillips, R. L., and Bérczi, I., 1985, Processes and depositional environments of Neogene deltaic-lacustrine sediments, Pannonian basin, southeast Hungary: core investigation summary; U.S. Geological Survey Open-file Report 85-360, 66 p.

Pickrill, R. A., and Irwin, J., 1983, Sedimentation in a deep glacier-fed lake-Lake Tekapo, New Zealand; Sedimentology, 30, 63-75.

Prior, D. B., and Bornhold, B. D., 1986, Sediment transport on subaqueous fan delta slopes, Britannia Beach, British Columbia; Geo-Marine Letters, 5, 217-224.

Prior, D. B., Bornhold, B. D. and Johns, M. W., 1984, Depositional characteristics of a submarine debris flow; Journal of Geology, 91, 707-727.

Prior, D. B., and Coleman, J. M., 1982, Active slides and flows in underconsolidated marine sediments on the slopes of the Mississippi delta; *in* Niuwenhuis, J.K. and Saxov, S. (eds.), Marine slides and other mass movements; Plenum Press, New York, 21-49.

Prior, D. B., Wiseman, W. J. and Bryant, 1981, Submarine chutes on the slopes of fjord deltas; Nature, 290, 326-328.

Royden, L. H., 1988, Late Cenozoic tectonics of the Pannonian basin system; *in* Royden, L. H., and Horváth, F., (eds.), The Pannonian Basin -- a study in basin evolution; Am. Assoc. Petrol. Geol., Mem. 45, 27-48.

Royden, L. H., Horváth, F., and Burchfiel, B. C., 1982, Transform faulting, extension and subduction in the Carpathian-Pannonian region; Geol. Soc. America, Bull., 73, 717-725.

Sturm, M., and Matter, A., 1978, Turbidites and varves in Lake Brienz (Switzerland): deposition of clastic detritus by currents; *in* Matter, A. and Tuckern, M.E., (eds.), Modern and ancient lake sediments, Int. Assoc. Sedimentologists, Spec. Publ. 2, 147-168.

Szentgyörgyi, K. and Teleki, P. G., Facies and depositional environments of Miocene sedimentary rocks; this volume, 83-98.

4 Facies and Depositional Environments of Miocene Sedimentary Rocks

Károly Szentgyörgyi[1] *and Paul G. Teleki*[2]

ABSTRACT

The area of the Békés basin began to subside in early Badenian time during the late Styrian orogenic phase. The initial phase of subsidence and transgression of the Paratethys Sea is marked by conglomerate and/or breccia that is present almost everywhere where Miocene deposits occur. The Miocene sequence is 0-275 m thick, where penetrated by wells in the Békés basin; however, in the deepest parts of the basin the sequence is inferred to be more than 1000 m thick based on interpretation of seismic data.

During the Badenian, the region was covered by a shallow, well-oxygenated, warm sea of normal salinity. Numerous, small islands, that exist now as buried basement highs, were fringed by *Lithothamnium* algal flats. Along the shorelines of the Badenian sea, dominantly coarse, sandy sediments, and in littoral zones, sandstones and carbonates were deposited. Shelf and offshore areas are represented by complex interfingering lithofacies, ranging from bioclastic limestones that make up the fabric of submarine plateaus, to dark-colored, turbiditic marls and shales in the deepest parts of the basin.

In early Badenian time an almost continuous transgression of marine waters took place; whereas, in the late Badenian almost continuous regression of the sea followed. The Sarmatian epoch started with a new transgression in the southern part of the basin, while in the northern part regression continued, indicating a tilting of the basin. From the beginning of Sarmatian time, the area became isolated from the Paratethys Sea, the salinity of the water gradually decreased, and an endemic fauna characterized by euryhaline living organisms developed in an anoxic environment.

Sarmatian sedimentation took place in a basin whose extent was smaller than its Badenian equivalent, as indicated by the deposition of lower Pannonian sediments on lower Badenian rocks, especially in the northern part of the Békés basin. In the deepest parts of the basin, however, sedimentation may have been continuous from middle Miocene through Pliocene time with the lower Pannonian section being highly condensed as a result of little or no deposition.

1 MOL Rt. (Hungarian Oil and Gas Co., Ltd.), H-1039, Budapest, Hungary
2 U.S. Geological Survey, Reston, Virginia 22092, USA

P. G. Teleki et al. (eds.), Basin Analysis in Petroleum Exploration, 83–97.

INTRODUCTION

In keeping with the terminology of the chronostratigraphic subdivision of the Central Paratethys (Steininger and others, 1985, 1988; Hámor and others, 1987), the term "Miocene" rocks, as used in this chapter, refers to pre-Pannonian, mid-to-late Miocene sedimentary rocks of the Badenian and Sarmatian stages (Fig. 1). The time interval represented by these stages is 5 million years (11.5 to 16.5 Ma) (Steininger and others, 1985; Vass and others, 1985; Hámor and others, 1987; Bérczi and others, 1988). The boundary between early and late Badenian is placed at 15 Ma, and the Badenian-Sarmatian boundary at 13.7 Ma as determined by K/Ar dating (Vass and others, 1985; Hámor and others, 1987). Comparison of this chronostratigraphic division to Mediterranean and Paratethys stages can be found in Berger (1992).

Miocene sedimentary rocks lie unconformably on Mesozoic and Paleozoic basement rocks and are overlain by late Miocene (Pannonian *sensu stricto* - Pontian) sedimentary rocks. The Miocene sequence, as defined, was deposited in a subsiding basin, initially under open-marine conditions in early Badenian time. These waters were part of the Paratethys Sea, which was connected by a narrow passageway to the Adriatic Sea in the southwest, and in the east the Paratethys extended beyond the present Caspian Sea (see Halmai and others, 1988, map no. 3).

The Paratethys is a relict arm of the east-west trending Tethys seaway present over Europe in the Paleogene. The Paratethys formed by isolation as the Alpine orogenies created land barriers between itself and the future Mediterranean Sea. The Mediterranean was closed off from the Indo-Pacific ocean by mid-Miocene time. The region of the Central Paratethys, thus, became confined to the Pannonian Basin, the Alpine-Carpathian foredeep and intramontane basins between present Austria on the west, and the Ukraine and Romania on the east (Rögl and Steininger, 1984).

The transgression of marine waters in the Central Paratethys region began in Karpatian (Langhian) time from the southwest (Rögl, 1985); however, no evidence exists to suggest that waters reached the Békés basin. By Badenian time, the epicontinental sea covered all of the Pannonian Basin and areas to the west, and warm-water foraminifera (*Amphistegina, Heterostegina*), large mollusks and corals invaded as Indo-pacific waters reflooded the Mediterranean and Paratethys Seas (Rögl and Steininger, 1984; Rögl, 1985). In mid-to-late Badenian a "salinity crisis" occurred in the Transylvanian Basin and in the Carpathian molasse foredeep (Steininger and others; 1989) as waters became cut off from the Eastern Paratethys, but it had no effect on sedimentation in the Békés basin. Renewed flooding from the east took place in late Badenian-early Sarmatian time.

The climate was subtropical in early Badenian time. On a global scale, a major climatic change, resulting in the formation of the Antarctic ice sheet, began at about 14 Ma (Kennett, 1982, p. 734), corresponding to the Badenian-Sarmatian boundary. Locally, however, the climate remained subtropical. The waters of this Sarmatian sea (termed the Pannonian Lake by Kázmér, 1990) turned gradually more brackish by late Sarmatian time, having become isolated from the Eastern Paratethys, as a result of thrusting and uplift of the Carpathians.

Well logs and core data were used to determine the extent and thickness of Miocene sedimentary rocks in most areas of the Békés basin. These data show that the Miocene section penetrated by wells is no more than 275 m thick, with an average thickness of about 100 m, and is absent on major basement highs (Fig. 2). Miocene sediments were deposited unconformably on Mesozoic and pre-Mesozoic rocks of variable composition. In the Békés-1 and Békés-2 wells, Upper Cretaceous sandstone and siltstone were penetrated beneath Miocene deposits (Szentgyörgyi, 1983). Mesozoic sedimentary rocks underlie Miocene deposits in the Doboz-I, Endröd (En)-7, Köröstarcsa (Köt)-I, Nagyszénás (Nsz)-2, Csanádapáca (Csa)-2, Csa-9, Medgyesbodzás (Med)-1, Med-2, and Magyardombegyháza (Domb)-1 wells. In most of the remaining wells, Paleozoic and/or older metamorphic rocks constitute the basement complex. The basement was not penetrated in the Hunya-1 and Kondoros (Kond)-1 wells.

In the east-central part of the basin, where well data were not available, the Miocene section is estimated to reach a thickness of 1000-1500 m based on the interpretation of seismic data (János Rumpler, MOL Rt. - Hungarian Oil and Gas Co., personal communication, 1992). Along the margins of the basin, numerous control points were provided by wells that reached basement. These were used to construct the isopach map of Figure 2. Basinward, the areal density of wells rapidly decreases and in the deep part of the basin, several wells did not penetrate the entire Miocene section, as in the Kond-1 and the Hunya-1 wells, which bottomed in the Miocene after penetrating 9 and 112 meters of the Miocene section, respectively. Elsewhere in the basin the Békés-1, Békés-2, En-1, En-N-1, Gyoma (Gyo)-2, Köt-1, Dévaványa (Déva)-7, and Sarkadkeresztúr (Sark)-NW-2

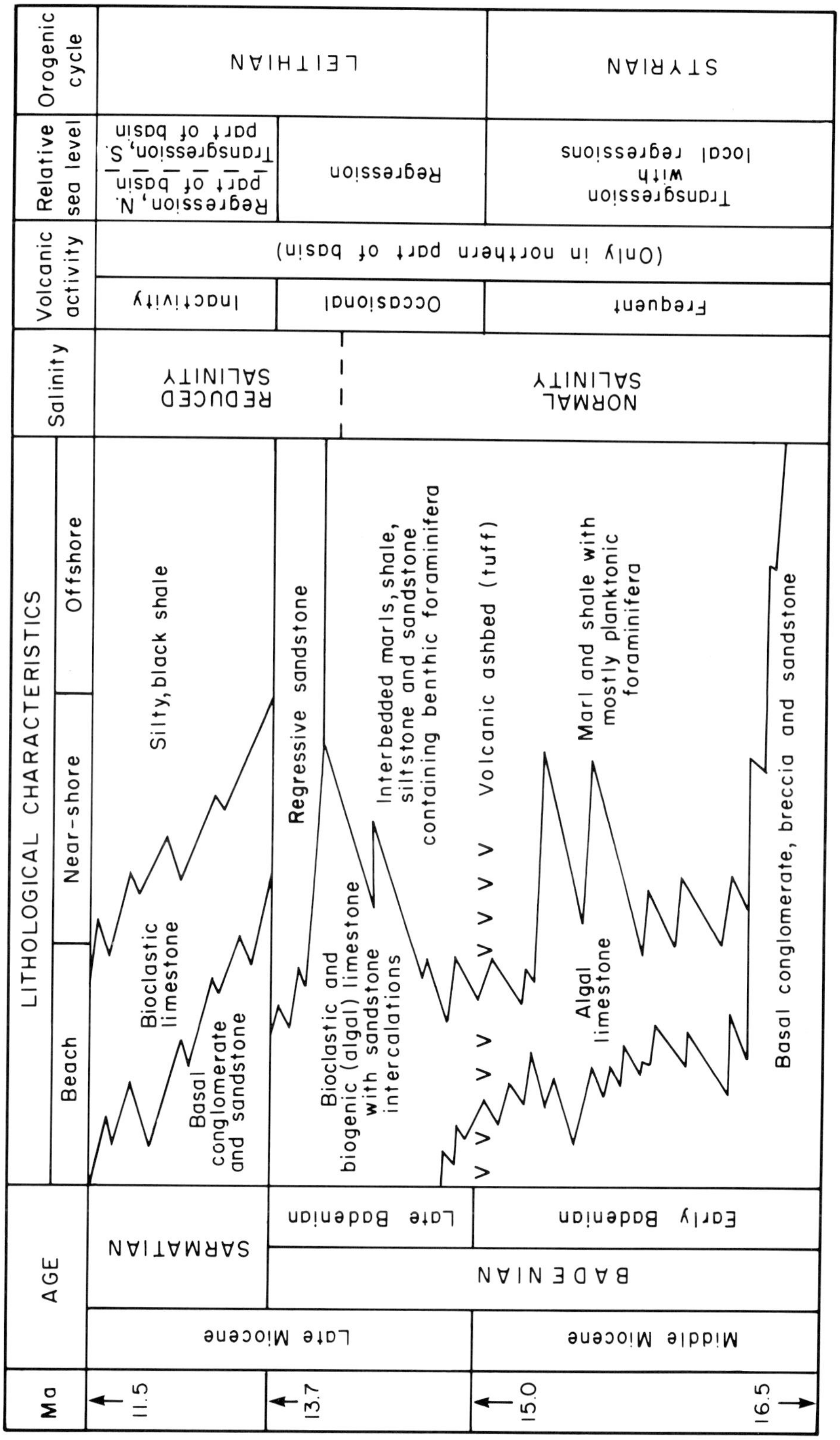

Figure 1. Chronostratigraphic and lithostratigraphic relations among Miocene facies and their depositional environments in the Békés basin.

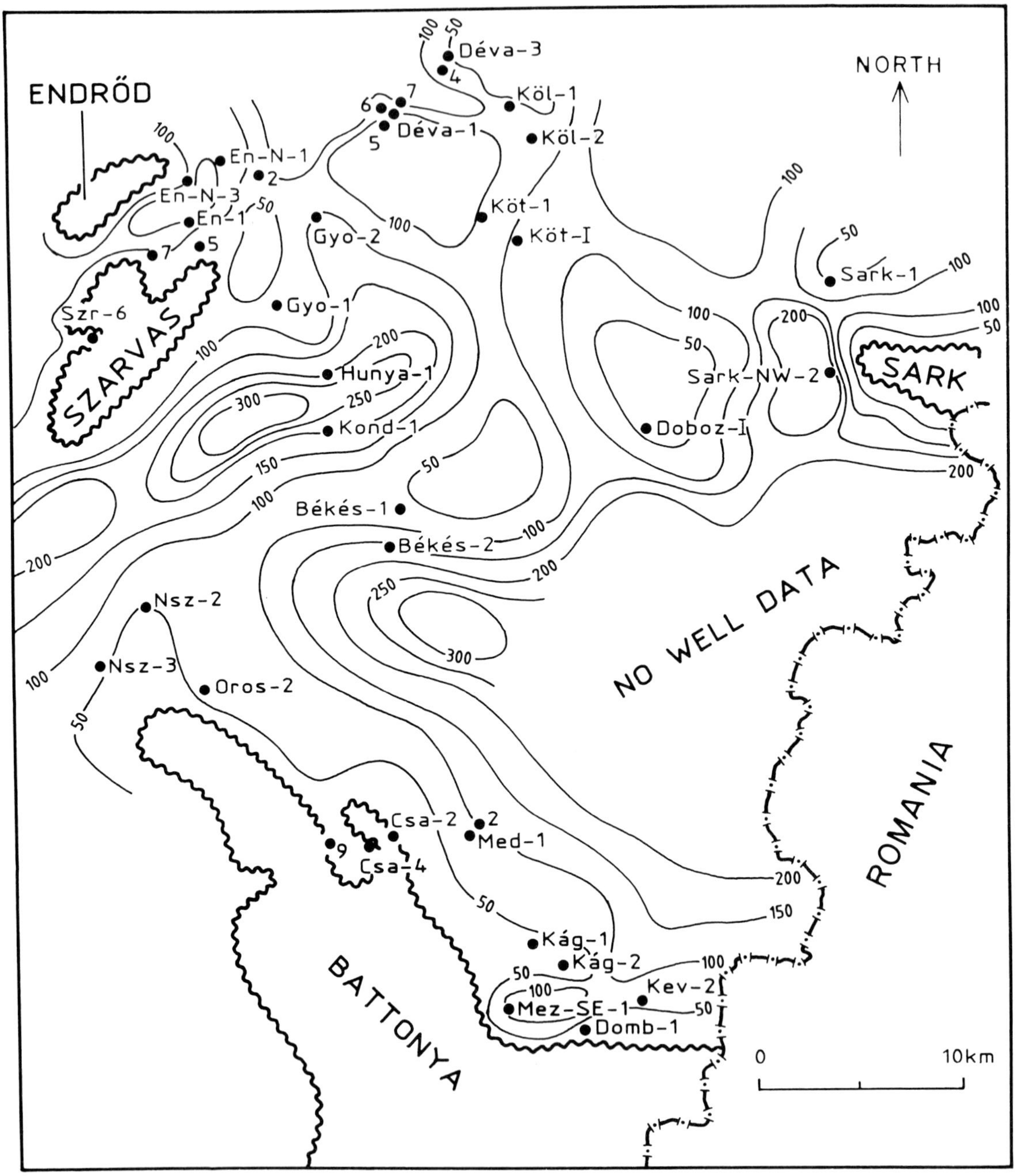

Figure 2. *Isopach map (in meters, 50 meter contour interval) of pre-Pannonian Miocene (Badenian and Sarmatian) sedimentary rocks and location of selected wells in the Békés basin. Wavy lines enclose areas where Miocene rocks are absent as a result of nondeposition; these areas were islands in the Miocene sea. Csa: Csanádapáca; Déva: Dévaványa; Domb: Magyardombegyháza; En: Endröd; Gyo: Gyoma; Kág: Kunágota; Kev: Kevermes; Kond: Kondoros; Köl: Körösladány; Köt: Köröstarcsa; Med: Medgyesbodzás; Mez: Mezöhegyes; Nsz: Nagyszénás; Oros: Orosháza; Sark: Sarkadkeresztúr; Szr: Szarvas.*

wells penetrated more than 100 m of Miocene sedimentary rocks and bottomed in basement. In the Dévaványa area (Déva wells, Fig. 2), abrupt changes in thickness occur in the Miocene section, from 34 m in the Déva-1 well to 118 m in the Déva-7 well. Similar abrupt changes in thickness occur in the Endröd area (En-1 and En-7 wells). In the Dévaványa and Endröd areas, the abrupt changes in thickness probably reflect paleogeographic relief of the basement's surface. In other areas, changes in thickness are inferred to be the result of erosion or non-deposition (Déva-2, Déva-3, En-1, Köt-1, Köt-I, Nsz-2 and Nsz-3 wells). In the area of Dévaványa-Endröd-Köröstarcsa, Sarmatian beds are missing and lower Pannonian (Miocene-Pliocene) sediments directly overlie lower Badenian marine deposits.

STRATIGRAPHIC RELATIONS OF THE MIOCENE SEQUENCE

Most of the marine deposits in the Békés basin contain identifiable microfauna, especially the lower Badenian deposits. Rocks of mid-Badenian age, that would correspond to the "salinity crisis" east of the basin, could not be clearly differentiated. However, the upper part of the upper Badenian sequence can be demarcated, as it contains beds of pyroclastics (tuffs) that were deposited over a wide region. Biostratigraphical studies of cores indicate that upper Badenian beds are absent in some areas of the basin. Index fossils near the Sarmatian-Pannonian contact in central parts of the basin, however, are lacking (Hunya-1 and Kond-1 wells). The stratigraphic relations near the contact are further complicated by the fact that the lithology of the lower part of the Pannonian section is similar to the lithology of the upper part of the Miocene sequence.

The absence of upper Miocene (Sarmatian) strata in some areas indicates that, an erosional or non-depositional interval had preceded the deposition of the Pannonian sequence. It is also possible, that in certain areas, the depositional sequences may have become condensed (e.g., Hunya-1 and Kond-1 wells). According to Pogácsás and others (this volume), the lowest part of the Pannonian section is highly condensed due to little or no deposition.

Miocene rocks are unconformably overlain by Pannonian rocks in many parts of the basin, although continuous deposition may have occurred in the deepest, east-central part of the basin (Phillips and others, this volume; Mattick and others, this volume).

LITHOLOGY AND FACIES OF MIOCENE SEQUENCES

Badenian Deposits

Unfossiliferous basal conglomerate and/or breccia is present everywhere in the basin where Miocene deposits occur, and the thickness of these beds ranges from a few meters to 30 m (Fig. 3). The coarse clastic material consists of debris flow and material originating from syntectonic landslides (gravity flow) deposited in channels and fans. The source of the mostly polymictic clasts is the rocks of the underlying pre-Neogene basement (Fig. 3 and 4). The conglomerate or breccia is mostly unstratified or unbedded, although, in places, it is interbedded with sandstone that is fluvial in origin. The matrix is composed commonly of sand or silt (Table 1).

Table 1. Mineral composition of basal, fluvial sandstone (Sark-NW-2 well, 3769-3779 m).

MINERAL	PERCENT
Quartz	17
Feldspar	3
Calcite	64
Dolomite	4
Illite/montmorillonite	5
Chlorite	5
Muscovite	3
Other	3

Cyclic, coarse-fine deposition has been observed in places (e.g., Sark-NW-2 well). At a few locations, the color of the matrix, at the base of the conglomerate, is red (caused by limonite staining), which indicates subaerial exposure during initial deposition of these sediments. The subsurface form of these deposits is likely to be similar to other Neogene fan-delta deposits produced by rapid synsedimentary tectonics (see, e.g., Dabrio and Polo, 1991). The synrift conglomerates and breccias are common adjacent to paleogeographic islands in the basin, as in the Szarvas (Szr)-6, Sark-2, Sark-NW-2, Békéssámson (Bés)-1, Csanádalbert (Csal)-1, Pitvaros (Pit)-S-1, Csa-9, Kúnágota (Kág)-2 and Mezöhegyes (Mez)-SE-1 wells (Fig. 5), but are also present in deeper parts of the basin, as in the Békés-2, Gyo-1, Gyo-2 (embedded in red sandstone) and Hunya-1 wells.

The basal conglomeratic sequence is succeeded by either carbonates, or siliciclastic sequences, or interbedded sequences of both.

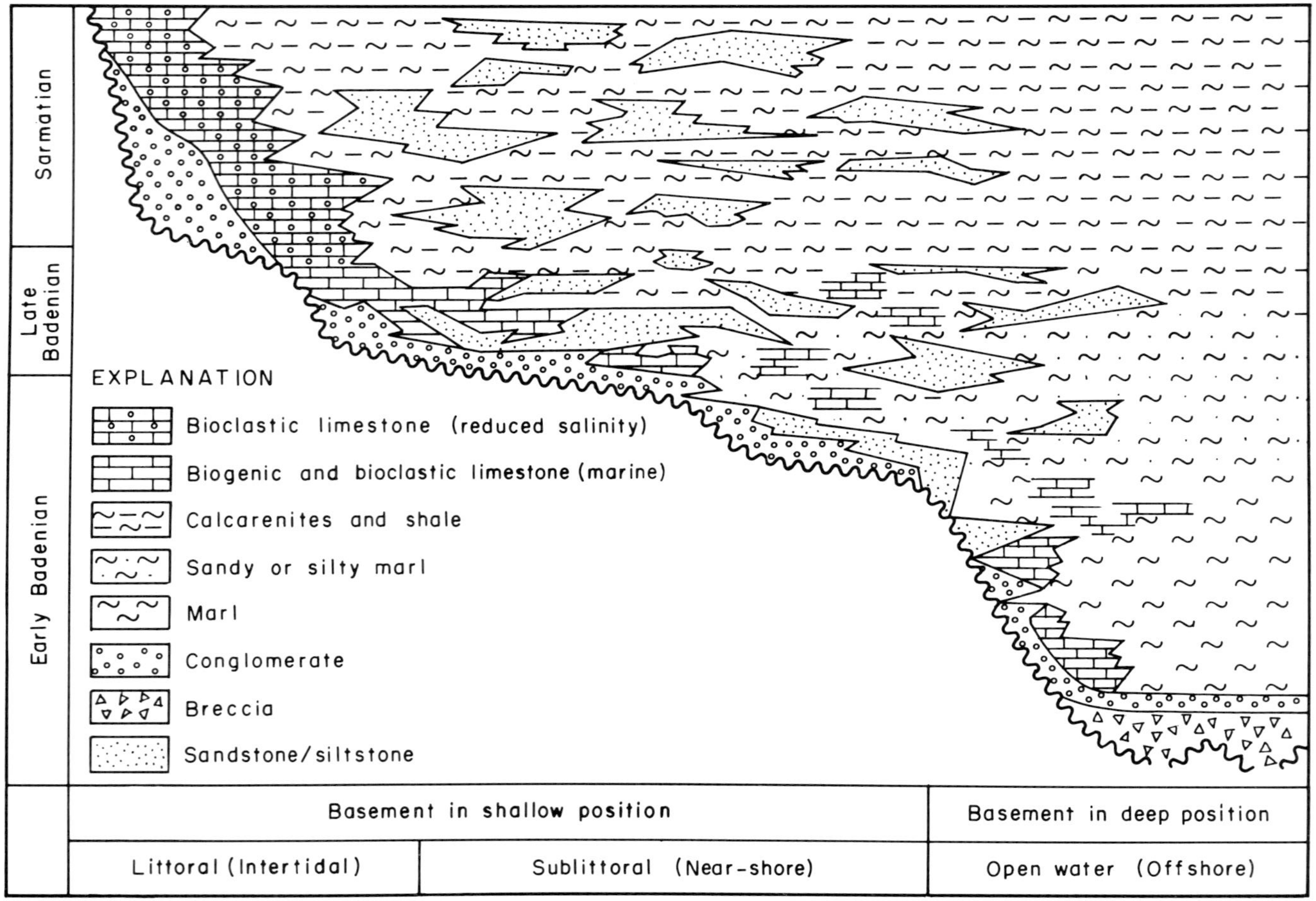

Figure 3. *Generalized diagram of Miocene (Badenian and Sarmatian) lithofacies in the Békés basin. Figure is not drawn to scale.*

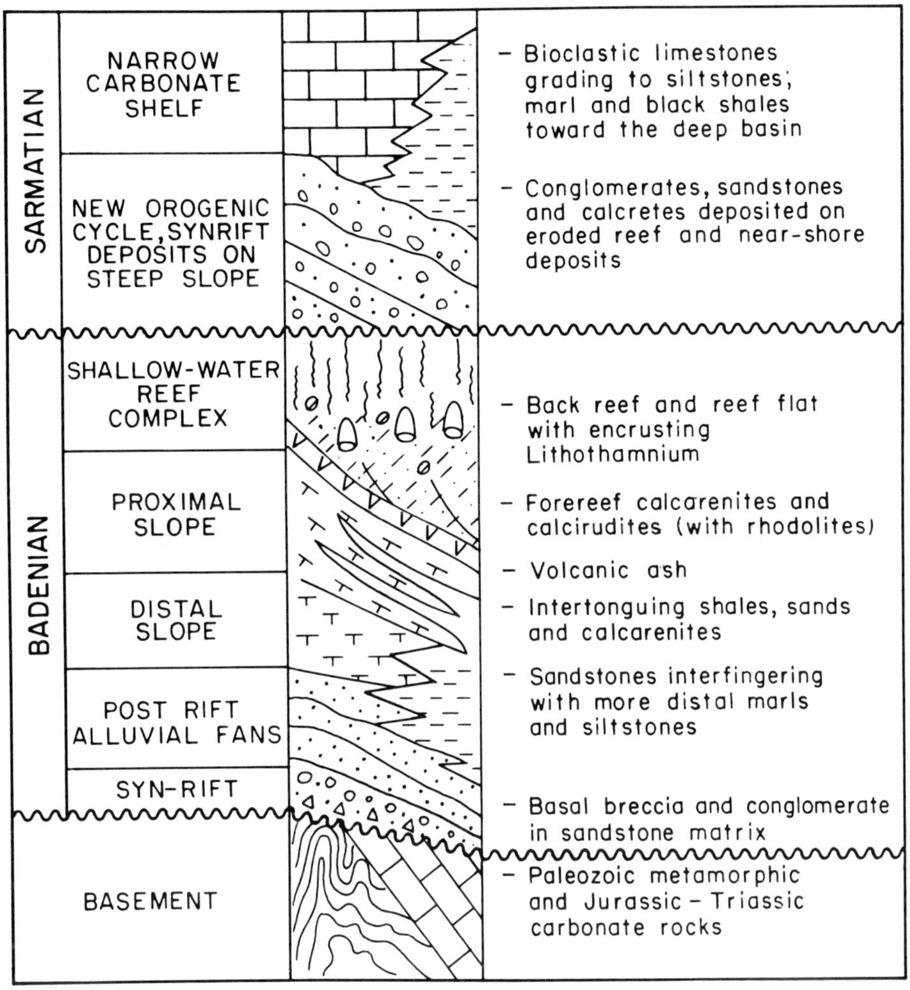

Figure 4. Generalized stratigraphic column of the Miocene sequence in the Békés basin.

Nearshore environments are characterized by sandstones of considerable thicknesses deposited postrift in alluvial-to-marine environments, constituting channel-fan systems, mainly on the gently dipping flanks of basement highs.

Benthic foraminifera, such as *Cibicides dutemplei, Nonion soldanii, Nonion umbilicatum, Pullenia bulloides, Eponides majzoni, Bolivina dilatata,* and *Uvigerina tenuistriata* are present in these deposits. The number, thicknesses of sandstone bodies, and grainsize decline basinward (toward the central and southeastern part of the basin). Toward the offshore, the sandstones become finer grained and dark gray in color, and are compacted and fractured. Offshore, these sandstones, that are distal slope deposits (Fig. 4), intertongue with turbidites, with interbedded marls, siltstones, sandstones and shales that exhibit load and flame structures, as in the Hunya-1, Kond-1 and Gyo-1 wells (Phillips and others, this volume).

The siltstones occur as thin intercalations in sandstone units or in shales and marls. Their bounding surfaces are micaceous, and both sharp and gradational contacts occur. Gray in color and well-sorted in grainsize, the siltstones are compacted, and the degree of diagenesis as well as carbonate content is higher than that of sandstones (Table 2).

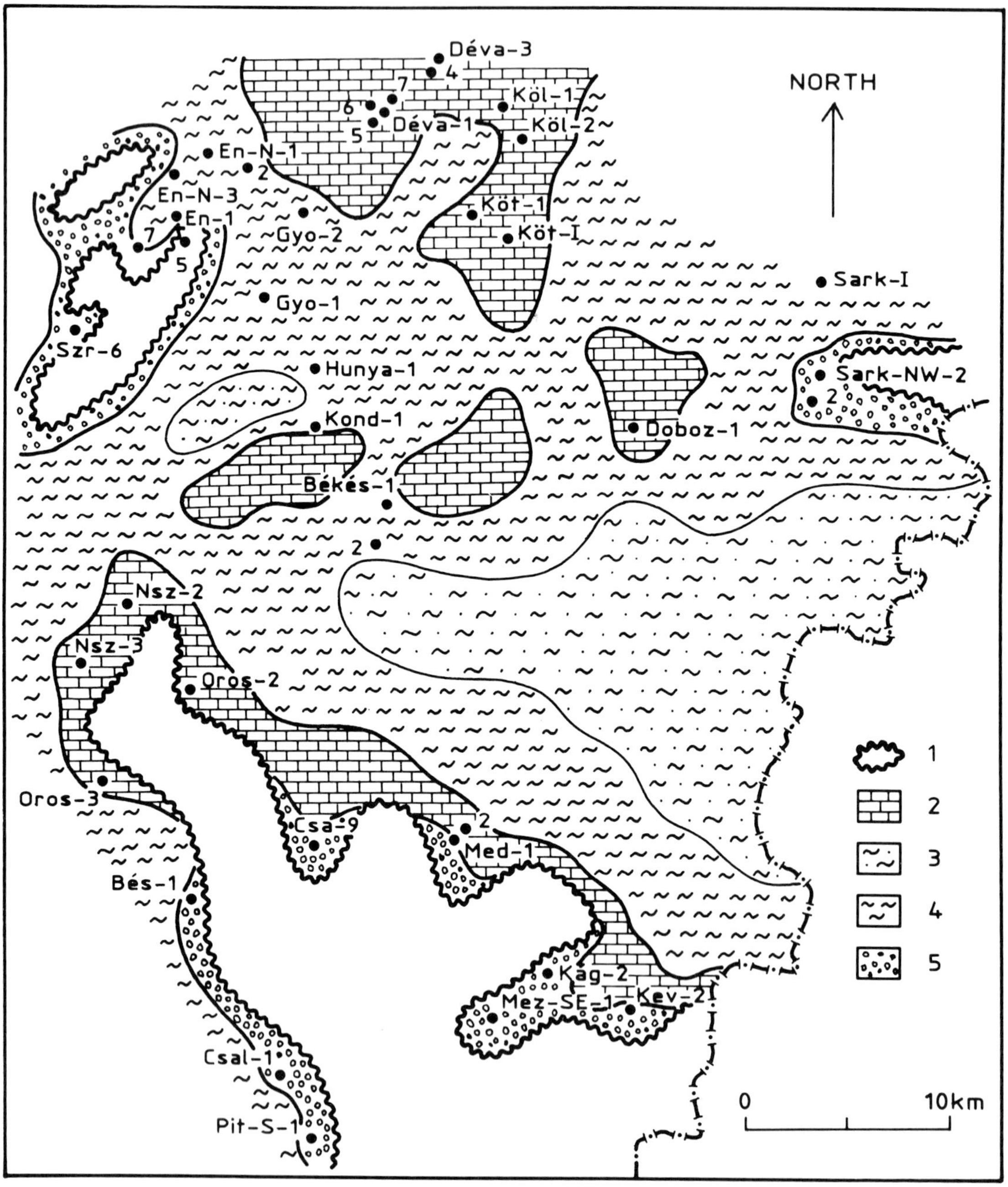

Figure 5. Distribution of lower Badenian lithofacies in the Békés basin. 1: paleoislands; 2: biogenic and bioclastic limestones; 3: marls; 4: calcarenites, shales and siltstones; 5: conglomerates and breccia. Well notations: Szr: Szarvas; En: Endröd; Déva: Dévaványa; Kl: Körösladány; Kt: Köröstarcsa; Sark: Sarkadkeresztúr; Gy: Gyoma; Kond: Kondoros; Oros: Orosháza; Nsz: Nagyszénás; Med: Medgyesbodzás; Kág: Kúnágota; Mez: Mezöhegyes; Kev: Kevermes; Bés: Békéssámsom; Csal: Csanádalberti; Pit: Pitvaros.

Upward in the sequence, the sandstones are interbedded with layers of shale and marl or limestone, interpreted to have been deposited in tidal to proximal slope environments (Fig. 4). Along the bedding planes, coalified plant fragments occur. In neritic depositional environments, sandstone is overlain by bioclastic, algal limestone (Nsz-3, Köt-I, Köt-1 and Déva-2 wells). In the Déva-4 well, 2-3-m-thick sandstone intercalations in the limestone may represent changes in the position of the shoreline and water depth.

Table 2. Mineral composition of siltstone (Sark-NW-2 well, 3642-3645 m).

MINERAL	PERCENT
Quartz	17
Feldspar	3
Calcite	64
Dolomite	4
Illite/montmorillonite	5
Chlorite	5
Muscovite	3
Other	3

The sandstones customarily fringe the paleogeographic highs. Benthic foraminiferal species, such as *Cibicides, Nonion, Pullenia, Eponides, Virgulina, Lagena* and *Uvigerina* indicate that the sandstones are marine in origin.

On the narrow, shallow shelf, marlstones and mudstones became interbedded with algal limestones. These sequences, illustrated in the lower half of Figure 3, are present in the southwestern and north-northeastern parts of the basin. Where open-marine (offshore) conditions were present, thin layers of calcareous marlstones are interbedded with and subordinated to shale and marl. Thick deposits of marl and shale form continuous sequences in the offshore zone, such as in the central (Hunya-1 well) and southeastern parts of the basin, signifying cyclic deposition alternating between clastic sediment influx and greater carbonate production. The marls occur as brownish-gray, indurated, unbedded to laminate deposits, containing 60-70% carbonate material as bioclasts, and occasional inclusions of small lenses of fine-grained sandstone. The bioclasts consist of nodules of Lithothamnium algae (rhodoliths), tests of benthic foraminifera and ostracoda, and Echinoidea and Bryozoa fragments in great abundance in a carbonate-mudstone matrix. The assemblage represents reef front deposits encrusted by algae.

Characteristic foraminiferal species found in the reef front deposits are *Amphistegina hauerina, Heterostegina costata, Gaudryina bradyi, Textularia abbreviata, Anomalina badensis, A. cryptomphala, Cibicides boueanus,* and *C. dutemplei.* The quantities of planktonic foraminifera in these deposits are commonly quite small.

The distinctive lithology of the Badenian series is bioclastic-biogenic algal limestone deposited on reef flats in shallow, clear, well-oxygenated waters of a nearshore zone. The thickness of these carbonate deposits is 20-50 m. Bioclastic limestone deposits, that have been encountered by drilling in the deeper parts of the basin, are interpreted to have formed on submarine plateaus, or banks, located in offshore areas where the water depths were relatively shallow, then became buried as the basin subsided. Bank carbonates were penetrated in the Doboz-I, Köt-I, Köt-1 and Körösladány (Köl)-2 wells. Near the shore, these carbonate facies (so-called "Leitha" limestones in Hungarian terminology) may succeed the basal conglomerates or alluvial sandstones directly. Toward the offshore, the carbonate deposits interfinger with turbiditic, distal slope deposits, that are composed largely of shales and siltstones.

In most places, two separate successions of shallow-marine, reef-forming limestone bodies can be recognized, isolated by thin shale or sand layers. The age of the lower unit is early Badenian, and the upper unit late Badenian. The two successions could be related to changes in the Paratethyan sea level. The sea-level curves of Steininger and others (1989) show a major transgression at the beginning of Badenian time, and a minor one during the late Badenian. There is no other evidence in the Békés basin to substantiate a new cycle starting in the mid-Badenian, as indicated by these sea-level curves. Instead, it can be argued that subareas of the basin were differentially subsiding throughout the Miocene.

Along the northern perimeter of the basin, that is, in the Körösladány-Dévaványa areas, only the upper Badenian unit has been deposited (Köl-2 and Déva-3 wells). This signifies that, during late Badenian, the sea advanced further northward, compared to early Badenian time, because subsidence of the northern part occurred later than that of the southern part. Elsewhere, the upper unit is missing, because of uplift and erosion at the end of the Badenian stage (see below), and only the lower unit is present (Déva-2, Köt-I, Köt-1 and Nsz-2 wells).

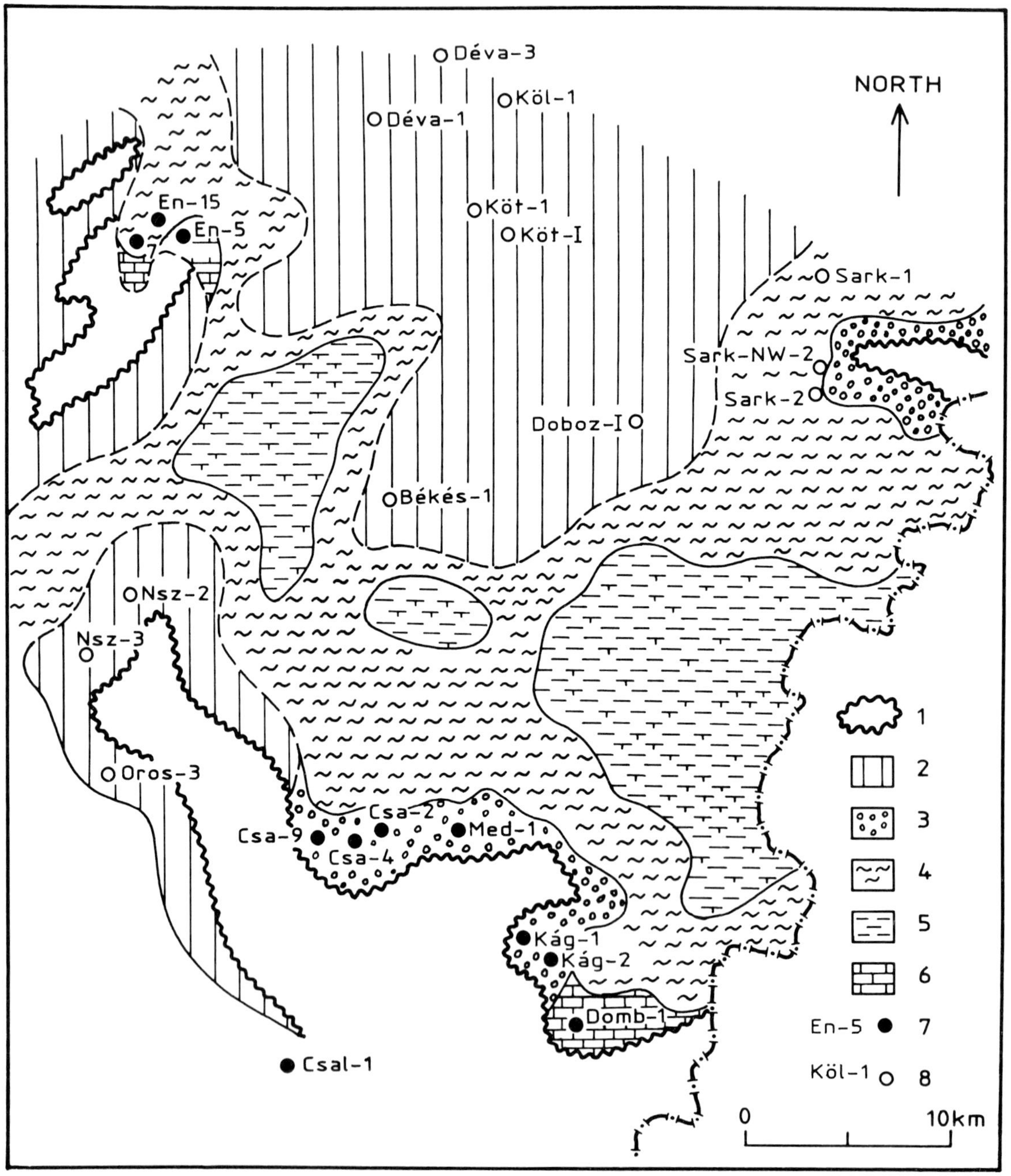

Figure 6. Distribution of Sarmatian lithofacies in the Békés basin: 1) paleoislands; 2) no deposition; 3) conglomerates and breccia; 4) marls; 5) shales and siltstones; 6) bioclastic limestones. Well names as in Fig. 5.

The grainsize of limestones ranges from calcilutite to calcarenite and 70-80% of the rock volume consists of benthic foraminifera, Bryozoans, and algal (*Lithothamnium* sp.) nodules (rhodoliths). Fragments of shells with grain-supported texture occur in a few samples. The cement is composed of microcrystalline calcite. The mineralogical composition of the bioclastic limestone, based on x-ray diffraction data for core from the Fábiánsebestyén-4 drillhole is shown in Table 3.

Table 3. Mineral composition of bioclastic limestone.

MINERAL	PERCENT
Quartz	35
Feldspar	13
Chlorite	20
Muscovite	23
Montmorillonite	5
Mg-Calcite	13
Other	4

The limestone deposits are rich in microfauna. Benthic, warm-water foraminifera predominate: *Amphistegina hauerina, A. vulgaris, Asterigerina planorbis, Heterostegina costata, Cibicides boueanus, C. lobatulus, C. dutemplei, Anomalina badensis, A. ammonoides, Textularia abbreviata, T. mayeriana, T. agglutinans, Sphaerogypsina globula, Eponides praecinctus, Elphidium fichtelianum, E. crispum, Borelis melo* and *Discorbis mira*. The presence of *Elphidium* and *Textularia* indicate waters of normal salinity (Bizon, 1985).

Significant volumes of the Badenian marine rocks consist of fine-grained sediments (shale, marl and calcareous marl) with variable carbonate content. Nevertheless, marl dominates these sequences in the central and southeastern parts of the basin (Fig. 5).

The shales, in the central part of the basin, are commonly interbedded with beds or laminae of fine-grained sandstone or siltstone. Coalified plant fragments and fine-grained mica are concentrated along bedding surfaces (Békés-2 well). The marls are generally dark gray, massive or sometimes slabby, but commonly also contain lighter colored, dense, calcareous sandstone beds or lenses a few centimeters thick. Quartz grains, carbonate bioclastic material, and muscovite flakes are scattered throughout the marls (Table 4).

The deposits are rich in foraminifera, the majority of which is planktonic. Typical foraminiferal assemblages consist of planktonic species such as *Orbulina suturalis, O. bilobata, Globigerinoides bisphaericus, G. quadrilobatulus, G. tarchanensis, G. glomerosus, Globigerina juvenilis, G. nepenthes, G. eggeri, Globoquadrina altispira globosa, Globorotalia scitula, G. praemenardii* and *G. obesa*. According to Iaccarino (1985), *Orbulina suturalis* is the index fossil used to demarcate the partition of early Miocene (ending with the Karpatian stage) from the middle Miocene (beginning with the Badenian stage).

Table 4. Mineral composition of shale/clay-marl (Békés-2 well, 3239-3240 m).

MINERAL	PERCENT
Quartz	38
Feldspar	7
Chlorite	10
Muscovite	18
Illite/montmorillonite	2
Calcite	13
Kaolinite	5
Other	7

Benthic forms frequently encountered are *Gyroidina, Elphidium, Textularia, Eponides, Cibicides, Amphistegina, Robulus, Epistomina, Reusella, Discorbis, Uvigerina, Nonion, Rotalia, Gaudryina, Anomalina* and other genera. These microfaunal suites indicate that the major part of the sequences are of early Badenian age (Rögl, 1985; Hámor and others, 1985, 1987; Nagymarosy and Müller, 1988). The entire Badenian section in the Nsz-2, Nsz-3, En-7, En-N-1, En-N-3 and Déva-4 wells is of early Badenian age. This suggests that in these areas the upper Badenian rocks were eroded. Elsewhere uneroded upper Badenian strata are present. Planktonic foraminifera, however, are rare in upper Badenian sediments. Where present, they consist of a few forms of *Globigerina* (Hunya-1, Gyo-2 and Kond-1 wells).

Thin layers of rhyodacite tuff occur in the Badenian section of the Békés basin (Fig. 4). Although the volume of these rocks is negligible, they are significant time markers. The pyroclastic rocks occur as thin interbeds in the Badenian section (Figs. 1, also see Pogácsás, Mattick, Elston and others, this volume), but are absent in overlying Sarmatian sediments. The tuff represents windblown particles deposited in marine waters; thus, the deposits are finely stratified, dense and sometimes contain microfauna as in the En-18 well. The beds contain calcite and silica nodules. Their mineral composition is dominated by feldspars replaced by calcite, biotite and zeolites. The most significant occurrences in the Békés basin are in the Szarvas (Szr)-6, Déva-3, En-1, En-3, En-18, Köt-1, Kond-1 and Nsz-2 wells, and near the basin at the Fábiánsebestyén (-1, -2, -4) wells.

Sarmatian Deposits

Two features differentiate the Sarmatian sequence from the underlying Badenian sequence. One is that paleontological evidence indicates that the salinity of Sarmatian sea waters became reduced gradually after Badenian time, because the Sarmatian Paratethys became isolated from the Eastern Paratethys (Rögl and Steininger, 1984; Halmai and others, 1988), the outcome of which was the development of a local endemic fauna. The other is the onset of a new cycle of sedimentation in the Békés basin, starting with the deposition of basal conglomerates. The climate was still subtropical-mediterranean in early Sarmatian time, but by late Sarmatian cooling became evident (Kennett, 1982; Bizon, 1985).

Nearshore Sarmatian deposits, 10-40 m thick, consist of two rock assemblages: (1) a basal conglomerate and sandstone, and (2) a basal conglomerate and bioclastic limestone (Figs. 1, 3, and 4). Basal units of the Sarmatian sequence are usually composed of conglomerates in a fine-grained, sandy matrix. The rock generally is poorly consolidated, unstratified, well sorted, and has a low carbonate content. The pebbles are well rounded and polymictic; the matrix is sandstone and the cement is calcite. In a few cases, sandstone overlies the conglomerate. Where the conglomerate is absent, sandstone overlies basement rocks. The sandstone is stratified and scattered coalified plant fragments and mica commonly occur on bedding surfaces. Some of the sandstone contains interbeds of siltstone and shale. As shown on Figure 6, the basal conglomerate and sandstone is present on the north side of the Battonya high (Csa-2, -4, -9 and Med-1 wells) and the basal conglomerate succeeded by limestone occurs along the flank of this high (Domb-1 well) and the Szarvas high (En-5 well). Conglomerates also flank the Sarkadkeresztúr high. The appearance of these coarse deposits signifies a new cycle of sedimentation in the Békés basin. According to Rögl (1985), discordances and transgressions mark the beginning of the Sarmatian epoch in the marginal areas of the Central Paratethys basin, and it appears that such movements affected the more central parts as well, where paleogeographic islands existed.

In the central part of the basin, Sarmatian sediments were deposited on upper Badenian strata without interruption; the Sarmatian sequence commences with sandstone deposition followed by marl (Békés-1 well). The marl is generally unstratified, and occasionally is found to include fossiliferous beds. These deposits are 50-90 m thick. In other locations, such as near paleogeographic islands (presently basement highs) Sarmatian deposits overlie Mesozoic and pre-Mesozoic rocks (Csa-2 and Domb-1 wells), or Badenian basal conglomerates (Kág-2 well) directly.

The absence of a section, representing about 2.5 million years, indicates that either the Badenian sea did not reach these areas, or, as evidenced by the occurrence of a new basal conglomerate section (Fig. 4), uplift and erosion could have taken place at the close of Badenian time, at least along the perimeter of the basin (Fig. 6). These Sarmatian-age basal deposits, composed of conglomerates, sandstones and calcretes, possibly signify the onset of another orogenic cycle (the Leithian orogeny), which was followed by the Sarmatian sea transgressing the flanks of basement highs. This event is also demarcated in those areas where Badenian-Sarmatian sedimentation remained continuous, in that, siliciclastic sediments were deposited in the form of an intercyclic sedimentary unit at the base of the Sarmatian sequence.

Limestone deposition in Sarmatian time took place in shallow-water. The limestone is commonly stratified, biogenic or rich in bioclasts, and contains euryhaline foraminifera (species of *Elphidium, Nonion, Nubecularia, Quinqueloculina, Triloculina* and *Miliolidea*), ostracoda, abundant shell fragments of molluscs (*Irus gregarinus* and *Mactra podolica*) and fragments of Bryozoa. In Sarmatian deposits, stenohaline foraminifera, molluscs and *Lithothamnium algae are absent. The grain size of bioclasts ranges between calcarenitic and micritic. The rock often contains large quantities of quartz grains or small pebbles. Oolitic limestone was observed only in cores from the En-5 well, otherwise it is absent in the Békés basin (Szentgyörgyi, 1978).*

In more central parts of the basin (Hunya-1, Kond-1 and Köt-1 wells), a non-fossiliferous shale sequence deposited in an offshore environment occurs between the biostratigraphically identifiable lower Badenian and lower Pannonian sequences, as observed in well cores. The thickness of the shale sequence is 50-90 m, and no break in sedimentation can be recognized between these beds and the underlying and overlying strata (Hunya-1, Kond-1, Köt-1 wells). The upward-coarsening shale sequence consists of dark-gray, black, unstratified, silty, micaceous marl and shale with calcareous marl laminae containing fish scales. Scattered coalified plant fragments and grains of primary pyrite commonly occur on bedding surfaces. The pyrite indicates that the unit may have been deposited in an anoxic environment. This sequence may represent the full upper Miocene (late Badenian and Sarmatian) section (Szentgyörgyi, 1978), and signify deposition in a basin of little circulation, as both the calcareous algae and the euryhaline foraminifera vanished.

A significant part of the Békés basin appears to be devoid of Sarmatian deposits, as shown in Figure 6, that is, identifiable Badenian sequences are overlain by Pannonian age rocks (Békés-1, Doboz-I, Köt-1, Köt-I, Köl-1, Déva-1, Déva-3 wells). This suggests that the magnitude of uplift and erosion at the close of the Badenian was higher in the northern part of the basin, and along the eastern rim near the northwest tip of the Battonya "island" than elsewhere. These factors imply a southeastward tilting of the basin.

PALEOGEOGRAPHY OF MIOCENE SEDIMENTATION

Those areas where Miocene sediments are absent (Szarvas, Endröd, Battonya and Sarkadkeresztúr highs) were probably islands in the Paratethys Sea during much of Miocene time, as inferred from spatial relations of sedimentary facies. For example, the dominance of nearshore facies in the southwestern part of the basin indicates that much of the Battonya high remained subaerial during Miocene time. Basinward from the Battonya high, nearshore facies grade to shelf facies dominated by bioclastic limestone that interfinger with deep-water (basinal) facies. Island coastlines were probably embayed, and the current-dominated, sand and gravel and gravel prisms fringing them occupied narrow zones.

The paleogeographic relation between islands and the sea changed little during Miocene time. For the basin as a whole, the position of shorelines varied only slightly through time; locally, however, shorelines prograded toward the basin, as, e.g., in the north and east in Sarmatian time. Local transgressions of the sea occurred, especially during Sarmatian time, as documented by the fact that, in the southeastern part of the basin (Kág-1 and Domb-1 wells, Fig. 6), Sarmatian clastic sequences are in contact with pre-Neogene basement rocks.

HISTORY AND EVOLUTION

The Békés basin was an emergent area from Late Cretaceous to Badenian time. Paleogene and older Miocene (Eggerburgian, Ottnangian and Karpatian) deposits have not been found to date (Fig. 4). Post-Mesozoic subsidence began in early Badenian time during the late Styrian orogenic phase and subsidence has continued to the present. In early Badenian time, basin subsidence was accompanied by a major transgression, and the region became engulfed by a shallow sea containing islands, namely the Central Paratethys Sea.

Paleoecological considerations of the microfaunal associations indicate well-oxygenated, warm water conditions were present (Gall, 1983). Circulation may have generated shelf currents; the occurrence of sandstone in sublittoral zones may signify their past existence. Complex interfingering lithofacies characterize the sequences in both the offshore and along the coast (Figs. 3 and 4). Beach sandstones are interbedded with marl, silty shale, and siltstone, indicating minor water-level fluctuations, and deposition in waters depths of 10-30 m. Algal flats, rich in Lithothamnium algae, fringed the islands (basements highs) in the Badenian sea.

Early-to-middle Badenian time is characterized by the continuous transgression of marine waters of normal salinity, as connection was established with Indo-pacific waters that also flooded the Mediterranean Basin. In the Békés basin, the shoreline of the sea moved landward during this time. Lower Badenian rocks can be biostratigraphically correlated throughout the basin, and the fauna is characteristically Indo-pacific in origin (Rögl and Steininger, 1984; Rögl, 1985). In addition, an acid tuff bed in the lower Badenian section at the lower-upper Badenian boundary is a basinwide chronostratigraphic marker bed (Figs. 1 and 4). The center of volcanism was not too distant; one such site has been shown by Halmai and others (1988) to have existed in the center of the Békés basin. Based on well data, the lower Badenian marine sequence is the most widespread pre-Pannonian, Miocene unit.

According to Rögl and Steininger (1984) and Steininger and others (1989, mid-to-late Badenian time is marked by a regression of the sea, resulting in the widespread deposition of evaporites in the Carpathian foredeep and the Transylvanian Basin. In addition, as a result of the northward movement of the Dinaride Mountains, located southwest of the Pannonian Basin, the connection between the Central Paratethys Sea and the Mediterranean Sea, in the west, was severed. In the east, uplift of the Carpathians curtailed access to the Eastern Paratethys.

The late Badenian reflooding from the east (Rögl and Steininger, 1984; Halmai and others, 1988) cannot be distinguished in the Békés basin, although this is the time of reef building under subtropical conditions and the invasion of Indo-pacific faunal assemblages into the Central Paratethys.

The first sedimentary cycle closed at the end of the Badenian.

From the beginning of Sarmatian time, the Pannonian Basin became permanently isolated from the Eastern Paratethys, and its paleogeographical development became independent. Locally, a new cycle of transgression-regression began (Steininger and others, 1989). In the Békés basin, a new transgression-regression cycle can also be recognized, accentuated by the southeastward tilting, therefore northeastern rising of the basin. As the regression of the Paratethys Sea in the Pannonian Basin continued, however, the sea became shallower, and, with increasing fresh-water inflow, the waters became less saline. In these brackish environments, an endemic fauna characterized by euryhaline living organisms developed. The rate of subsidence gradually moderated, and at the end of Sarmatian time it was minimal except for the southeastern and central parts of the basin.

Sedimentation was replaced by erosion at the beginning of Pannonian time near the periphery of the Békés basin, where lower Pannonian sediments were deposited on lower Badenian rocks. However, in the deepest part of the Békés basin, sedimentation may have been continuous from middle Miocene through Pliocene time (Mattick and others, this volume), although the early Pannonian section was highly condensed as a result of little or no deposition (Pogácsás and others, this volume).

SUMMARY AND CONCLUSIONS

The lithostratigraphic record of the Békés basin represents excellent examples of marine-lacustrine, confined sedimentation patterns and facies developments in a tectonically mobile terrain. The record of the Miocene sea in this area is particularly fascinating, because it embodies major changes in sea-level, climate, water chemistry and temperature, sediment sources, and faunal assemblages within a brief 5 million years.

Two transgressive-regressive cycles characterize the Miocene sequence in the Békés basin. The first cycle is marine and Badenian in age, resting on the denuded pre-Neogene basement. Rocks of the basement served as source material for the Badenian deposits once the Styrian orogeny uplifted the terrain. Sediments, at first coarse, synrift conglomerates, then finer siliciclastics poured into a subsiding basin created by oblique extension and strike-slip faulting (Grow and others, this volume). Paleorelief varied; basement-high supported islands remained on the basin's perimeter. The paleo-islands were fringed by sands derived from alluvial fans, and further offshore, reef-forming algae built shallow submarine carbonate banks. The sedimentary sequence fines upward in the lower Badenian section, signifying a transgressive cycle. The upper part of the cycle is regressive, carbonate-dominated (reefs), closing with deposition of a regressive sand body.

The second cycle is separated from the first by an erosional unconformity near the paleo-islands, nondeposition in the northern and eastern part of the basin, and at least a lithologic disconformity in the interior of the basin. Appearance of conglomerates and sandstones above this boundary demarcate the onset of renewed tectonic movements, which may be related to the Leithian orogeny. Sediments of the second cycle fine upward to limestones and shales, in shallow and deep waters, respectively. Regression of the marine waters may be signified by the shale sequence, which coarsens upward and includes increasing amounts of mica and organic matter. The waters, at this time, had begun to become brackish and anoxic upon closure of seaways to the world's oceans caused by the uplift of the Carpathian Mountains. Fluctuations in lake-level occurred throughout the second cycle, probably in response to local subsidence. The erosional-nondepositional record in the Sarmatian suggests that, the basin was tilted, along a northwest-southeast axis, upward toward the northwest.

Only the Styrian orogenesis is associated with substantive and frequent volcanism that deposited ash in the northern part of the basin. A notable marker bed is the rhyodacite tuff beds in the upper part of the Lower Badenian section (15 Ma), although other layers are also present. The source of pyroclastics may have been in the basin itself (Halmai and others, 1988).

ACKNOWLEDGEMENTS

The authors wish to express their gratitude to Professor Géza Hámor, past director of the Hungarian Geological

Survey (MÁFI), for his valuable suggestions and reasoning, and to Robert Mattick, Kay Molenaar and N. Terence Edgar for their helpful reviews.

REFERENCES

Bérczi, I., Hámor, G., Jámbor, A., and Szentgyörgyi, K., 1988, Neogene sedimentation in Hungary, in Royden, L.H. and Horváth, F., (eds.), The Pannonian Basin: A study in basin evolution; Amer. Assoc. Petrol. Geol., Memoir 45, 57-67.

Berger, J.-P., 1992, Correlation chart of the European Oligocene and Miocene: Application to the Swiss Molasse Basin; Eclogae Geol. Helv., 85 (3), 573-609.

Bizon, G., 1985, Mediterranean Foraminiferal changes as related to paleoceanography and paleoclimatology; in Stanley, D.J., and Wezel, F.-C. (eds.) Geological evolution of the Mediterranean Basin; Springer Verlag, New York, 453-603.

Dabrio, C.J., and Polo, M.D., 1991, Fan-delta slope deposits and sequences in the Murcia-Carrascoy Basin (Late Neogene, S.E. Spain); Cuadernos de Geología Ibérica, Madrid, 15, 49-71.

Gall, J.C., 1983, Ancient sedimentary environments and the habitats of living organisms; Springer Verlag, Berlin, 1-167.

Grow, J.A., Pogácsás, G., Bérczi-Makk, A., Hajdú, D., Várnai, P., Varga, E., and Péró, C., Mesozoic and basement structures of the Békés basin inferred from seismic reflection profiles, drillhole, and gravity data; this volume, 1-38.

Halmai, J. (compiler), and Steininger, F.F., Kojumdgieva, E., Cicha, I., Vass, D., Berthelt, D., Hámor, G., Bocaletti, M., Gelati, R., Moratti, G., Slaczka, A., Marinescu, F., Berger, J.P., Babak, E.V., Goncharova, I.A., Ilyina, L.B., Nevesskaya, L.A., Paramonova, N.P., Popov, S.P., Eremija, M., and Marinovic, D. (eds.), 1988, Neogene palaeogeographic atlas of Central and Eastern Europe; Hung. Geol. Inst., 7 map sheets.

Hámor, G., Báldi, T., Bohn-Havas, M., Hably, L., Halmai, J., Hajós, M., Kókai, J., Kordos, L., Korecz-Laky, I., Nagy, E., Nagymarosy, A., and Völgyi, L., 1985, The biostratigraphy of the Hungarian Miocene (abs.); 8th Cong., Reg. Comm. Mediter. Neogene Stratigraphy, Budapest, 252-257.

Hámor, G., Báldi, T., Bohn-Havas, M., Hably, L., Halmai, J., Hajós, M., Kókai, J., Kordos, L., Korecz-Laky, I., Nagy, E., Nagymarosy, A., and Völgyi, L., 1987, The bio-, litho-, and chronostratigraphy of the Hungarian Miocene; Proc. 8th Cong. Reg. Comm. Mediter. Neogene Stratigraphy, Budapest, Ann. Inst. Geol. Publ., 70, 351-354.

Iaccarino, S., 1985, Mediterranean Miocene and Pliocene planktic foraminifera; in Bolli, H.M., Saunders, J.B. and Perch-Nielsen, K. (eds.), Plankton stratigraphy; Cambridge Univ. Press, 283-314.

Kázmér, M., 1990, Birth, life and death of the Pannonian Lake; Palaeogeog., Palaeoclim., Palaeoecology, 79, 171-188.

Kennett, J., 1982, Marine geology; Prentice-Hall, Englewood Cliffs, New Jersey, 813 p.

Mattick, R.E., Rumpler, J., Ujfalusy, A., Szanyi, B., and Nagy, I., Sequence stratigraphy of the Békés basin; this volume, 39-65.

Nagymarosy, A. and Müller, P., 1988, Some aspects of Neogene biostratigraphy in the Pannonian Basin, in Royden, L.H. and Horváth, F. (eds.), The Pannonian Basin; A study in basin evolution; Amer. Assoc. Petrol. Geol. Memoir 45, 69-75.

Phillips, R.L., Révész, I., and Bérczi, I., Lower Pannonian deltaic-lacustrine processes and sedimentation, Békés basin; this volume, 67-82.

Pogácsás, Gy., Mattick R.E., Elston, D.L., Hámor, T., Jámbor, Á., Lakatos, L., Lantos, M., Simon, E., Vakarcs, G., Várkonyi, L., and Várnai, P., Correlation of seismo- and magnetostratigraphy in southeastern Hungary; this volume, 143-160.

Rögl, F., 1985, Late Oligocene and Miocene planktic foraminifera of the Central Paratethys; in Bolli, H.M., Saunders, J.B. and Perch-Nilesen, K. (eds.), Plankton stratigraphy; Cambridge Univ. Press, 315-328.

Rögl, F. and Steininger, F.F., 1984, Neogene Paratethys, Mediterranean and Indo-pacific seaways - implications for the paleobiogeography of marine and terrestrial biotas; in Brenchley, P. (ed.), Fossils and climate; John Wiley & Sons Ltd., 171-200.

Steininger, F.F., Müller, C., and Rögl, F., 1988, Correlation of Central Paratethys, Eastern Paratethys and Mediterranean Neogene stages in the Pannonian Basin, in Royden, L.H. and Horváth, F., (eds.), The Pannonian Basin: A study in basin evolution; Amer. Assoc. Petrol. Geol. Memoir 45, 79-87.

Steininger, F.F., Rögl, F., Hochuli, P. and Müller, C., 1989, Lignite deposition and marine cycles - the Austrian Tertiary lignite deposits, a case history; Österreich. Akad. der Wissenschaften, Biol. Wissenschaften u. Erdwissenschaften, 197, (5-10), 309-332 (Springer Verlag, Wien).

Steininger, F.F., Senes, J., Kleemann, K., and Rögl, F., 1985, Neogene of the Mediterranean Tethys and Paratethys; Int. Union Geol. Sci., University of Vienna, 25, 59-60.

Szentgyörgyi, K., 1978, The Sarmatian formations in the Tiszántúl area (Eastern Hungary) and their stratigraphic position; Acta Mineral. Petrog., Szeged, Hungary, 23, 279-297.

Szentgyörgyi, K., 1983, Lithostratigraphic units of the epicontinental Senonian in the Hungarian Plain; Acta Geol. Hung., 26, 197-211.

Vass, D., Repcok, I., Halmai, J., and Balogh, K., 1985, Contributions to the improvement of the numerical time scale for the Central Paratethys, (abs.); 8th Cong. Reg. Comm. Mediter. Neogene Stratigraphy, Budapest, 595-596.

5 Stratigraphic Framework and Sandstone Facies Distribution of the Pannonian Sequence in the Békés Basin

C. M. Molenaar[1], I. Révész[2], I. Bérczi[3], A. Kovács[3], Gy. K. Juhász[2], I. Gajdos[4], and B. Szanyi[3]

ABSTRACT

The Békés basin is a late Tertiary structural basin containing as much as 6500 m of Neogene sandstone, siltstone, claystone, and marl, most of which comprise the Pannonian sequence or group. Interpretation of well data and seismic reflection data indicates that these rocks were deposited in a wide range of depositional environments ranging from deep-basinal brackish/lacustrine to basin slope to shallow lake and delta plain to alluvial plain. Seismic reflection profiles clearly show topset, foreset (clinoform), and bottomset (basinal) depositional architecture and that the different facies are highly diachronous. Correlations of seismic reflections with well logs indicate that the reflection patterns within each of the three main reflection groups contain characteristics that can be related to lithologies and depositional environments, which in turn reflect formational terminologies. In ascending order, these formations are (1) the Tótkomlós and Nagykörü Formations, a basal clayey marl and marl unit; (2) the Szolnok Formation, a basinal unit consisting of thick amalgamated beds of very fine grained turbidite sandstone and interbedded marl; (3) the Algyö Formation, consisting of prodelta-slope clay and marl and lesser amounts of turbidite or slumped sandstone in the lower part; (4) the Törtel Formation, consisting of delta-plain and delta-front very fine- to medium-grained sandstone and interbedded clay and clayey marl; and (5) the Zagyva and Nagyalföld Formations, consisting of upper delta-plain or alluvial-plain and inland-swamp sandstone and clay. The lower three formations are referred to as lower Pannonian, and the upper formations as upper Pannonian. Sandstones in the lower Pannonian are more compacted, and the reservoir properties are not as good as the generally friable sandstones in the upper Pannonian.

INTRODUCTION

The Békés basin, a subbasin of the Pannonian Basin, is a late Tertiary structural basin that was filled by as much as 6500 m of Neogene sandstone, siltstone, claystone, and marl (Fig. 1). These rocks, known as the Pannonian sequence or group, are oil and gas productive in structural traps around the flanks of the basin. Recent reports by Jámbor and others (1987), Pogácsás and Révész (1987), and Pogácsás (1987) describe the general stratigraphy

[1] U.S. Geological Survey, Denver, Colorado 80225, USA
[2] MOL Rt. (Hungarian Oil and Gas Co., Ltd.), Szeged, Hungary
[3] MOL Rt. (Hungarian Oil and Gas Co., Ltd.), H-1062, Budapest, Hungary
[4] MOL Rt. (Hungarian Oil and Gas Co., Ltd.), H-5001, Szolnok, Hungary

P. G. Teleki et al. (eds.), Basin Analysis in Petroleum Exploration, 99–110.

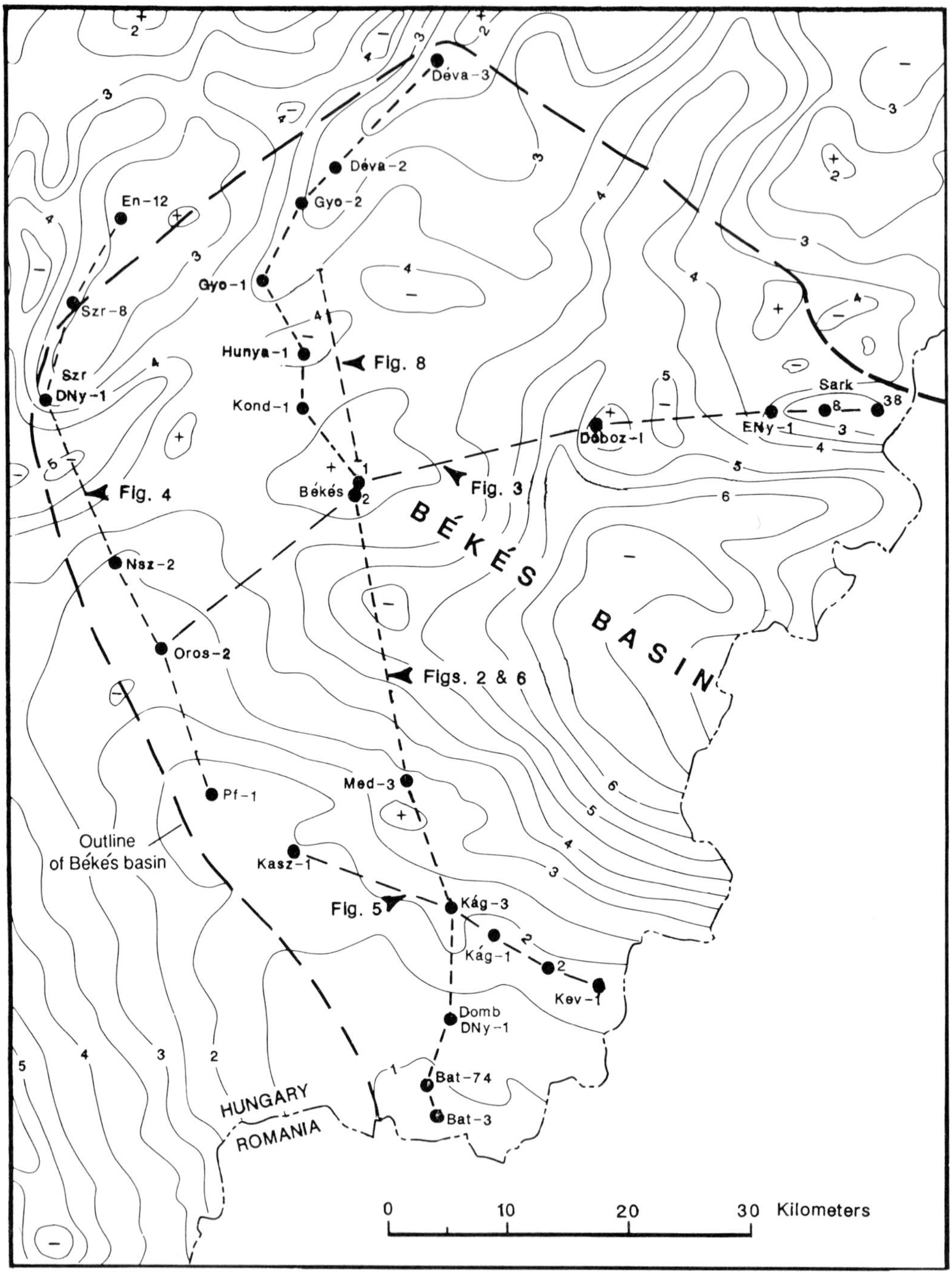

Figure 1. Index map of the Békés basin showing locations of cross sections (Figures 2-6 and 8). Isopachs are of Neogene rocks, most of which are of Pannonian age. Thickness is in kilometers, with a 500 m contour interval. Isopachs essentially represent the structural configuration of the top of the Mesozoic or older basement rocks measured from a datum 100 m above sea level. Well names: En: Endröd; Déva: Dévaványa, Gyo: Gyoma, Szr: Szarvas, Szr-DNy: Szarvas SW, Nsz: Nagyszénás, Kond: Kondoros, Oros: Orosháza, Pf: Pusztaföldvár, Kasz: Kaszaper, Med: Medgyesbodzás, Kág: Kúnágota, Domb: Dombegyháza, Bat: Battonya, Sark: Sarkadkeresztúr, Sark-ENy: Sarkadkeresztúr NW.

and seismic characteristics of the Pannonian sequence in eastern Hungary. This report describes the stratigraphic framework and facies distribution of these rocks in the Békés basin as interpreted from seismic reflection profiles and four well-log lithostratigraphic cross sections correlated with the seismic data.

The Pannonian sequence was deposited in a wide range of depositional environments, ranging from deep-basinal brackish/lacustrine to basin slope to shallow lake and delta plain to alluvial plain. Essentially one major depositional cycle is represented; however, on the basis of magnetostratigraphy, Elston and others (this volume) and Pogácsás and others (this volume) recognized several hiatuses or disconformities associated with lake-level changes and a disconformity between the alluvial-plain Zagyva Formation and overlying Nagyalföld Formation (Fig. 2). Much of the deposition of the Pannonian sequence resulted from deltas rapidly prograding into and across the basin from the northwest and northeast. Because of the rapid deposition of sediments and the lack of lacustrine or nonmarine index fossils and pollen, attempts to establish biostratigraphic zones within the thick sequence have not been successful. The large amount of high-quality, multichannel, common-depth-point (CDP) seismic reflection data, however, provides an excellent method of establishing time correlations between wells and provides a means to decipher depositional environments.

Four well-log-controlled lithostratigraphic cross sections were correlated with seismic data to establish the stratigraphic framework of the Pannonian sequence in the Békés basin (Figs. 2-5). Figure 6 shows the interpreted chrono- or time-stratigraphic relations of the north-south section.

In order to correlate seismic reflection profiles to wells, it is important to have good velocity data on or near the stratigraphic section, because velocity changes occur both laterally and vertically. Velocity control for correlating seismic reflections to the wells along the four cross sections is derived from velocity measurements at 12 wells. A few of these wells are shown on the cross sections. The correlated seismic reflections (time lines) shown on the cross sections were correlated with an accuracy of about 50 m (vertically).

The four seismically correlated well-log cross sections and observations made on the adjacent seismic profiles provide the stratigraphic framework for describing the depositional sequences, lithologies, and the general distribution of the different lithofacies within the Békés basin.

DEPOSITIONAL RELATIONS OF THE PANNONIAN SEQUENCE AS INDICATED BY SEISMIC PROFILES AND WELL-LOG CROSS SECTIONS

Seismic reflections as shown on seismic profiles almost always represent time-bounded surfaces (Mitchum and others, 1977). Exceptions, such as reflections from fault planes and diffraction patterns, can usually be distinguished from true bedding reflections in the Pannonian sequence. In the Békés basin, the reflections can be separated vertically into three groups: (1) topset, (2) foreset, and (3) bottomset reflections or beds (Fig. 7). The topset reflections represent beds deposited on an almost flat surface. Foreset reflections (often referred to as clinoforms) represent beds deposited on the inclined slope of the basin (or the prodelta slope), and the bottomset reflections represent beds deposited on the bottom of the basin, whose original surface of deposition was nearly horizontal.

Correlations of well logs and cores with seismic profiles indicate that the reflection patterns within each of these three main reflection groups contain characteristics that can be related to lithologies and/or depositional environments. In general, the seismic facies patterns correspond to the different depositional environments, which in turn reflect established formational or lithogenetic terminologies.

Topset (Shallow Lake-Delta Plain-Alluvial Plain) Reflections

The lower 300-450 m (200-300 milliseconds (ms) two-way time on seismic profiles) of the 1000-2000 m (1000-1900 ms)-thick section of topset reflections generally consists of fairly high-amplitude, fairly continuous reflections (Fig. 8). Well logs indicate this to be a section of sandstone and shale, or clayey marl, in which the sandstone units are as much as 10 to 20 m thick and constitute about 50 or 60% of the section. Based on position in the sequence, this part of the section is interpreted to represent lower delta-plain and shallow-lake deposits. The sandstone units are interpreted as distributary-channel-mouth bars and distributary-channel sandstones. Reflections above this part of the section are more discontinuous, of moderate and variable amplitude, and many show a hummocky pattern (Fig. 8). Well logs

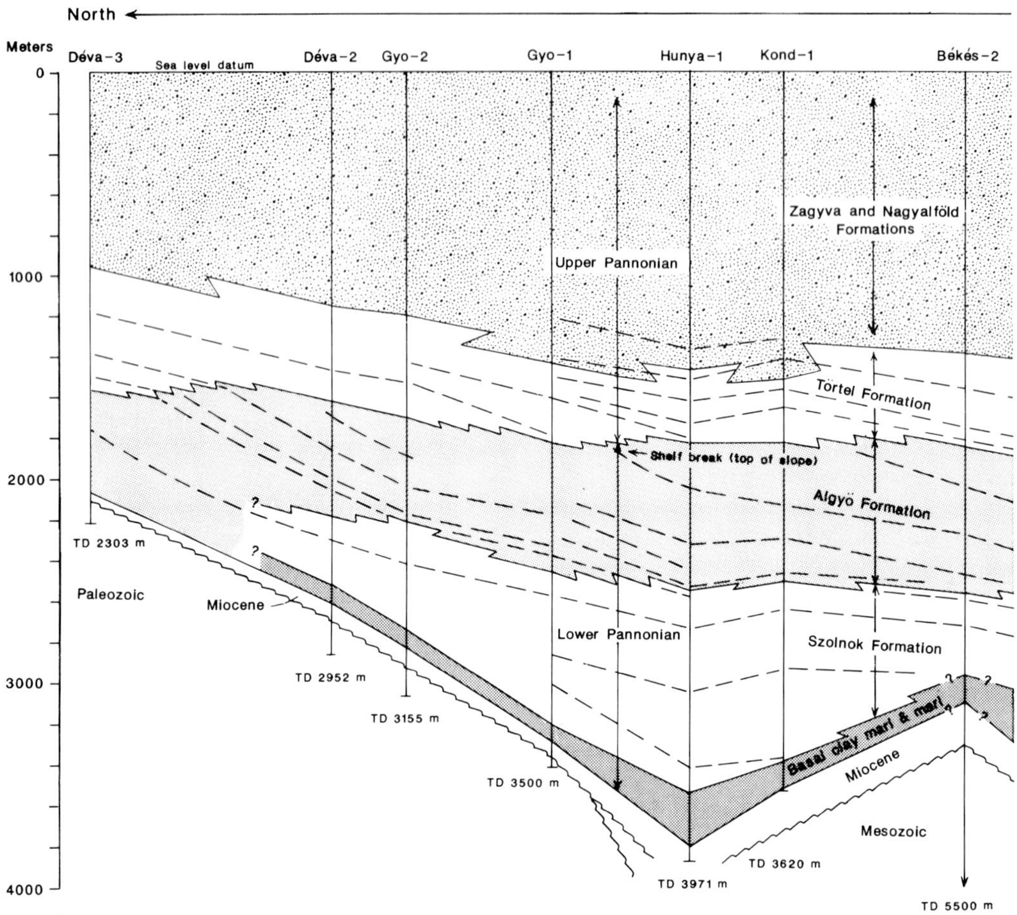

Figure 2. North-south lithostratigraphic cross section across the Békés basin showing depositional relations of the Pannonian sequence. Base of stippled pattern corresponds to the top of the Törtel Formation as determined from well logs. Location of section is shown in Figure 1. Gyo: Gyoma; Kond: Kondoros; Déva: Dévaványa; Med: Medgyesbodzás; Kág: Kunágota; Domb: Magyardombegyháza; Bat: Battonya. Figure is continued on the next page.

indicate an interbedded sandstone and clayey marl section in which the sandstone beds are less than 5 or 10 m thick. This section is interpreted to represent upper delta-plain or alluvial-plain deposits. Inland swamp deposits are probably also present.

The topset interval corresponds to the upper Pannonian (Pa_2) lithogenetic unit of Szalay and Szentgyörgyi (1979), which includes, in ascending order, the Törtel, Zagyva, and Nagyalföld Formations. Elston and others (this volume) recognized a disconformity between the Zagyva and overlying Nagyalföld Formations on the basis of magnetostratigraphy. The differentiation of individual formations, using seismic data within this part of the section, is difficult if not impossible. The lower 200-300 ms of more continuous reflections corresponds to at least part of the Törtel Formation. As shown in Figures 2-5, the contact between the Törtel and overlying formations does not necessarily follow a pattern of prograding facies.

Clinoform (Prodelta Slope) Reflections

The clinoform or prodelta-slope reflection package covers an interval of 300 to 600 ms (500 to 900 m thick). The thinner intervals occur on, or adjacent to basement highs. These reflections range from moderate to low

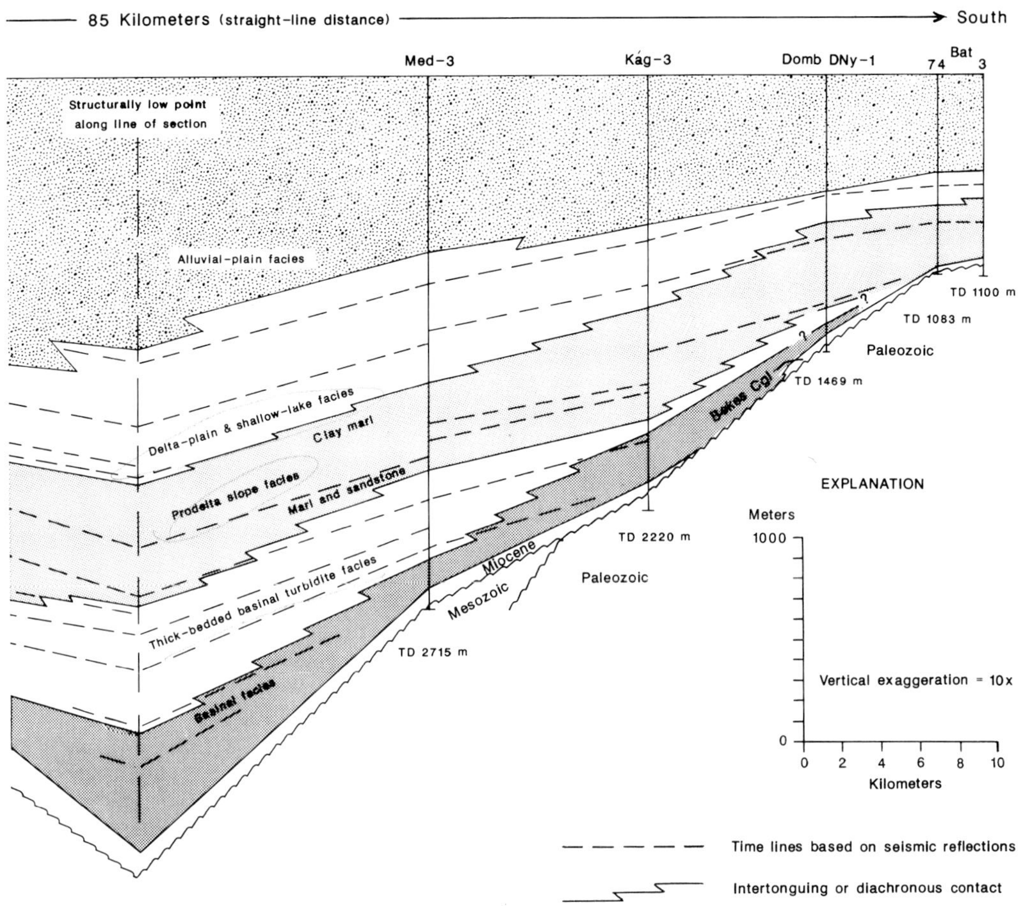

Figure 2. Continued from previous page.

amplitude and are continuous to discontinuous to locally chaotic (Fig. 8). On seismic profiles oriented within 50 or 60 degrees of the direction of depositional dip, the reflections appear as dipping clinoforms. However, on seismic profiles oriented normal to depositional dip, the clinoform reflections appear roughly parallel with the overlying topset and underlying bottomset reflections. The internal reflection pattern within the clinoform sequence, however, is one of discontinuous, hummocky reflections. The clinoform beds decrease in steepness from top to bottom, grading asymptotically at their base into bottomset reflections (Fig. 8). Well logs indicate the upper half of the section to be dominantly shale or clayey marl with a few thin sandstone interbeds. The lower half of the section contains increasing amounts of thicker units of sandstone, although it is generally mostly shale or marl. This part of the spontaneous potential (SP) log response is weak or reversed.

The clinoform reflections are interpreted to represent a combination of prodelta hemipelagic clayey and marl, and slump and proximal turbidity current deposits of sand and clay marl. The lesser amount or absence of sand in the upper part is interpreted to be due to the sand having been transported down the slope and into deeper parts of the basin by turbidity currents; the sand bypassed the steeper, upper part of the slope. Chaotic or disrupted reflections represent slumped deposits.

The clinoform part of the section approximately corresponds to the Algyö Formation, the upper part of the lower Pannonian, or part of the $Pa_{1(2)}$ lithogenetic unit of Szalay and Szentgyörgyi (1979).

In most places the reflections show an abrupt change from topset bedding above to foreset bedding below. This type of pattern is called an oblique progradational

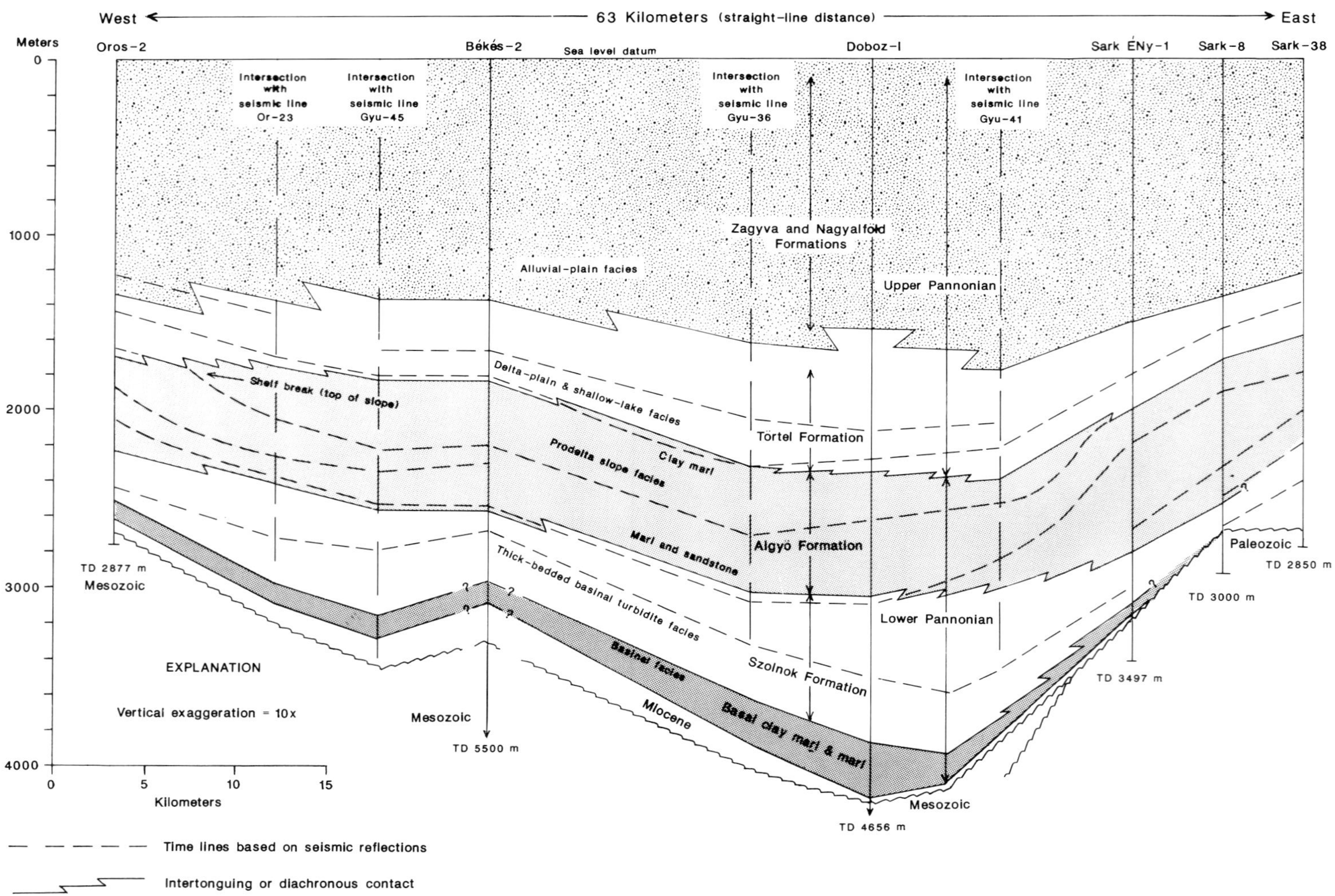

Figure 3. East-west lithostratigraphic cross section across the Békés basin showing depositional relations of the Pannonian sequence. Base of stippled pattern corresponds to the top of the Törtel Formation as determined from well logs. Location of section is shown in Figure 1. Sark: Sarkadkeresztúr; Oros: Orosháza.

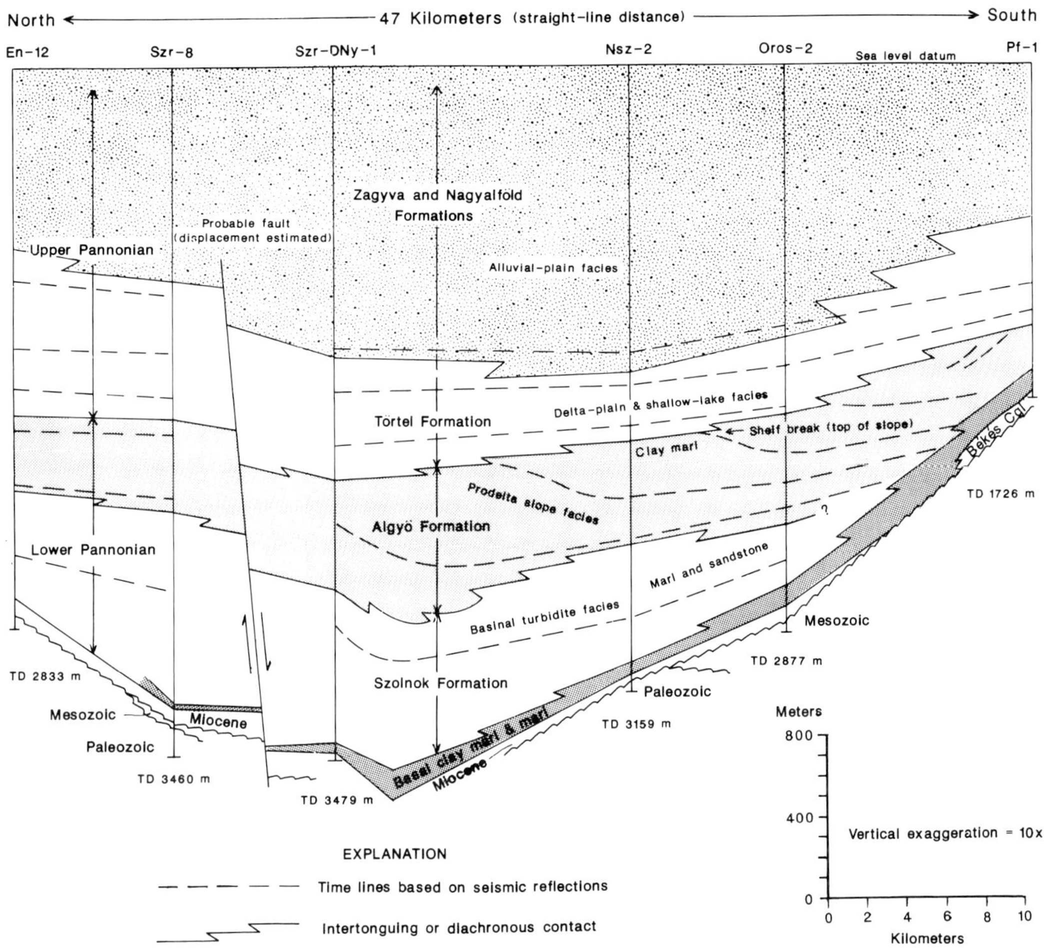

Figure 4. North-south lithostratigraphic cross section along west side of the Békés basin showing depositional relations of the Pannonian sequence. Base of stippled pattern corresponds to the top of the Törtel Formation as determined from well logs. Location of section is shown in Figure 1. Wells: En: Endröd; Szr: Szarvas; Nsz: Nagyszénás; Oros: Orosháza; Pf: Pusztaföldvár.

configuration and implies rapid outbuilding of the delta (Mitchum and others, 1977, p. 128). A more gradual change from topset to clinoform reflections, where reflections can be traced from a topset position to a clinoform position is called a sigmoidal progradational configuration and implies a low sediment supply (an interdeltaic area) and relative basin subsidence (a transgression) (Mitchum and others, 1977, p. 125). Both types of configurations are present in the Békés basin (Fig. 7).

Significance of the Shelf Break

The inflection point that marks the change from topset reflections to clinoform reflections represents the shelf break (or shelf-slope break) at the time of deposition (Figs. 7 and 8). This inflection point is important because shelf breaks are related to a critical water depth, which is probably related to the storm wave base of the body of water in which the sediments were deposited. Because of a limited wind fetch and probable limited current

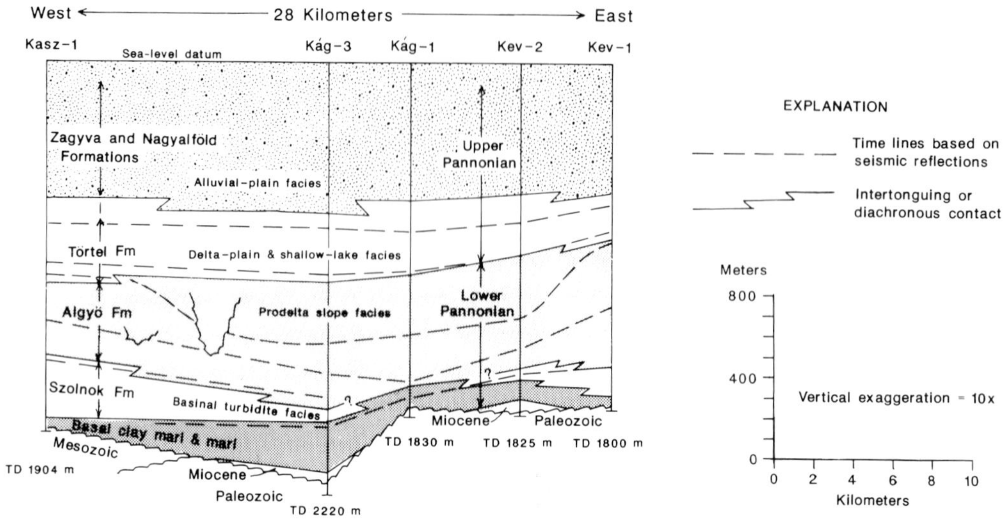

Figure 5. East-west lithostratigraphic cross section across the southern part of the Békés basin showing depositional relations of the Pannonian sequence. Base of stippled pattern corresponds to the top of the Törtel Formation as determined from well logs. Location of section is shown in Figure 1. Kasz: Kaszapér; Kág: Kunágota; Kev: Kevernes wells

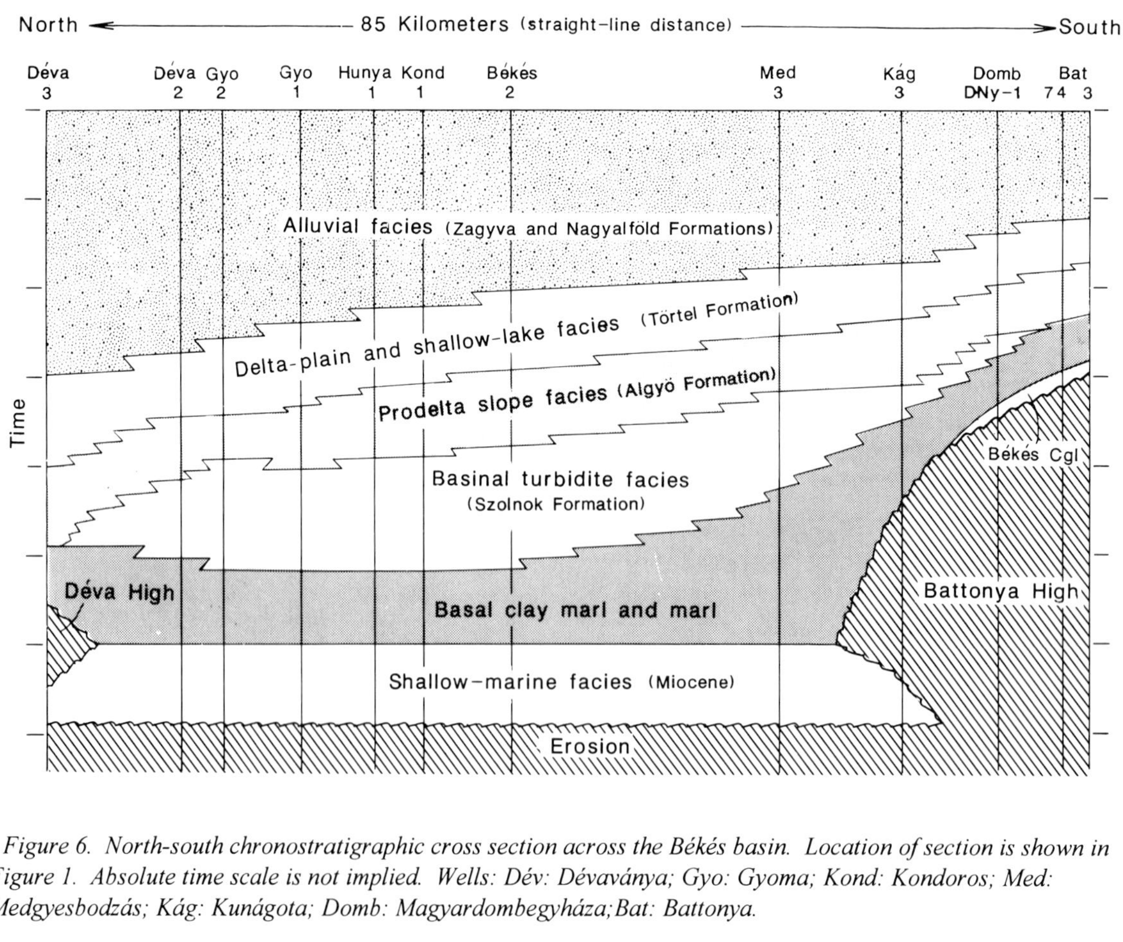

Figure 6. North-south chronostratigraphic cross section across the Békés basin. Location of section is shown in Figure 1. Absolute time scale is not implied. Wells: Dév: Dévaványa; Gyo: Gyoma; Kond: Kondoros; Med: Medgyesbodzás; Kág: Kunágota; Domb: Magyardombegyháza; Bat: Battonya.

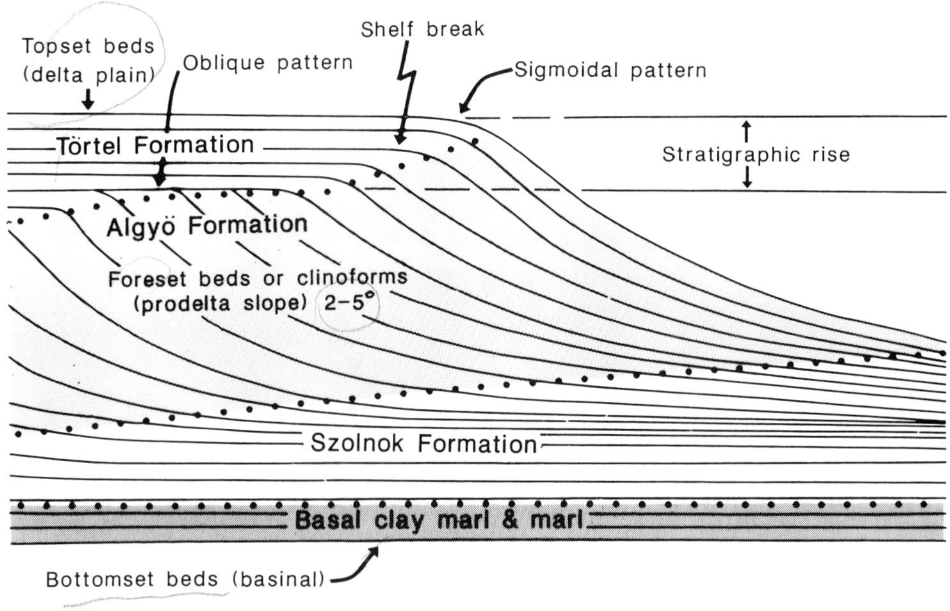

Figure 7. Schematic diagram showing depositional patterns indicated by seismic reflections, relationship of formational units to depositional patterns, and stratigraphic terminologies used in the text.

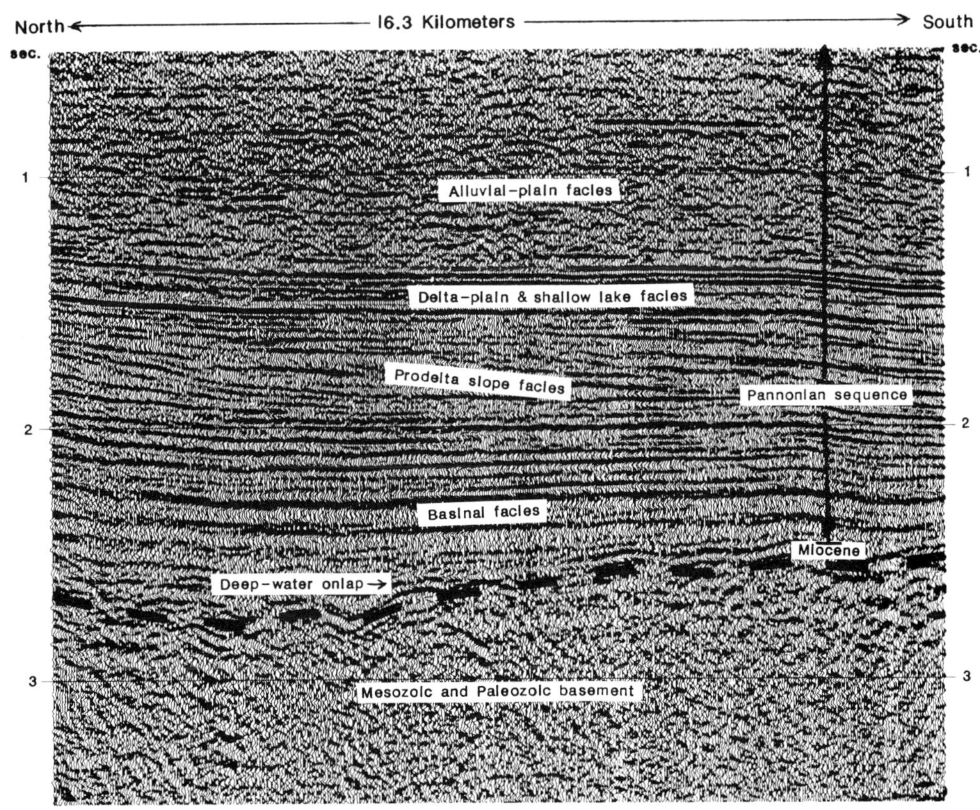

Figure 8. Seismic profile Gyu-30 (north half) showing reflection characteristics of different facies of the Pannonian sequence. Location of section is shown in Figure 1.

action in an inland lake, the depth of water at the shelf break in the Pannonian Lake was probably not great, say less than 35 m. The shelf break separated beds deposited in shallow lake and subaerial environments (Törtel Formation) from those deposited below storm wave base and more basinal environments (Algyö Formation). Because the sands (or sandstones) above the shelf break (Törtel Formation) are unconsolidated or friable, core recovery has been poor (Phillips and others, this volume). Nevertheless, these sands are presumed to have been deposited by traction currents in streams or rivers and as mouth bars or sheet sands in shallow-lake environments. Sands below the shelf break were deposited by slumping and turbidity currents (Phillips and others, this volume). Therefore geometries of sand bodies above the shelf break are different from those below.

Based on the correlation of seismic profiles with well logs, the position of the shelf break on seismic profiles is at, or a short distance below, the base of the Törtel Formation. At the Doboz-I well (Fig.3), however, the shelf break, as indicated by the seismic data, seems to be as much as 200 m below the base of the Törtel Formation. It may be that the contact at the base of the Törtel Formation was placed too high, owing to poor or attenuated-response on SP and resistivity logs, or possibly a greater than normal thickness of shale deposits could have existed above the shelf break. A velocity survey was made in this well, thus, there cannot be a significant error affecting the calculated position of the shelf break as determined from seismic data.

The seismic profiles show gradual to abrupt upward shifts in stratigraphic position of the shelf break in a down-depositional dip direction. This is called the stratigraphic rise (Fig. 7) and indicates relative rising of the lake level (possibly due to local basinal subsidence). The more abrupt rises probably represent minor transgressions. These upward shifts or stratigraphic rises also cause the formational boundaries to shift upward in the direction of progradation, thus accentuating the diachronous nature of the depositional units (Figs. 2 and 6).

Unlike sea level, lake level is controlled by a spill point (unless the lake is totally enclosed) which in turn is affected by erosion, local tectonics or igneous activity, or even depositional barriers. Additional factors affecting the lake level may be climatic variations in the drainage basin, or sediment volume replacing water volume. Thus, the lake level can vary, through time, as a result of many either independent or interrelated factors.

Basin Slope and Water Depth Measurements

The upper part of the clinoform (prodelta slope) reflection is usually its steepest part (Fig. 8). Calculations made on a few of the steeper reflections indicate inclinations of 3.0 to 4.5 degrees relative to topset beds. Because these beds have been compacted, the slope angle at the time of deposition was somewhat steeper than that measured, possibly by a degree or more.

The relief or height of the clinoform beds is an indication of the depth of the water into which the deltas prograded. Calculations of the relief (or height) of some of the largest clinoform beds, which are about 10 km southeast of the Doboz-I well, show relief of about 475 m. Allowing for compaction and the water depth at the top of the clinoform beds (the shelf break), the water depth had to be greater than 500 m, and probably closer to 600 m in the deeper part of the basin at the time of deposition. The water depth during deposition of lower basinal beds may have been greater than 600 m, but the approximate depth is not possible to determine within the Békés basin because these basinal beds cannot be traced up the clinoform beds to a shelf break. The clinoforms and shelf break for these basinal beds are north of and outside the Békés basin.

Bottomset (Basinal) Reflections

The bottomset or basinal reflection intervals along the lines of well-log cross sections range in thickness from zero on basement structural highs, which are onlapped by basinal and clinoform beds, to about 1200 m in the Hunya-1 well. The interval is thicker in the structurally low areas that are not crossed by the cross sections. The reflections are parallel, range from low to high amplitude, and are continuous for distances of several kilometers, although their amplitudes may change (Fig. 8). The lowermost part of the bottomset reflection zone commonly contains one or two high-amplitude reflections. Well logs and cores indicate this part of the section consists of shale or clayey marl, and marl, about 50 to 300 m thick, and is designated as such on the cross sections. This part of the section corresponds to the Nagykörü and/or Tótkomlós Formations, or to the $Pa_{1(1a)}$ lithogenetic unit of Szalay and Szentgyörgyi (1979).

The remaining thick overlying bottomset reflections are mostly of very low amplitude in the lower part and medium to high amplitude in the upper part of the section. Well logs indicate this entire interval to be interbedded sandstone and marl; the lower half or two-thirds of the interval consists of 50 to 70% sandstone in units as thick as 10 to 30 m, and the remaining upper part contains slightly less sandstone in beds less than 10 m thick. Analyses of cores indicate that the sandstone is mostly very fine-grained and of distal turbidite origin (Phillips and others, this volume). The thick sandstone units are composed of amalgamated thinner beds.

Resistivities of the sandstones are fairly high, and many formation tests and core analyses indicate that they are of low porosity and permeability. Recoveries are mostly small amounts of water and traces of gas.

The thick basinal sandstone and marl section approximately corresponds to the Szolnok Formation (Figs. 2-5) or to part of the $Pa_{1(2)}$ and the $Pa_{1(1b)}$ lithogenetic units of Szalay and Szentgyörgyi (1979).

The contact between the Szolnok Formation and overlying Algyö Formation on Figures 2-5 was determined from well logs and placed at the point where clayey marls and thinner bedded sandstone (Algyö Formation) grade downward into thicker bedded sandstone and clayey marl (Szolnok Formation). In many wells, the contact is poorly defined on well-logs.

The Basal Clayey Marl and Marl

The basal clayey marl and marl comprise an important unit because it is a fair-to-good, oil-prone source rock in some parts of the basin (Clayton and others, this volume). This unit is diachronous and laps upward onto and over topographic highs on the lake floor. In the deeper parts of the basin, the clayey marl is black, owing to lack of circulation and a euxinic environment, whereas on the highs the clayey marl is yellowish gray owing to more oxygenated conditions (Phillips and others, this volume). As shown in Figure 6, the lower part of this unit is interpreted to be a distal pelagic deposit that conformably overlies marine Miocene beds; the upper part is interpreted to be more diachronous because it represents initial turbidite deposition. The basement highs were exposed to subaerial erosion in early Pannonian time and in some areas, a basal conglomerate, called the Békés Conglomerate, was deposited (Figs. 2, 4, and 6). The conglomerate probably accumulated as a result of wave action along the transgressing shoreline. The fact that the Békés Conglomerate is best developed on the north sides of the highs suggests that the north sides were exposed to greater wave action. As the highs subsided below wave base, clayey marl was deposited before the turbidites overtopped the highs. Thus, the clayey marls on the highs are younger than those deeper in the basin. However the younger and older clayey marls probably form a continuous deposit (Fig. 6).

SUMMARY OF SANDSTONE FACIES AND DISTRIBUTION

Sandstone occurs in many intervals in the sedimentary section, but the main concentrations occur as turbidites in the basinal facies, excluding the basal clayey marl, and in the shallow-lake, delta-plain, and alluvial-plain facies. The prodelta slope contains much less sandstone. The following discussion of sandstone facies and distribution, in ascending order, is based on the interpretation of well logs and on observations made during this study of seismic correlations of the four well-log lithostratigraphic cross sections.

Basinal Turbidites (Szolnok Formation)

The sandstone in the lower two-thirds of the basinal facies (Szolnok Formation or $Pa_{1(1b)}$) occurs as thick, amalgamated units of mostly very fine grained turbidite sandstone (Phillips and others, this volume). Well logs indicate that several of the sandstone units are as thick as 30 m and that sandstone comprises 50 to 70 percent of the section, the remainder being shale or marl. The upper third of the formation contains slightly less sandstone (40-50 percent) in beds generally less than 10 m thick.

The upper third of the Szolnok Formation and all of the formations on the flanks of the basement highs can be traced by seismic reflections updip into clinoform beds. However, the lower two-thirds, which occur in more basinal positions, cannot be traced to clinoform beds within the Békés basin. The delta front or fronts from which these basinal facies were derived was located farther to the northwest or northeast. Thus, the lower beds are more distal in origin than the upper beds. The great percentage of sand in the lower part of the section suggests that the delta system at that time was either more rich in sand, as compared to later delta systems, or more likely, the turbidites were deposited by down-slope currents in troughs between basement highs. Such a process would relatively rapidly fill the low areas with turbidite sands.

Seismic reflection data indicate that the basinal turbidite sandstone beds, both individually and as a unit, are continuous and widespread throughout low areas of the

Békés basin. The sandstone beds onlap and pinchout against flanking highs.

Prodelta-Slope Sandstones (Algyö Formation)

Sandstone beds in the prodelta-slope environment (Algyö Formation) are generally less than 10 m thick and constitute only 10 to 20% of the section. The percentage of sand is variable, but it is almost always higher in the lower part of the formation. Because of the variations of current intensity across the slope during deposition, together with the steep inclination of the depositional surface, these sand bodies are probably discontinuous and many were likely deposited by slumping or gravity sliding off the delta front.

Delta-Plain/Delta-Front Sandstones (Törtel Formation)

The delta-plain and delta-front (shallow-lake) sandstones constitute about 50 to 60% of the delta-plain deposits. The sand bodies occur as distributary-mouth bars and distributary-channel sands in units as thick as 20 m, although commonly the sand bodies are only about 10 m thick. Because of their friability, few cores are available for examination; nonetheless, the sands are reported to be very fine to medium grained (Phillips and others, this volume). Their reservoir properties are much better than those of the basinal sandstones.

Distributary-mouth bars probably thicken and thin along depositional strike, depending on their proximity to the distributary channel. Distributary-channel sand bodies probably had limited lateral continuity originally, but they probably merge laterally with younger or older sand bodies. Channel sands may be separated from the main sand bodies by another channel that was later abandoned and filled with clay. Unless these features are large, they cannot be recognized on seismic records.

Alluvial-Plain Sandstones (Zagyva and Nagyalföld Formations)

The alluvial-plain deposits contain a variable amount of sand or sandstone. (Because of lack of compaction and cementation, many of the arenaceous clastics in this part of the section are sands rather than sandstones). Examination of the well logs along the cross sections indicates that sand or sandstone comprises not more than 50 percent of the section and bed thickness is generally less than 5 or 10 m, although a few thicker beds are locally present. Because of their shallow burial depth, sands or sandstones of the alluvial-plain facies probably have good reservoir properties.

REFERENCES

Clayton, J.L., Koncz, I., and King, D.J., and Tatár, É., Organic geochemistry of crude oils and source rocks, Békés basin; this volume, 161-185.

Elston, D.P., Lantos, M., and Hámor, T., High resolution polarity records and the stratigraphic and magnetostratigraphic correlation of late Miocene and Pliocene (Pannonian s.l.) deposits of Hungary; this volume, 111-142.

Jámbor, Á., Balázs, E., Balogh, K., Bérczi, I., Bóna, J., Horváth, F., Gajdos, I., Geiger, J., Hajós, M., Kordos, L., Korecz, A., Korecz-Laki, I., Korpás-Hódi, M., Kőváry, J., Mészáros, L., Nagy, E., Németh, G., Nusszer, A., Pap, S., Pogácsás, Gy., Révész, I., Rumpler, J., Sütö-Szentai, M., Szalay, Á., Szentgyörgyi, K., Széles, M., and Völgyi, L., 1987, General characteristics of Pannonian s.l. deposits in Hungary; Proc. 8th Cong., Reg. Comm. on Mediterr. Neogene Stratigraphy, Budapest, September 15-22, 1985; Hung. Geol. Inst. Ann. Rept., 70, 155-167.

Mitchum, R.M., Vail, P.R., and Sangree, J.B., 1977, Stratigraphic interpretation of seismic reflection patterns in depositional sequences; *in* Payton, C.E. (ed.), Seismic stratigraphy --applications to hydrocarbon exploration; Amer. Assoc. Petrol. Geol. Memoir. 26, 117-133.

Phillips, R.L., Révész, I., and Bérczi, I., Lower Pannonian deltaic-lacustrine processes and sedimentation, Békés basin; this volume, 67-82.

Pogácsás, Gy., 1987, Seismic stratigraphy as a tool for chronostratigraphy -- Pannonian basin; Proc. 8th Cong. Reg. Comm. on Mediterr. Neogene Stratigraphy; Budapest, September 15-22, 1985; Hung. Geol. Inst. Ann. Rept., 70, 55-63.

Pogácsás, Gy., and Révész, I., 1987, Seismic stratigraphic and sedimentological analysis of Neogene delta features in the Pannonian Basin; Proc. 8th Cong. Reg Comm. on Mediterr. Neogene Stratigraphy, Budapest, September 15-22, 1985; Hung. Geol. Inst. Ann. Rept., 70, 267-273.

Pogácsás, Gy., Mattick, R.E., Elston, D.P., Hámor, T., Jámbor, Á., Lakatos, L., Lantos, M., Simon, E., Vakarcs, G., Várkonyi, L., and Várnai, P., Correlation of seismo- and magnetostratigraphy in southeastern Hungary; this volume, 143-160.

Szalay, A., and Szentgyörgyi, K., 1979, Subdividing the lithology of Pannonian deposits based on well data, and using trend analysis; Hung. Acad. Sci., Rept. of the Earth and Mining Science Div., 12, (4), 401-423.

6 High Resolution Polarity Records and the Stratigraphic and Magnetostratigraphic Correlation of Late Miocene and Pliocene (Pannonian, s.l.) Deposits of Hungary

Donald P. Elston[1], *Miklós Lantos*[2] *and Tamás Hámor*[2]

ABSTRACT

Stratigraphic records from four widely spaced holes, continuously cored from the surface of the Great Hungarian Plain to depths of 1.2 to 2 km, have been correlated and placed in relative stratigraphic positions by means of seismic profiles. Polarity zonations for two 2-km-thick cored sections were correlated with the polarity time scale for much of late Miocene time, and two cored sections, 1.2 km thick, were correlated with the polarity time scale for much of Pliocene and Pleistocene time. The high-resolution magnetostratigraphic records from the four cored sections contain many more polarity reversals than the accepted polarity time scale for the late Miocene and Pliocene; because of this, only the broader polarity intervals in the drill cores are correlated with the polarity time scale. The new seismic-stratigraphic and magnetostratigraphic correlations have led to revised correlations of Pannonian stratigraphic units in the subsurface, and to a new chronostratigraphic framework and model for the manner and timing of accumulation of late Miocene and Pliocene deposits in the Pannonian Basin. The high resolution polarity zonations also provide new information on the detailed character of the polarity time scale for parts of late Miocene, Pliocene, and Pleistocene time.

INTRODUCTION

Pannonian Basin

The Pannonian Basin (Fig. 1) is a complex depression, partly bordered by the Carpathian Mountains, containing deposits of middle Miocene to Pleistocene age. The Carpathian Mountains, orogenically active during early and middle Miocene time, developed as a result of compression arising partly from encroachment of the African plate on the Eurasian plate on the south, and partly from interaction of the Eurasian plate with the Anatolian plate on the east (see, for example, Hámor, 1984). The Pannonian Basin began to subside following deformation during Oligocene and early Miocene time, with subsidence continuing through the remainder of Miocene, Pliocene, and Pleistocene time. Some geoscientists have referred to the entire intra-Carpathian late Paleogene and Neogene basin as the Pannonian Basin (e.g., Balla, 1987a). Others workers (e.g., Jámbor and others, 1987) have applied the general term Pannonian (*sensu lato*) to

[1] U.S. Geological Survey, Flagstaff, Arizona 86001, USA
[2] Hungarian Geological Institute, H-1442, Budapest, Hungary

P. G. Teleki et al. (eds.), Basin Analysis in Petroleum Exploration, 111–142.

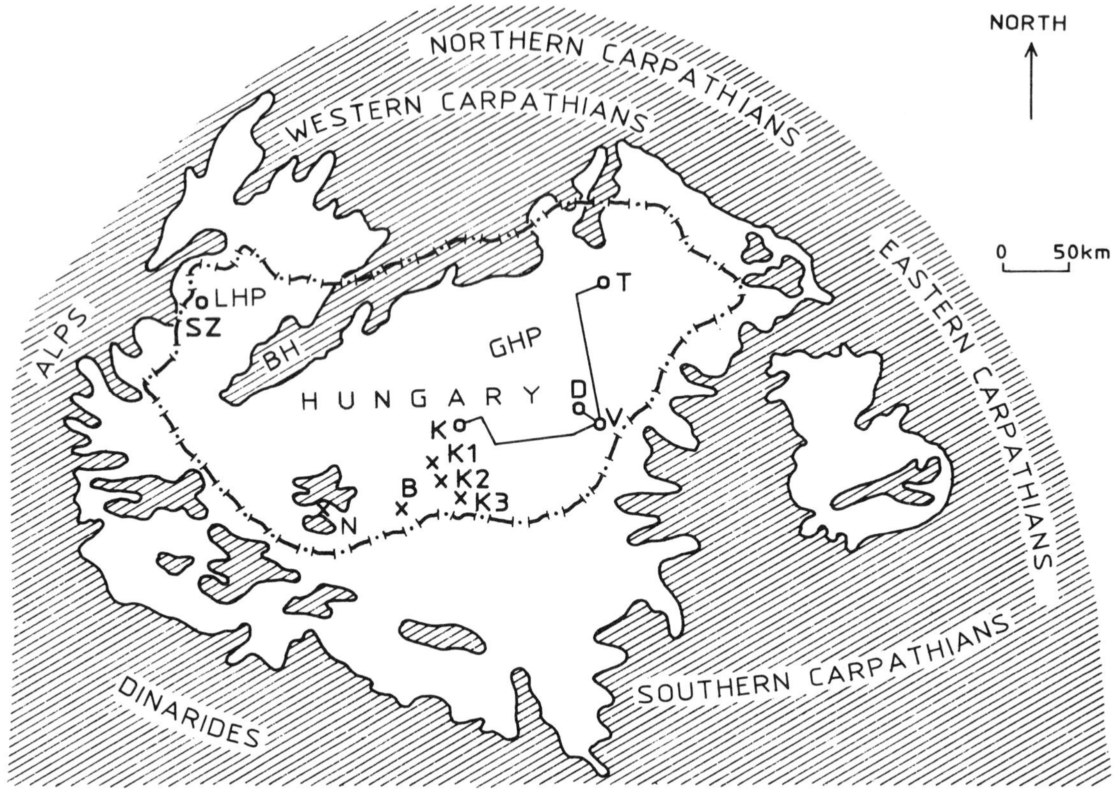

Figure 1. Map showing distribution of late middle and late Miocene to Pliocene (Pannonian) deposits (without pattern) and outline of Hungary (dash-dot symbol). Great Hungarian Plain: GHP; Little Hungarian Plain: LHP; Balaton Highlands area: BH. Open circles show location of core holes where geomagnetic polarity studies were performed. T: Tiszapalkonya; K: Kaskantyú-2; D: Dévaványa; V: Vésztö; Sz: Szombathely. Drill hole from which isotopic data are available is shown by X (N: Nagykozár-2; B: Bácsalmás-1; K3: Kiskunhalas-3; K1: Kecel-1; K2: Kecel-2). Lines connecting the Tiszapalkonya, Vésztö, Dévaványa, and Kaskantyú wells show locations of seismic profiles.

deposits that accumulated in the basin during middle and late Miocene and Pliocene time.

The Pannonian Basin contains a thickness of 0.2 to 7 kilometers of Miocene-Pliocene sediments. The dominantly clastic deposits accumulated initially in a shallowing, semi-enclosed to closed inland sea, - an environment that changed gradually from brackish water to lacustrine, and then to subaerial, fluvial conditions of deposition (Bérczi and Phillips, 1985). Subsurface study of the Pannonian sedimentary section in Hungary has been particularly intense because of its hydrocarbon resources (Dank, 1987; Molnár and others, 1987). However, in spite of extensive study, the age and correlation of many of the Pannonian stratigraphic units encountered in the subsurface remained uncertain because, (1) time-diagnostic fossils are lacking (Cicha and Senes, 1968; Steininger and Senes, 1971), and (2) only sparse volcanic deposits amenable for isotopic dating are present (see Balogh and Jámbor, 1985; and a compilation by Vass and others, 1987).

Assumptions Underlying Previous Correlations in the Pannonian Basin

The general lack of temporal control in pre-Pleistocene deposits underlying the Great Hungarian Plain required that the discrimination and correlation of units identified in drill cores, and their assignment to discrete parts of the geologic time scale, be based mainly on lithofacies (rock units representing the same depositional environments).

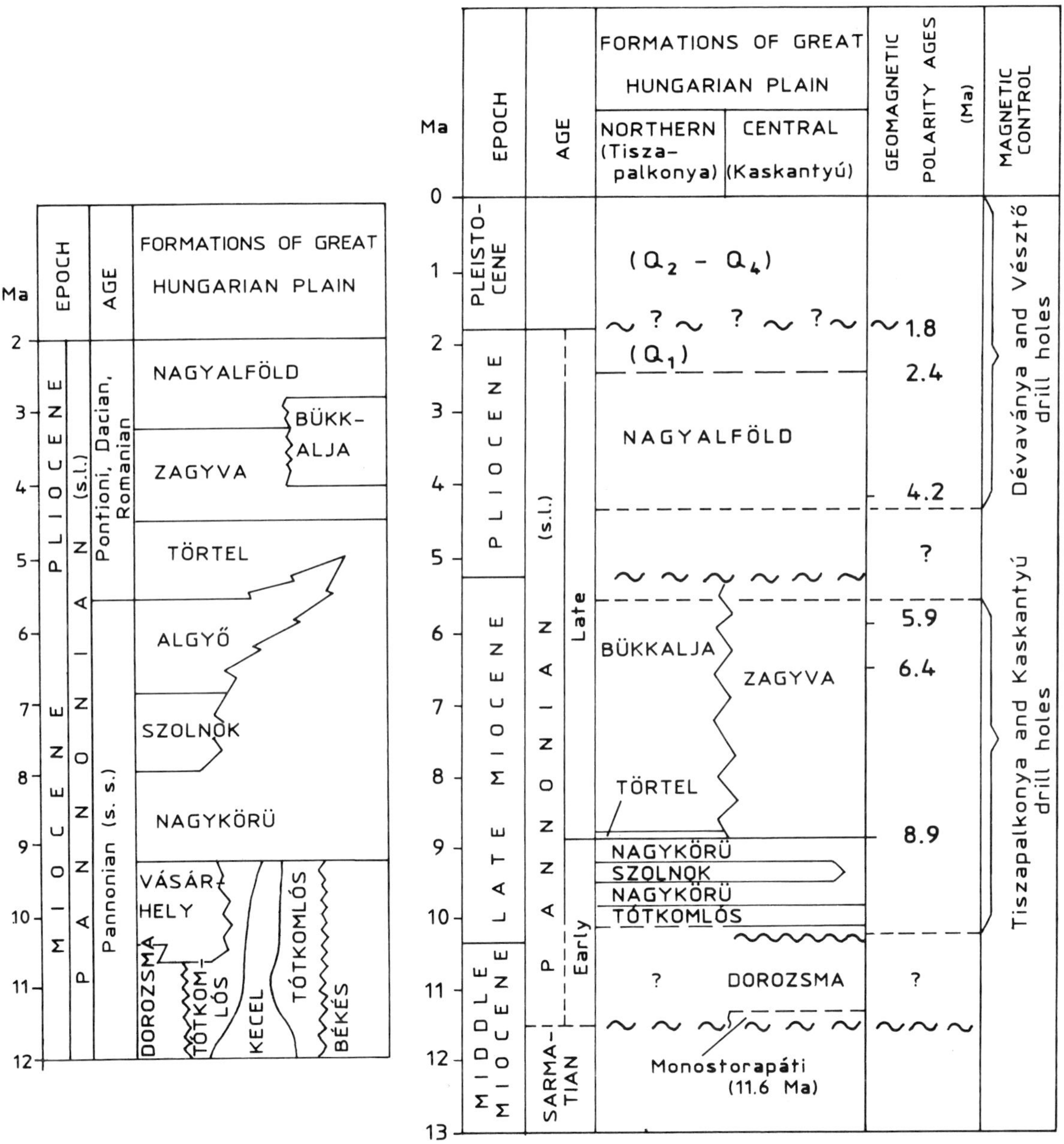

Figure 2. Lithostratigraphic and temporal correlations of Pannonian formations underlying the Great Hungarian Plain: (a) accepted by the Stratigraphic Commission of Hungary, 1983 (after Jámbor and others, 1987); and (b) current stratigraphic framework developed from a correlation of formations penetrated in four deep core holes with geomagnetic time scale. Neogene time scale is from Berggren and others (1985).

By working down-section and backwards in time from the present across the Neogene time scale, implicit assumptions as to age and contemporaneity of units were employed for regional correlations. These included, (1) that thicknesses of deposits could be generally equated with time, (2) that the deposits accumulated more or less uniformly across the basin, and (3) that no significant regional unconformities were present. Subsurface correlations employing the above assumptions also appealed to facies changes to account for differing lithologic characteristics lying at similar depths beneath the surface of the Great Hungarian Plain. The foregoing assumptions led to the correlations and inferred ages of deposits shown in Figure 2a. In support of this model, we note that thickness appears to generally equate with time for the Pleistocene and late Pliocene parts of sections penetrated in the Dévaványa and Vésztö core holes (Cooke and others, 1979) drilled in the east-central part of the Great Hungarian Plain (Fig. 1). In view of its apparent validity, this approach was used in an initial attempt to place the Kaskantyú drill core and its polarity zonation in the geologic time scale (Elston and others, 1985; Hámor and others, 1985). This approach proved unsatisfactory and was abandoned once seismic-stratigraphic profiles became available.

Advent of Seismic-Stratigraphic Profiles

With the acquisition of high-resolution seismic-stratigraphic profiles across the Great Hungarian Plain (Pogácsás, 1987; see Pogácsás and others, this volume), the regional character and distribution of Pannonian (*s.l.*) stratigraphic deposits became known in considerable detail. However, the depositional framework deduced from the seismic records could not be dated in an absolute sense until magnetostratigraphic records correlated with the polarity time scale in four deep holes were applied to the seismic records. Even general correlations of polarity zonations in the cores with the polarity time scale allowed geomagnetic time lines to be assigned to stratigraphic horizons in the seismic profiles, and these time lines then were traced in the subsurface across the basin, allowing the ages of different delta-slope and delta-plain deposits defined in the seismic profiles to be determined. This approach assumed that laterally traceable and apparently correlative seismic reflectors represented the same depositional or non-depositional events. The traces of seismic profiles connecting the core sections containing polarity information in the Great Hungarian Plain are shown in Figure 1.

From an integration of stratigraphic, seismic-stratigraphic, and magnetostratigraphic data, geologic and paleomagnetic correlations arose that bear on the accumulation history of late Miocene and Pliocene deposits in eastern subbasins of the Pannonian Basin, i.e., those underlying the Great Hungarian Plain. In addition to providing an improved understanding of the history and environments of deposition, the polarity zonations also provide high-resolution polarity records of the geomagnetic field for parts of late Miocene, Pliocene, and Pleistocene time. Paleomagnetic records from four Hungarian cored sections, if accepted at face value, indicate that the polarity time scale for this interval of time is highly generalized.

GEOLOGIC SETTING

Structure

The Pannonian Basin began to subside in early Miocene time, during the Styrian phase of the Alpine orogeny. Paleozoic and Mesozoic units comprising the basement were subjected to strong extensional, compressional, and rotational tectonic movements that broke the Pannonian Basin into several subbasins, each of which subsided at a different rate. These basins became filled with late Miocene, Pliocene, and Quaternary sediments totaling as much as 7 km in thickness (Kilényi and Rumpler, 1985). Except for a brief episode of uplift in the Bakony Mountains near the end of Pliocene time, locally amounting to as much as 250-300 m (Jámbor, 1980), the Pannonian Basin appears to have subsided rapidly although not everywhere uniformly during late Miocene and Pliocene time. From the analysis presented in this report, an unconformity may lie at or near the Miocene-Pliocene boundary as well as at or near the Pliocene-Pleistocene boundary. The former conceivably could represent a pre-Pleistocene episode of uplift and erosion that interrupted subsidence of the basin.

The Pannonian Basin is underlain by relatively narrow depressions, 4-8 km deep, separated by relatively high-standing ridges having 0.5-2 km of structural relief (e.g., see Körössy, 1980; Pogácsás, 1980; Kilényi and Rumpler, 1985). As interpreted from regional stratigraphic studies refined by seismic stratigraphy, the elongated depressions (or troughs) developed during early and middle Miocene time. These Miocene events set the stage and were responsible for the accumulation of Pannonian (*s.l.*) deposits. Detrital sediments derived from nearby sources accumulated in the troughs, followed by the accumulation of upper Miocene and Pliocene deposits accompanying synsedimentary listric faulting (Hámor, 1984; Pogácsás, 1984). It was a time marked by a progressive deepening of central parts of the Pannonian Basin.

Stratigraphy

Nomenclature

The term Pannonian (*s.l.*) has been applied by Hungarian geologists to strata in the Pannonian Basin that accumulated during late middle Miocene, late Miocene, and Pliocene time. The term Pannonian (*sensu stricto*) has been applied to deposits considered to be only of late Miocene age, whereas deposits considered to be Pliocene in age variously have been assigned to the Pontian, Dacian, or Romanian stages of deposition (Fig. 2). The diversity of classifications proposed by Hungarian and other workers has served to aggravate rather than resolve problems concerned with the age and subsurface correlation of stratigraphic units. A representative nomenclatural framework and grouping of units as depicted by Jámbor and others (1987) is shown in Figure 2a. The correlations shown in Figure 2a are based on lithostratigraphy, inferred facies changes, and inferred ages. They differ markedly from the ages and correlations shown in Figure 2b, arrived at from a correlation of seismic and magnetic stratigraphy. Although the position of the Miocene-Pliocene boundary in the Pannonian section has been narrowed as a result of our study, its position remains to be identified with certainty in the subsurface.

The chart shown in Figure 2b results from the stratigraphic and seismic stratigraphic correlation of formations penetrated in four deep core holes, and the correlation of their polarity zonations with the geomagnetic time scale. These results suggest that major changes are needed in the correlation and age assignments for several units with respect to correlations shown in Figure 2a. In particular, strata of the Nagyalföld Formation appear to occupy a major part of the Pliocene, and have a maximum age of at least 4.25 Ma. The possibility exists that the base of the Nagyalföld Formation may be marked by an unconformity, and that the Miocene-Pliocene boundary may lie near (if not correspond with) the base of this formation. If so, discrepant thicknesses and correlations of units underlying the Nagyalföld Formation may be resolved by appealing to loss of section by erosion at the presumed unconformity. In the classification proposed here, the base of the upper Pannonian Nagyalföld Formation lies near or at the Miocene-Pliocene boundary.

The lignite-bearing upper Pannonian Bükkalja Formation, rather than middle Pliocene in age as depicted in Figure 2a, is interpreted from its polarity zonation and seismic stratigraphic position as having accumulated during a substantial interval of late Miocene time (from about 8.9 to less than 6.4 Ma). The nominal boundary between the lower and upper Pannonian is drawn at the base of the Törtel Formation, which in the northern part of the Great Plain has an age of about 8.9 Ma as determined from magnetostratigraphy. A similar boundary (and age) is recognized on magnetostratigraphic grounds at the base of the Zagyva Formation in the central part of the basin. Lastly, lower Pannonian strata penetrated in the Tiszapalkonya and Kaskantyú core holes, on the basis of their normal polarity, appear to have accumulated entirely during the early part of late Miocene time. Evidence for the correlations and ages is reviewed later.

Geologic characteristics of the core sections

The four core holes studied and evaluated for this report are located in three different sub-basins of the Pannonian Basin (Fig. 1). These continuously cored holes were drilled by the Hungarian Geological Institute as part of a program of regional stratigraphic studies designed to elucidate stratigraphic relations in the Pannonian Basin.

Each basin of the Pannonian Basin system had somewhat different depositional and subsidence histories during late Miocene, Pliocene, and Pleistocene time. Nonetheless, the lithologies of three stratigraphic intervals are common to all four drill-core sections. These are, (1) the late Miocene Zagyva and Bükkalja Formations, (2) the Pliocene Nagyalföld Formation, and (3) the Pleistocene sequence. Additionally, Lower Pannonian stratigraphic units encountered in the lower parts of the Tiszapalkonya and Kaskantyú drill-core sections also correlate lithologically and temporally. The new stratigraphic and temporal correlations are depicted in Figures 2b, 3, and 4.

Dévaványa and Vésztö

The Dévaványa-1 and Vésztö-1 core holes (Fig. 3) were drilled in the central part of the Békés basin, which contains about 3200 m of Pannonian strata in its deepest part. The two core holes, each about 1200 m deep, intersected a fairly complete Pleistocene sequence, 430-500 m thick. The underlying Pliocene Nagyalföld Formation was encountered in both drill holes, but its base was penetrated only in the Dévaványa drill hole. The Vésztö core hole was drilled to somewhat greater depth,

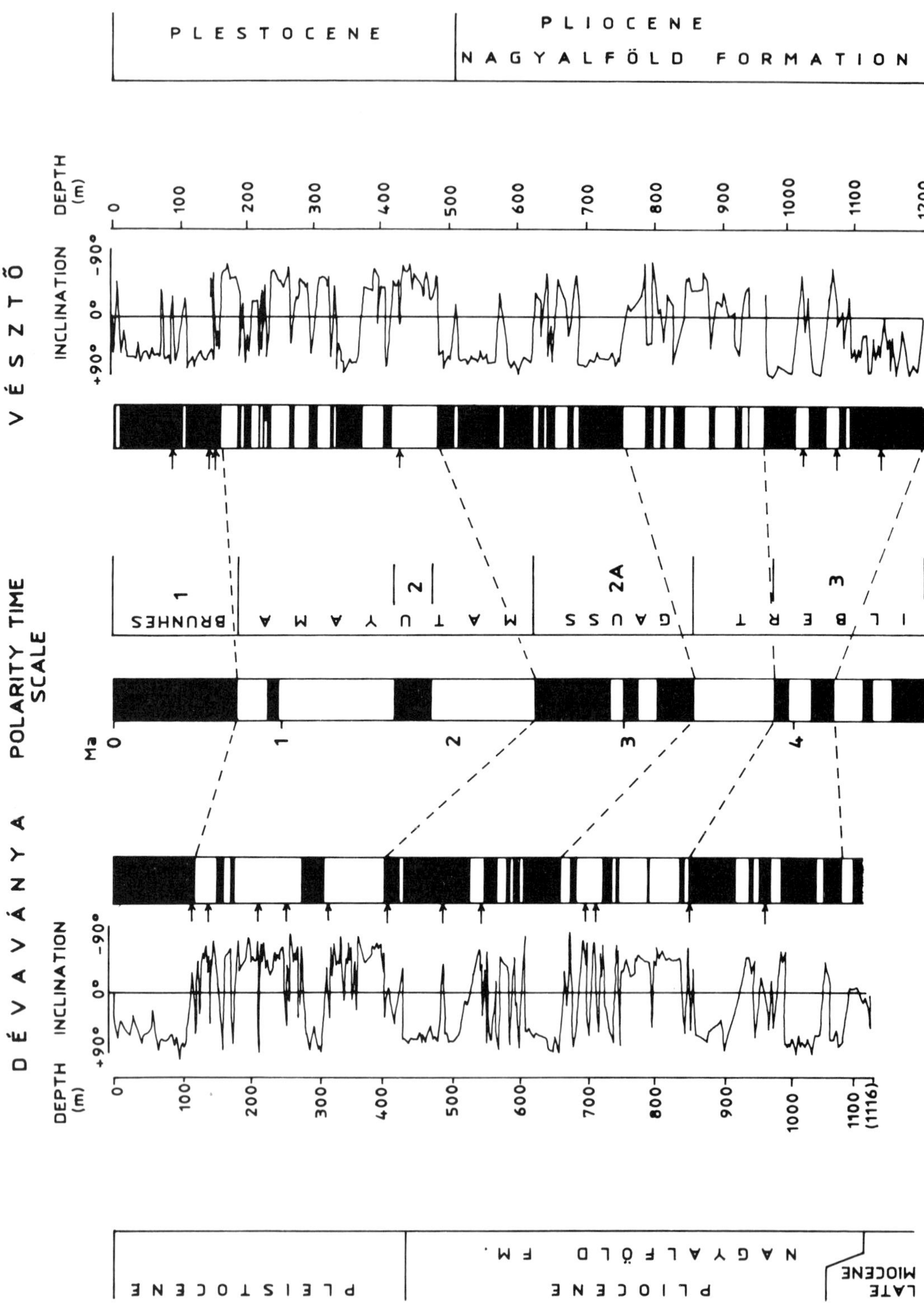

Figure 3. Lithostratigraphy, polarity zonations, and correlations with the polarity time scale for the Dévaványa and Vésztő core holes (modified from Cooke and others, 1979). Black: normal polarity; white: reversed polarity; slashed: no control; arrow: short reversal. Polarity time scale is from Berggren and others (1985).

but because it is located in a deeper part of the Békés basin, the base of the Nagyalföld was not reached. The greater thickness of the Nagyalföld Formation in Vésztö implies that the central part of the Békés basin continued to subside differentially during Pliocene time.

The Nagyalföld Formation contains chiefly mottled claystone that accumulated in a cyclic, fluviolacustrine environment. The deposits characteristically consist of unstratified yellowish-gray (rarely reddish), variegated clay and silt that contain root remnants, desiccation cracks, lime concretions, and subordinate layers of fine- to medium-grained sand. The Nagyalföld Formation and the overlying Pleistocene sequence are similar in composition and appearance, and their alternating clays and silts are distinguished from one another with difficulty. In the Dévaványa and Vésztö cores, sedimentological analysis served to identify the lowest fluvial cycle, interpreted by geologists of the Hungarian Geological Institute to mark the base of the Pleistocene; it is an interpretation supported by palynological and paleontological (Ostracoda zonation) analyses (Cooke and others, 1979).

Strata characteristic of the Pannonian (*s.l.*) were encountered beneath the Nagyalföld Formation in the Dévaványa drill-core section (Fig. 3). In Dévaványa, gray interstratified clayey silt and silty sand that accumulated in a lacustrine environment and contain a molluscan fauna characteristic of the pre-Nagyalföld Pannonian, are stratigraphically low in the section. Pannonian strata in Dévaványa are overlain by a relatively thick layer of sand at the base of the Nagyalföld Formation. The contact between the Pannonian Zagyva Formation and the Nagyalföld Formation may be unconformable.

Tiszapalkonya and Kaskantyú

The Tiszapalkonya-1 core hole (Fig. 4) was drilled in the northeastern part of the Great Hungarian Plain, and the Kaskantyú-2 core hole was drilled in the west-central part of the Great Hungarian Plain (Fig. 1). In Tiszapalkonya, nearly 2 km of late Miocene strata are overlain disconformably by a relatively thin Pliocene and Pleistocene section. In Kaskantyú, a relatively thick late Miocene section also is disconformably overlain by a thin Pliocene and Pleistocene section.

The youngest stratigraphic units in the Tiszapalkonya and Kaskantyú drill-holes comprise thin (128 and 151 m-thick) sections of middle and upper Pleistocene periglacial deposits that consist of fluvial sand, gravel, and loess. These deposits disconformably overlie the (upper Pannonian) Pliocene Nagyalföld Formation, whose characteristics are similar to those observed in the Nagyalföld Formation in the Dévaványa and Vésztö drill cores. In Kaskantyú, the Nagyalföld Formation either conformably or disconformably overlies the Zagyva Formation, whereas in Tiszapalkonya an erosional surface (a disconformity) is inferred from the character of the contact in the core and from the presence of relatively coarse reworked and weathered materials in core above the disconformity.

The late Miocene section in Kaskantyú is markedly thinner than the section of the same age encountered in the Tiszapalkonya drill hole (1 km in contrast to 2 km thick). The thin section at Kaskantyú overlies a structural high in the pre-Pannonian basement. This area subsided considerably less than nearby parts of the basin. Except for the differences in thickness, the paleoenvironments of upper Miocene deposition at Tiszapalkonya and Kaskantyú were similar, and generally typical of upper Miocene sediments in the area of the Great Hungarian Plain.

Upper Miocene strata underlying the Pliocene Nagyalföld Formation consist of the Bükkalja Formation in the Tiszapalkonya drill hole and the correlative Zagyva Formation in the Kaskantyú drill hole. These fine-grained units accumulated during a time of moderate to abundant plant growth in the region, characteristic of swampy and fluviatile environments of deposition. Each formation occupies major parts of the cored sections. The Bükkalja Formation (1311 m thick) consists of a cyclical series of gray, in part cross-bedded sand, silt, clay, and lignitic layers that accumulated under swampy and shallow water (1-20 m deep) lacustrine conditions, and that contain a brackish water fauna consisting of molluscs and ostracods. These deposits represent deposition in a delta plain environment as inferred from seismic stratigraphic studies (Mattick and others, this volume). The Zagyva Formation (625 m thick) consists of alternating layers of gray, laminated sandy silt, interbedded laminated and cross-bedded well-sorted sand, and unstratified clayey marl. These deposits accumulated in part under shallow-water, lagoonal, and nearshore conditions, and in part under fluvial conditions. These sediments were also deposited in a delta plain environment as inferred from seismic stratigraphy.

The basal unit of the upper Pannonian is the Törtel Sandstone, which underlies the Bükkalja Formation in the Tiszapalkonya drill-core section. The Törtel Sandstone, about 25 m thick, consists of gray, fine-grained cross-bedded sand, sandstone, and clayey marl. These

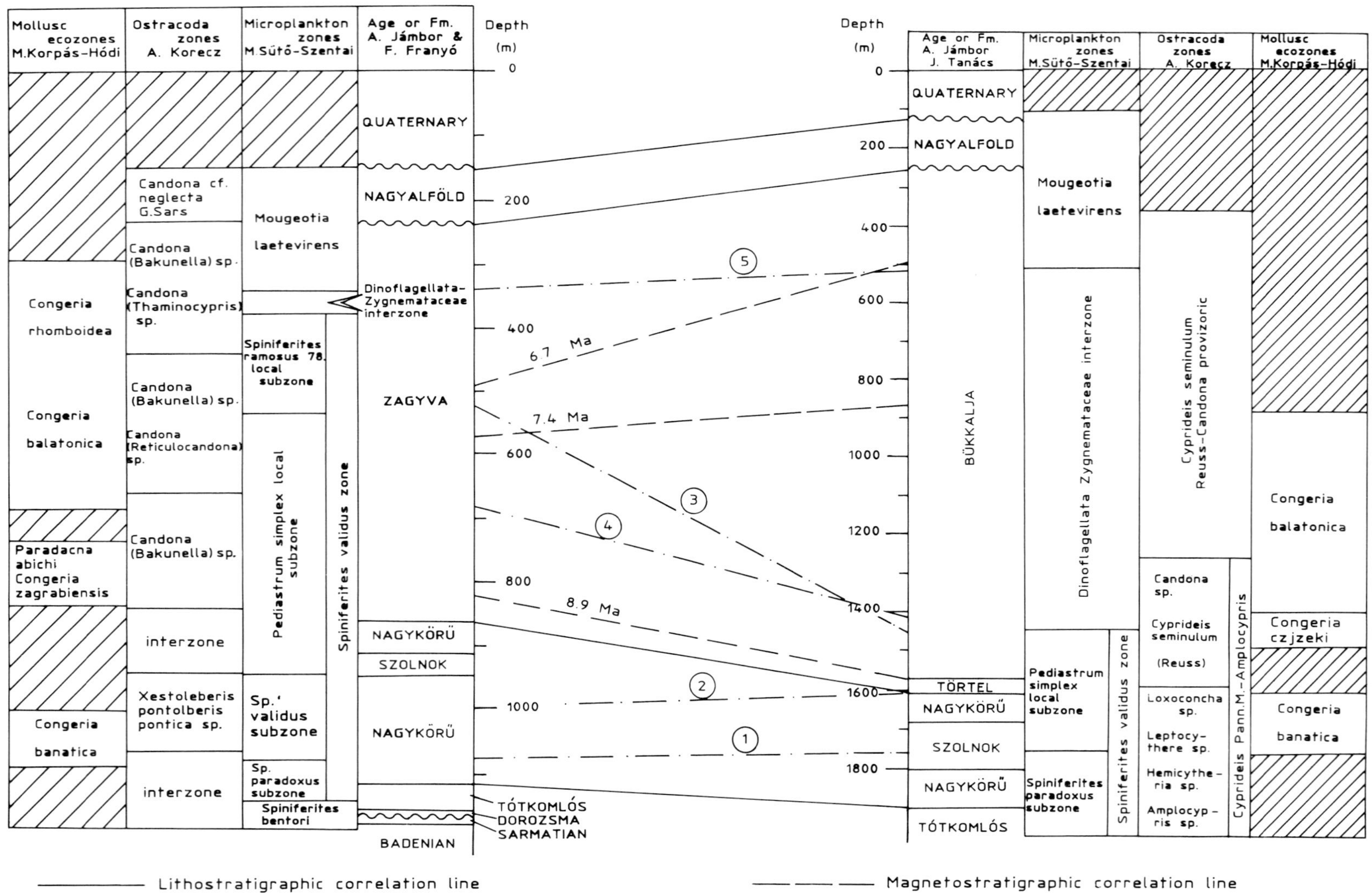

Figure 4. Diagrams showing lithostratigraphy and biostratigraphy from the Tiszapalkonya and Kaskantyú core holes. Key correlations between the holes are also shown. Depths are referenced to ground surface. 1: upper boundary of Spiniferites paradoxus subzone; 2: upper boundary of Congeria banatica zone; 3: upper boundary of Pediastrum simplex local subzone; 4: lower boundary of Congeria balatonica zone; 5: lower boundary of Mougeotia laetevirens zone.

represent the initial deposits of the progradational deltaic sequence of the Pannonian Basin. In Kaskantyú, interbedded sandstones and marls of the Zagyva Formation appear to represent the same stratigraphic interval.

Lower Pannonian strata underlying the Törtel and Zagyva Formations in the Tiszapalkonya and Kaskantyú drill-core sections, respectively (Fig. 4), consist of units that accumulated in deep (600-800 m), quiet, near-anoxic and brackish water conditions. These deposits include the Nagykörü Formation (a clayey marl), the Tótkomlós Formation (a calcareous marl), and the Dorozsma Formation. The Dorozsma consists of dark gray, organic-rich, well laminated clayey marl, calcareous marl, and silty clay, which contain authigenic pyrite, molluscs, ostracods, and fish scales. These formations, and sandstone of the interbedded Szolnok Formation, belong to an inferred prodelta facies (Mattick and others, this volume; Phillips and others , this volume).

Conglomerates and turbidites that characterize lower Pannonian deposits in the deep parts of the basin are absent in the drill cores. The turbidites can be identified on seismic records (Mattick and others, this volume) by their seismic character. In the Kaskantyú section, an unconformity separates strata at the base of the Pannonian from underlying strata of Sarmatian (late middle Miocene) age (Fig. 2b). An hiatus and loss of record at this horizon also is inferred from paleontological information (discussed later).

Biostratigraphy

The distribution of floral and faunal elements across the Pannonian sequence have been summarized in Jámbor and others (1987). The various fossil assemblages include molluscs, microplanktons, vertebrates, ostracods, theacamoebans, diatoms, foraminifera, nannoplanktons, spores and pollen, ichnofossils, and sponge spicules. These fossils have allowed the environments of deposition to be discriminated. However, because all of the fossils are characterized by low specific diversity, they have proven to be of minimal value for placing the beds within specific parts of the late Neogene time scale.

Isotopic Ages

Great Hungarian Plain.

Only a few isotopic dates, determined mainly from the K-Ar analysis of basaltic flows in the sedimentary section, serve to constrain the age of the Pannonian (*s.l.*) sequence. All ages were determined near the lower boundary, or at somewhat higher levels, in the lower Pannonian (Fig. 2b). Vass and others (1987) assigned an age of 11.5 Ma (plus or minus 0.5) for the base of the Pannonian, reporting K-Ar ages ranging from 10.9 to 12.3 Ma. Additionally, K-Ar ages of 10.7 to 11.0 Ma have been reported from alunite crystals from veins in an uppermost Sarmatian tuff in northeast Hungary, considered by the authors to correspond with the Pannonian (*s.l.*) (Balogh and others, 1984). The apparently most reliable age for the lower Pannonian has come from unweathered and unaltered biotite from a dacite tuff that is interbedded in marl of the Monostorapáti Formation in southern Hungary, penetrated in the Nagykozár-2 drill hole (Fig. 1). The biotite tuff, encountered at a depth of 263 m, has an age of 11.6 Ma (plus or minus 0.5) (Vass and others, 1987). This, and the foregoing ages, thus indicate that accumulation of Pannonian (*s.l.*) deposits began in latest middle Miocene time. (The nominal boundary between the middle and late Miocene, drawn at the base of Anomaly 5, is 10.42 Ma in the time scale of Berggren and others, 1985).

A few younger K-Ar ages have been reported from strata higher in the Pannonian section. An age of 9.6 Ma (plus or minus 1.0) (S. Lenner, written communication, 1985) has been determined from the analysis of biotite from a rhyolite tuff in the Nagykörü Formation (upper part of lower Pannonian). This tuff, encountered in the Bácsalmás-1 drill hole (Fig. 1), was sampled at a depth of 486 m. A virtually identical age has been reported for basalt penetrated at a depth of 1167 m in the Kiskunhalas-3 drill hole, also located in southern Hungary (Fig. 1). Its age, 9.61 Ma (plus or minus 0.38)(Balogh and others, 1984), was determined from whole-rock K-Ar analysis of basalt. The horizon containing the basalt, mapped in seismic reflection profiles, correlates with strata falling within the long interval of normal polarity in the Kaskantyú-2 drill hole (Pogácsás and others, this volume). This interval of normal polarity is correlated (in a later section of this report) with marine magnetic Anomaly 5, which has a nominal age range of 10.42 to 8.92 Ma. Although the possibility exists that some Ar loss occurred as a consequence of burial and heating, resulting in an age somewhat younger than the true time of emplacement of the flow, the reported isotopic age of 9.6 Ma in the Kiskunhalas-3 drill hole is reasonable in view of the seismic correlation of the horizon with Anomaly 5 in Kaskantyú.

Still younger K-Ar ages of 8.47 Ma(plus or minus 0.77) and 8.13 Ma (plus or minus 0.71) (Balogh and others,

1984) have been reported from basalt encountered in the Kecel-1 and -2 drill holes, also located in southern Hungary (Fig. 1). As in Kiskunhalas-3, these basalt flows are interbedded in strata containing fauna whose age is considered to lie in the middle part of the lower Pannonian (*s.l.*) (Cserepes-Meszéna, 1978). If no Ar loss has occurred, the slightly younger ages would suggest that the basalt flows in the Kecel-1 and -2 drill holes are in a somewhat higher part of the Pannonian sequence. However, such an age assignment is not in accord with ages greater than 8.9 Ma assigned to lower Pannonian (*s.l.*) strata in Tiszapalkonya and Kaskantyú on the basis of the age of the upper boundary of Anomaly 5. The true stratigraphic relations with respect to the sections in the Kecel-1 and -2 drill holes remain unclear. No other isotopic ages exist from Pannonian strata beneath the Great Hungarian Plain.

Little Hungarian Plain.

Strata beneath the Little Hungarian Plain, northwest of the Transdanubian Mountains (Balaton Highlands; Fig. 1), have been classed as upper Pannonian (*s.l.*) and dated from intercalated basaltic layers, some of which also crop out in the Balaton Highlands area. The basalts, which are believed to have been extruded in three stages, range in age from 5.5 to 2.9 Ma as determined from K-Ar analysis (Balogh and others, 1985). These ages, extrapolated by some workers to upper Pannonian deposits beneath the Great Hungarian Plain, do not correspond with the magnetostratigraphically determined estimates of a late Miocene age for lower parts of the upper Pannonian sections in the Tiszapalkonya and Kaskantyú drill-cores.

Recent drilling, stratigraphic, seismic-stratigraphic, and magnetostratigraphic studies in the Little Hungarian Plain have shown that deposition of a very thick (1.25 km) section of upper Pannonian deposits of early late Miocene age (8.8 to 8.5 Ma) occurred following deposition of lower Pannonian strata in the interval 9.2 to 8.8 Ma (Szombathely well, Fig. 1). The section in the Szombathely well is correlated with the Tiszapalkonya well in the Great Hungarian Plain at the level of the top of Anomaly 5. Correlations of these stratigraphic sections with the polarity time scale indicate that pronounced sinking of both basins occurred in the early to middle parts of late Miocene time.

MAGNETOSTRATIGRAPHY

A joint magnetostratigraphy study between the Hungarian Geological Institute and the U. S. Geological Survey was undertaken to develop geomagnetically controlled time lines in Neogene sections of the Pannonian Basin. It was hoped that the polarity time lines could help refine stratigraphic correlations, and that the polarity zonations would serve to place the stratigraphic units in the geomagnetic time scale for the Neogene.

A parallel objective, having global implications, was to develop high-resolution polarity zonations for substantial parts of Neogene time, thereby obtaining new information on the frequency and character of polarity reversals for comparison with the zonation of the commonly accepted polarity time scale. The joint work was begun in 1982. Preliminary results for the Kaskantyú drill core were reported in 1984 (Elston and others, 1985). Since then, second-stage samples from the Kaskantyú core hole, and samples from a second deep core hole (Tiszapalkonya), were collected and analyzed. Results are summarized below.

Sampling

Kaskantyú-2 drill hole

The Kaskantyú-2 hole was cored continuously to a depth of 1.2 km, employing a 5.9 m-long core barrel. Approximately 1200 samples, at spacings of about 1/2 m, were collected at the drill site from the moderately indurated parts of the section. However, unconsolidated and well-indurated materials were not sampled at the time of drilling, which resulted in many small and several large gaps in magnetostratigraphic coverage. The moderately indurated cores that were sampled were sliced longitudinally, following which plastic boxes were inserted into the cut surfaces. Although this is a routine procedure for soft marine cores contained in plastic liners, this procedure may have been responsible for the generation of several one-sample reversals arising from the inadvertent inversion of samples during handling.

Approximately 200 additional (second-stage) samples were collected from the better-indurated rocks at a later date to provide paleomagnetic data for the major unsampled intervals, and also to check on several of the one-sample reversals observed in the initial data set. One-sample reversals not resolved by the second stage sampling were arbitrarily discarded from the data set. Although little in the way of declination information was developed, a useful paleomagnetic data set nonetheless was obtained from the Kaskantyú core section.

Tiszapalkonya-1 drill hole

The Tiszapalkonya-1 hole was cored continuously to a depth of 2 km, also employing a 5.9 m-long core barrel. Samples were collected at 1/2 m intervals at the drill site immediately following removal of core from the core barrel. Slightly more than 3000 oriented samples were collected. Each "string" of core was scribed longitudinally as the core was extracted from the core barrel, providing an arbitrary reference direction to which the individual samples from the run were oriented. The cores commonly were extracted as largely unbroken cylinders during the scribing process, and where breaks occurred, attempts were made to match the broken sections and extend the reference line to include as much of the "core run" as possible. The forgoing procedure allowed a series of declinations as well as inclinations to be determined. The core was transversely sliced at each sampling horizon, a procedure less likely to produce accidentally inverted samples (and spurious reversals) than the sampling of longitudinally cut faces. The samples were taken from the central parts of the core, away from core that had been in contact with the core barrel. Following cutting with a diamond saw, the samples were immediately placed in cubical plastic boxes, which then were sealed and stored in a cool refrigerator at the drill site to inhibit desiccation. The samples also were refrigerated at the paleomagnetics laboratory following their shipment to the United States. Only a few intervals of poor core recovery were encountered in the Tiszapalkonya section, and core recovery for the entire section was essentially 100%.

Laboratory Procedures

Samples from the Kaskantyú and Tiszapalkonya drill holes were processed at the U.S. Geological Survey's paleomagnetics laboratory at Flagstaff, Arizona. The natural remanent magnetization (NRM) of the samples, prior to and following demagnetization treatment, was measured in a cryogenic magnetometer (noise level 10^{-7} emu or 10^{-10} Am^2). Following this, a number of pilot samples representing various lithologies, and of both normal and reversed polarity, were selected for progressive alternating-field (AF) demagnetization. Treatment was carried out in a tumbling AF demagnetizer of commercial manufacture having a maximum peak demagnetizing field of 100 mT (milliTesla; = 1000 Oersted); the pilot samples were routinely demagnetized to the limits of stability of their magnetization. Response of the pilot samples to AF demagnetization indicated that the dominant carrier of the magnetization was magnetite. Following analysis of the pilot samples, the remaining samples were partially demagnetized either at single, optimal steps for given lithologies and depths, or, for many samples, at two and more demagnetizing steps to assure that the samples had undergone cleaning treatment adequate to remove at least the softer components of potential secondary magnetizations. For Kaskantyú, most samples were partially demagnetized at 5 mT (50 Oe) fields. The Tiszapalkonya samples were partially demagnetized in 2.5 to 20 mT (25 to 200 Oe) fields, mostly near and at the high end of this range. Demagnetization levels generally increased with sample depth. Samples that did not appear to respond to demagnetization at the lower AF levels were routinely subjected to higher levels of alternating fields, but this rarely resulted in appreciable changes in direction. Thermal demagnetization was not conducted on the moderately to poorly indurated sediments.

Thermal demagnetization was carried out on pilot samples from different lithologic units from the lower part of the Pannonian section in Szombathely well (Fig. 1) using a commercial thermal demagnetizer and a cryogenic magnetometer at the joint laboratory of the Eötvös Lóránd Geophysical Institute and the Hungarian Geological Institute at Budapest. The objective was to compare results with respect to AF demagnetization treatment of another suite of pilot samples. No significant differences in directions were observed between the two methods for approximately 70% of the samples, but as a group, directions of samples subjected to thermal treatment were somewhat more scattered. For the remaining samples, the AF treatment produced less scattering. There is no reason to expect that thermal treatment of the samples from the Tiszapalkonya and Kaskantyú wells would have produced different or improved results with respect to directions obtained from AF treatment.

Demagnetization Analysis

The stability of magnetization, and the behavior during AF demagnetization, of representative pilot samples from the Tiszapalkonya and Kaskantyú drill cores is depicted on orthogonal vector diagrams (Figs. 5a-5h). Some samples exhibited little or no secondary overprinting, only a single component of magnetization being removed during much of the range of stepwise treatment (Figs. 5a - 5c); such samples were not abundant. More commonly, samples of the various rock types contained two (and in a few cases even three) distinct components of magnetization (Figs. 5d - 5e). A very few samples responded to demagnetization treatment to reveal an

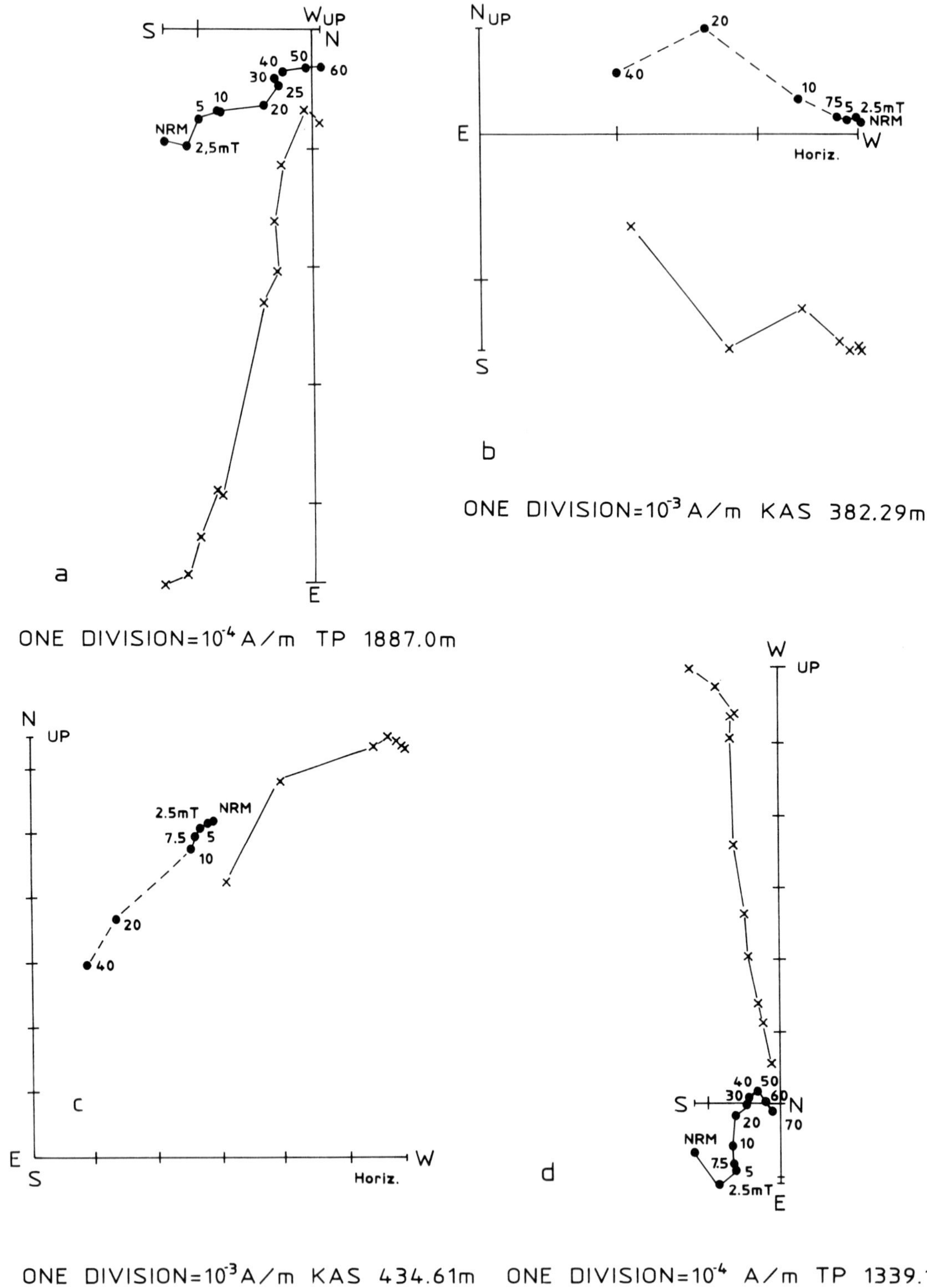

Figure 5 a/h. Orthogonal vector diagrams showing stability of magnetization during alternating-field (AF) progressive demagnetization treatment for representative normal and reversely polarized samples from the Tiszapalkonya (TP) and Kaskantyú (KAS) drill holes. Dots: horizontal plane; X: vertical plane; mT: milliTesla; NRM: natural remanent magnetization. Declinations are arbitrary. Demagnetization characteristics are discussed in text. Figure continued on next page.

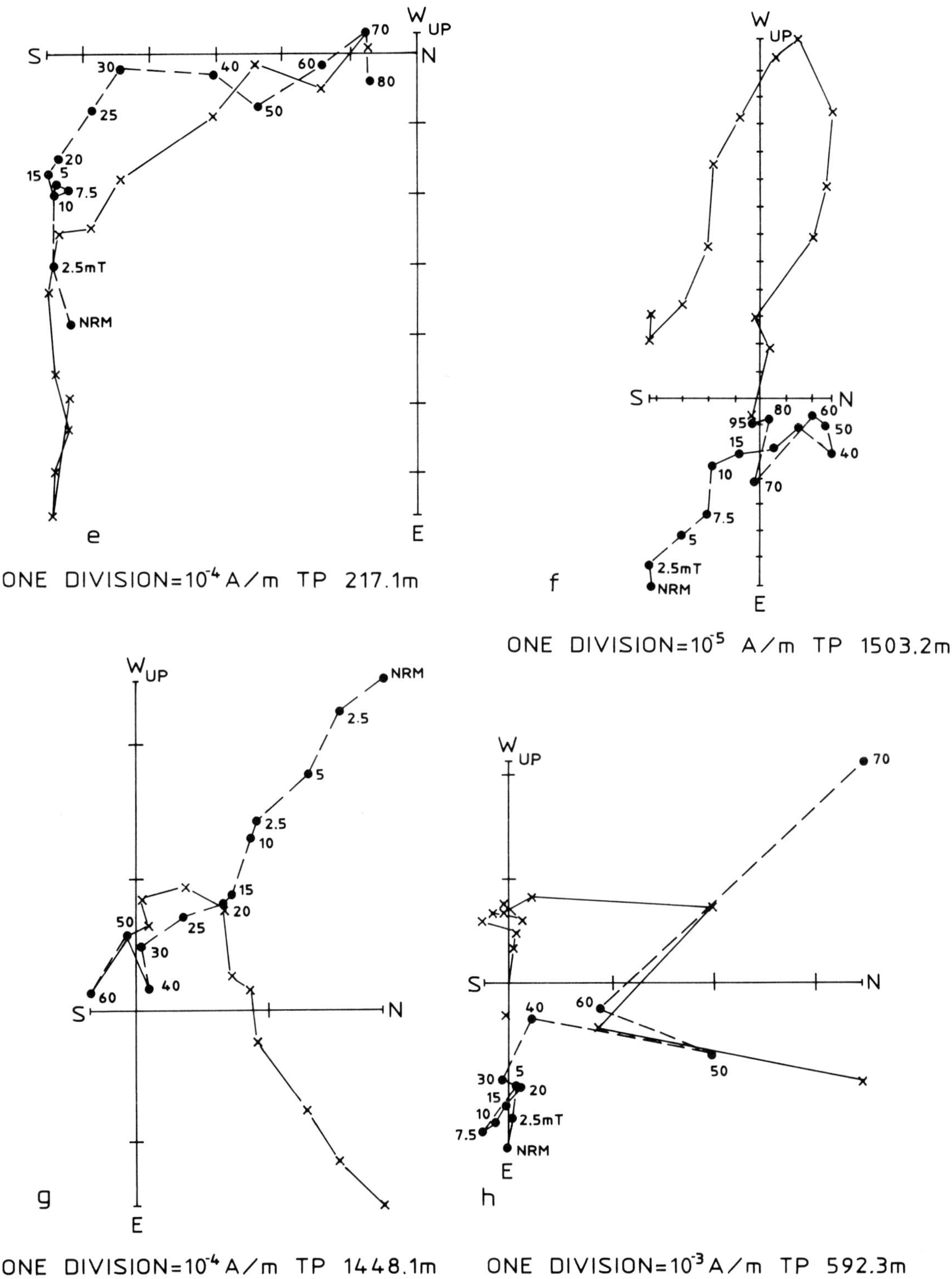

Figure 5. Continued from previous page.

inclination opposite to that observed prior to cleaning (Figs. 5f - 5g). Single (blanket) demagnetization steps were chosen after a review of the behavior of all progressively demagnetized pilot samples, examined with respect to a combination of lithology and depth. The selected demagnetization steps, ranging from 2.5 to 20 mT, served to reduce the scatter in inclination values, but rarely resulted in appreciable changes in inclination or to changes in the polarity zonation seen prior to cleaning treatment.

Multicomponent magnetizations particularly characterized yellowish, fine- to coarse-grained sand, a subordinate lithologic type that (not unexpectedly) contained the least reliable magnetization. With the exception of yellow sand, and gray coarse-grained sand and sandstone, no correlations were observed between demagnetization behavior and rock type, or with depth of burial. Although the quality of the demagnetization records is not superb, most samples of the different lithologic types failed to display appreciable changes in direction with demagnetization. Except for the few samples that exhibited reversals in inclination with cleaning, the cleaning revealed no hint that different polarities resided in remanences near the threshold levels of stability. The directions (inclinations) obtained after treatment thus were considered to reflect original directions. We note that this accords with the history of rapid accumulation and burial of the sediments, the lack of evidence for alterations due to the passage of solutions, or alterations resulting from exposure and weathering. Secondary magnetizations arising from geologic factors therefore were expected to be minor, an expectation supported from the results of the AF demagnetization. No magnetizations related to drilling were recognized.

Although rare in the sections, one rock type (coarse-grained gray sand and sandstone) proved troublesome because it exhibited erratic demagnetization behavior. Up to certain demagnetization levels, the behavior was similar to that seen in the other rocks. However, above this, at levels in the range of 10 to 30 mT, intensities increased and all semblance of order in direction became lost as secondary magnetizations were acquired in the demagnetizer (Fig. 5h). Such sand and sandstone samples thus were routinely demagnetized at the lowest demagnetization levels that served to remove minor, soft, potentially secondary components of magnetization, to provide more stable, presumably original directions. Minimum levels used for this rock type also were found to be suitable for the partial demagnetization of samples of other rock types.

Intensities of Magnetization

Intensities of magnetization commonly were reduced by 20-40% by the blanket demagnetization. Intensities of the Kaskantyú samples following demagnetization ranged from about 10^{-2} to 10^{-3} A/m. Intensities of samples from Tiszapalkonya following cleaning ranged from 10^{-2} to 10^{-4} A/m above a depth of 1200 m, and from 10^{-3} to 10^{-4} A/m below that depth. Except for a very few samples, the intensities of magnetization of the samples following demagnetization comfortably exceed the noise level of the cryogenic magnetometer.

Polarity Zonations

Polarity zonations for the Tiszapalkonya and Kaskantyú drill cores are shown in Figures 6 and 7. Results for samples collected above a depth of 266 m in Tiszapalkonya are not shown. Inadequate section exists above the unconformity at 266 m to allow a correlation to be made with the polarity time scale. Only about one-third of the individual sample points are plotted for each of these drill holes, allowing individual points to be discriminated in the greatly reduced stratigraphic plots of inclination. Thus, only about 1000 points are shown for Tiszapalkonya and about 700 data points are shown for Kaskantyú. Narrow intervals of normal and reversed polarity shown on the plots are based on multiple samples. Still narrower reversals (commonly consisting of two-sample reversals) are shown by arrows along the left sides of the columns. Intervals of alternating normal and reversed polarity, occurring on a scale of 1 to 2 samples and repeated at least twice consecutively, are classed as intervals of mixed polarity and designated by M on the plots. Where inclinations appear to progress from one stable polarity state into the other, and where declinations (for Tiszapalkonya) also show a systematic progression, the behavior has been classed as a transition (designated as T). Places where only partial or incomplete polarity reversals are observed are classed as excursions (designated as E). As shown on Figures 6 and 7, intervals of mixed polarity, transitions, and excursions are found across the Tiszapalkonya and Kaskantyú sections.

Polarity Reversals, Excursions, and Oscillations

Examples of the character of intervals of mixed polarity (exhibiting an oscillating inclination record), transitions, and excursions are shown in Figures 8a-8c. An interval of mixed polarity, documented by 12 data points in about 6 m of section, is found at about 830 m depth in the

Kaskantyú core (Fig. 8a) where it directly overlies a long interval of normal polarity. The top of the normal polarity interval (correlated with marine magnetic Anomaly 5), is marked by a series of narrow reversals and oscillations, rather than by a single polarity switch. In the Tiszapalkonya drill core, the top of the same long interval of normal polarity is similarly characterized by an interval of narrow polarity reversals. The record in Tiszapalkonya appears to document this field behavior more completely because of a higher-resolution stratigraphic record. A transition from reversed to normal polarity is recorded by 8 samples at about 775 m depth in Tiszapalkonya (Fig. 8b), where a progression in values of declination, as well as inclination, and a switch in polarity are observed within a single core-run. (Core-run boundaries are indicated by dashed lines; declinations across core-run boundaries have not been adjusted.) Lastly, an example of an excursion within a single core-run at a depth of about 595 m in Tiszapalkonya is shown on Figure 8c. Here, two samples exhibit an excursion in declination followed by a shallow reversed polarity inclination.

The fine scale structure of the inclination records in Tiszapalkonya and Kaskantyú display regularities and progressions, suggesting that they have recorded field behavior. Independent evidence supporting this interpretation has been obtained from the polarity zonation and inclination record in a well drilled recently in the Little Hungarian Plain (Szombathely, Fig. 1). The record in Szombathely correlates with part of the polarity zonation and inclination record in the Tiszapalkonya section (Fig. 9). Geological analysis of the 2-km-thick Szombathely core record, sampled on 1/2 m intervals, indicates that a lower and upper Pannonian sequence accumulated with no significant break in deposition (R.L. Phillips, written communication, 1989). This section, generally placed in the time scale on stratigraphic grounds, was placed in the polarity time scale from correlation with Anomaly 5 of the polarity time scale. Similar transitions from normal to reversed polarity at the top of Anomaly 5 are present in both Szombathely and Tiszapalkonya. Additionally, approximately 30 oscillations of the inclination record, in part correlatable on character of the oscillation intervals, are present in both Szombathely and Tiszapalkonya for an interval correlated with 8.8 to 8.5 Ma of the polarity time scale. Not surprisingly, the record of oscillations is less completely represented in Tiszapalkonya than in Szombathely. In Szombathely, individual oscillation sequences commonly are characterized by substantial intervals of very stable magnetization of a given polarity (mostly reversed); these stable inclinations progress into oscillations of increasing amplitude, some passing into excursions or even reversals, which in turn are followed by either abrupt or oscillating returns to stable inclinations. The Szombathely inclination record is interpreted to reflect a very high resolution record of secular variation, with control points at considerably less than 1000 year intervals. The character of the oscillating inclination records for Szombathely and Tiszapalkonya, and their correlations, are to be presented in another report.

Recognition that the fine scale inclination record in Tiszapalkonya correlates with a still higher resolution record that is attributable to secular variation, has answered questions concerning the timing of magnetization in the Tiszapalkonya section, the abundance of polarity reversals, and the presence or absence of secondary magnetizations that might compromise the Tiszapalkonya polarity record. Discussions concerning these questions, written before drilling of the Szombathely well and analysis of its samples, have been retained in the following section because such questions still are germane to any magnetostratigraphic study.

Timing of Magnetization

Many stratigraphically narrow intervals of reversed and normal polarity not represented in the commonly accepted geomagnetic time scale for the late Miocene and Pliocene are present in the cored sections of Tiszapalkonya, Kaskantyú, Dévaványa, and Vésztö. The records display a number of apparent transitions, locally abundant intervals of mixed polarity, and a number of more prominent, antiparallel or near-antiparallel intervals of normal and reversed polarity inclinations, all of which lie within the framework of a much simpler polarity time scale. Some of these "extra" reversals might be arbitrarily rejected as arising from complex overprinting reflecting a delayed acquisition of remanence, particularly where the narrow reversals closely overlie a long interval of single polarity. However, it is difficult to explain the origin of the complex polarity zonations and transitions as arising from secondary magnetizations where these intervals are considered to correlate with parts of the time scale for which no or only few short reversals are shown, i.e., how can fine scale structures (oscillations, transitions) and differing polarities in the core records across the sections reflect secondary magnetizations if the ambient field is broadly stable and only infrequently reversed itself?

The multiple narrow reversals from about 860 to 950 m in Tiszapalkonya, when compared with a less complex interval in Kaskantyú, might be pointed to as evidence

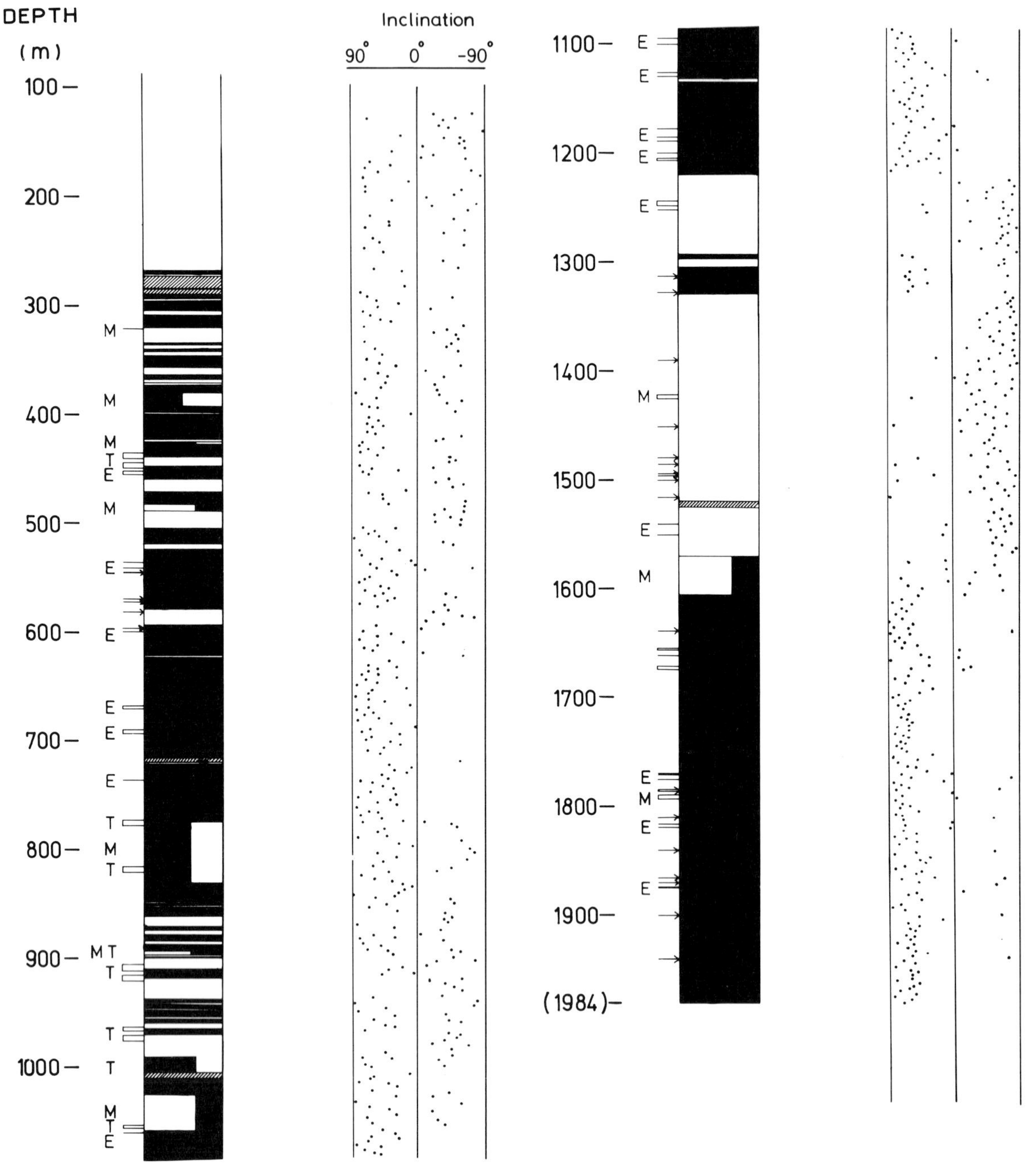

Figure 6. Polarity zonation for the Tiszapalkonya core hole showing locations of intervals of mixed polarity (M), transitions (T), and excursions (E). Black: normal polarity; white: reversed polarity; black and white: mixed polarity with dominant polarity on left; arrow: short reversal; slashed: no control. The inclinations of about 1/3 of the 3000+ samples analyzed are shown. Depths are referenced to ground surface.

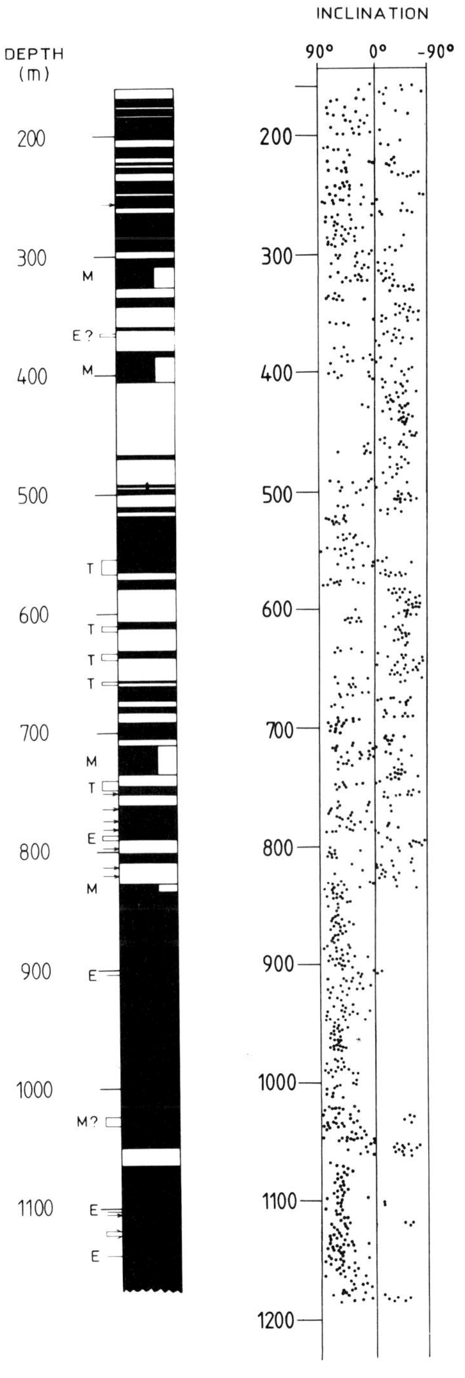

Figure 7. Polarity zonation for the Kaskantyú-2 core hole showing locations of intervals of mixed polarity (M), transitions (T), and excursions (E). Explanation for other symbols is given in Figure 6. Depths are referenced to ground surface.

for complex overprinting. However, a similar discrepancy exists between the frequency of polarity reversals for the Matuyama Reversed-Polarity Chron in Dévaványa and Vésztö (Fig. 3), with the higher frequency record in Vésztö being considered the more complete and reliable (Cooke and others, 1979; Cooke, 1985). We note that the several intervals of normal polarity in the Matuyama Chron in Vésztö, correlate better with the polarity time scale than the less abundant reversals in Dévaványa. Even if all of the fine-scale reversals are discarded, the polarity records in the Hungarian drill cores still appear to be the most complete records of the geomagnetic field yet obtained for parts of late Miocene, Pliocene, and Pleistocene time.

Additionally, with few exceptions, progressive AF demagnetization of large numbers of pilot samples, and progressive demagnetization of many of the remaining samples, did not reveal or suggest the existence of secondary magnetizations masking directions that could be considered to be earlier. Secondary overprinting recognized in many of the samples only appears to have degraded the quality of the magnetostratigraphy record in the sections, not mask an original magnetization, and the quality of the records improved with demagnetization treatment.

Geologic considerations also support the conclusion that original magnetizations reside in the core records. Most of the cored strata are very fine grained, dominantly fine-grained argillaceous sand, silt, and clay. With the exception of infrequent beds of yellowish sand, most beds are an even shade of gray, are extremely fresh in appearance, and display no hint of weathering or alteration arising from the movement of fluids. The deposits are the product of rapid accumulation and burial of sediments, without exposure during or since accumulation, and without apparent alteration arising from the movement of connate or ground waters. In view of this, appreciable secondary overprinting arising from weathering at the surface, such as commonly complicates paleomagnetic records in samples collected from outcrops and from near-surface cores, was not expected.

Lastly, minor secondary magnetizations in the Tiszapalkonya, Kaskantyú, and Szombathely cores, irregularly distributed in the sections, do not appear to have been acquired as a consequence of pervasive overprinting arising from relatively deep burial of the magnetite-bearing sediments. Overprinting has not been overly severe because (1) long intervals of normal polarity (Anomaly 5) in the lower parts of the Tiszapalkonya, Kaskantyú, and Szombathely sections contain narrow

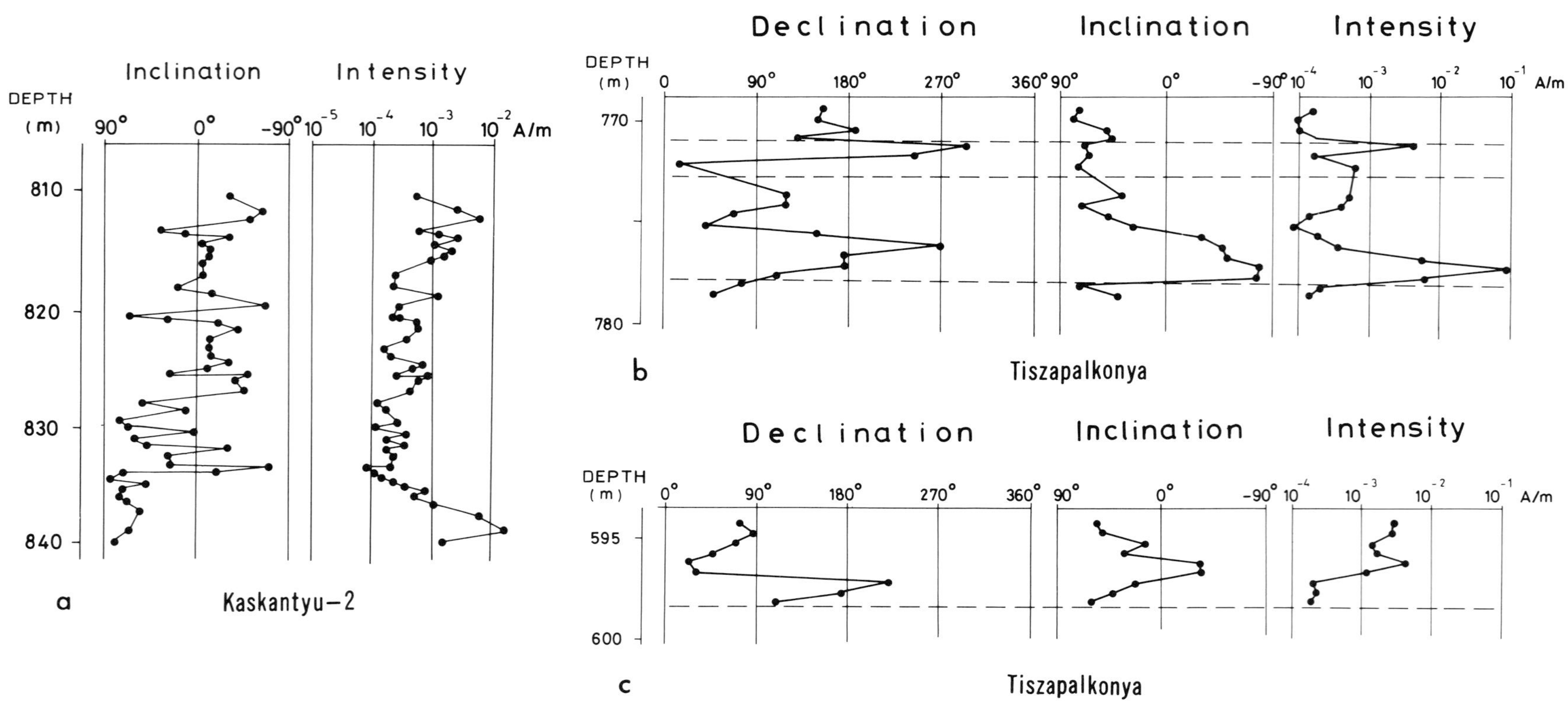

Figure 8 a/c. Examples of intervals of (a) mixed polarity (with oscillations), (b) transition, and (c) excursion for core samples from the Kaskantyú and Tiszapalkonya drill holes. Dashed lines indicate core-run boundaries in the Tiszapalkonya core hole. Within each core run declinations are oriented with respect to one another. Depths are rerefenced to ground surface.

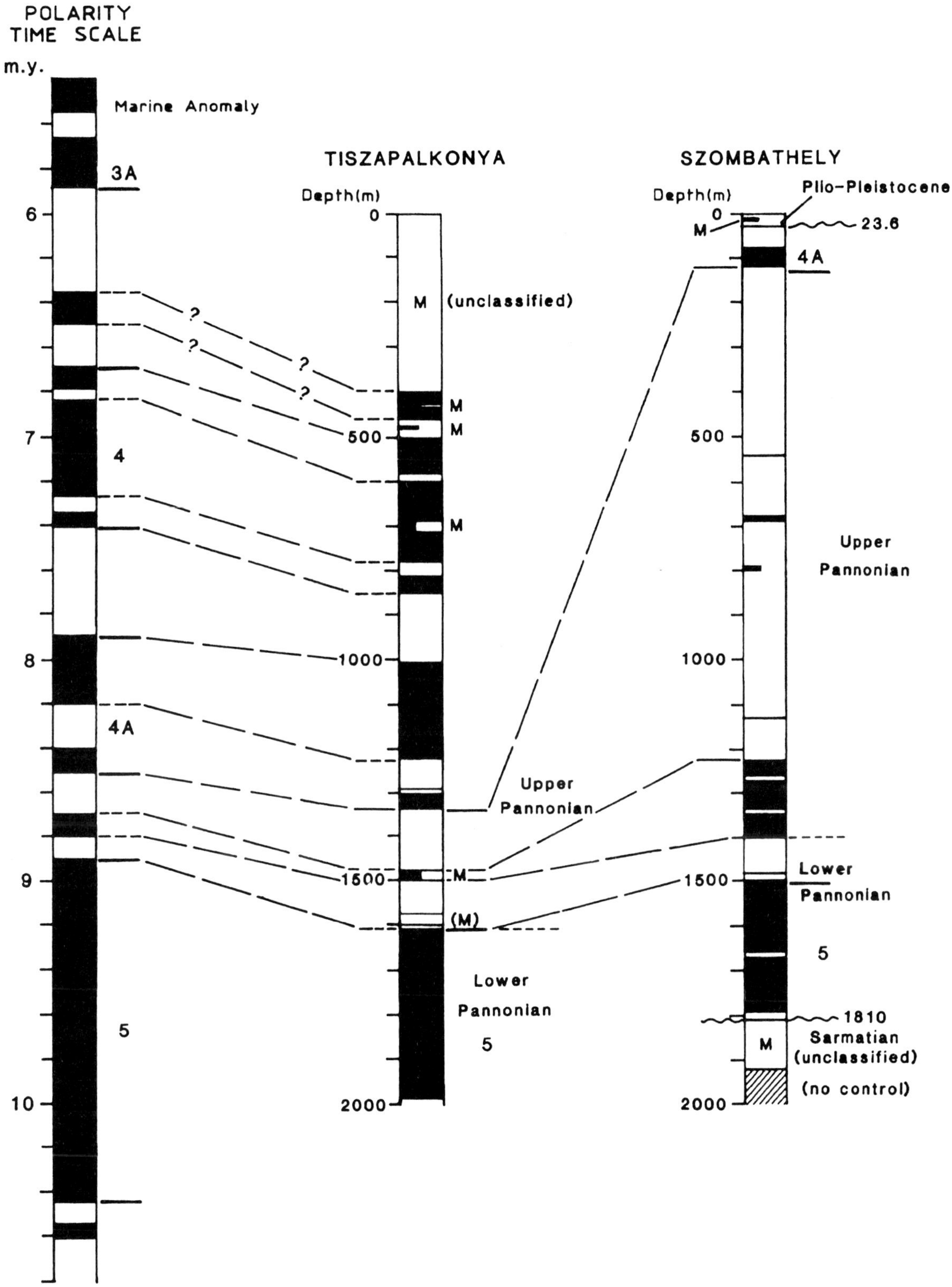

Figure 9. Magnetostratigraphic correlation of Szombathely and Tiszapalkonya well sections with the polarity time scale. Black: normal polarity; white: reversed polarity; M: mixed polarity. Polarity time scale from Berggren and others (1985). Depths are referenced to ground surface.

intervals of reversed polarity, (2) similar polarity structures are found at the top of the long interval of normal polarity, (3) similar frequencies in oscillations of the inclination record occur in strata overlying Anomaly 5 and correlated with 8.8 to 8.5 Ma of the polarity time scale, and (4) multiple polarity reversals are preserved in strata of Tiszapalkonya and Kaskantyú for upper Pannonian strata overlying Anomaly 5. These characteristics attest to the general stability of magnetization, and validity of at least the coarser intervals of normal and reversed polarity in these drill cores.

CORRELATIONS WITH POLARITY TIME SCALE

Assumptions and Tie Lines

The Pannonian (*s.l.*) sections penetrated in deep drillholes beneath the Great Hungarian Plain lack fossils that could have provided temporal control, and the core samples contain only a few volcanic deposits in the lower part of the section dated by isotopic methods. Prior to the development of seismic stratigraphic techniques, dating of the sections from magnetostratigraphy relied on correlation of polarity zonations in the subsurface sections with the polarity time scale employing some unverifiable assumptions. Correlations of the drill core records formerly assumed that the Pannonian sections accumulated without major interruption and that their polarity zonations thus could be correlated with the polarity time scale by simply working backward from the present. This approach included the assumption that accumulation rates for the Pleistocene and late Pliocene also could be applied to the Miocene, and that no significant hiatuses in deposition existed in the basin. Assignment of polarity time lines to seismic reflections at drill holes, and the tracing of these reflections in the subsurface across the basin, now has shown that assumptions about uniform and uninterrupted deposition beneath the Great Hungarian Plain are invalid (Pogácsás and others, this volume).

Dévaványa and Vésztö

Two core holes in the Pannonian Basin (Dévaványa and Vésztö; Figs. 1 and 3), each approximately 1200 m deep, penetrated sections extending from the Bruhnes Normal-Polarity Chron at the top into the Gilbert Reversed-Polarity Chron near the base (Cooke and others, 1979). Their polarity zonations were matched by Cooke and others (1979) with the polarity time scale from the Bruhnes to the base of the Gauss. This correlation implied fairly uniform rates of accumulation across the Pleistocene and the later part of Pliocene time. Because the Vésztö hole was drilled to verify the zonation previously obtained from Dévaványa, its core was sampled and processed with greater care than the Dévaványa core, and its samples yielded a more refined polarity record. The zonation later led Cooke (1985) to note that the record in Vésztö contains a number of polarity reversals not reported in the polarity time scale. From the correlations, the Dévaványa and Vésztö core holes appear to have penetrated essentially complete magnetostratigraphic records as far back as the beginning of the Gauss Normal-Polarity Chron, although a gap may exist at the level of the Pliocene-Pleistocene boundary.

In spite of the apparently satisfactory correlation for the interval Bruhnes to Gauss, a question remained about the validity of correlations with earlier parts of the polarity time scale that Cooke and others (1979) proposed. The polarity zonations in the lower parts of Dévaványa and Vésztö core sections were inferred to correlate with the polarity time scale for the early Pliocene and late Miocene by assuming a uniform rate of accumulation for the entire section. Not taken into account was the possibility that a higher-resolution record may have been obtained in the lower parts of Dévaványa and Vésztö arising as a consequence of a higher rate of accumulation, and that the lower parts of the sections might contain considerably more reversals than are shown in the polarity time scale.

Subsurface correlations based on seismic stratigraphy (Pogácsás and others, this volume) have led to a new correlation of Pliocene units in the basin (Fig. 2b). Results of the correlation indicate that pre-Gauss strata of Pliocene age in the lower parts of Dévaványa and Vésztö must have accumulated at higher rate than younger strata. Additionally, correlation of the late Miocene polarity records in Tiszapalkonya and Kaskantyú, related to a common datum at the top of Anomaly 5, also led to a revised correlation of Miocene stratigraphic units in the subsurface (Fig. 2b), and to accumulation rates comparable to those for the early Pliocene in the Dévaványa and Vésztö drill-core sections.

Kaskantyú

The earlier-held assumption that sedimentation was more or less continuous in the Pannonian Basin, and that deposits accumulated rather uniformly and could be correlated stratigraphically in a horizontal and temporal sense, was employed for the initial correlation of the polarity zonation in the Kaskantyú-2 core hole with the polarity time scale. Following the first-stage sampling, a polarity zonation existed which, although containing major gaps, exhibited sufficient reversals to allow a possible match with the polarity time scale for the past 10+ m.y. (Fig. 3 in Elston and others, 1985). The apparent match seemed to support the assumption of a uniform rate of sedimentation, although a rate somewhat less than that calculated for the Dévaványa and Vésztö sections. This initial correlation arose because of the existence of several major unsampled intervals in the lower part of the Kaskantyú drill-core section. However, second-stage sampling subsequently allowed a large interval of normal polarity to be identified in the lower part of the section (Fig. 7), which then led to the interpretation that this part of the section correlated with marine magnetic Anomaly 5 (8.9 - 10.30 m.y.). Such an interpretation was presented at the 1985 Neogene Conference held at Budapest (Hámor and others, 1985), although the new results and interpretation are not given in the abstract for the meetings volume.

The new paleomagnetic results for the Kaskantyú core hole (Fig. 7) presented a dilemma. The thick interval of normal polarity invalidated the previously employed assumptions used for correlating the section with the polarity time scale. If the earlier correlation were to be valid above the level of Anomaly 5, then a very large decrease in the accumulation rate had to occur for strata above Anomaly 5, a feature not supported from the stratigraphy. Additionally, no rationale existed for correlating the Kaskantyú polarity zonation with a smaller interval of the polarity time scale, nor for choosing the size of the "window" for attempting such a correlation. Clearly, far too many polarity reversals are present in the Kaskantyú section to permit a correlation with the polarity time scale by simple pattern matching alone.

Tiszapalkonya

Only a cursory examination of the polarity zonation from the Tiszapalkonya drill hole (Fig. 6) is needed to see that a major problem also exists for correlation of its polarity zonation with the polarity time scale. As is the case for Kaskantyú-2, a thick interval of normal polarity strongly suggests the existence of Anomaly 5. Again, as for Kaskantyú, an independent method was needed that would serve to objectively place the cored section in a discrete part of the Pannonian sequence, thus limiting the size of the temporal window used for correlation with the polarity time scale. Subsurface correlations derived from seismic stratigraphic studies provided the much needed evidence for placing the Kaskantyú and Tiszapalkonya sections within a discrete part of the Pannonian sequence and within a general part of the time scale.

Seismic Stratigraphy

Mattick and others (1985) and Pogácsás (1987) have shown that stratigraphic time lines in the Pannonian sequence have great lateral continuity, and that they follow seismic reflections. The seismic profiles provided abundant subsurface marker horizons (time lines) that clearly indicate that the Pannonian deposits did not accumulate uniformly in the basin and that many of the depositional units are time-transgressive. The seismic correlations, thus, disproved the assumption that time-stratigraphic correlations could be made from inferring contemporaneity of deposition on the basis of horizontal deposits. The correlations also revealed that depositional rates in the Pannonian Basin were highly variable, and depended on environment of deposition. However, "absolute" ages remained to be assigned to the time lines.

The seismic correlations of Pogácsás and others (this volume) indicate that the Pliocene sections cored in the Dévaványa and Vésztö drill holes correspond stratigraphically to the upper part of the upper Pannonian, a conclusion already deduced from magnetostratigraphy correlations with the Bruhnes Normal-, Matuyama Reversed-, and Gauss Normal-Polarity Chrons. In contrast, substantial parts of the sections penetrated by the Tiszapalkonya and Kaskantyú core holes were shown by Pogácsás and others (this volume) to correlate with strata that stratigraphically belong in the lower part of the upper Pannonian.

The stratigraphic position of the major part of the Tiszapalkonya section was determined by tracing seismic horizons from Tiszapalkonya southward into the Békés basin, and then, using other profiles, extending the correlations to Kaskantyú (Pogácsás and others, this volume). Specifically, a seismic horizon at the Tiszapalkonya drill site, at 280 m depth, was traced to the Vésztö drill site where it lies at a depth of between 2240 and 2910 m beneath the surface, about 1000 to 1700 m below the deepest strata penetrated in Dévaványa and Vésztö. Moreover, seismic horizons in the upper part of the Kaskantyú drill section, at 250 m depth, correlate with horizons at a depth of 1,100 m at the Vésztö drill site. This correlation indicates a substantially greater age for much of the Kaskantyú section with respect to strata of the Dévaványa and Vésztö sections. A relatively great age for the long intervals of normal polarity found in the lower parts of the Tiszapalkonya and Kaskantyú sections thus is indicated by the correlations. Lastly, seismic markers near the top of the interval of normal polarity in Kaskantyú can be traced laterally to basaltic rocks dated at 9.6 Ma in the Kiskunhalas-3 drill hole (Fig. 1). This correlation verifies that the long interval of normal polarity in Kaskantyú correlates with Anomaly 5 of early late Miocene age.

The stratigraphic framework and age assignments for the Tiszapalkonya and Kaskantyú sections, (Figures 2b and 4) were derived from seismic stratigraphic correlations. The correlations allowed the polarity zonations for the four core holes to be placed in two different, non-overlapping parts of the time scale (late Miocene, and Pliocene to Pleistocene). For Tiszapalkonya and Kaskantyú, the correlations established the temporal window that allowed their complex polarity records to be correlated with the polarity time scale for the late Miocene.

Comparison with Marine Core Records

In view of the broadly dated seismic stratigraphic correlations, the polarity records of Tiszapalkonya and Kaskantyú can be examined with respect to a polarity record obtained from upper Miocene marine sediments cored in the South Atlantic Ocean (Fig. 10; Site 519, Deep Sea Drilling Project Leg 73; Tauxe and others, 1984, their Fig. 2). DSDP Site 519 was located to encounter sediments whose polarity zonation would correlate with, and overlie, Anomaly 5. Tauxe and others (1984, p. 611) report the existence of a one-sample polarity reversal in the lower part of Anomaly 5, which they believed to be a real event. Additionally, they report the existence of an "extra" sub-chron in chron 8 (interval of normal and reversed polarity above Anomaly 5), noting that an extra sub-chron also has been observed in rocks of the Siwalik Group of northern Pakistan. For an interval correlated with the reversed polarity interval between Anomalies 4 and 4A, plots for most samples from Site 519 show shallow inclinations, and a general shift from low inclinations to steeper, reversed polarity inclinations from the bottom to the top of the interval. Lastly, Tauxe and others (1984, their Fig. 2) show four one-sample reversals in and adjacent to this interval of reversed polarity, and they speculate that these also may represent brief reversals of the field.

The distribution of polarity reversals from DSDP Site 519 section is shown on Figure 10 for an interval that includes and overlies Anomaly 5. For purposes of comparison, polarity zonations from Tiszapalkonya and Kaskantyú, more detailed than shown in Figures 6 and 7 (but that still are highly generalized), are also shown on Figure 10. A correlation between the terrestrial and marine records is proposed in this figure.

The one-sample reversal in the lower part of Anomaly 5 at Site 519 may correlate with the interval of reversed polarity at a depth of 1049 to 1066 m in Kaskantyú. From a general matching of the broader intervals of reversed and normal polarity above Anomaly 5, the interval of reversed polarity in the upper part of the Tiszapalkonya section, which contains many intervals of normal polarity, is provisionally correlated with the interval of reversed polarity lying between Anomalies 4A and 4. Beneath this complex reversed-polarity interval, several one-sample R-polarity reversals are present in the Site 519 section; these presumably correlate with some of the narrow reversals in the Tiszapalkonya and Kaskantyú sections. Evident in Figure 10, even with exclusion of the interval of highly mixed polarity in the upper part of the Tiszapalkonya section, is the lack of obvious correspondence of fine-scale polarity structure for the three sections.

From the differences in thickness of the stratigraphic sections, and from the time interval suggested by the correlations shown on Figure 10, the sedimentation rate in the deltaic environment at Tiszapalkonya appears to have been about 50 times greater than the rate of sedimentation in the open-marine environment at Site 519. For this reason, and because numerous hiatuses almost certainly exist in the Hungarian terrestrial sections, at least above basement highs where most wells have been drilled, a lack of correspondence in fine detail of the polarity zonations of Hungarian cores and the marine sediments is to be expected. Additionally, even the marine section may have problems with non-uniform deposition rates and structural disturbances. The Site 519 record does not correlate with the geomagnetic time scale in a way to support the idea that the thickness of that section can be generally equated with time. Tauxe and others (1984) suggest that the part of the section they correlate with chrons 5 (Anomaly 3A) and 6 may have been attenuated during the Messinian salinity crisis, and the section they correlate with chron 2 (Matuyama) may have been thickened as a consequence of slumping.

In summary, problems exist when attempting correlations of polarity records from the Pannonian Basin (and from the marine record), with the standard polarity time scale. The high-resolution records from Tiszapalkonya and Kaskantyú can not be correlated either with each other, or with the polarity time scale, by simple pattern matching alone. Although the lack of correlation might be assumed, ad hoc, to stem from overprinting, a major part of the problem would seem to be attributable to fragmented high-resolution records that arise from still-less-than-complete recordings of polarity history at hiatuses that occur irregularly across these sections. The lack of correlations on a fine scale is mitigated if correlations are proposed mainly on the basis of the principal N and R intervals of the polarity time scale.

Correlations of Hungarian Sections with Polarity Time Scale

Late Miocene

A correlation of the Tiszapalkonya and Kaskantyú sections with the geomagnetic polarity time scale of Berggren and others (1985) for the late Miocene is proposed in Figure 11. The correlation is anchored to the top of marine magnetic Anomaly 5. Above the level of 8.9 m.y., correlations with the accepted time scale are drawn only at boundaries separating relatively large intervals of normal and reversed polarity. The many narrow intervals of normal and reversed polarity in the two core sections are unrepresented in the accepted time scale. In particular, an interval of mixed polarity in Tiszapalkonya is correlated with the interval of reversed polarity between normal polarity Anomalies 4A and 4 (Fig. 11). This allows the coarse-scale polarity zonation in Tiszapalkonya to be correlated with the accepted polarity time scale for the late Miocene in a manner compatible with a relatively uniform average rate of deposition, including deposition during the long interval of normal polarity (Anomaly 5).

The Kaskantyú section displays a somewhat different polarity zonation, somewhat compressed with respect to that observed in Tiszapalkonya, although their zonations appear to embrace the same general interval of time. The differences between the zonations presumably result from an attenuated Kaskantyú section, arising from

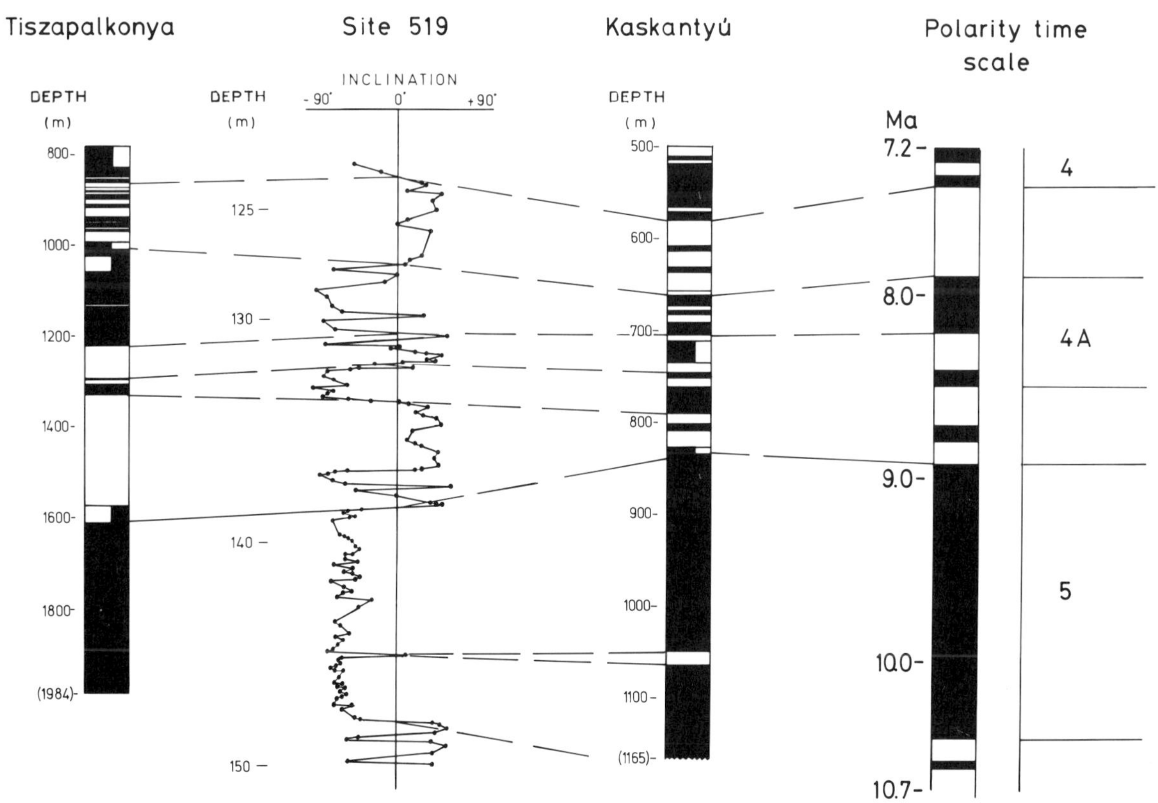

Figure 10. Comparison of polarity records from the Tiszapalkonya and Kaskantyú drill holes with the polarity record from part of Deep Sea Drilling Project Leg 73, Site 519 (after Tauxe and others, 1984, Fig. 2). Depths are referenced to ground surface.

deposition above a structurally elevated basement block. From seismic stratigraphic analysis (Pogácsás and others, this volume), the Kaskantyú area does not appear to have subsided as uniformly as adjacent, deeper parts of the basin, explaining why deposits of the Kaskantyú section do not contain as uniform a polarity record as those at Tiszapalkonya. In spite of this, all major polarity zones for the interval 8.9 to 5.9 Ma appear to be recorded at Kaskantyú (Fig. 11).

Pliocene and Pleistocene

Polarity zonations for the Dévaványa and Vésztö drill holes reported by Cooke and others (1979) correlate broadly with the principal intervals of normal and reversed polarity for the Bruhnes Normal-, Matuyama Reversed-, and Gauss Normal-Polarity Chrons of the geomagnetic polarity time scale (Fig. 3). However, these drill core records display more reversals than are present in the polarity time scale. Cooke and others (1979), employing the assumption of a uniform rate of accumulation, proposed that the reversals in the lower part of the sections correlate with the time scale for the Gilbert on an essentially one-to-one basis. This correlation served to place the base of the drill-core sections near or below the Miocene-Pliocene boundary, a correlation not supported by paleontological evidence. An alternate interpretation arose following analysis of seismic profiles and development of subsurface correlations that led to the stratigraphic framework depicted in Figure 2b. These correlations indicated that (as for the late Miocene deposits of Tiszapalkonya and Kaskantyú), only the major polarity intervals in Dévaványa and Vésztö should be correlated with the major polarity intervals of the polarity time scale (Fig. 3). In such a correlation, the base of the Dévaványa and Vésztö sections is inferred to lie at about 4.25 Ma rather than 5+ Ma as had been proposed previously by Cooke and others (1979). The new interpretation implies that, as for the Miocene, either there were significantly more reversals of the geomagnetic field than is recorded for the accepted polarity time scale or that the many "extra" reversals are the result of unrecognized overprinting. The differences in polarity zonation between Dévaványa and Vésztö for the well-known Matuyama Reversed-Polarity Chron suggests that the former is the case because, (1) the six or more intervals of normal polarity in Vésztö accord better with the six normal polarity intervals for the Matuyama shown on a recent version of the polarity time scale (Seward and others, 1986), and (2) the Vésztö cores were the more carefully collected and processed, and its record is considered the more reliable (Cooke and others, 1979; Cooke, 1985).

DISCUSSION

Strata penetrated in four deep core holes are placed in the time scale for the late Miocene and Pliocene on the basis of their polarity zonations (Fig. 12). In this figure, the thicknesses of the individual sections and their polarity zonations have been adjusted to provide a proportionally uniform fit with the time scale. This temporally-controlled stratigraphic framework, tied to seismic-reflection data, has allowed the timing of accumulation of delta plain and delta slope deposits in the Pannonian Basin to be assessed (Pogácsás and others, this volume). A brief summary of the regional stratigraphic relations developed from this assessment, and a discussion of generalized accumulation rates inferred from correlations with the polarity time scale, follow.

Regional correlations

The time lines (Fig. 12) indicate that three rock units (or "packages" of rock) occupy discrete parts of the time scale and record three discrete episodes involved with deposition in the Pannonian Basin. The units represent (1) a deep water facies or sequence (lower Pannonian); (2) a lacustrine and terrestrial sequence characterized by abundant plant life (lower part of upper Pannonian); and (3) terrestrial deposits characterized by mottled claystone containing a lesser abundance of plant life (upper part of upper Pannonian). Unit 1 is late-middle Miocene and early-late Miocene in age, accumulating mainly in the interval 10.4 - 8.9 Ma. Unit 2 is late Miocene in age, accumulating in the interval 8.9 to <5.9 Ma; in the deep parts of the basin, its upper boundary may lie near or at the Miocene-Pliocene boundary. Unit 3 appears to be entirely Pliocene in age, and its lower boundary may lie near or coincide with the Miocene-Pliocene boundary. This interpretation applies and may be restricted to the southeastern part of Hungary. According to Pogácsás (1984), Mattick and others (1985), Mattick and others (this volume), and Pogácsás and others (this volume), the boundaries between these Pannonian depositional facies are strongly time-transgressive when the entire Pannonian Basin is considered. Characteristics and ages of the three units follow.

Lower Pannonian deposits accumulated in deep water, an environment representing a vestige of the Paratethys sea that once was connected to the Tethys on the east.

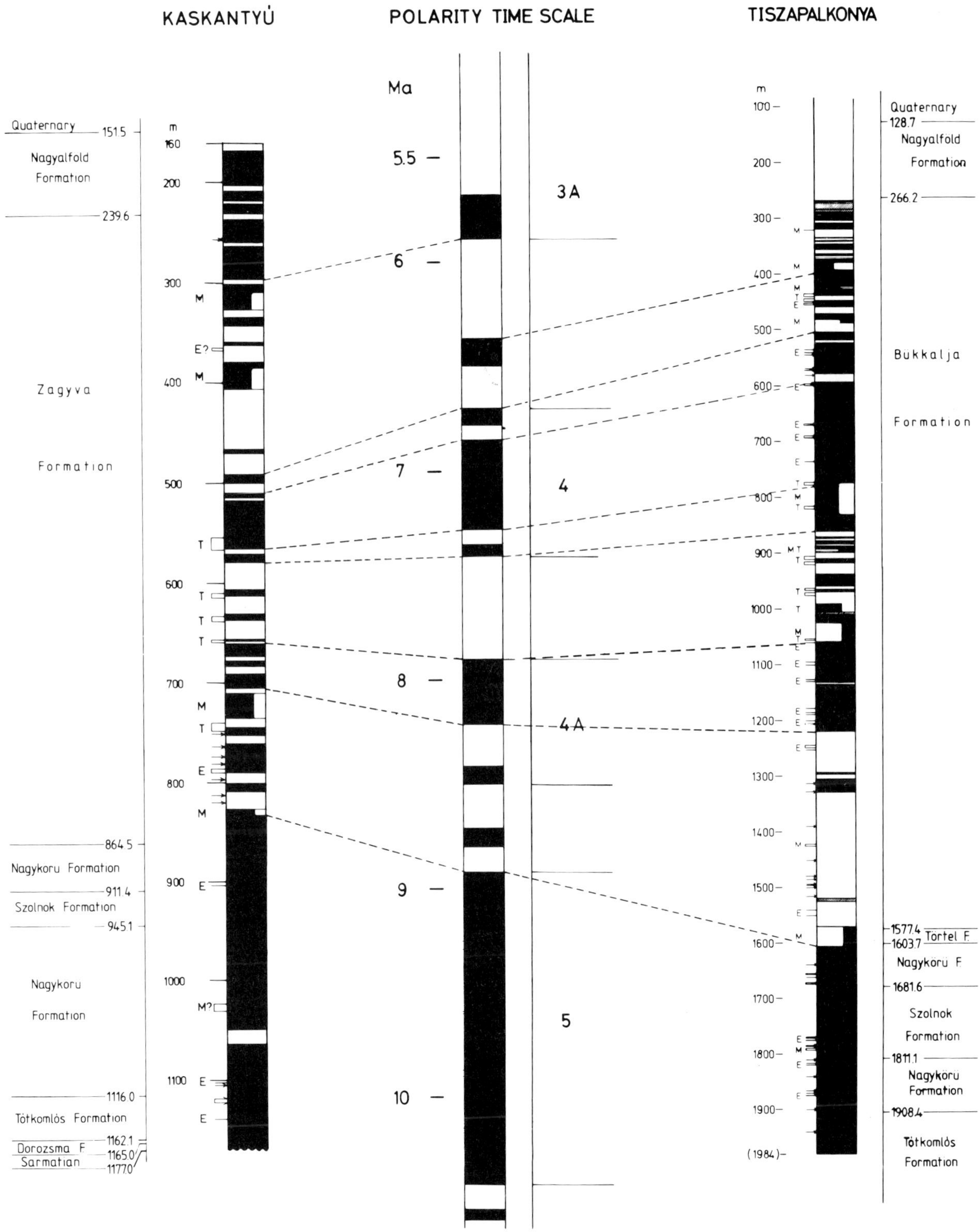

Figure 11. Proposed correlation of polarity zonations from the Tiszapalkonya and Kaskantyú core holes with the geomagnetic polarity time scale for part of late Miocene time. Depths are referenced to ground surface.

Although the base of the Pannonian is dated isotopically at about 11.5-12 Ma, none of the lower Pannonian units penetrated in the Tiszapalkonya and Kaskantyú cores display intervals of reversed polarity. This indicates that the switch from reverse to normal polarity at the base of the long interval of normal polarity (Anomaly 5, 8.9-10.4 Ma in the time scale of Berggren and others, 1985) probably is not represented in the cored lower Pannonian deposits. This also suggests that the base of the Pannonian section in the drill holes is early late Miocene in age rather than late middle Miocene as indicated by the isotopic dating of lower Pannonian rocks elsewhere in the Pannonian basin. Therefore, the lower parts of the Pannonian sections in Tiszapalkonya and Kaskantyú are considered here to be late Miocene in age, suggesting that the unconformity separating Pannonian from underlying Sarmatian strata in Kaskantyú may represent more than one million years.

The foregoing interpretation assumes (1) that the base of Anomaly 5 is properly placed in the time scale, and (2) that isotopic dates obtained for igneous rocks present locally in the lower Pannonian section are reliable. Among geologic and paleomagnetic professionals, discussions continue regarding ages that may be assigned to the polarity zonation for the Miocene. Berggren (1987) recently has proposed a revised time scale for an interval of late Neogene time that includes Anomaly 5. The revised estimates are derived from a series of K-Ar whole-rock isotopic dates from basalt flows related to a polarity zonation reported from northwest Iceland (McDougall and others, 1984). If the Icelandic age data are accepted, they provide an estimated age of 11.12 Ma for the base of Anomaly 5 in the revised time scale proposed by Berggren (1987, his Fig. 8). However, as Berggren points out, the Icelandic isotopic ages may not be reliable due to alteration (loss of potassium). A similar suggestion also had been made by Tauxe and others (1985), who favor a data set based on mineral ages from Kenya that indicate a considerably younger age for the base of Anomaly 5. Berggren (1987, p. 29) notes that reasons for the discrepancies may be more complicated, involving unresolved problems in isotopic dating, the identification of the magnetic polarity stratigraphy in the Icelandic data set, or a combination of both. For the moment, the arguments of Tauxe and others (1985) appear to be the more sound, and the polarity zonations from the Hungarian core sections have been referred to the earlier time scale of Berggren and others (1985).

An isotopic age of 11.6 Ma (plus or minus 0.5) determined by K-Ar analysis of fresh biotite from a tuff interbedded in the lower Pannonian, penetrated by the Nagykozár-2 drill hole in southern Hungary (Fig. 1), is accepted as a reliable age for the lower part of the lower Pannonian. Thus, we follow the assignment of Vass and others (1987) and recognize the base of the Pannonian as late middle Miocene (11.5 Ma (plus or minus 0.5)). The anomaly between the polarity "ages" from Tiszapalkonya and Kaskantyú and the isotopic ages from near the base of the Pannonian at other localities may have a simple explanation. Initiation of Pannonian deposition across the Pannonian Basin may not have been synchronous (Lóczy, 1913; Körössy, 1968). From the paleomagnetic data, onset of Pannonian deposition perhaps occurred over a 1 to 1.5 m.y. span of time. Additionally, from their normal polarity, accumulation of lower Pannonian deposits occurred mainly within the interval of time marked by marine magnetic Anomaly 5 (10.4 - 8.9 Ma), and, the boundary between the lower and upper Pannonian lies close to the top of Anomaly 5.

Upper Pannonian deposits of late Miocene and Pliocene age mark the time of filling of the Pannonian Basin with delta plain and delta slope deposits, identified from a combination of seismic and drill-core records. Lignite-bearing and lignitic deposits, although given different formation names at different places, accumulated from about 8.9 to at least 6.4 Ma, and probably to less than 5.9 Ma (the highest polarity interval in the Kaskantyú core provisionally correlated with the polarity time scale). From the stratigraphic records, the abundance of plant life decreased during the latter part of late Miocene time, presumably coinciding with a time of increased aridity elsewhere (e.g., with the Messinian "salinity crisis" of the western Mediterranean; Rouchy, 1982; Moissette and Pouyet, 1987). The Messinian is identified by Berggren and others (1985) as having occurred in the interval 6.4 to 5.2 Ma. In northern Greece, southeast of the Pannonian Basin, red beds are present in upper Miocene, Pliocene, and Pleistocene strata (Psilovikos and others, 1987), and these also may reflect regional conditions of increased aridity near the end of the Miocene and during Pliocene time.

The upper part of the upper Pannonian is represented by the Pliocene Nagyalföld Formation, a lithologic unit recognized across the region. It is represented in Dévaványa and Vésztö by a mottled claystone unit that may record somewhat drier conditions.

Accumulation rates

Average rates of accumulation for the four deep drill holes can be estimated from plotting thicknesses of the

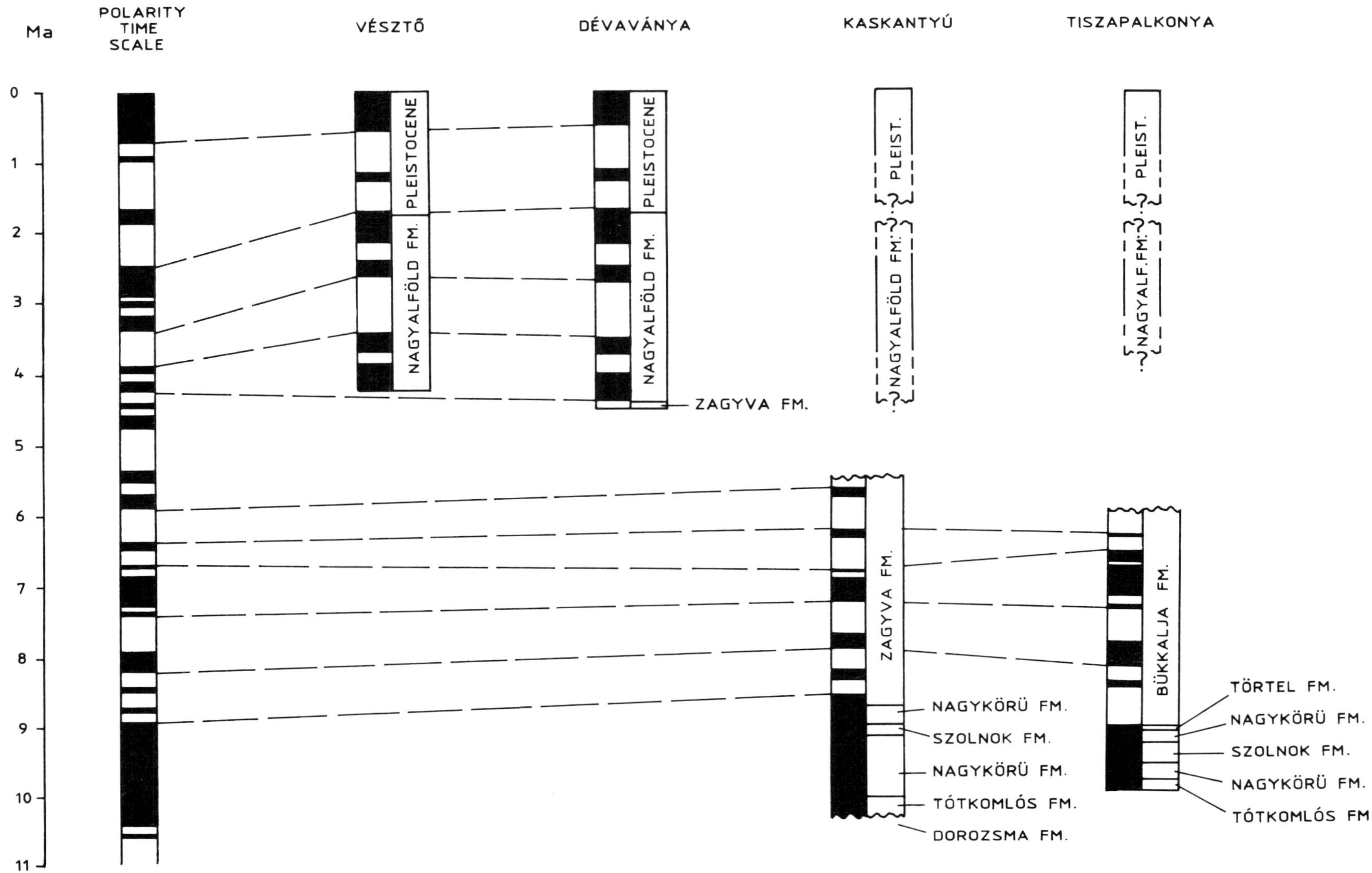

Figure 12. Correlation of generalized polarity records and stratigraphic units of late Miocene and Pliocene age from four deep core holes in the Pannonian Basin. Thicknesses of individual sections have been adjusted to provide general fits with correlated intervals of the polarity time scale.

sections against time (Fig. 13). In Tiszapalkonya, a 1+ km-thick interval accumulated at an average rate of 460 m/m.y., whereas a 500 m-thick interval in the Kaskantyú core accumulated at an average rate of 175 m/m.y. These estimates are average minimum rates of accumulation, and do not take into account details of the depositional history that may be inferred from the seismic stratigraphy. The appreciably lower accumulation rate for the Kaskantyú section is explained by its position on a structurally elevated block that did not sink as rapidly as adjoining parts of the basin (Pogácsás and others, this volume). The revised correlations with the polarity time scale for Dévaványa and Vésztö (Fig. 3) lead to accumulation curves that change with time (Fig. 13). These curves display accumulation rates for the middle Gilbert (600 to 620 m/m.y.) that are distinctly greater than the 460 m/m.y. rate for Tiszapalkonya. The greater rates may be explained by the location of Dévaványa and Vésztö, which are in one of the deepest, most actively subsiding parts of the Pannonian Basin. The accumulation curves flatten for the younger parts of the Dévaványa and Vésztö sections, averaging 200 to 220 m/m.y. from the Gauss to the Bruhnes Chrons.

Character of Geomagnetic Field

The geomagnetic polarity time scale for the late Miocene, when viewed with respect to the high resolution paleomagnetic records from Tiszapalkonya and Kaskantyú (Fig. 11), and from Szombathely (Fig. 9), appears to be a generalized representation of field behavior. A similar conclusion is reached when the polarity zonations of Dévaványa and Vésztö are compared with the polarity time scale for the Pliocene and Pleistocene (Fig. 3).

For the late Miocene, Anomaly 5 exhibits stable normal polarity, although locally containing a few excursions and brief reversals. The upper boundary of Anomaly 5 is closely overlain by a series of brief reversals rather than a single reversal in Tiszapalkonya, Kaskantyú, and Szombathely, apparently recording a change to less stable geomagnetic conditions. This assumes that delays in acquisition of stable remanence did not suddenly take place at this horizon across the basins. Intervals of normal and reversed polarity overlying this lower interval of mixed polarity contain abundant subordinate reversals. At places in the Tiszapalkonya and Kaskantyú sections, the reversals have frequencies near the limit of resolution of the 1/2 m sampling intervals (approximately 1000 yr for Tiszapalkonya and 3000 yr for Kaskantyú).

Some systematic patterns in inclination and declination are recognized in the high-resolution records that are suggestive of field behavior (Figs. 8a-8c), an interpretation now supported by detailed correlations of oscillations of the inclination record in the Szombathely well in the Little Hungarian Plain with the inclination record in Tiszapalkonya in the Great Hungarian Plain (Fig. 9). On a broad scale, the interval between Anomalies 4 and 4A may not be a well-defined interval of reversed polarity, but rather may be an interval of mixed polarity (Fig. 11). A few well-defined intervals of normal and reversed polarity are found above and below this interval in Tiszapalkonya and Kaskantyú, but even these better-defined polarity intervals contain multiple reversals. It is difficult to explain the differing number of multiple reversals in the sections as arising from complex magnetizations from an ambient field having the stability that is portrayed in the accepted polarity time scale. Except for the Szombathely section, it is simpler to explain discrepancies in fine detail between the records in the drill cores, and the differences with the polarity time scale, as arising from hiatuses in the sections and a considerably less than complete recording of the field.

If the records in Tiszapalkonya and Kaskantyú are broadly accepted at face value, the geomagnetic field proceeded from a state of relative stability during the time of Anomaly 5 into a state of increasing instability, with instability peaking during a time that correlates with the interval of reversed (but highly mixed) polarity lying between Anomalies 4 and 4A. Within this interval, regularities seen at a fine scale may reflect complex transitions from one polarity state to the other. For this mixed polarity interval, and for underlying and overlying broad intervals of normal and reversed polarity as well, transitions from one polarity state to the other appear to range from simple switches to complex, multiple switches of polarity. Complexity in the reversal pattern cannot be related to details of the lithology or stratigraphy.

The drill-core records from Tiszapalkonya and Kaskantyú suggest that some reversals, portrayed in the accepted geomagnetic time scale as consisting of a single polarity state, contain multiple polarity reversals. Linear magnetic anomalies in the ocean-floor record have been modelled as short subchrons (Blakely, 1974) or as fluctuations of intensity (Cande and LaBrecque, 1974). Tauxe and others (1984) note that, for DSDP Site 519 (see Fig. 10, this report), at least some of the "tiny wiggles" in the magnetic anomaly pattern could be caused by short-term reversals of the magnetic field. Tauxe and others (1984) also concluded that the DSDP

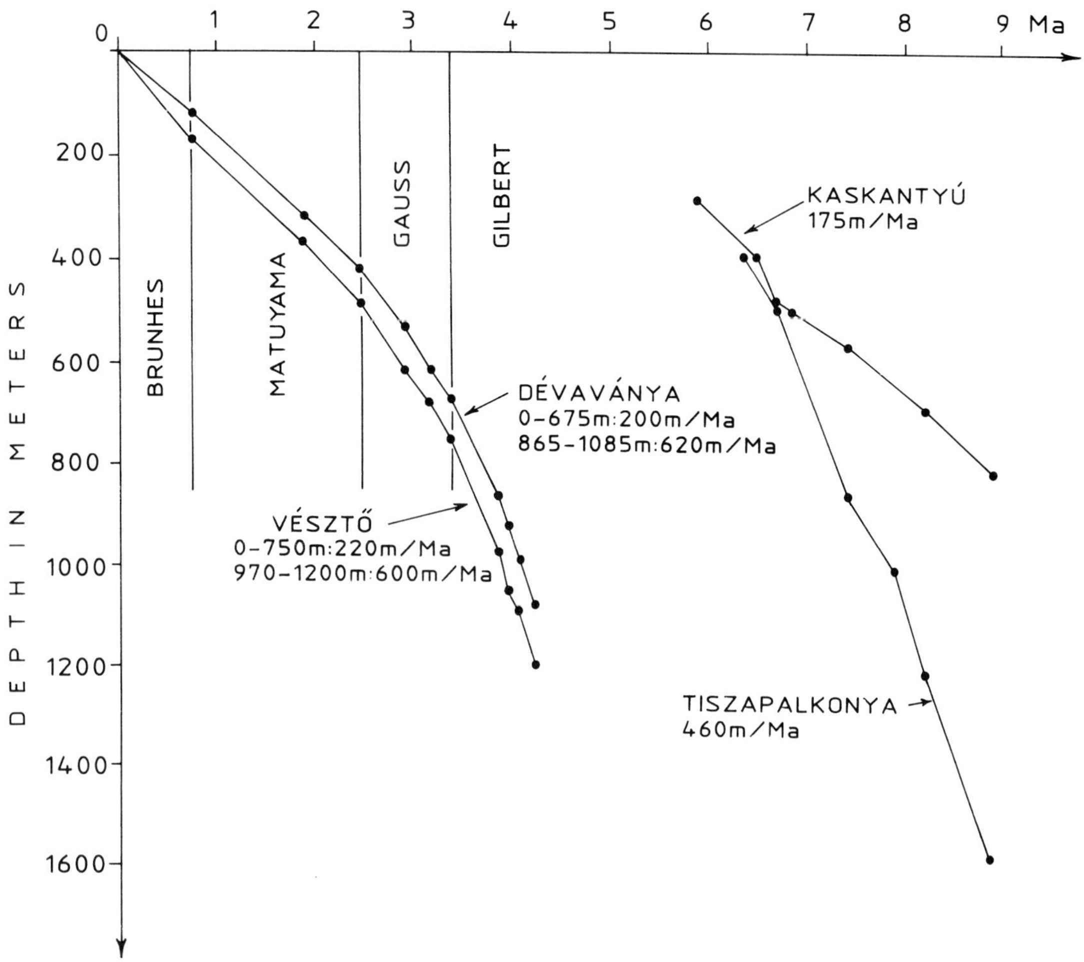

Figure 13. Accumulation curves for parts of the Tiszapalkonya and Kaskantyú core sections lying above marine magnetic Anomaly 5, and for the Dévaványa and Vésztö core sections. Depths are referenced to ground surface. Brunhes: Normal polarity chron; Matuyama: reversed polarity chron; Gauss: normal polarity chron; Gilbert: reversed polarity chron.

Leg 73 paleomagnetic data set constrains the length of undetected subchrons to less than 20000 years. This would represent about 10 m of record in the Tiszapalkonya drill-core and 4 m of record in the Kaskantyú drill core.

Records in the four Hungarian cores indicate that appreciably more short reversals occurred during late Miocene and Pliocene time than is depicted in the polarity time scale. The high-resolution Hungarian records result in the disquieting thought that correlations based on polarity zonations for the late Miocene to Pleistocene, obtained either from relatively narrow stratigraphic intervals or from temporally "coarse" records, may be less certain than would appear from a simple matching of polarities using the accepted polarity time scale as reference.

SUMMARY

High-resolution polarity records were obtained from two continuously-cored holes (2 km- and 1.2 km-deep), drilled at Tiszapalkonya and Kaskantyú in the northern and central parts of the Great Hungarian Plain. Correlated with the late Miocene geomagnetic field for the interval 10 to 6 Ma, the 1/2 m sampling intervals in these holes have provided temporal resolutions of about 1000 and 3000 years, respectively. The stratigraphic sections and their polarity records were placed in the polarity time scale by correlation of long intervals of normal polarity, present in the lower parts of both holes, with marine magnetic Anomaly 5 (early late Miocene). Stratigraphic horizons mapped (in time units) on seismic reflection profiles across the Pannonian basin (Pogácsás and others, this volume) indicate that the lower stratigraphic units of the upper Pannonian sequences in Tiszapalkonya and Kaskantyú are of late Miocene age, and that they correlate with strata underlying Pliocene and Pleistocene deposits penetrated in the Dévaványa and Vésztö core holes drilled in the east-central part of the Great Hungarian Plain. Above Anomaly 5 in Tiszapalkonya and Kaskantyú, only the broad intervals of normal and reverse polarity can be correlated with the geomagnetic polarity time scale for the interval 8.9 to 6.4 and 5.9 Ma. The correlations lead to calculated average sedimentation rates of 460 and 175 m/m.y., respectively, for the Tiszapalkonya and Kaskantyú sections.

Magnetostratigraphic records in the Dévaványa and Vésztö core holes, 1.2 km in depth and correlated by Cooke and others (1979) with the geomagnetic time scale for the Pleistocene, Pliocene, and early Miocene, were re-evaluated in light of seismic stratigraphic studies, and the new stratigraphic and geomagnetic correlations were developed with respect to Tiszapalkonya and Kaskantyú. The analyses resulted in assignment of upper Pannonian strata near the base of Dévaványa and Vésztö, and their polarity records, to the middle part of the Gilbert Reversed-Polarity Chron (late early Pliocene), rather than to the polarity zonation for the late Miocene (5+ Ma) and early Pliocene as had been previously proposed. Strata at the base of the Dévaványa and Vésztö sections are now provisionally assigned a maximum age of about 4.25 Ma. The new correlations suggest an accumulation rate during the early Pliocene of about 600 m/m.y. for the east central part of the Pannonian basin. Pannonian strata in the age range 4.25-5.9 Ma are not recognized paleomagnetically or paleontologically in the four deep drill cores.

ACKNOWLEDGEMENTS

The magnetostratigraphic studies were carried out as part of a bilateral cooperative agreement between the U.S. Geological Survey (U.S.A.) and the Central Office of Geology (Hungary). Laboratory work and associated costs in the U.S. were supported by the Geologic Framework and Synthesis Program, Geologic Division, U.S. Geological Survey. We acknowledge the careful work of geologists and technicians who carried out the sampling of the Tiszapalkonya drill hole. We thank Géza Hámor, former Director of the Hungarian Geological Institute (MÁFI), for the broad range of support provided this project.

Lithologic, biostratigraphic, and stratigraphic analyses of the Tiszapalkonya and Kaskantyú core holes were under the supervision of Áron Jámbor, Hungarian Geological Institute, who freely provided pertinent information bearing on regional stratigraphic correlations derived from paleontological and sedimentological studies of his group, and on the existence and extent of unconformities recognized in the subsurface sections. Á. Jámbor not only provided important information bearing on the stratigraphy of the deposits studied but he also reviewed an early draft of this manuscript. However, he is not responsible for interpretations presented here concerning the age and correlation of stratigraphic units. We thank Richard Blakely, Edward Mankinen, Paul Teleki, and Robert Mattick of the U.S. Geological Survey for their careful and helpful reviews of the manuscript.

REFERENCES

Balla, Z., 1987a, Neogene kinematics of the Carpatho-Pannonian region, Proc. 8th Cong., Reg. Comm. on Mediterr. Neogene Stratigraphy, Budapest, 15-22 September, 1985: Ann. Inst. Geol. Hung., 70, 193-199.

Balla, Z., 1987b, The middle section of the Alpine-Mediterranean belt in the Neogene; Proc. 8th Cong., Reg. Comm. on Mediterr. Neogene Stratigraphy, Budapest, 15-22 September, 1985: Ann. Inst. Geol. Hung., 70, 301-306.

Balogh, K., and Jámbor, Á., 1985, Radiometrische Daten zur Charakteristik postsarmatischer Ablagerungen in Ungarn, *in* Papp, F. (ed.), Chronostratigraphie und Neostratotypen Miozän (M6), Pannonien, Akadémiai Kiadó, Budapest, 177-179.

Balogh, K., Jámbor, Á., Partényi, Z., Ravasz-Baranyai, L., Solti, G., and Nusszer, A., 1983, Petrography and K/Ar dating of the Tertiary and Quaternary basaltic rocks in Hungary; Anuarul Institutului. Geologie si Geofizika, Bucharest, 61, 365-373.

Bérczi, I., and Phillips, R L., 1985, Processes and depositional environments within Neogene deltaic-lacustrine sediments, Pannonian Basin, southeast Hungary; Eötvös Lóránd Geophys. Inst., Geophysical Transactions, 31, (1-3), 55-74.

Berggren, W.A., 1987, Neogene chronology and chronostratigraphy-new data; Proc. 8th Cong., Reg. Comm. on Mediterr. Neogene Stratigraphy, Budapest, 15-22 September, 1985: Ann. Inst. Geol. Hung., 70, 19-41.

Berggren, W.A., Kent, D.V., Flynn, J.J., and Van Couvering, J.A., 1985, Cenozoic geochronology: Geol. Soc. Am. Bull., 96, (11), 1407-1418.

Blakely, R.J., 1974, Geomagnetic reversals and crustal spreading rates during the Miocene; Jour. Geophys. Res., 79, 279-300.

Cande, S.C. and LaBrecque, J.L., 1974, Behavior of the Earth's paleomagnetic field from small scale marine magnetic anomalies; Nature, 247, 26-28.

Cicha, I. and Senes, J., 1968, Sur la position du Miocène de la Paratethys centrale dans le cadre du Tertière de l'Europe; Geol. Zbornik (Geol. Carpatica), Bratislava, 19, (1), 95-116.

Cooke, H.B.S., 1985, Potential of Great Hungarian Plain drilling for paleomagnetic theory and climatic records; Abs. vol., 8th Cong., Reg.l Comm. on Mediterr. Neogene Stratigraphy, Hung. Geol. Inst., Budapest, 153-155.

Cooke, H.B.S., Hall, J.M., and Rónai, A., 1979, Paleomagnetic, sedimentary, and climatic records from boreholes of Dévaványa and Vésztö, Hungary; Acta Geol. Hung., 22, (1-4), 86-109.

Cserepes-Meszéna, B., 1978, On the Lower Pannonian basalts and migmatites uncovered by the hydrocarbon-exploratory borehole Kiskunhalas-Ny-3 (in Hungarian); Földtani Közlöny, 108, (1), 53-64.

Dank, V., 1987, The role of Neogene deposits among the mineral resources in Hungary; Proc. 8th Cong., Reg. Comm. on Mediterr. Neogene Stratigraphy, Budapest, 15-22 September, 1985: Annal. Inst. Geol. Hung., 70, 9-17.

Elston, D.P., Hámor, G., Jámbor, Á., Lantos, M., and Rónai, A., 1985, Magnetostratigraphy of Neogene strata penetrated in two deep core holes in the Pannonian Basin: Preliminary results; Eötvös Lóránd Geophysical Institute, Geophysical Transactions, 31, (1-3), 75-88.

Hámor, G., 1984, Paleogeographic reconstruction of Neogene plate movements in the Paratethyan realm; Acta Geol. Hung., 27, (1-2), 5-21.

Hámor, G., Jámbor, Á., Lantos, M., Rónai, A., and Elston, D.P., 1987, Magnetostratigraphic correlations of some Neogene strata of the Pannonian Basin, Hungary; Abs., 8th Cong., Reg. Comm. on Mediterr. Neogene Stratigraphy, Hung. Geol. Inst., Budapest, 257.

Jámbor, Á., 1980, Pannonian in the Transdanubian Central Mountains; Ann. Inst. Geol. Hung., 62, 243 p.

Jámbor, Á., Balázs, E., Balogh, K., Bérczi, I., Bóna, J., Horváth, F., Gajdos, I., Geiger, J., Hajós, M., Kordos, L., Korecz, A., Korecz-Laky, I., Korpás-Hódi, M., Kőváry, J., Mészáros, L., Nagy, E., Németh, G., Nusszer, A., Pap, S., Pogácsás, Gy., Révész, I., Rumpler, J., Sütö-Szentai, M., Szalay, A., Szentgyörgyi, K., Széles, M., and Völgyi, L., 1987, General characteristics of Pannonian s.l. deposits in Hungary; Proc. 8th Cong., Reg. Comm. on Mediterr. Neogene Stratigraphy, Budapest, 15-22 September, 1985: Ann. Inst. Geol. Hung., 70, 155-167.

Kilényi, É. and Rumpler, J., 1985, Pre-Tertiary basement relief map of Hungary: Eötvös Lóránd Geophysical Institute, Geophys. Trans.,30, (4), 425-428.

Körössy, L., 1968, Entwicklungsgeschichtliche und paleogeogräphische Grundzüge der ungarische Unterpannon: Acta Geol. Hung., 12, 199-217.

Körössy, L., 1980, Investigation into Neogene paleogeography in the Carpathian Basin (in Hungarian): Földtani Közlöny, 110, (3-4), 473-484.

Lóczy, L., Sr., 1913, A Balaton környékének geológiai képződményei és ezek vidékek szerinti telepedése (Geological formations and their regional distribution in the vicinity of Lake Balaton: in Hungarian): Balaton monografia 1 (1) (1), 173-581.

Mattick, R.E., Rumpler, J., and Phillips, R.L., 1985, Seismic stratigraphy of the Pannonian Basin in southeastern Hungary: Eötvös Lóránd Geophysical Institute, Geophysical Transactions, 31, (1-3), 13-54.

Mattick, R.E., Rumpler, J., Ujfalusy, A., Szanyi, B. and Nagy, I., Sequence stratigraphy of the Békés basin; this volume, 39-65.

McDougall, I., Krijansson, L., and Saemundsson, K., 1984, Magnetostratigraphy and geochronology of Northwest Iceland; Jour. Geophys. Res., 89, (B8), 7029-7060.

Moisette, P, and Pouyet, S., 1987, Bryozoan faunas and the Messinian salinity crisis; Proc. 8th Cong., Reg. Comm. on Mediterr. Neogene Stratigraphy, Budapest, 15-22 September, 1985; Ann. Inst. Geol. Hung., 70, 447-453.

Molnár, K., Pogácsás, Gy., and Rumpler, J., 1987, Seismic reflection investigations in the Hungarian part of the Pannonian Basin: Application to hydrocarbon exploration; Proc. 8th Cong., Reg. Comm. on Mediterr. Neogene Stratigraphy, Budapest, 15-22 September, 1985; Ann. Inst. Geol. Hung., 70, 593-599.

Phillips, R.L., Révész., I., and Bérczi, I., Lower Pannonian deltaic-lacustrine processes and sedimentation, Békés basin; this volume, 67-82.

Pogácsás, Gy., 1980, Evolution of Hungary's Neogene depressions in the light of geophysical surface measurements (in Hungarian); Földtani Közlöny, 110, (3-4), 485-497.

Pogácsás, Gy., 1984, Seismic stratigraphic features of Neogene sediments in the Pannonian Basin; Eötvös Lóránd Geophysical Institute, Geophysical Transactions, 30, (4), 373-410.

Pogácsás, Gy., 1987, Seismic stratigraphy as a tool for chronostratigraphy: Pannonian Basin; Proc. 8th Cong., Reg.l Comm. on Mediterr. Neogene Stratigraphy, Budapest, 15-22 September, 1985; Ann. Inst. Geol. Hung., 70, 55-63.

Pogácsás, Gy., Mattick, R.E., Elston, D.P., Hámor, T., Jámbor, A., Lakatos, L., Simon, E., Vakarcs, G., Várkonyi L., and Várnai, P., Correlation of seismo- and magnetostratigraphy in southeastern Hungary; this volume,143-160.

Psilovikos, A., Koufos, G. and Syrides, G., 1987, The problem of red-beds in Northern Greece; Proc. 8th Cong., Reg. Comm. on Mediterr. Neogene Stratigraphy, Budapest, 15-22 September, 1985; Ann. Inst. Geol. Hung., 70, 509-516.

Rouchy, J.M., 1982, La genèse des evaporites messiniennes de Mediterranée: Mem. Mus. Hist. Nat., Paris, ser. C., 50, 267.

Senes, J., 1971, Einleitung, *in* Senes, J. and Steininger, F., Chronostratigraphie und Neostratotypen: 2,: (M/1), Eggenburgien, Die Eggenburger Schichtengruppe und ihr Stratotypen; Vidavel'stvo Slovenskej akadémie vid, Bratuslava, 21-27

Seward, D., Christofel, D.A., and Linert, B., 1986, Magnetic polarity stratigraphy of a Plio-Pleistocene marine sequence of North Island, New Zealand; Earth and Planetary Science Letters, 80, 353-360.

Tauxe, L., Tucker, P., Petersen, N.P., and LaBrecque, J.L., 1984, Magnetostratigraphy of Leg 73 sediments; Initial reports of the Deep Sea Drilling Project, 73, 609-621.

Tauxe, L., Monaghan, M. Drake, R., Curtis, G. and Staudigel, H., 1985, Paleomagnetism of Miocene East African Rift sediments and the calibration of the geomagnetic reversal time scale; Jour. Geophys. Res., 90, 4639-4646.

Vass, D., Repcok, I., Balogh, K., and Halmai, J., 1987, Revised radiometric time-scale for the Central Paratethyan Neogene; Proc. 8th Cong., Reg. Comm. on Mediter. Neogene Stratigraphy, Budapest, 15-22 September, 1985; Ann. Inst. Geol. Hung., 70, 423-434.

7 Correlation of Seismo- and Magnetostratigraphy in Southeastern Hungary

György Pogácsás[1], Robert E. Mattick[2], Donald P. Elston[3], Tamás Hámor[4], Áron Jámbor[4], László Lakatos[1], Miklós Lantos[4], Ernö Simon[1], Gábor Vakarcs[1], László Várkonyi[1] and Péter Várnai[1]

ABSTRACT

Correlation of results from magnetostratigraphic and seismic-reflection studies indicate that the Pannonian Basin, during the postrift phase of its evolution (middle Miocene to present), became filled by sediments of southward and eastward prograding deltaic wedges.

In the Békés basin (a subbasin of the Pannonian Basin), the Badenian-Sarmatian age (16.5 to12 Ma) section, where explored by drilling, is relatively thin (generally less than 275 m) and represents principally shallow-water, nearshore, marine-brackish water environments. During latest Sarmatian and early Pannonian time (12-9 Ma), the Békés basin was starved, as other subbasins located along its margins captured most of the sediment load carried by rivers. During this time interval, a combination of relatively low deposition rates and high subsidence rates produced great water depths (about 900 m) in the Békés basin. By middle Pannonian time (6-7 Ma), subbasins on the margins of the basin had become filled with sediments as deltas gradually prograded across them. As a result of this filling process, a platform was constructed across which rivers transported their sediment loads into the Békés basin. Thereafter, deltaic filling of the basin proceeded rapidly and rates of sediment accumulation reached 1000 m/million years. Lacustrine sediments, more than 6000 m thick, were deposited in the deeper parts of the basin.

Indirect evidence suggests that lake levels in the Pannonian inland sea (a remnant of the Paratethys), although isolated from the world's oceans, were affected by eustatic sea level changes. Three hiatuses were identified on seismic profiles from the northern margin of the Pannonian Basin, and are inferred to represent non-deposition. These hiatuses (7.9-7.6, 6.8-5.7, and 5.4-4.6 Ma) appear to correlate with global eustatic sea level minima. This suggests that the Pannonian inland sea became gradually isolated from the world oceans, and fluctuated in phase with global sea level. The hiatus between 6.8 and 5.7 Ma is tentatively correlated with the Messinian global stage, during which time evaporite deposition in the Mediterranean was widespread - the so-called "Messinian salinity crisis" (Hsü and others, 1973).

[1] MOL Rt. (Hungarian Oil and Gas Co., Ltd.), H-1068, Budapest, Hungary
[2] U.S. Geological Survey, Reston, Virginia 22092, USA
[3] U.S. Geological Survey, Flagstaff, Arizona 86001, USA
[4] Hungarian Geological Survey, H-1144, Budapest, Hungary

P. G. Teleki et al. (eds.), Basin Analysis in Petroleum Exploration, 143–160.

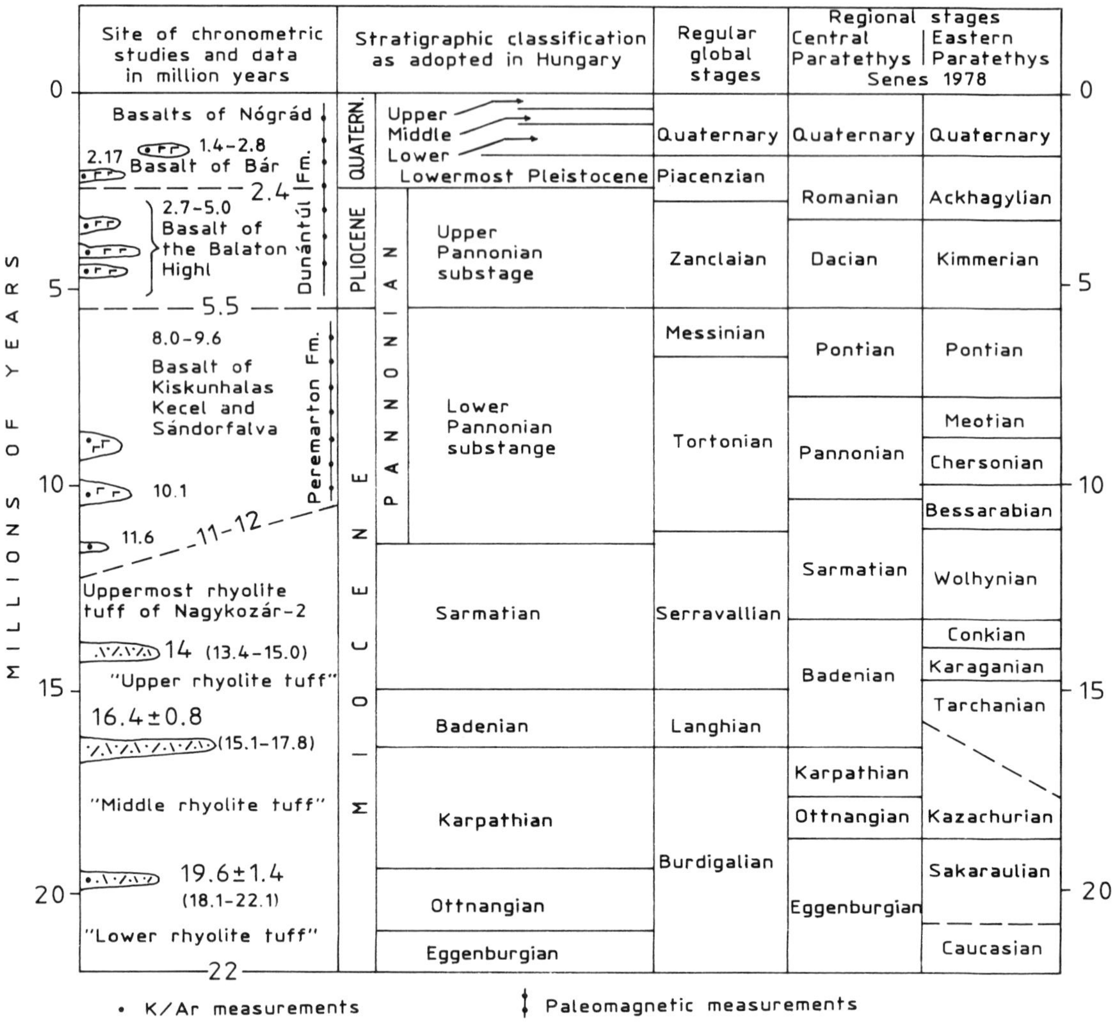

Figure 1. Stratigraphic classification used in Hungary with comparable global stages and regional stages. Ages, established for volcanic rocks in Hungary by K/Ar measurements, are shown in the left-hand column (after Hámor and others, 1980). Figure modified from Hámor and others (1987).

INTRODUCTION

During the post-rift phase of its evolution (from the middle Miocene to the present time), the Pannonian Basin was filled primarily by southward and eastward prograding sediment wedges (Pogácsás, 1984; Mattick and others, 1985; Jámbor and others, 1987; Pogácsás and others, 1987). Based principally on the interpretation of seismic-reflection data, these authors inferred that, as a consequence of this progradation, the post-rift sediments of the Pannonian basin become progressively younger from the basin's margins toward the central depocenters. Until recently, however, this hypothesis could not be tested because the bulk of the volume of post-rift sediments is of Pannonian age and the chronostratigraphic scale of the Pannonian section in Hungary had not been defined with sufficient resolution to confirm that the sediments become younger in a basinward direction. In this paper, the results of magnetostratigraphic studies of the Pannonian section (Elston and others, this volume) are combined with seismic stratigraphic interpretations, with the aim of improving time-stratigraphic correlations among post-rift sedimentary rock units in the Pannonian Basin.

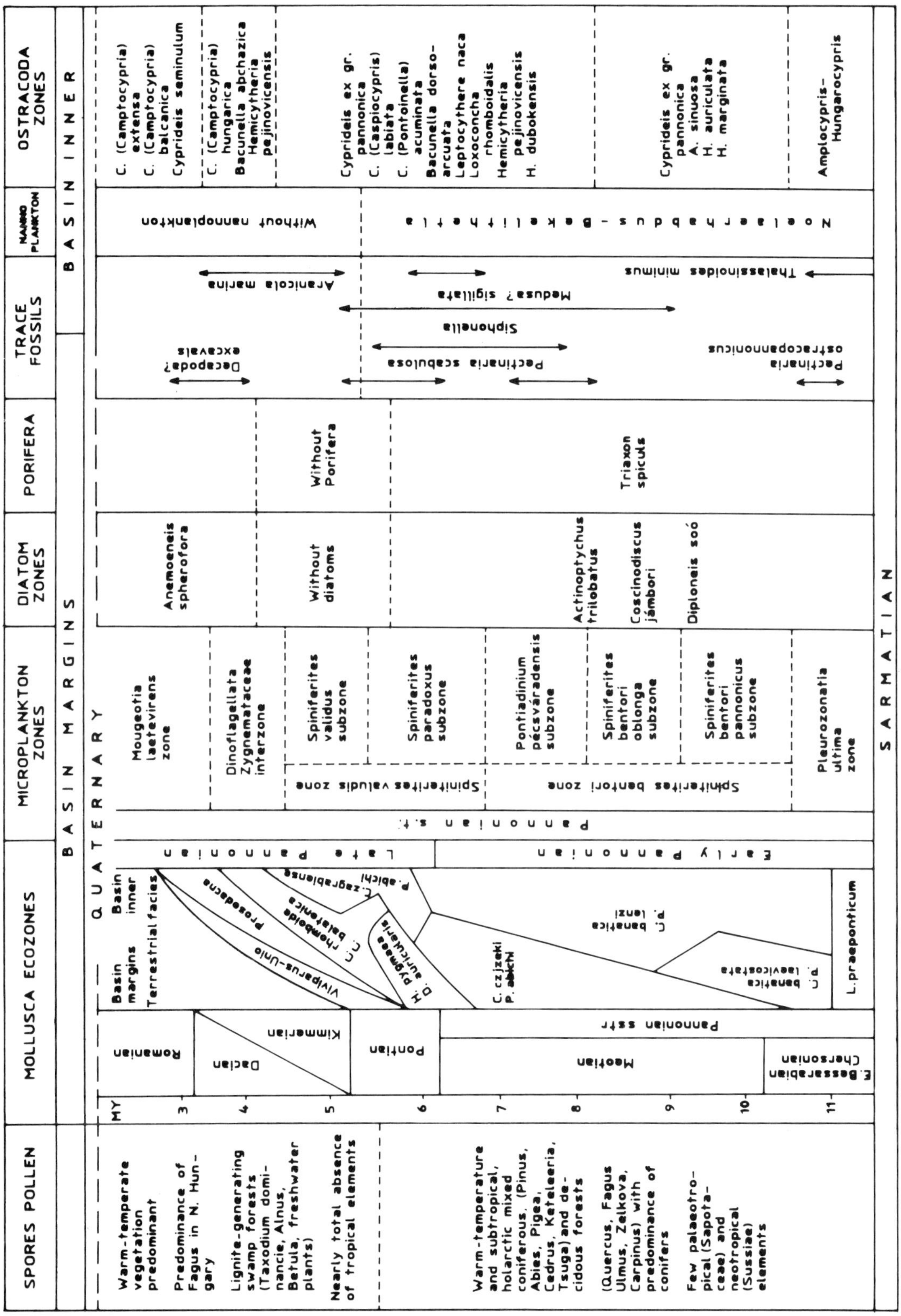

Figure 2. Biostratigraphic subdivisions of the Pannonian stage in Hungary (after Jámbor and others, 1987).

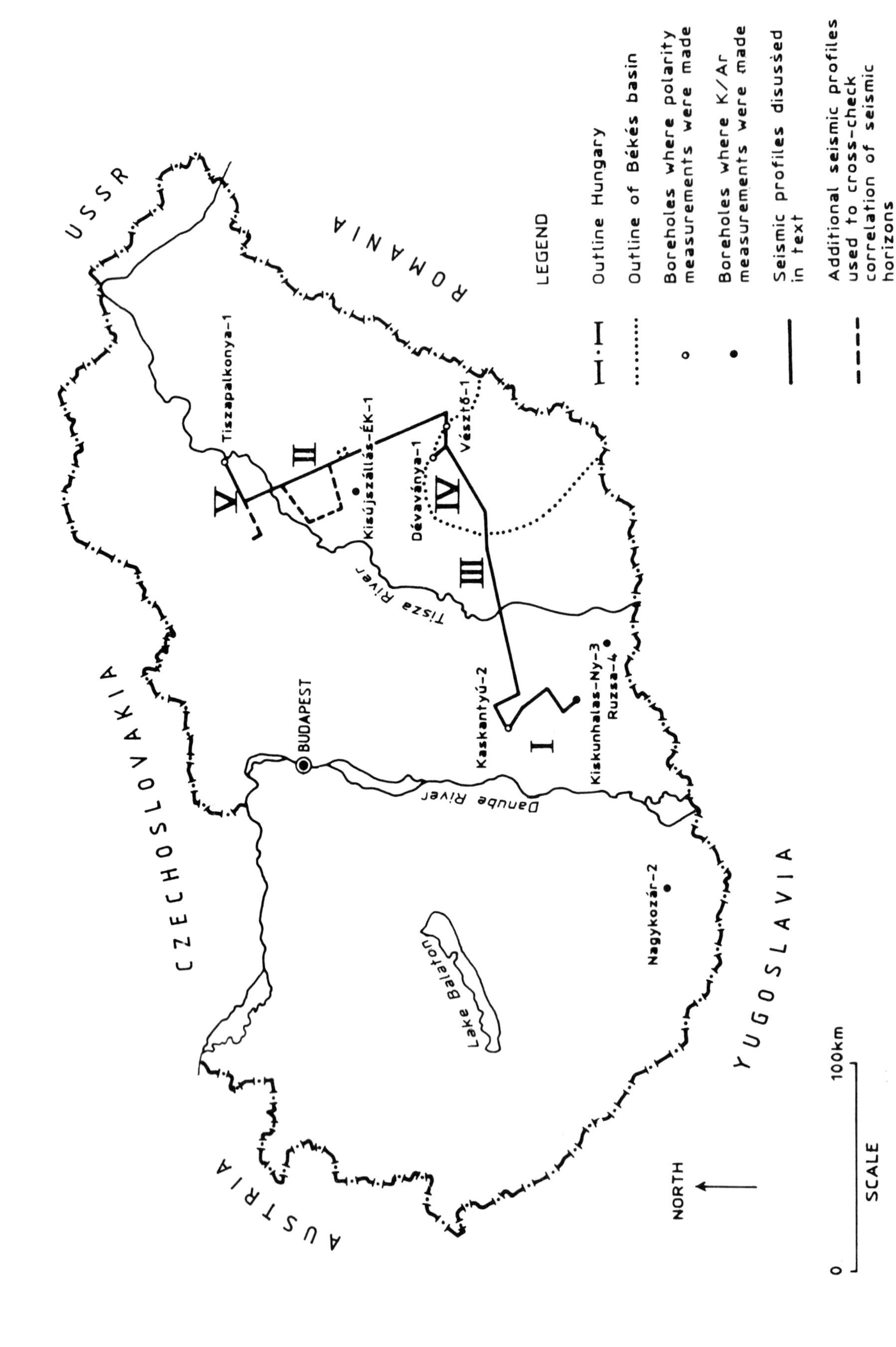

Figure 3. Map of Hungary showing location of the Békés basin, and seismic profiles and boreholes discussed in the text.

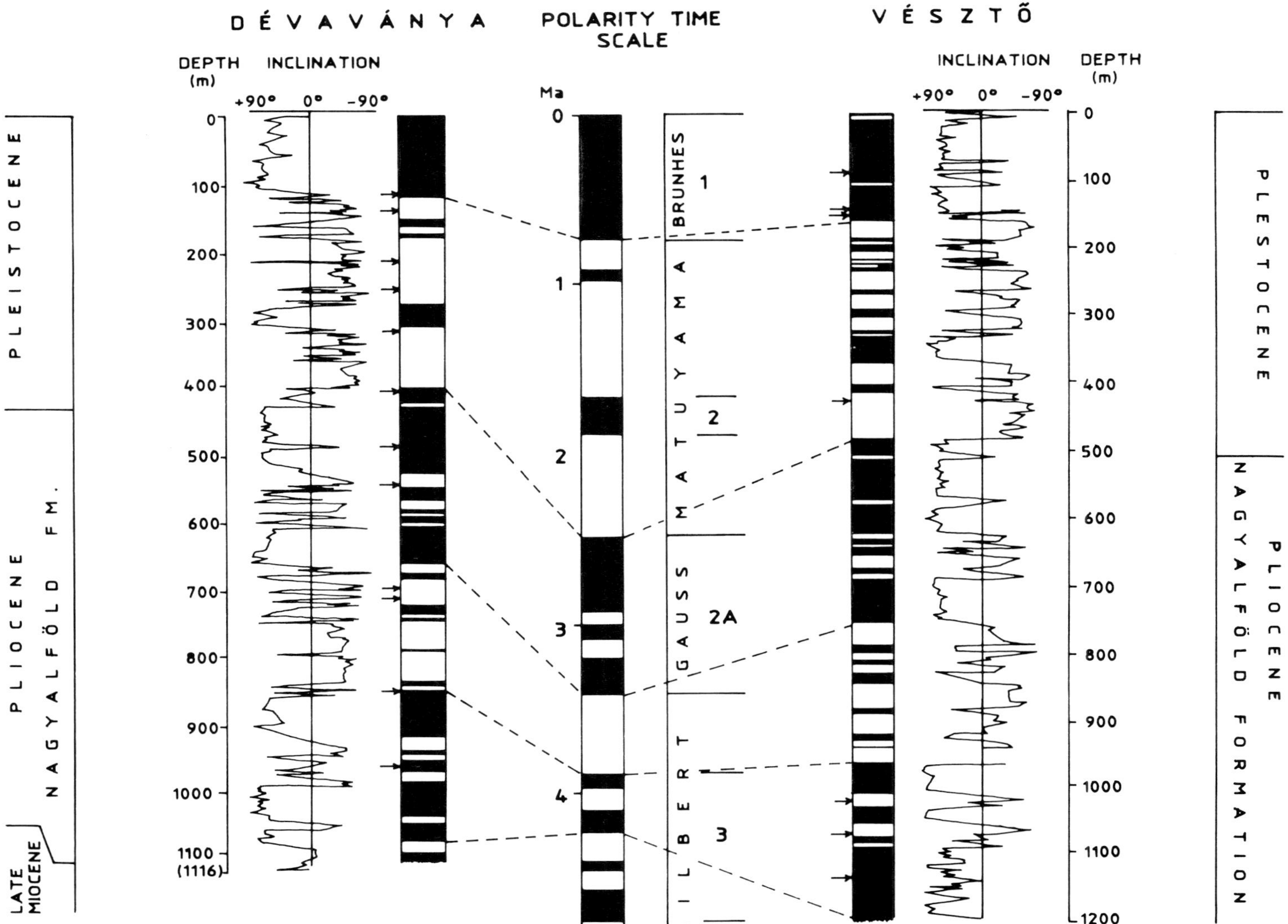

Figure 4.. Correlation of generalized polarity records and lithostratigraphic units in the Dévaványa-1 and Vésztö-1 boreholes. M, T and E show locations of intervals of mixed polarity (M), transitions (T), and excursions (E). Black: normal polarity; white: reversed polarity; black and white: mixed polarity with dominanat polarity on left; arrow: short reversal; slash: no control. Depth in meters from the ground surface.

Table 1. Depths and inferred ages of magnetic polarity boundaries measured in southeastern Hungary. N: normal

Magnetic Polarity in Boreholes	Age (Ma) Berggren and others (1985)	Age (Ma) Lowrie and Alvarez (1981)	Depth (m)	Two-way travel time (sec)
Tiszapalkonya-1				
R/N	6.37	6.34	401	435
R/N	6.70	6.68	506	530
N/R	7.41	7.38	863	915
M/N	7.90	7.77	1059	1085
N/R	8.21	8.09	1221	1195
M above N	8.92	8.80	1571	1435
Total depth			2000	1785
Kaskantyú-2				
N/R	5.89	5.87	295	330
M/R	6.50	6.47	406	430
R/N	6.70	6.68	491	515
R/N	6.85	6.83	510	530
N/R	7.41	7.38	579	590
N/R	8.21	8.09	705	690
R/N	8.92	8.27	827	795
Total depth			1185	1040
Vésztő-1				
N/R		0.75	162	110
R/N		2.47	480	435
N/R		3.40	750	705
R/N		3.87	970	960
N/R		3.97	1050	1010
R/N		4.10	1090	1055
N/R(?)		4.25	1200	1125
Total depth			1200	1125
Dévaványa-1				
N/R		0.75	120	70
R/N		2.47	420	370
N/R		3.40	675	620
R/N		3.87	865	825
N/R		3.97	925	935
R/N		4.10	985	985
N/R(?)		4.25	1085	1100
Total depth			1120	1125

PREVIOUS STUDIES

For Neogene rocks, the relation between the time-stratigraphic classification adopted in Hungary and the time-stratigraphic classification for other regions has been difficult to establish because the sedimentary and tectonic history of the Pannonian Basin, unlike that of surrounding European areas, was not well established. Presently recognized subdivisions of the Miocene-Holocene epochs in Hungary and their relation to global and regional stage subdivisions are shown in Figure 1. According to Hámor and others (1987), the Miocene marine faunal assemblages from the Pannonian Basin show varying affinities with those outside the basin; an Atlantic-boreal affinity in the lower Miocene (Eggenburgian -Ottnangian stages), a Mediterranean affinity in the

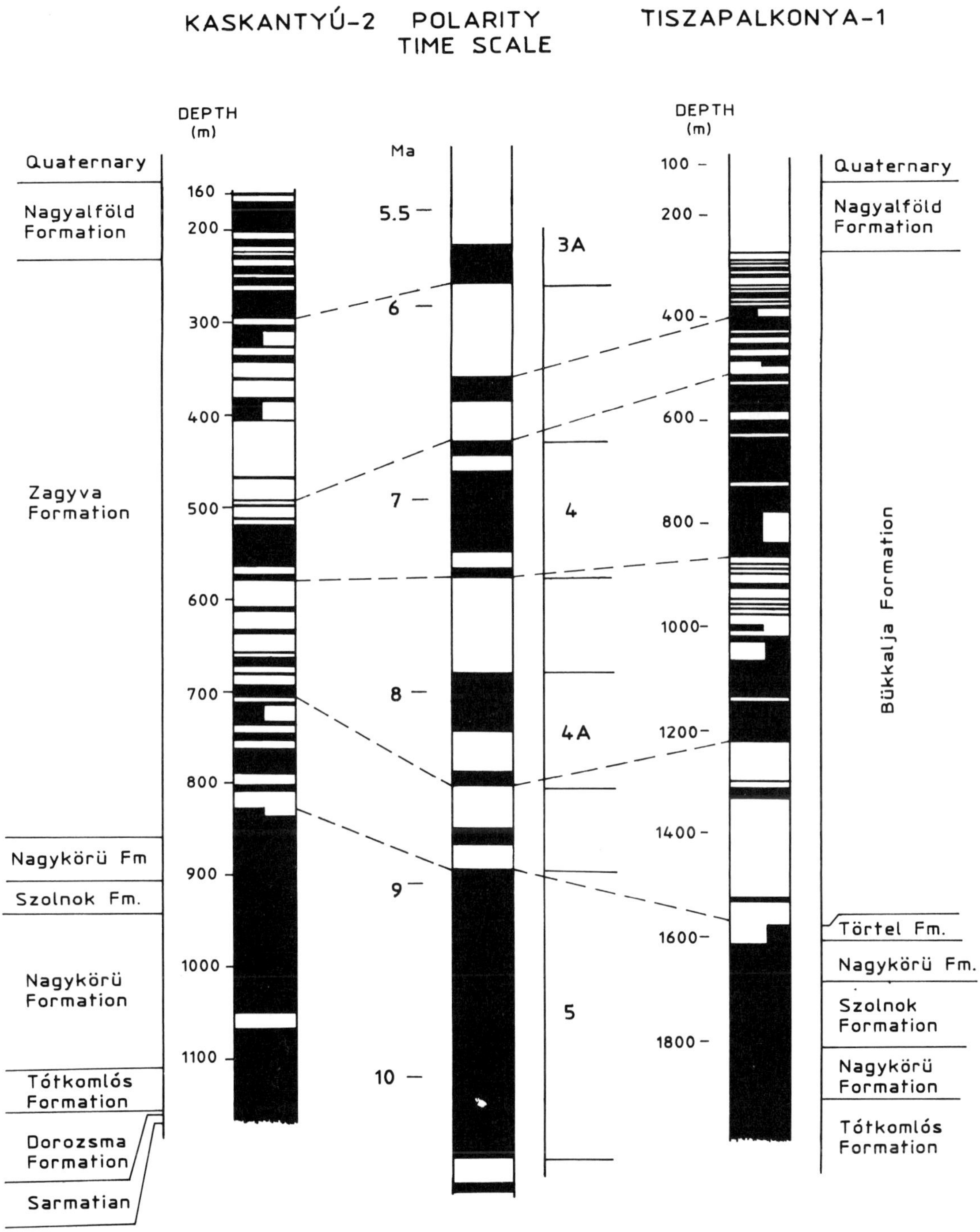

Figure 5. Correlation of generalized polarity records in the Kaskantyú-2 and Tiszapalkonya -1 boreholes. Locations of boreholes are shown in Figure 3. Black: normal polarity; white: reversed polarity; black and white: mixed polarity.

lower-middle Miocene (Carpathian-lower Badenian stages), and Caspian affinity in the middle-upper Miocene (upper Badenian-Pannonian stages).

Attempts to subdivide the Pannonian section using K/Ar methods or biostratigraphy have not been very successful in Hungary. A few absolute age determinations have been made using K/Ar dating methods for basalts and rhyolite tuff beds, both of which are widespread throughout Hungary (Fig. 1). Biostratigraphic dating of strata in the Pannonian basin has resulted in the chart shown in Figure 2. Extending these horizons to the deeper parts of the basin has not met with success (Nagymarosy, 1981; Jámbor and others, 1987). In the deeper parts of the basin, biostratigraphic subdivision, aside from the availability of few data points, is further hampered by the fact that the endemic, lacustrine Pannonian fauna are indicative mostly of depositional environments rather than age (Magyar and Révész, 1976). In addition, the sparseness of boreholes in the deeper (4-7 km) parts of the basin, the incompleteness of core samples, and the lack of Pannonian rhyolite tuff intercalations suitable for K/Ar analysis compound the difficulty of age dating.

Another factor that confuses the issue of subdivision of the Pannonian section is the usage, in the Hungarian literature, of the terms "Upper and Lower Pannonian" to subdivide the Pannonian section into lithostratigraphic units; delta plain and deep-water facies, respectively (Dank and Kókai (1969, p.138). Pogácsás (1984) and Mattick and others (1985) have shown that this boundary, as mapped on the basis of well-log response, is strongly time-transgressive.

Recently Pogácsás (1984, 1987) and Horváth and Pogácsás (1988) correlated seismic-reflection, magnetostratigraphic, and radiometric (K/Ar) data from the Pannonian Basin. These authors demonstrated that absolute age determinations based on K/Ar measurements in the Kiskunhalas-Ny-3 borehole are in general agreement with ages determined from paleomagnetic-polarity reversal studies made at the Kaskantyú-2 borehole. These authors also demonstrated that, if seismic-reflection horizons are projected between the Kiskunhalas-Ny-3, Kaskantyú-2, Vésztö-1 and Dévaványa-1 boreholes, individual seismic horizons at the borehole sites correlate with rock units inferred to be of similar age based on K/Ar and paleomagnetic-polarity reversal studies. A recent K/Ar determination has shown that the age of the uppermost rhyolite tuff penetrated in the Nagykozár-2 borehole is 11.6 Ma. Comparison with biostratigraphic data from this borehole and the Kaskantyú-2 borehole indicate that 11.6 Ma is equivalent to the pre-Pannonian Miocene/Pannonian hiatus. The locations of the above-mentioned boreholes are shown in Figure 3.

PRESENT STUDIES

In an attempt to correlate seismic-reflection and magnetic polarity data further in the Pannonian Basin, the Hungarian Geological Institute and the U.S.Geological Survey undertook a cooperative magnetic polarity study of core samples from the Tiszapalkonya-1 borehole (Elston and others, this volume). polarity reversals measured at the Tiszapalkonya-1 site (Fig. 3) were fitted to the paleomagnetic time scales of Berggren and others (1985) and Lowrie and Alvarez (1981). The estimated ages of polarity epochs measured at the Tiszapalkonya-1 site and their respective depths are shown in Table 1. As an extension of that study, four seismic-reflection sections were composited from individual seismic profiles of the Hungarian Oil and Gas Co. (MOL Rt.). The four composite profiles, labeled I-IV in Figure 3, connect the following boreholes: (I) Kiskunhalas-Ny-3 and Kaskantyú-2; (II) Tiszapalkonya-1 and Vésztö-1; (III) Kaskantyú-2 and Vésztö-1; and (IV) Dévaványa-1 and Vésztö-1. The investigations (Elston and others, this volume) also included re-evaluation of results reported earlier (Rónai and Szemethy, 1979; Rónai, 1981), including the correlation of measured polarity reversals between the Dévaványa-1 and Vésztö-1 sites (Fig. 4), and the Kaskantyú-2 and Tiszapalkonya-1 sites (Fig. 5).

PROFILE I

Composite seismic-reflection profile I, between the Kaskantyú-2 and the Kiskunhalas-Ny-3 boreholes, and an interpretation of the primary seismic events are shown in Figure 6. K/Ar dating indicates that basalts, penetrated at a depth of 1162-1167 m in the Kiskunhalas-Ny-3 borehole, were extruded 9.61 (plus or minus 0.38) Ma (Balogh and others, 1983). The depth to the basalts was converted to two-way travel time and plotted on the seismic profile (Fig. 6). It appears that the basalts correlate with a seismic horizon that terminates by onlap against a middle Miocene (Badenian) unconformity. The ages determined by polarity reversal techniques (Elston and others, this volume) are displayed at the north end of the section (Kaskantyú-2 site). At least three seismic horizons at the Kaskantyú-2 site, corresponding to 8.9, 7.4 and 6.7 Ma can be followed on the seismic section to the Kiskunhalas-Ny-3 borehole site (Fig. 6). Although few, if any, major conclusions can be drawn from this extrapolation of seismic horizons between the

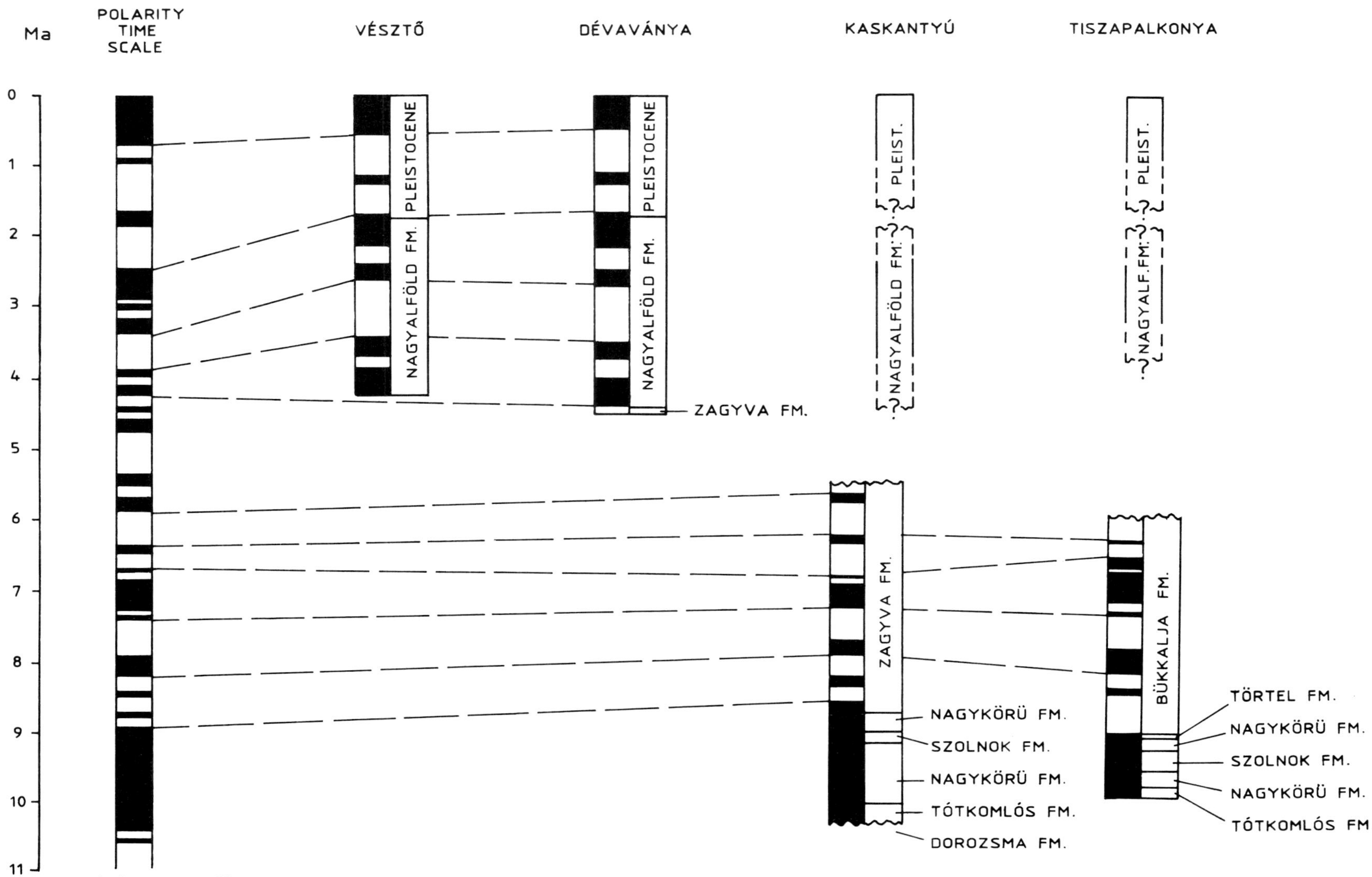

Figure 6. Seismic profile (above) between the Kiskunhalas-Ny-3 borehole and the Kaskantyú-2 borehole. Interpretation of seismic profile is shown below. Labels on seismic horizons indicate inferred age (in millions of years) of corresponding magnetic polarity epoch at the Kaskantyú-2 borehole. Horizons are dashed where quality of seismic record is poor. Location of profile is shown in Figure 3.

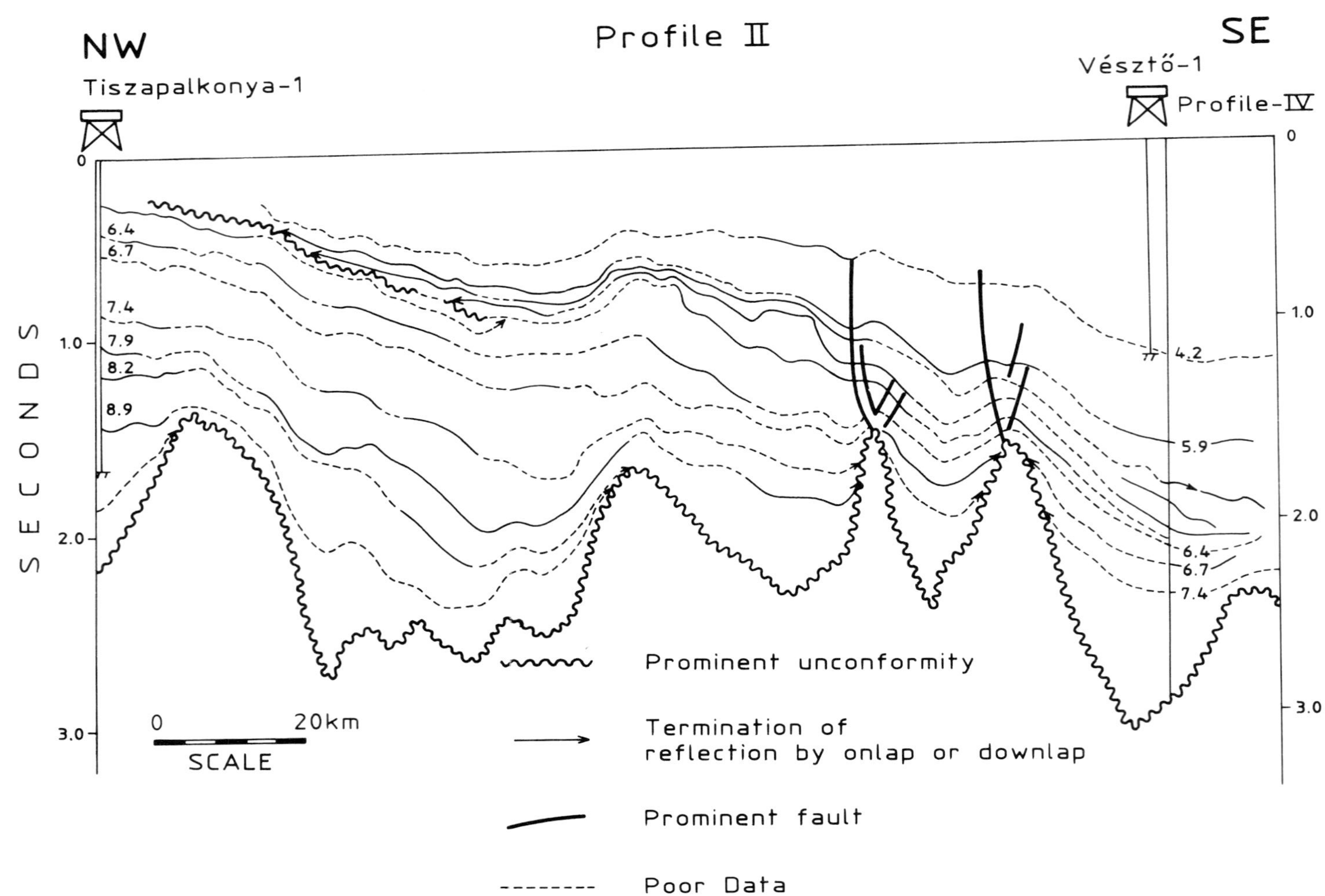

Figure 7. Interpretation of seismic profile II between the Tiszapalkonya-1 and the Vésztö-1 boreholes. With the exception of those marked 4.2 and 5.9, labels on the seismic horizons indicate inferred ages (in millions of years) of corresponding magnetic polarity epochs at the Tiszapalkonya-1 borehole. The horizons labeled 4.2 and 5.9 refer to polarity epochs identified in the Vésztö-1 and Kaskantyú-1 boreholes, respectively. Horizons are dashed where the quality of the seismic record is poor. Location of profile is shown in Figure 3.

two boreholes, the results do show that there are no major contradictions between the results derived from K/Ar age dating methods, and those based on the correlation of polarity reversals with the polarity time scale.

PROFILE II

Elston and others (this volume) infer that the rocks penetrated near the bottom of the Vésztö-1 borehole are about 4.2 million years old and those below the Nagyalföld Formation penetrated in the Tiszapalkonya-1 borehole are mainly between 6.4 to 8.9 million years old. The results of correlating seismic horizons between these boreholes confirm these findings (Fig. 7). Seismic horizons, corresponding to 7.9, 8.2, and 8.9 Ma at the Tiszapalkonya-1 site, are seen to terminate near the center of the profile in a downlap configuration against an unconformity that marks the contact between rocks of the basement complex and sedimentary rocks that lie above them. Seismic horizons corresponding to 6.4, 6.7 and 7.4 Ma at the Tiszapalkonya-1 site dip steeply to the southeast, from depths corresponding to 400-1000 milliseconds two-way travel time on the seismic section at the Tiszapalkonya-1 site, to depths corresponding to 2200--2600 milliseconds at the Vésztö-1 site. The seismic horizon corresponding to 4.2 Ma at the Vésztö-1 site could not be traced on the seismic profile to the Tiszapalkonya-1 site; it is assumed that this horizon, in the vicinity of the Tiszapalkonya-1 site, terminates against a shallow unconformity by onlap, or offlap, as do other, shallower seismic horizons (Fig. 7).

PROFILE III

Seismic profile III extends from the Kaskantyú-2 borehole to the Vésztö-1 borehole. An interpretation displaying the main seismic horizons is shown in Figure 8. Several horizons could be correlated to polarity reversals noted by Elston and others (this volume). Six seismic horizons, corresponding to ages of 5.9, 6.5, 6.7, 7.4, 8.2 and 8.9 Ma, could be mapped starting at the Kaskantyú-2 borehole site. From the Vésztö-1 site, three seismic horizons, corresponding to ages of 4.2, 6.4 and 7.4 Ma, also could be mapped. The seismic horizon labeled 4.2 is inferred to correspond to an age of 4.2 Ma based on magnetostratigraphic analysis of rock cores from the Vésztö-1 borehole; whereas, the seismic horizons labeled 6.4 and 7.4 Ma are inferred to correspond to ages of 6.4 and 7.4 Ma based on extrapolation of seismic horizons from the Tiszapalkonya-1 borehole.

Based on results shown in Figures 6-8, the authors conclude that seismic-stratigraphic techniques can be used to correlate paleomagnetic polarity reversal data obtained from widely-spaced boreholes in the Pannonian Basin. The results of such correlations are, at least, internally consistent. Magnetostratigraphic results obtained from the Kaskantyú-2 and the Tiszapalkonya-1 sites were extrapolated for distances of approximately 250 and 140 km, respectively, along seismic profiles to the Vésztö-1 site where earlier studies (Cooke and others, 1979) had established magnetostratigraphic subdivisions of the Pannonian sequence. As seen in Figure 8, the inferred ages of sedimentary beds (or actually, the equivalent seismic horizons) are consistent with the stratigraphic model of the Pannonian Basin proposed by Mattick and others (this volume) and Molenaar and others (this volume); that is, the Pannonian Basin in northeastern Hungary was filled, for the most part, by massive sediment wedges that prograded southward.

It should be noted that the apparently "good" correlation between the results of seismic-stratigraphic and magnetostratigraphic interpretations shown in Figure 7 resulted partly from "circular reasoning", because the two analytical approaches were not performed independently of one another. Prior to performing the magnetostratigraphic interpretations, seismic studies had already indicated that the upper part of the sedimentary section at the Tiszapalkonya-1 site is much older than the upper part of the sedimentary section at the Vésztö-1 site (Fig. 7). Elston and others (this volume) used this knowledge to fit their polarity reversal data measured at the Tiszapalkonya-1 site to the time scale of Berggren and others (1985), thus leading to the correlations they show between the Tiszapalkonya-1 and Vésztö-1 sites.

PROFILE IV

Seismic profile IV connects the Dévaványa-1 and Vésztö-1 borehole sites. An interpretation depicting the principal seismic horizons is shown in Figure 9. At the two boreholes, depths corresponding to polarity epochs dated by Elston and others (this volume) were converted to two-way seismic travel times, and these are plotted on the section. Although the reflections in this area, especially in the upper 500 milliseconds of record section, are discontinuous, it appears that the principal seismic horizons correspond well with sedimentary horizons that have been dated. These results indicate that seismic reflections parallel geologic time lines, or depositional surfaces, at least in this area of the Pannonian Basin.

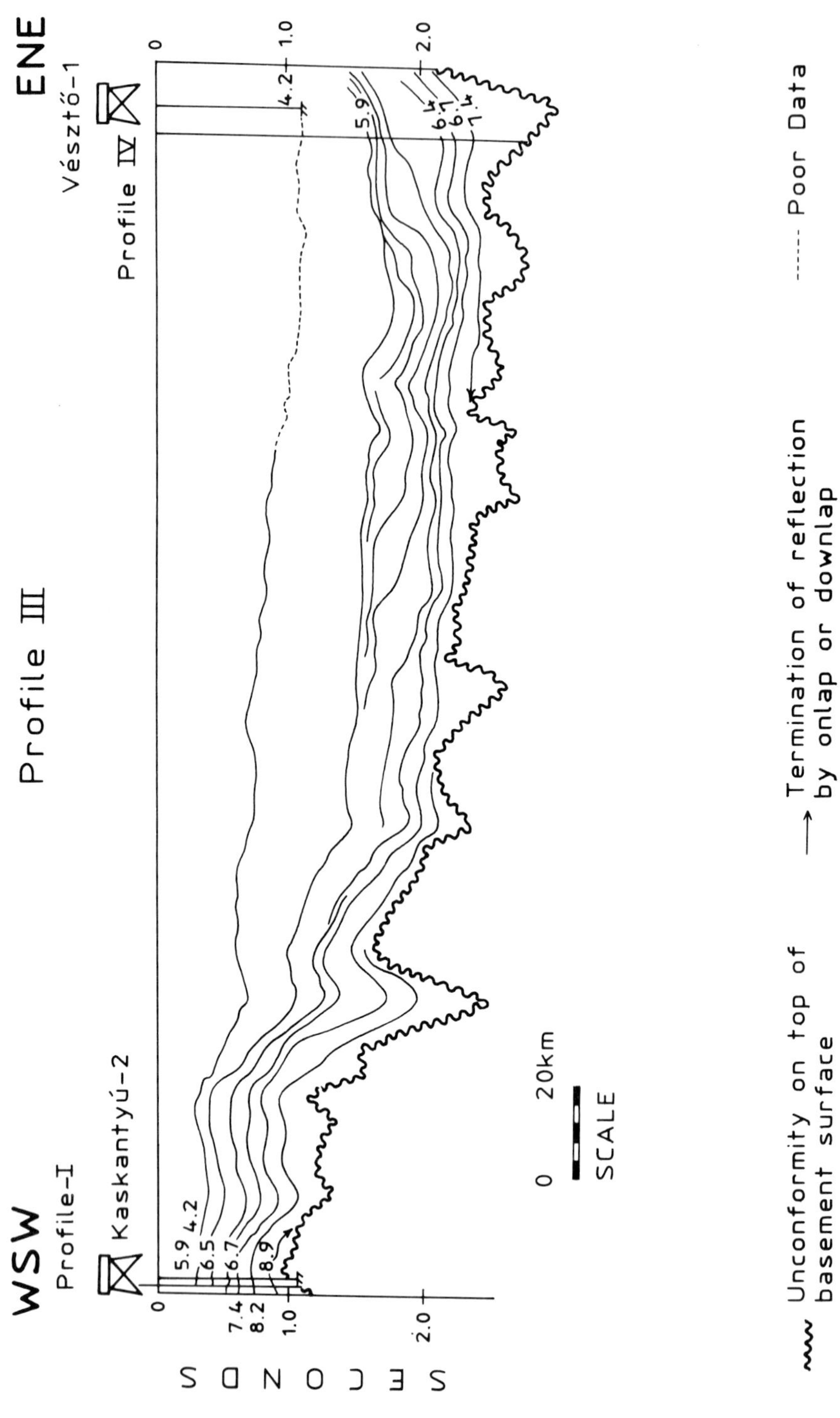

Figure 8. Interpretation of seismic profile III between the Kaskantyú-2 and Vésztö-1 boreholes. Labels on seismic horizons indicate inferred age (in millions of years) of corresponding magnetic polarity epochs. The 4.2 Ma polarity epoch was identified in the Vésztö-1 borehole. Polarity epochs 5.9, 6.5, 6.7, 8.2, and 8.9 Ma were identified in the Kaskantyú-2 and Tiszapalkonya-1 boreholes. Horizons are dashed where the quality of the seismic data is poor. Location of profile is shown in Figure 3.

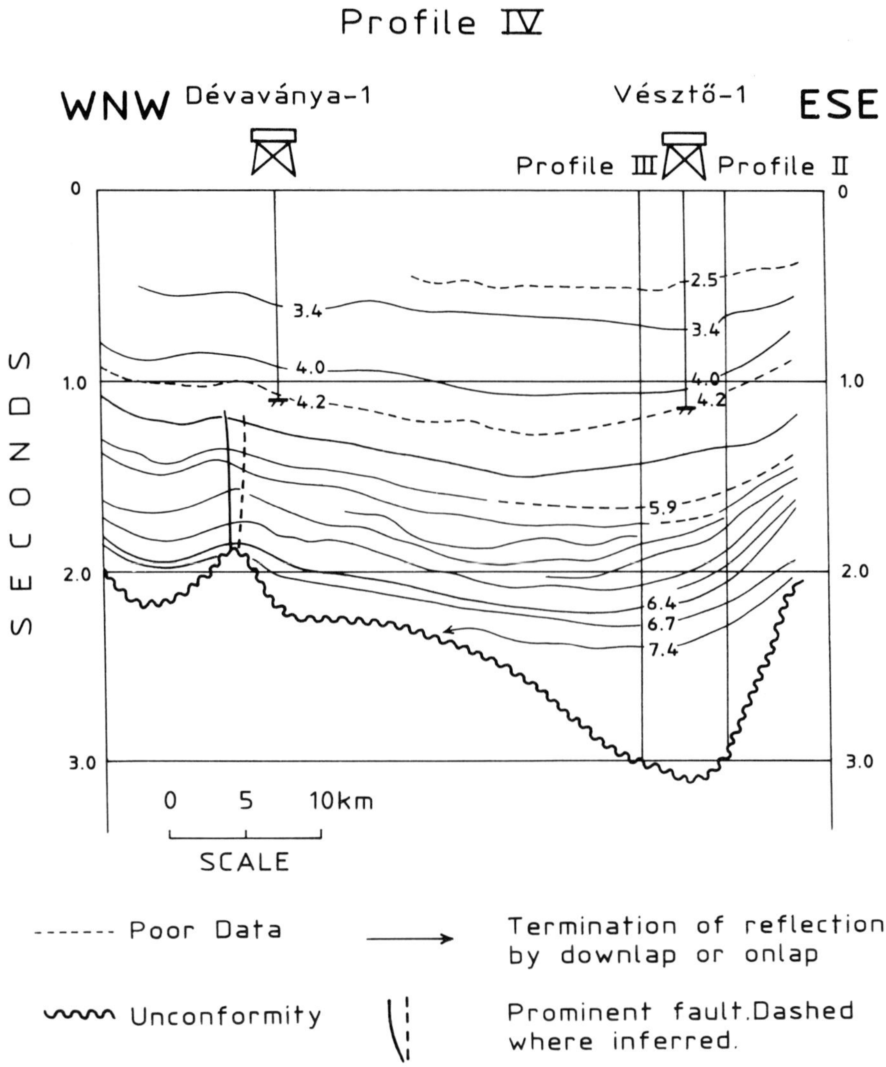

Figure 9. Interpretation of seismic profile IV between the Dévaványa-1 and Vésztő-1 boreholes. Numbers at borehole sites indicate inferred ages (in millions of years) of polarity epochs represented by rocks penetrated in the boreholes. Horizons are dashed where the quality of seismic data is poor. Location of profile is shown in Figure 3.

DISCUSSION OF RESULTS

Filling of Békés Basin

According to Szentgyörgyi and Teleki (this volume), the oldest Neogene sedimentary rocks penetrated in wells in the Békés basin (Fig. 5.) are middle Miocene in age, and correspond to the Badenian and Sarmatian stages. The combined Badenian and Sarmatian stages represent deposition from 16.5 to 11.5 million years ago (Hámor and others, 1985; Steininger and others, 1985). The thickness of the combined Badenian and Sarmatian section in the Békés basin ranges from 0-275 m (Szentgyörgyi and Teleki, this volume). These sedimentary strata represent shallow-water, near-shore, marine to brackish water deposition and unconformably overlie Mesozoic and older rocks of the basement complex.

Sarmatian strata in the Békés basin are overlain by lacustrine sedimentary strata of Pannonian age. At the

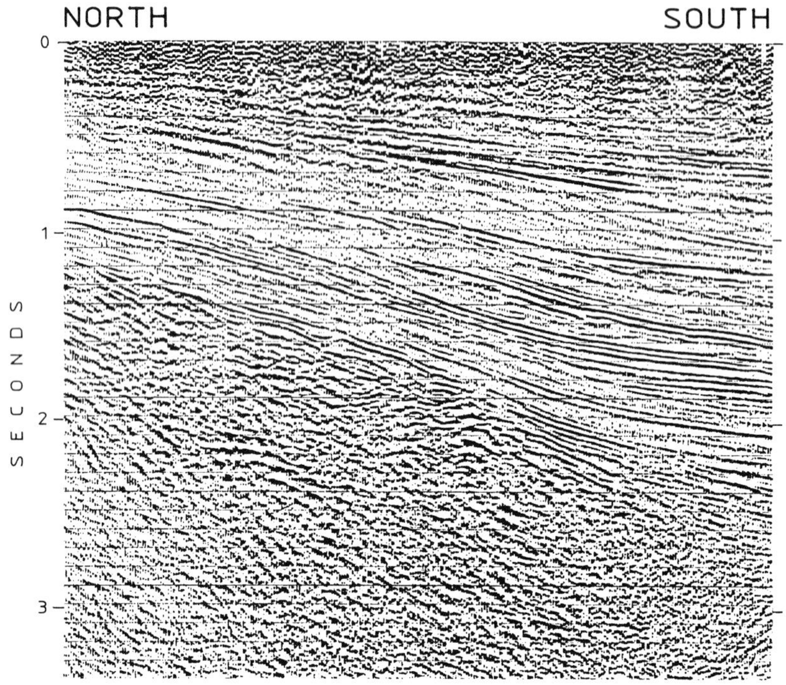

Figure 10. Seismic profile V. Location of this profile is shown in Figure 3 and its interpretation is shown below in Figure 11.

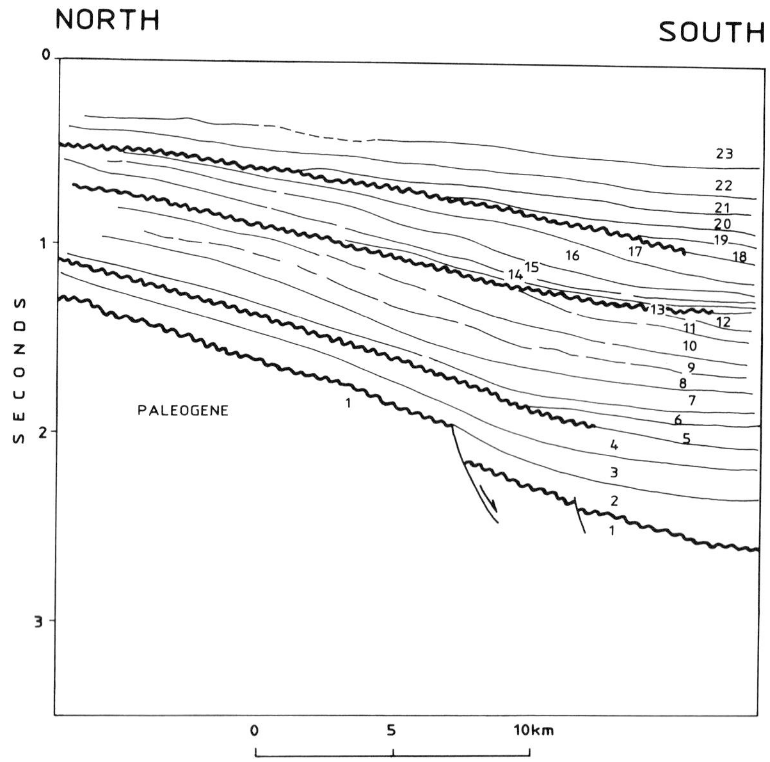

Figure 11. Interpretation of seismic profile V shown in Figure 10. The intervals, representing rock strata between the principal seismic events, are labeled sequentially from oldest to youngest. Interval 1 represents rocks of the basement complex.

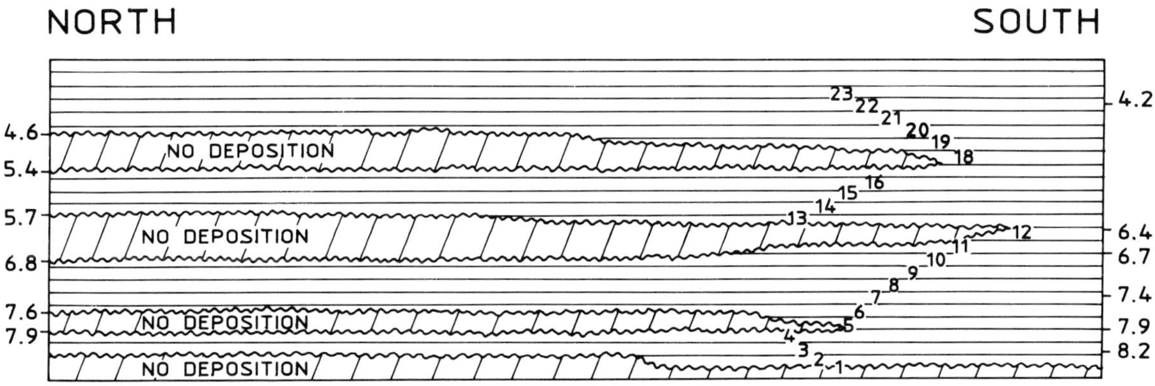

Figure 12. Chronostratigraphic chart constructed from data of Figure 11. Geologic time is the ordinate and the time intervals of individuals strata are assumed to be equal. Numbers on the right side of the figure indicate ages of each stratum (in millions of years) as inferred from magnetostratigraphic studies and extrapolated along seismic profiles. Numbers on the left side show inferred ages of periods of non-deposition.

basin's margins, the contact between Sarmatian and Pannonian strata is unconformable, whereas in deeper parts of the basin deposition was probably uninterrupted (Mattick and others, this volume). The Sarmatian (*sensu stricto*)-Pannonian boundary is placed at 11.5 million years ago by Hámor and others (1985).

A controversy exists concerning water depths in the Békés basin during latest Sarmatian-earliest Pannonian time. This controversy affects interpretations of the time that the major phase of subsidence of the basin took place. According to Szentgyörgyi and Teleki (this volume), the Békés basin opened during the Styrian orogenic cycle (lower Badenian), and Badenian and Sarmatian sediments were deposited in near-shore, shallow-water environments. However, Lukács-Miksa and others (1983) and Mattick and others (this volume) argue that early Pannonian turbidites in the Békés basin represent deep-water conditions (depths of about 900 m). All of these authors concede that sedimentation was probably uninterrupted from Badenian through Pannonian time, at least in central parts of the Békés basin. The data of Szentgyörgyi and Teleki(this volume) indicate that the rate of sediment deposition during Badenian-Sarmatian time ranged from 0 to 55 m/million years. During Pannonian time, however, rates of sediment deposition had increased to as much as 1000 m/million years (Mattick and others, this volume).

The change in water depths and sedimentation rates in the Békés basin from Sarmatian to Pannonian time perhaps can be explained as follows. During latest Sarmatian-early Pannonian time, the Békés basin was largely starved of sediments because detritus transported by rivers was being captured by subbasins located along the margins of the basin. Concurrently, however, subsidence accelerated and water depths increased to about 900 m. The consequence of this hypothesis, put forth by Mattick and others (this volume), is that sedimentary filling of the Békés basin during Pannonian time was one of the last episodes in the evolution of the larger Pannonian Basin.

The foregoing hypothesis is corroborated by evidence from the seismic profiles described in this paper. The best examples supporting this sequence of events is exhibited on seismic profiles II and III (Figs.7 and 8.). The Vésztö-1 borehole, located on the northern boundary of the Békés basin, is shown on both profiles. The Tiszapalkonya-1 borehole, located about 115 km to the north of the basin, is shown on Figure 7. On this figure, the oldest horizons corresponding to polarity reversals at 8.9, 8.2 and 7.9 Ma (Table 1, column 1) are seen to terminate against basement highs near the center of the profile. Horizons representing polarity reversals at 6.4 and 7.4 Ma dip steeply southeastward and, at the southeastern end of the profile, they approach the basement surface. On the seismic section, it can be seen that the sedimentary section younger than 4.2 Ma (value at the bottom of the Vésztö-1 borehole) thickens considerably in a northwest to southeast direction. Correlation of seismic horizons and polarity measurements shown in Figure 8 indicate that sedimentation processes were similar west of the Békés basin.

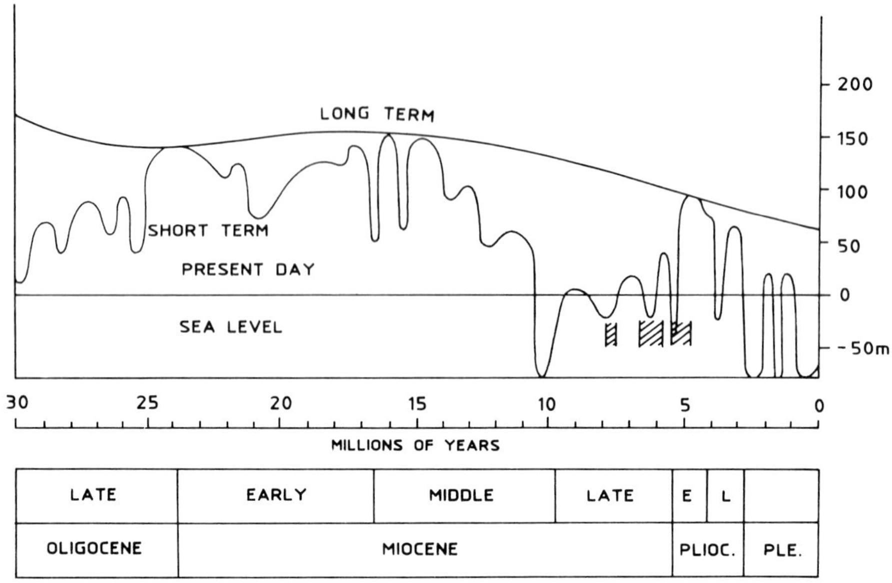

Fugure 13. Tertiary and Quaternary eustatic fluctuations of sea level. Meters above or below present-day sea level are approximate. Figure modified from Haq and others (1987). Hachured areas (7.9-7.6, 6.8-5.7, and 5.4-4.6 Ma) show non-depositional periods on the flanks of the Pannonian basin. These periods of non-deposition appear to correlate with the eustatic sea level minima of 7.8, 6.3, and 5.2 Ma. PLE: Pleistocene.

On the strength of these interpretations, it can be concluded that between 12 and 9 Ma the Békés basin was starved as a result of sediments being captured by subbasins located along the basin's margins. Subsidence, however, continued and water depths increased rapidly. By about 6-7 Ma, subbasins near the Békés basin became filled by southward and eastward prograding sediment wedges and rivers were able to reach the Békés basin. Thereafter, deltas began to build along the edges of the basin and sediments began to accumulate. During the course of deposition, sedimentation rates reached 1000 m/million years.

Lake Level Changes

Mattick and others (this volume) attribute the origin of unconformities in the Pannonian section to delta lobe switching, based on interpretation of seismic records from the Békés basin. However, there is evidence that some of the unconformities may be related to lake-level changes, caused by fluctuations in eustatic sea level.

Seismic-reflection profile V, located about 40 km southeast of the Tiszapalkonya-1 borehole near the northern margin of the Pannonian Basin, is shown in Figure 10. An interpretation of this profile is given in Figure 11. The intervals between prominent seismic reflectors are sequentially numbered from oldest to youngest. At least four major unconformities can be identified: (1) between units 1 (the basement complex) and 2, (2) between units 4 and 5, (3) between units 12 and 13, and (4) between units 17 and 18.

The stratigraphic interpretation of Figure 11 is transposed to a chronostratigraphic section with geologic time shown as the ordinate in Figure 12. The methods used to construct this chronostratigraphic section have been outlined by Mitchum and others (1977). The geologic time interval of each unit (or stratum) is assumed to be the same. The seismic sequences are bounded by hiatuses (or periods of non-deposition). The time intervals of these hiatuses decrease toward the south in a basinward direction. To place the stratigraphic intervals in a standard chronostratigraphic framework, the results of magnetostratigraphic studies at the nearby Tiszapalkonya-1 borehole site (Elston and others, this volume) were extrapolated along a connecting seismic profile (dashed line in Figure 3) and are displayed along the right margin of Figure 12. With the exception of the lowermost unconformity that marks the top of the basement surface, inferred maximum and minimum ages of the prominent unconformities are shown on the left side of Figure 12.

In the Mediterranean, the end of the Miocene is characterized by widespread evaporites (Hsü and others, 1973) - the so-called "Messinian salinity crisis". According to Moisette and Pouyet (1987), the increase in salinity was related to either evaporation-desiccation of the Mediterranean Sea in the late Messinian or, more likely, a partial interruption of the connection between the Mediterranean Sea and the Atlantic ocean. The eustatic sea-level curve of Haq and others (1987) shows significant sea-level decreases in the late Miocene at approximately 7.8, 6.3 and 5.2 million years ago (Fig. 13). By comparison, lake level decreases in the Pannonian inland sea, although isolated from the world's oceans, occurred between 7.9-7.6, 6.8-5.7 and 5.4-4.6 Ma (Fig. 13). This suggests that lake levels in the Pannonian inland sea rose and fell in phase with global sea level changes.

REFERENCES

Berggren, W.A., Kent, D.U., Flynn, J.J., and van Couvering, J.A., 1985, Cenozoic geochronology; Geol. Soc. Am. Bull., 96, (2), 1407-1418.

Balogh, K., Jámbor, Á., Partényi, L., Ravasz-Baranyai, L., Solti, G. and Nusszer, A., 1983, Petrography and K/Ar dating of Tertiary and Quaternary basaltic rocks in Hungary; Ann. Inst. Geol. Geophys., Bucharest, 61, 365-373.

Cooke, H.B.S., Hau, J.M., and Rónai, A., 1979, Paleomagnetic sedimentary and climatic records from boreholes of Dévaványa and Vésztö, Hungary; Acta Geol. Hung., 22, (1-4), 89-109.

Dank, V., and Kókai, J., 1969, Oil and gas exploration in Hungary, *in* Hepple, P. (ed.), The exploration for petroleum in Europe and North Africa; Inst. of Petroleum, London, 131-145.

Elston, D.P., Lantos, M., and Hámor, T., High resolution polarity records, and the stratigraphic and magnetostratigraphic correlation of late Miocene and Pliocene (Pannonian, *s.l.*) deposits of Hungary; this volume, 11-142.

Hámor, G., Ravasz-Baranyai, C., Balogh, K., Árva-Sóos, E., 1980, Radiometric age of Hungarian Miocene age rhyolite tuffs, (in Hungarian); Ann. Rept. for 1978, Hung. Geol. Inst., 65-73.

Hámor, G., Jámbor, Á., Lantos, M., Rónai, A., and Elston, D.P., 1985, Magnetostratigraphic correlations of some Neogene strata of the Pannonian Basin Hungary; Preprints, 8th Cong. Reg. Comm. on Mediterr. Neogene Stratigraphy., (abs.), Budapest, 257.

Hámor, G., Báldi, T., Bohn-Havas, M., Hably, L., Halmai, J., Hajós, M., Kókai, J., Kordos, L., Korecz-Laky, I., Nagy, B., Nagymarosy, A., and Völgyi, L., 1987, The bio-, litho-, and chronostratigraphy of the Hungarian Miocene; Proc. 8th Cong. Reg. Comm. on Mediterr. Neogene Stratigraphy, Hung. Geol. Inst. Ann. Rept., 70, 351-353.

Haq, B.U., Hardenbol, J., and Vail, P.R., 1987, Chronology of fluctuating sea levels since the Triassic; Science, 235, 1156-1167.

Horváth, F., and Pogácsás, Gy., 1988, Contribution of seismic reflection data to chronostratigraphy of the Pannonian Basin; *in* Royden, L.H. and Horváth, F. (eds.), The Pannonian Basin, a study in basin evolution, Amer. Assoc. Petrol. Geol., Mem. 45, 97-106.

Hsü, K.J., Cita, M.B., and Ryan, W.B.F., 1973, The origin of the Mediterranean evaporites, *in* Ryan, W.B.F. and others (eds.), Initial Reports of the Deep Sea Drilling Project; U.S. Govt. Printing Office, Washington, 13, 1203-1231.

Jámbor, Á., Balázs, E., Balogh, K., Bérczi, I., Bóna, J., Horváth, F., Gajdos, I., Geiger, J., Hajós, M., Kordos,L., Korecz, A., Korecz-Laky, L., M. Korpás-Hódi, J. Kőváry, L., Mészáros, L., Nagy, E., Németh, G., Nusszer, A., Pap, S., Pogácsás, Gy., Révész, I., Rumpler, J., Sütö-Szentai, M., Szalay, A., Szentgyörgyi, K., Széles, M., and Völgyi, L., 1987, General characteristics of Pannonian (*s.l.*) deposits in Hungary; Proc. 8th Cong. Reg. Comm. Mediterr. Neogene Stratigraphy, Hung. Inst. Geol. Ann. Rept., 70, 155-168.

Lukács-Miksa, M., Pogácsás, Gy., and Varga, F., 1983, Seismic facies analysis and stratigraphic interpretation of the unconformably dipping Pliocene features in the Pannonian Basin (in Hungarian); Paper 2874 of the Geophysical Institute of Hungary, Proc. Balatonszemes symposium, 173-186.

Lowrie, W. and Alvarez, W., 1981, One hundred million years of geomagnetic polarity history; Geology, 9., 392-397.

Magyar, L. and Révész, I., 1976, Data on the classification of Pannonian sediments of the Algyö area; Acta Mineral.-Petrog., Szeged, Hungary, 22, 267-283.

Mattick, R.E., Rumpler, J., and Phillips, R.L., 1985, Seismic stratigraphy of the Pannonian Basin in southeastern Hungary; Geophys. Trans., Eötvös Lóránd Geophys. Inst., Budapest, 31, (1-3), 13-54.

Mattick, R.E., Rumpler, J., Újfalusy, A., Szanyi, B., and Nagy, I., Sequence stratigraphy of the Békés basin; this volume, 39-65.

Mitchum, R.M., Jr., Vail, P.R. and Thompson, S., III, 1977, Seismic stratigraphy and global changes of sea level, part 2: The depositional sequence as a basic unit for stratigraphic analysis, *in* Payton, C.E., (ed.), Seismic stratigraphy: application to hydrocarbon exploration, Amer. Assoc. Petrol. Geol., Mem., 26, 53-62.

Moisette, P. and Pouyet, S., 1987, Bryozoan faunas and the Messinian salinity crisis; Proc. 8th Cong. Reg. Comm. on Mediterr. Neogene Stratigraphy, Hung. Geol. Inst., Ann. Rept., 70, 447-453.

Molenaar, C.M., Révész, I., Bérczi, I., Kovács, A., Juhász, Gy., Gajdos, I., and Szanyi, B., Stratigraphic framework and sandstone facies distribution of the Pannonian sequence in the Békés basin; this volume, 99-110.

Nagymarosy, A., 1981, Chrono- and biostratigraphy of the Pannonian Basin: A review based mainly on data from Hungary: Earth Evolution Sciences, 1, (3-4), 183-194.

Pogácsás, Gy., 1984, Results of seismic stratigraphy in Hungary, Acta Geol. Hung., 27, (1-2), 91-108.

Pogácsás, Gy., 1987, Seismic stratigraphy as a tool for chronostratigraphy: Pannonian Basin; Proc. 8th Cong. Reg. Comm. on Mediterr. Neogene Stratigraphy, Hung. Geol. Inst., Ann. Rept., 70, 55-64.

Pogácsás, Gy., Jámbor, Á., Mattick, R., Elston, D., Várkonyi, L., Lantos, M., Simon, E., Várnai, P., Hámor, T., and Lakatos, L., 1987, Correlation of seismo-and magnetostratigraphy (abs.), Proc. of the COGEODATA Int. Workshop on Computerized Basin Analysis with Special Emphasis on Hydrocarbon Exploration; Szeged, Hungary, Sept. 1987, 1-13.

Rónai, A., and Szemethy, A., 1979, Latest results of lowland research in Hungary on paleomagnetic measurement on unconsolidated sediments (in Hungarian); Hung. Geol. Inst., Ann. Rept. for 1977, 67-83.

Rónai, A., 1981, Magnetostratigraphy of Miocene-Quaternary sediments in the Great Hungarian Plain, Earth Evol. Sciences, 1 (3-4), 265-267.

Steininger, F.F., Senes, J., Kleemann, K., and Rögl, F., 1985, Neogene of the Mediterranean Tethys and Paratethys (Stratigraphic correlation tables and sediment distribution maps); Inst. of Paleontology, Univ. Press of Vienna, Austria, 1, 189 p.; 2, 524 p.

Szentgyörgyi, K. and Teleki, P.G., Facies and depositional environments of Miocene sedimentary rocks; this volume, 83-97.

Seneš, I., 1978, Géochronologie des stratotypes des étages du Miocène inférieure et moyen de la Paratethys Centrale utilisables pour la correlation globale; Geol. Zbornik, Geol. Carpathica, 30, (1), 99-100.

8 Organic Geochemistry of Crude Oils and Source Rocks, Békés Basin

J. L. Clayton[1], I. Koncz[2], J. D. King[1] and E. Tatár[3]

ABSTRACT

Crude oil and rock samples in the Békés basin were analyzed to determine the genetic relationships of crude oils, estimate thermal conditions under which the oils were generated, establish crude oil-source rock correlations and infer secondary migration routes in the basin. The aim of the study was to provide a geochemical underpinning which could be used to predict the occurrence of undiscovered oil accumulations.

Organic geochemical characterization of the crude oils revealed that three major genetic oil-types or "oil-families" are present in the basin. This finding implies the existence of three effective source rocks, possibly different facies within the same age rock unit. The first oil type is typified by very high saturated hydrocarbon content (mostly 90%), commonly a predominance of odd-carbon-numbered *n*-alkanes in the C_{25}+ fraction, and generally high pristane/phytane ratios. In contrast, the second oil type is characterized by relatively higher contents of aromatic hydrocarbons, predominances of even-carbon-numbered *n*-alkanes, and lower pristane/phytane ratios. A third oil-group includes oils which contain relatively more nonhydrocarbons (resins + asphaltenes). Biological marker distributions are also useful in determining genetic relationships among the oils. Oleanane/hopane, moretane/hopane, Tm/Ts, and norhopane/hopane ratios are all useful in distinguishing genetic oil groups. Some of the oils contain marker distributions of more than one type; this may indicate that oils derived from more than one source were mixed.

Thermal maturities of the oils are variable. API gravities (26° to 55°) and maturation-sensitive geochemical parameters (biological marker distributions and *n*-alkane odd-even-carbon preferences) indicate that oils were expelled over a range of thermal maturities. The least-thermally altered oils have been derived from source rocks with thermal maturities corresponding to present-day burial depths of approximately 2000-3000 m [vitrinite reflectance (R_o) <0.4-0.6%]. More mature oils and condensates have maturities corresponding to present-day depths for the source rock burial of about 3000-4000 m (R_o > 0.6-0.9+%).

Middle Miocene rocks, containing as much as 5 weight (wt.) % organic carbon, are the major source rocks in the basin. Rocks in the lower part of the Pannonian (late Miocene-Pliocene) section also contain sufficient organic matter (as much as about 2 wt. % organic carbon) to be possible source rocks. Substantial secondary migration has occurred, resulting in oils from more than one source (and different thermal histories) occurring in a single field or producing structure.

[1] U.S. Geological Survey, Denver Federal Center, Denver, Colorado 80225, USA
[2] MOL Rt. (Hungarian Oil and Gas Co., Ltd.), H-2443, Százhalombatta, Hungary
[3] MOL Rt. (Hungarian Oil and Gas Co., Ltd.), H-5001, Szolnok, Hungary

P. G. Teleki et al. (eds.), Basin Analysis in Petroleum Exploration, 161–185.

INTRODUCTION

Organic geochemical studies have an important role in basin analysis aimed at petroleum exploration. An understanding of the timing and depths of oil and/or gas generation and expulsion are critical elements in the evaluation of the overall potential of a basin, and particularly for predicting secondary migration routes and possibilities for trapped hydrocarbons in specific exploration plays or prospects. Thermal models or indirect indicators of oil generation (e.g. vitrinite reflectance) can be used as general guides to the oil generation history of a basin (Horváth and others, 1988; Szalay, 1988), but determination of the actual amounts and molecular distribution of organic compounds in oils and source rocks is required to demonstrate specific oil-source bed relationships in a basin. The reason for this is that the processes of oil generation/migration and thermal evolution of vitrinite are not directly linked and may vary independently depending on specific conditions (e.g. type of organic matter, mineral matrix effects, heating rate, etc.) within a basin or at different locations in the same basin. Accordingly, in this chapter, oil-source rock correlations, based on a multiparameter approach are emphasized, and migration histories of the oils are determined in a general sense based on the current stratigraphic and structural relationships between the effective source rocks and oils in reservoirs. The question of secondary migration of oil and gas is treated in more detail by Spencer and others (this volume).

Oil (including condensate) in the Békés basin occurs mostly at depths between about 1000 and 3300 m, in rocks ranging in age from Precambrian through late Pannonian. This study addresses only the question of the origin and migration of crude oil in the basin. The geochemistry of natural gas is discussed in our companion papers (Clayton and others, 1990; Clayton and Koncz, this volume).

Nearly all of the oil fields found to date in the basin are associated with structural and combination structural-stratigraphic traps located on structural highs around the margins of the basin (Fig. 1). Oils with a wide range of API gravities are commonly produced from the same broad structural feature, or even from the same depth within a field. No systematic relationship exists between depth and API gravity of oil in the basin. Often, a wide range of gravities (and chemical composition, discussed later) occurs among oils produced from about the same depth but at different locations within a single field (Table 1). Chemical and isotopic analyses of gases in the basin indicate that significant vertical and lateral migration of gas has occurred (Clayton and Koncz, this volume). In this study, assessment of the thermal maturation of oils and rocks, oil-oil correlations, and oil-source correlations are used to infer secondary migration routes of oils and to predict possibilities for occurrence of additional undiscovered accumulations of oil.

ANALYTICAL PROCEDURES

Whole-rock samples were pulverized to about 100 mesh and analyzed by pyrolysis using a Delsi Model II Rock-Eval equipped with a total organic carbon module. Rock samples with the best source rock potential determined from pyrolysis were extracted with chloroform for 24 hours using a Soxhlet apparatus. Activated copper was used to remove elemental sulfur from the extracts obtained. An aliquot of the extract thus obtained and crude oils were separated by elution chromatography using 0.5 x 8 cm columns packed with silica gel and alumina. Saturated hydrocarbons, aromatic hydrocarbons, and resins were obtained by successive elution with iso-octane, benzene, and benzene:methanol (1:1 volume/volume (v/v)). An aliquot of each of the three eluates and the total extract ("bitumen") was evaporated under nitrogen and concentrations of the fractions were determined gravimetrically.

Gas chromatography-mass spectrometry (GC-MS) analyses were performed on whole bitumens and crude oils (diluted 10% v/v with chloroform). GC-MS analyses were performed with an HP 5880A GC equipped with a 50 m x 0.32 mm SE-54 fused silica capillary column, coupled with a Kratos MS 30 mass spectrometer at a mass resolution of 3000 (5% valley). The GC was programmed from 50 to 340°C at 4°C/minute using hydrogen as the carrier gas at a flow rate of 25 cm/sec. The MS source was operated in electron impact mode at 70 eV at a pressure of 10^{-6} torr at 250°C. Multiple ion detection was accomplished by switching the accelerating voltage with a constant magnetic field. Relative abundances of biological marker compounds were determined from peak areas of diagnostic ions. No attempt was made to determine response factors or absolute abundances. GC analysis of saturated hydrocarbon fractions was performed on the HP 5880A using a flame-ionization detector. Peak integrations were performed by digitizing the detector signal.

Aliquots of the saturated and aromatic hydrocarbon fractions were combusted and the carbon dioxide formed was purified and analyzed for $^{13}C/^{12}C$ using a Finnigan 251 isotope ratio mass spectrometer. Replicate analyses,

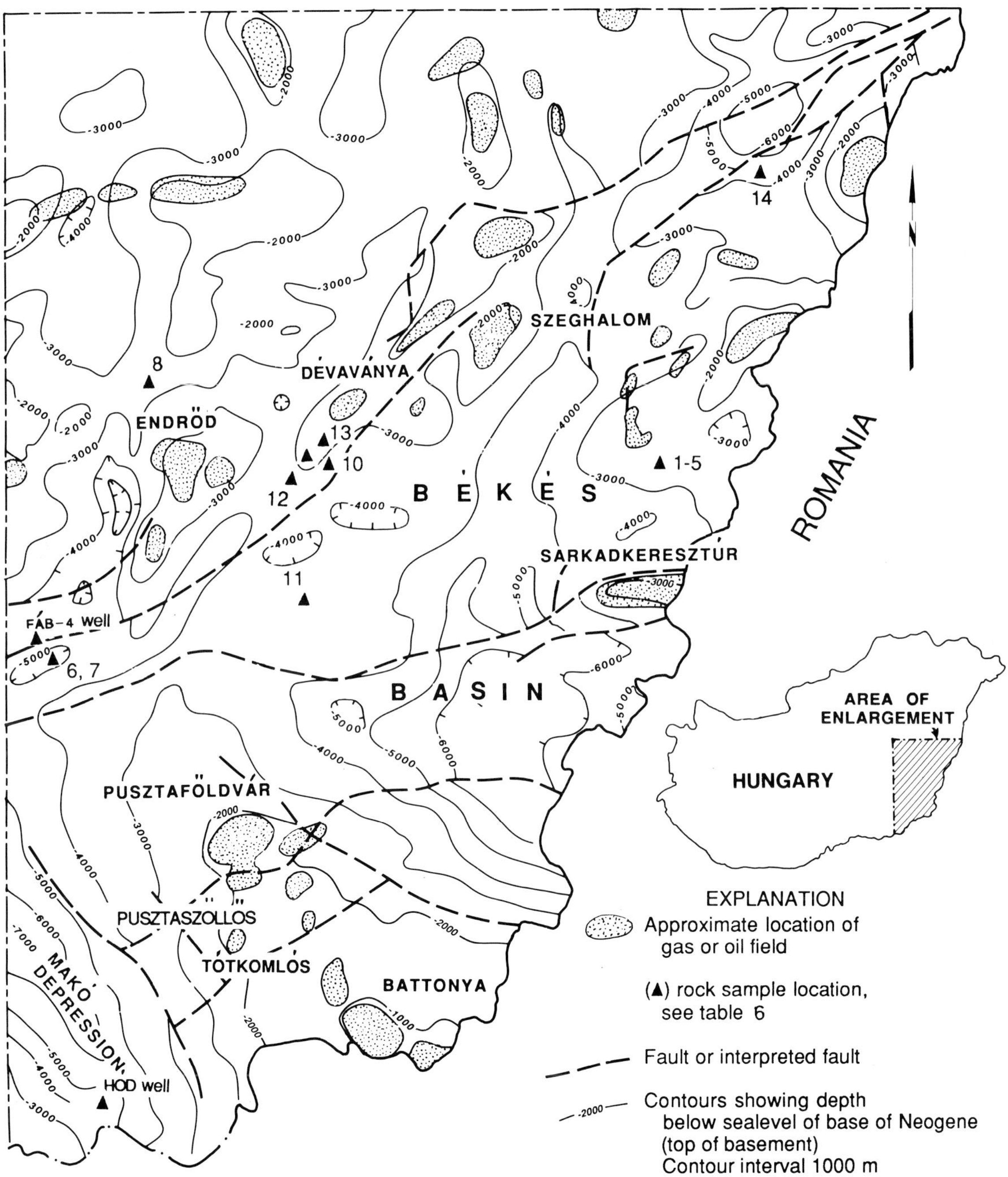

Figure 1. Index map of the Békés basin and surrounding areas showing locations of studied oil and rock samples. Oil and gas fields discussed in text are labeled with names. Fáb: Fábiánsebestyén; Hód: Hódmezövásárhely.

including sample preparation, have a standard deviation of 0.2-0.4 ‰.

RESULTS

Crude Oil Geochemistry

Oil sample locations and geochemical results are summarized in Tables 1 and 2. The triangular plot of Figure 2 shows that the oils are highly aliphatic, although the samples can be subdivided into three groups based on relative amounts of saturated hydrocarbons, aromatic hydrocarbons, and non-hydrocarbons. This compound class basis for grouping oils is convenient for refinery applications, but does not coincide precisely with the genetic classification of the oils and has, therefore, limited use for exploration purposes. Geochemical parameters more sensitive to the nature of the parent organic source material for the oils are discussed in subsequent sections.

Representative gas chromatograms of the saturated hydrocarbon fractions of the oils are shown in Figures 3 and 4. Group 1 oils (Fig. 3) all have pristane/phytane ratios

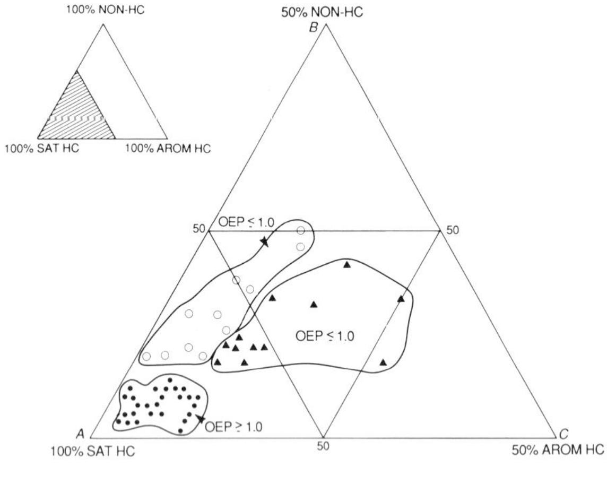

Figure 2. Normalized amounts (percent) of saturated hydrocarbons (SAT HC), aromatic hydrocarbons (AROM HC), and non-hydrocarbon (NON-HC) components in Békés basin oils determined by gravimetric analysis. Solid dots: "aliphatic" oils; open circles: "aromatic" oils; triangles: "asphaltic" oils; OEP: odd-even preference for n-alkanes.

(peaks labeled A and B) greater than one. The overall boiling point distribution among this group varies widely from fairly high wax oils to condensates consisting of predominantly low molecular weight alkanes (<C_{15}).

All oils which have pristane/phytane ratios less than one are in groups 2 and 3. However, pristane/phytane ratios much greater than one also occur in these two oil groups (Fig. 4; Table 2). As in the case of the group 1 oils, the overall molecular distribution of alkanes is variable, ranging from high wax oils (Pusztaföldvár-114 and Tótkomlós-I wells) to low wax oils (Szeghalom-28 and Pusztaszöllös-29 wells) (Figs. 3 and 4).

The group 1 oils generally have CPI (n-C_{25}-C_{31}) values greater than or equal to one. Oils with CPI values less than one have higher aromatic hydrocarbon or non-hydrocarbon contents. Odd-even preference (OEP) values (C_{15}-C_{33}) for oils are compared in Figure 5. The data are presented as OEP values rather than as carbon preference index (CPI) values because the odd/even preferences vary considerably over the C_{25} to C_{35} carbon number range, and in some cases, reversals occur in the odd or even preference. Two oil-groups with markedly different OEP values are evident in the Pusztaföldvár field and in the Komádi field, but a more limited distribution of OEP values occurs in the other oil fields.

The biological marker composition of the oils is summarized in Table 2 . No biological marker data are presented for very low density oils and condensates because the biological marker concentrations in these samples were too low for analysis. Two of the oils (sample nos. 42 and 66, Table 2) were included in Sajgó's (1984) study of oils from southeastern Hungary. In that study, no oleanane was reported for these two oils; however, we have tentatively identified oleanane in both of these oils using medium resolution (3000) GC-MS. This identification is based on retention time. Only one oil analyzed (sample no. 61, Szeghalom-28 well) contained no detectable oleanane. The 17 (H) α C_{27} trisnorhopane/18 α (H) C_{27} trisnorneohopane (Tm/Ts) and oleanane/hopane ratios are quite variable for the entire sample set (0.27 to 3.38 and 0 to 1.63 respectively; Table 2). With the exception of a high gammacerane content in three oils (samples 14, Endröd-4 well; 16, Gyoma-2 well; 28, Komádi-19 well), gammacerane and the C_{29} and C_{30} moretanes are present in only minor quantities in all of the Békés basin oils.

Among the steroids, the 20S/20S+20R ratio (in percent) for 24-ethylcholestane (C_{29}) varies from about 33-55+% and the triaromatic/triaromatic + monoaromatic ratio

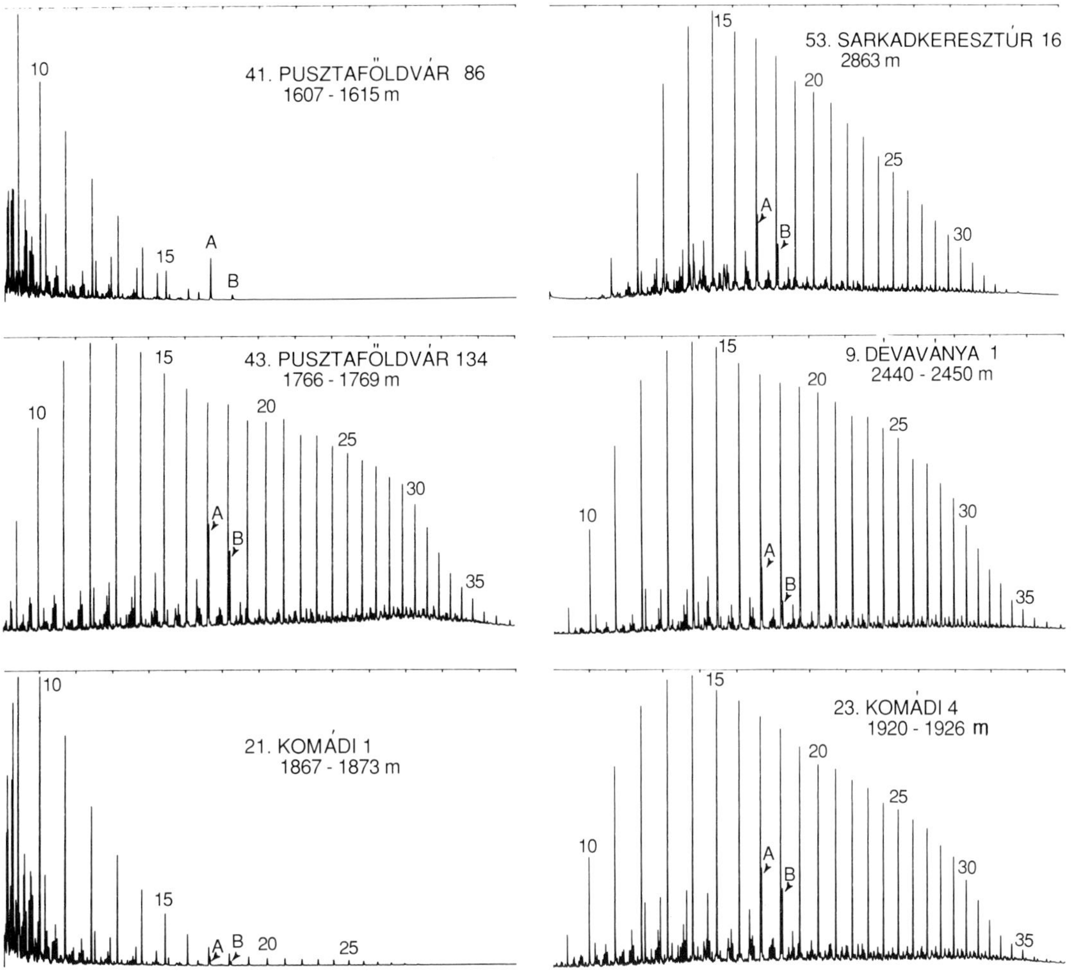

Figure 3. Representative gas chromatograms of saturated hydrocarbons from "aliphatic" type oils (group 1). Numbers proceeding well names refer to sample number shown in Table 2. All samples have pristane/phytane ratios greater than one (peaks are labeled A and B).

ranges from 0.55 to 0.95 (Table 2). These ratios were determined for a limited number of oil samples because many of the oils, especially the condensates, contained insufficient concentrations of steroids for analysis.

Source Rock Geochemistry

Distribution of organic carbon content (mg/g) for rocks in the Békés basin is shown in Tables 3 and 4. The upper Pannonian delta-plain sequence (Törtel Formation) contains the highest amounts of organic carbon, although analyzed samples of rock cuttings came from only two wells (Fábiánsebestyén-4 and Békés-2). Most of the upper Pannonian (Törtel Formation) samples from the Fábiánsebestyén-4 well contain greater than 5 mg/g organic carbon. More than 50 percent of these rocks contain in excess of 20 mg/g carbon and a few contain more than 80 mg/g.

Miocene ($M_{4,5}$ of Badenian and Sarmatian age) through lower Pannonian rocks from the basal clay and marl deposits in the Tótkomlós Formation (Molenaar and others, this volume) contain lower organic carbon contents than rocks of the upper Pannonian Törtel Formation. Organic carbon values are generally low for carbonate rocks and shales in the Neogene section as a whole (most contain less than 10 mg/g), but the basal clay and

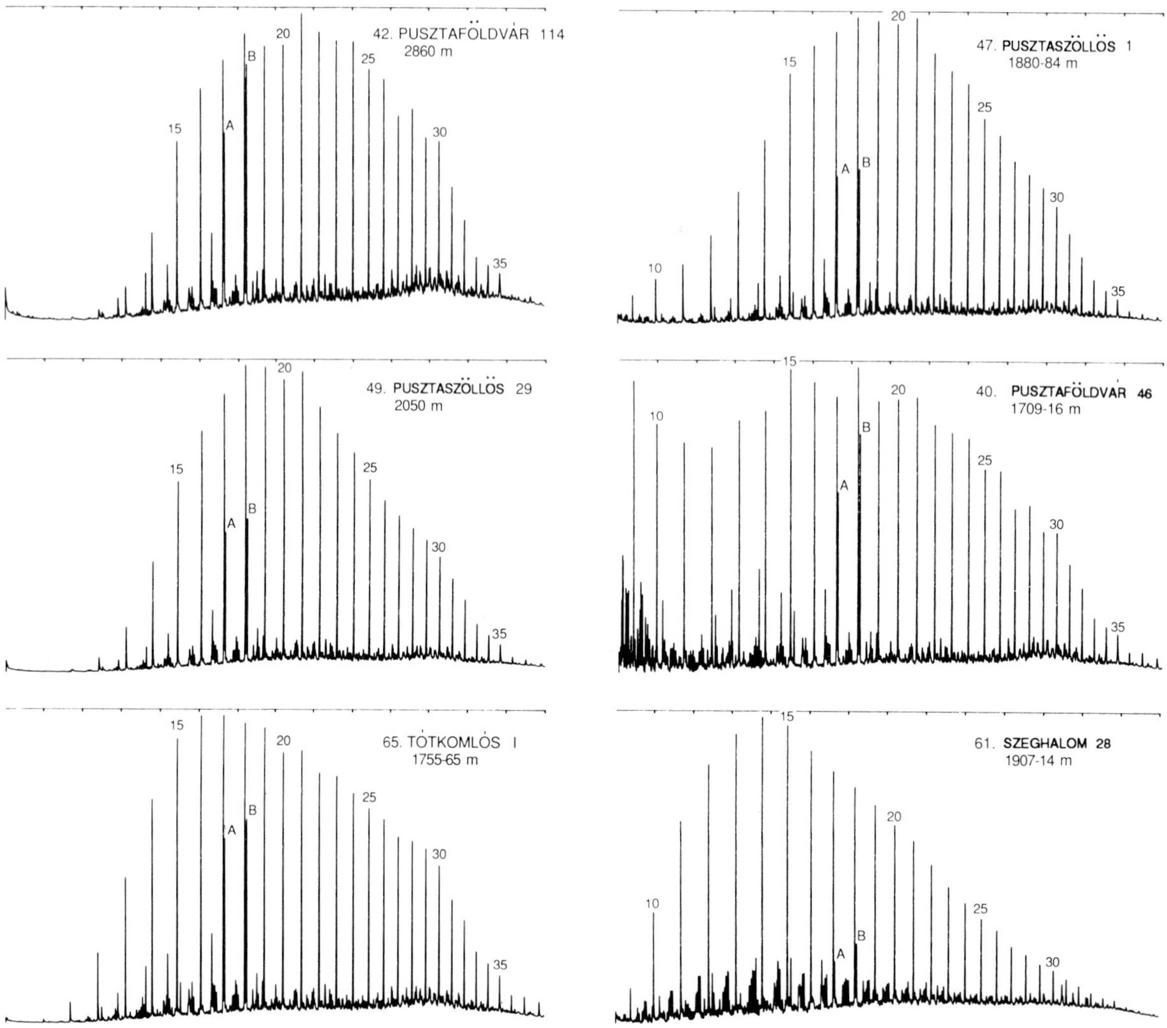

Figure 4. Representative gas chromatograms of saturated hydrocarbons from "aromatic" and "asphaltic" oils (genetic groups 2 and 3, respectively, in Table 7). See Table 2 and text for discussion. Numbers preceeding well names refer to sample numbers shown in Table 2. Pristane/phytane ratios are labeled A and B.

marl of the Tótkomlós Formation contain significant numbers of samples with 5-40 mg/g.

Most of the rocks from the uppermost part of Szolnok Formation and the Algyö Formation (delta-slope deposits) are organically lean. Seventy-nine percent of the samples analyzed contain less than 10 mg/g, and over 90% contain less than 20 mg/g. Sixty-five and 63% of the rocks of the Szolnok and Algyö Formations, respectively, contain less than 5 mg/g organic carbon. Rocks of pre-Pannonian Miocene age ($M_{4,5}$) contain between about 1.5-20 mg/g carbon, although core samples selected for detailed analysis contain as much as 50 mg/g. Sixty-seven percent of the $M_{4,5}$ rocks contain greater than 10 mg/g.

Extractable organic matter (bitumen) content ranges from 0.25 to 1.0 mg/g for most of the Neogene samples, although no rock samples representing upper Pannonian delta-plain deposition (Törtel Formation) were analyzed (Table 5). Samples selected from the lower Pannonian and pre-Pannonian Miocene ($M_{4,5}$) sections for detailed analyses (liquid chromatography, gas chromatography, mass spectrometry, etc.) contain above-average amounts of bitumen in absolute amounts (1-8 mg/g). Bitumen content normalized to organic carbon varies from about

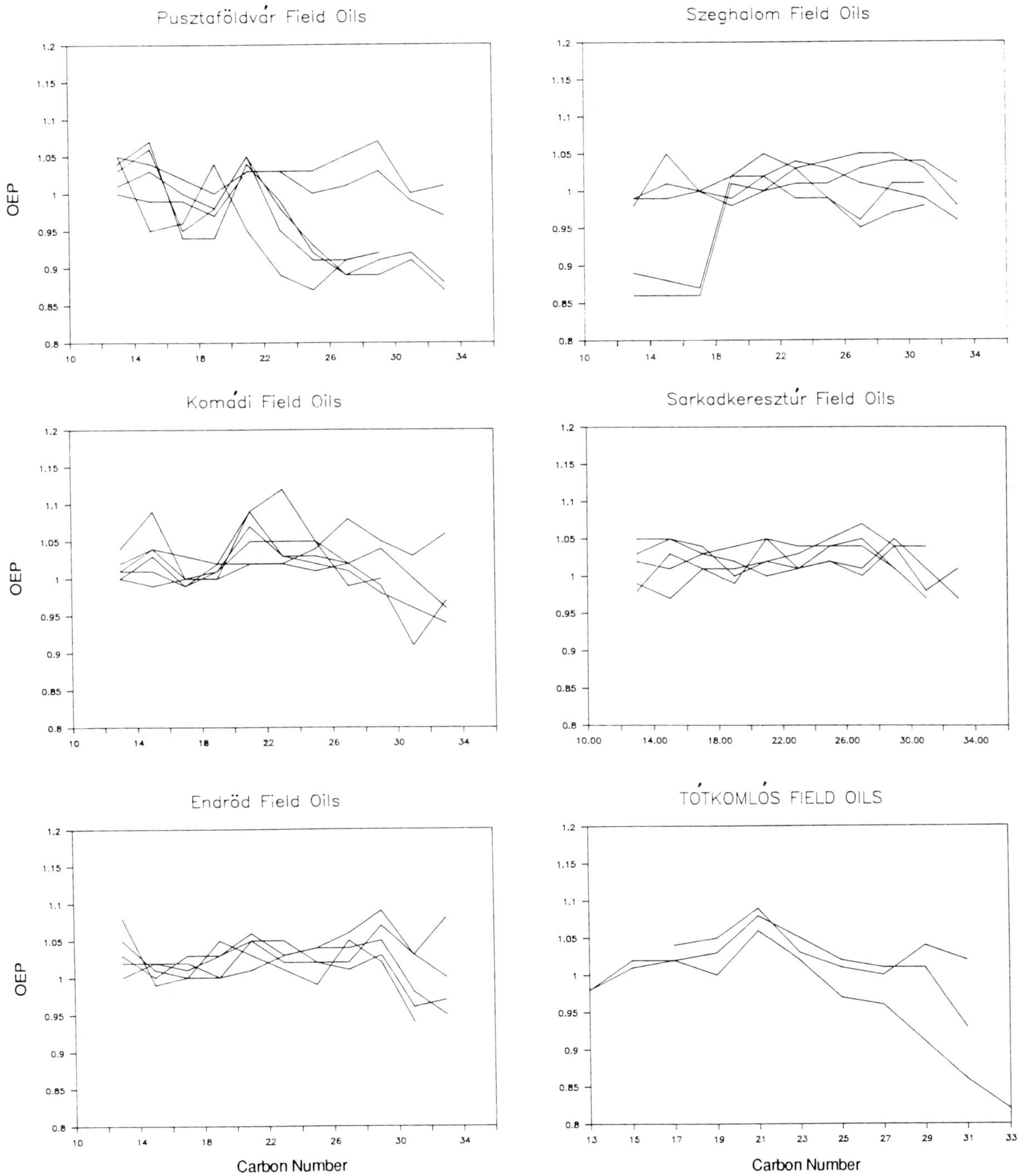

Figure 5. Comparison of odd-even preference (OEP) values for n-alkanes in Békés basin oils. OEP calculated by method of Scalan and Smith (1970).

3 to 30%. Values greater than 20% are common in rocks containing non-indigenous hydrocarbons (oil staining). All of the core samples extracted in the present study consist of fine-grained lithologies (non-reservoir lithologies) with no obvious evidence of hydrocarbon staining. Therefore, no definite conclusions can be made regarding the possibility of non-indigenous hydrocarbons, but we assume that all of the core samples except those from the Gyoma-2 well and possibly the Mezötúr-3 well (samples 8 and 10; Table 6) contain indigenous bitumen.

Like the oils, the hydrocarbons extracted from the cores consist predominantly of saturated compounds. The saturated/aromatic hydrocarbon ratios range from about 3 to more than 16. The total hydrocarbon fraction generally constitutes between about 40-70% of the total bitumen obtained, although higher relative hydrocarbon yields were obtained for four of the samples discussed above (samples 6, 7, 8 and 10; Table 6).

Rock-Eval pyrolysis data for samples of cuttings of Triassic through late Pannonian (Törtel Formation) age rocks from the Fábiánsebestyén-4 well are shown in Figure 6. The volatile hydrocarbon yield (S_1 peak) is correlative to bitumen content obtained by solvent extractions, although the S_1 yield is often lower than bitumen content because high molecular-weight bitumen constituents are not volatilized during the initial, low temperature (250°C) pyrolysis step. S_1 yields range from values of less than 10 mg/g in upper Pannonian rocks to typical values of 50-100 mg/g in lowermost Pannonian and Miocene ($M_{4,5}$) rocks. High S_1 yields are also common in the Mesozoic rocks, although most of the Mesozoic rock samples from the Fábiánsebestyén-4 well contain very low overall amounts of organic matter (organic carbon mg/g) and high S_1 yields occur only in the samples containing migrated hydrocarbons.

Figure 7 shows Hydrogen Index (HI) values for samples of rock cuttings collected from the Fábiánsebestyén-4 well and for core samples collected from 31 wells in the Békés basin. HI values are generally less than 300 for Neogene samples at thermal maturities corresponding to vitrinite reflectance values of about 0.4-1.5% or

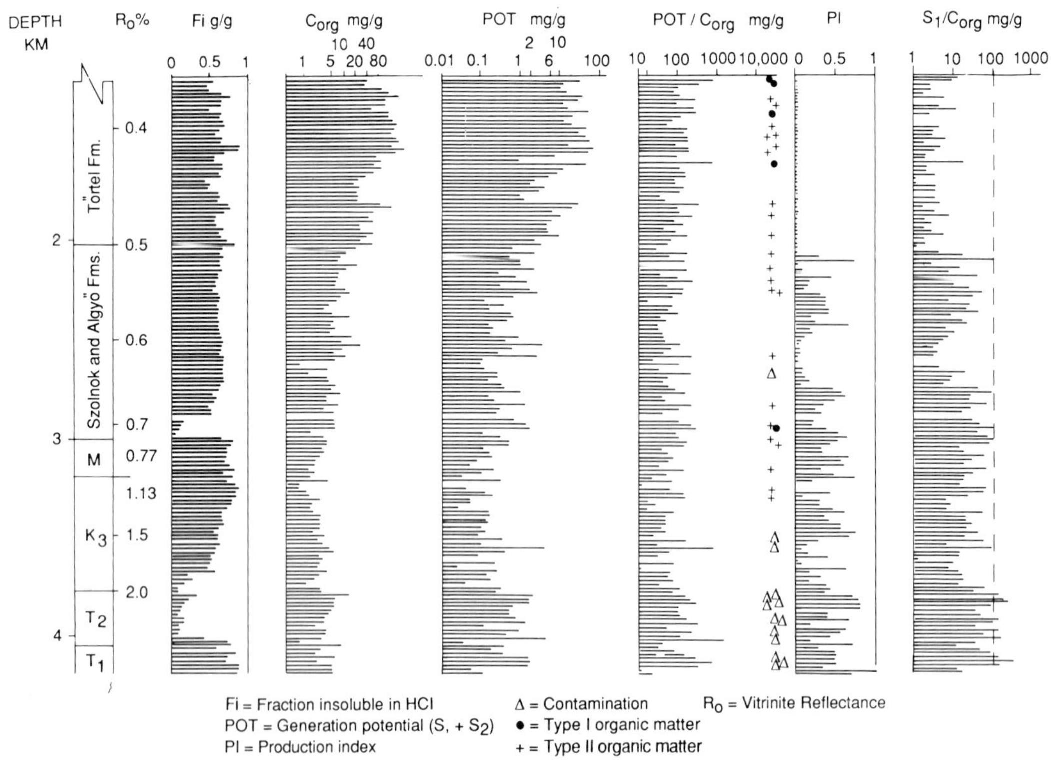

Figure 6. Rock-eval pyrolysis results from Fábiánsebestyén-4 well cuttings. M: Miocene; K_3: Upper Cretaceous; T_2: Middle Triassic, T_1: Lower Triassic.

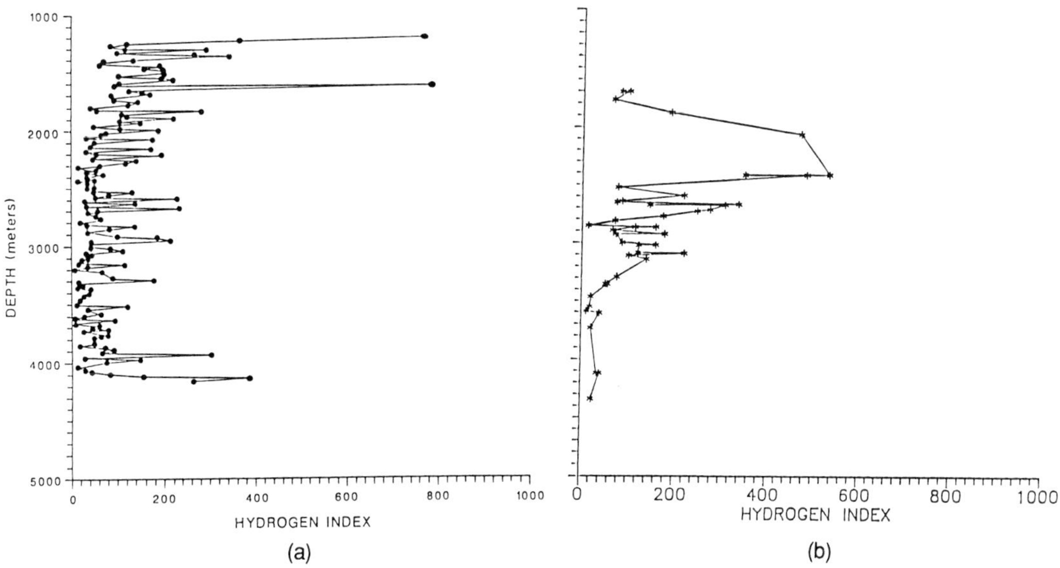

Figure 7. (a) Hydrogen Index (HI) in mg hydrocarbons/g organic carbon for samples of rock cuttings collected from the Fábiánsebestyén-4 well and (b) for core samples from 31 wells in the Békés basin.

greater. Rocks with notably higher HI values occur in delta-plain deposits of the upper Pannonian Törtel Formation (1200-1700 m depth in the Fábiánsebestyén-4 well; <2500 m depth throughout the Békés basin). With the exception of some of these shallow Törtel Formation rocks, the organic matter contained in Neogene rocks would be classified as type II and III according to the scheme of Tissot and others (1974) and Espitalié and others (1977).

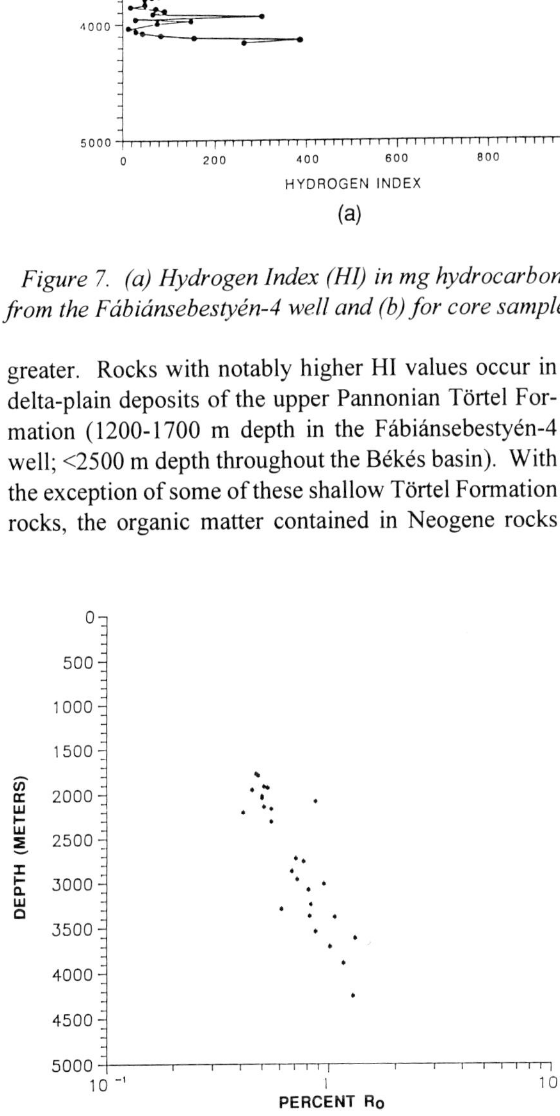

Figure 8. Distribution of vitrinite reflectance with depth for Neogene rocks in the Békés basin based on core samples from 10 wells. Data scatter largely reflects the effects of varying heat flow at different locations in the basin.

Vitrinite reflectance data for core samples of Neogene rocks from the Békés basin are shown in Figure 8. Vitrinite reflectance, T_{max}, and production index data for rocks from the Fábiánsebestyén-4 well are compared in Figure 9. Young rocks that represent delta-plain deposition (Törtel Formation) have vitrinite reflectance values less than about 0.5%, corresponding to T_{max} values between 420 and 430°C. Rocks that represent delta slope deposition (Algyö Formation) have vitrinite reflectance values between about 0.5-0.8% with occasionally higher values. Rocks from the lower part of the Pannonian section (including the Szolnok Formation and basal clays) and pre-Pannonian Miocene ($M_{4,5}$) rocks have variable maturities depending upon their location in the basin (i.e., depending on burial depth and heat flow). Vitrinite reflectance values for the pre-Pannonian Miocene rocks range from abut 0.7-2.0% or more in the deeper parts of the basin.

DISCUSSION

Oil Correlations

At least three genetic oil-types are present in the basin. The oils are listed according to interpreted genetic groups in Tables 2 and 7. We classified the oils according to

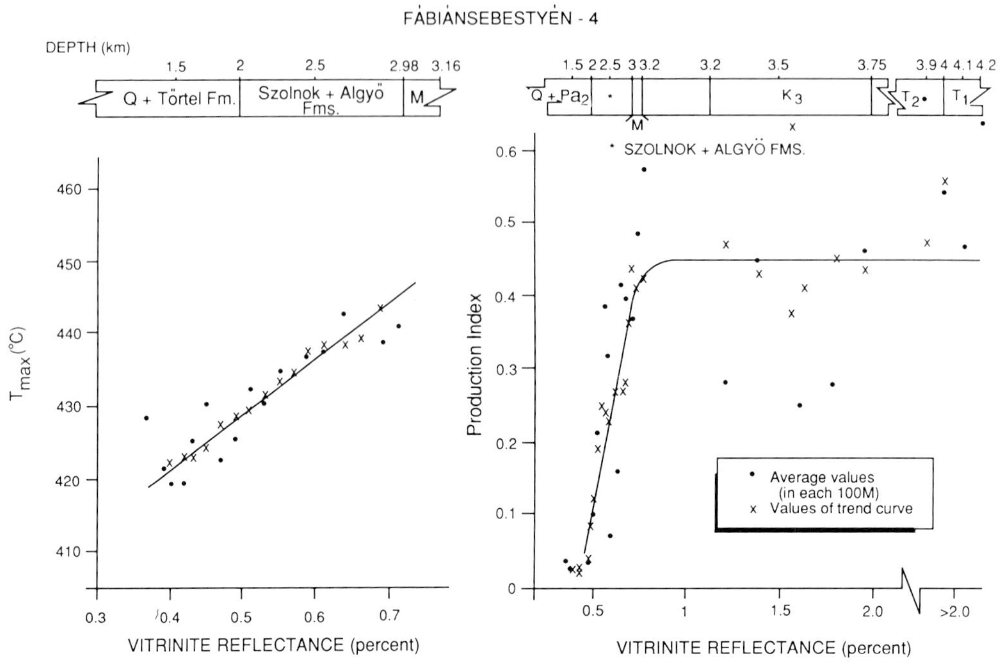

Figure 9. (left) T_{max} and (right) production index (PI) versus vitrinite reflectance for samples of rock cuttings from the Fábiánsebestyén-4 well. Dots show measured data points (averaged over 100 m intervals) and x's give trend or moving average. Q: Quaternary; M: Miocene; Pa_2: Upper Pannonian; K_3: Upper Cretaceous; T_2: Middle Triassic; T_1: Lower Triassic.

n-alkane distributions, pristane/phytane ratios, biological marker compositions, and carbon isotope ratios. These genetic classifications are based on consideration of all of the geochemical data, but differences in molecular distributions and carbon isotope ratios are given more weight than compound class distributions in the correlations. The data suggest that additional, volumetrically minor oil-types may be present within these major groups. Similarly, Sajgó (1984) correlated oils from the Battonya structure on the western flank of the Békés basin, including oils from the adjacent Hódmezövásárhely-Makó (Makó) depression (Fig. 1), and concluded that three genetic types are present. Sajgó's correlations were based mainly on relative amounts of oleanane and the amount of hopanes and moretanes relative to steranes. The oleanane, Tm/Ts, carbon isotope ratios, and OEP values (Fig. 10) indicate two oil-types at Pusztaföldvár field, and the virtual absence of oleanane in some of the other fields on the Battonya structure (Sajgó, 1984) is consistent with the interpretation of at least three genetic oil-types. In the case of these Pusztaföldvár field oils, a clear genetic distinction between the oil-groups is also indicated by the carbon isotope ratios. The group 1 oils are somewhat depleted in ^{13}C compared to the oils from groups 2 and 3. In general, the saturated and aromatic hydrocarbons in the group 1 oils (Table 7) have lower $^{13}C/^{12}C$ ratios than the other oil groups.

The two principal differences in alkane composition between the group 1 and groups 2 and 3 oils are: (1) the former generally have OEP or CPI values greater than one (odd carbon preference) while the group 2 and 3 oils commonly have values less than one (even carbon preference); and (2) the pristane/phytane ratios are greater than one for all of the group 1 oils but are less than one in some of the group 2 and 3 oils. Pristane/phytane ratios are believed to depend mainly on the origin of the organic matter (Powell and McKirdy, 1973), although the relationship between specific environments of deposition or contributing organisms and pristane/phytane ratio is imperfectly understood (ten Haven and others, 1987). Likewise, odd- or even-carbon preferences are thought to be a primary feature of the original organic matter. In contrast, secondary processes, most notably maturation of bitumen in a source rock prior to expulsion of the oils or in-reservoir maturation of oil, can significantly affect the relative amounts of compound classes (saturated hydrocarbons, aromatic hydrocarbons, resins, asphaltenes). Therefore, biological marker composition, alkane molecular distributions, and carbon isotope ratios are more sensitive than compound class composition to the organic matter source. Accordingly, the genetic classifications of Table 7 are based primarily on these source-dependent parameters and only secondarily on compound class composition.

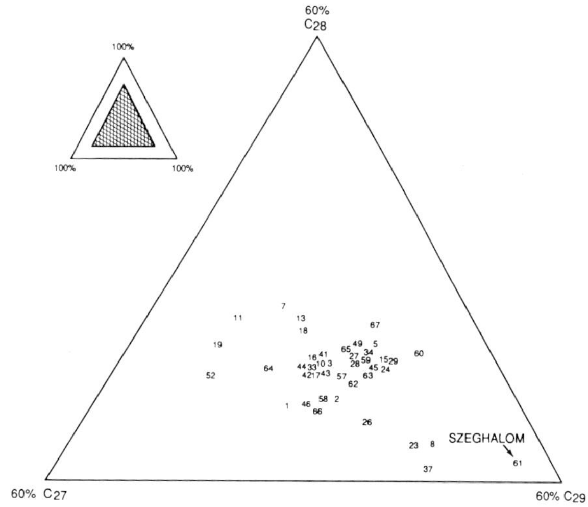

Figure 11. Normalized C_{27}-C_{29} sterane content (20R configuration) for Békés basin oils. Numbers refer to oil sample numbers listed in Table 1.

Thermal maturity

Superimposed on the source-related differences among the oils are effects of thermal maturity. For example, condensate samples 21 and 41 (Fig. 3) contain most features of the genetic group 1 oils (high saturated hydrocarbon content, high pristane/phytane ratio etc.), but contain predominantly low molecular weight alkanes ($< C_{15}$) owing to effects of thermal processes. Among

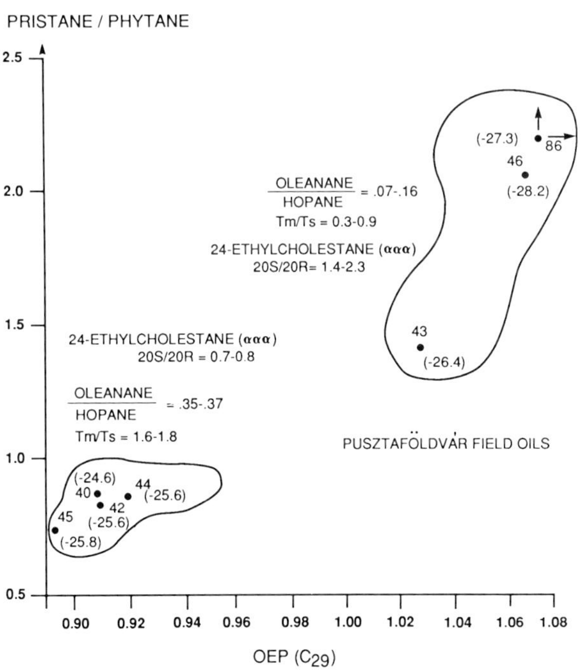

Figure 10. Comparison of selected correlation parameters for Pusztaföldvár field oils. Carbon isotope ratios (^{13}C) for saturated hydrocarbons are shown in parenthesis. OEP: odd-even preference for n-alkanes.

oil-types 2 and 3 (Table 7), the overall boiling-point distribution of alkanes also varies significantly. The Szeghalom-28 oil (sample 61, Szeghalom-28 well, Fig. 4) has a much lower wax content than the other related oils. However, in contrast to the group 1 oils, the lower wax content and lower isoprenoid/*n*-alkane ratio for the Szeghalom-28 oil is probably related more to source differences than to differences in thermal maturity. Unlike the other group 2 and 3 oils, the Szeghalom-28 oil contains no oleanane and has an anomalously high C_{29} sterane (24-ethylcholestane) content (Fig. 11). These features indicate that this oil represents a sub-type within group 2 and that the differences in alkane distribution (especially boiling point distribution and OEP) are related mainly to source rather than thermal maturity.

The ratios of 20S/20S+20R diastereomers of 24-ethylcholestane and triaromatic to monoaromatic steroids are shown in Figure 12. Some oil samples are omitted from this plot because some samples, particularly the high gravity oils and condensates, contained insufficient biological marker concentrations for analysis. The triaromatic/triaromatic + monoaromatic ratio, calculated according to the method of Mackenzie (1984), varies between 0.55-0.95. The 20S/20S + 20R ratio of 24-ethylcholestane range from 39-59%. Equilibrium for this latter reaction (20R-20S) is about 55%, and the mono- to triaromatic conversion is complete at a value of 1.00 (Mackenzie, 1984). It is noteworthy that these reactions have not reached completion (or equilibrium) in some of the oils. This finding, like the observation of odd-even

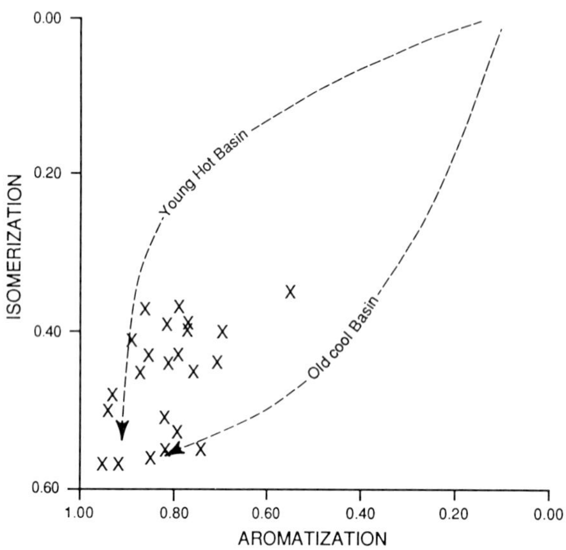

Figure 12. Comparison of steroid aromatization (monoaromatic/mono- + triaromatic; Mackenzie, 1984) and 24-ethylcholestane isomerization at C-20 (20S/20S + 20R) for Békés basin oils. Analyses are of whole oil samples (see analytical procedures section for description of method).

carbon preference in some of the oils, indicates that in some cases, oil expulsion from source rocks occurred at quite low temperatures before these biological marker reactions reached completion.

Oleanane Content

Oleanane is present in unusually high relative amounts (normalized to hopane) in some of the oils (Table 2). Although high amounts of oleanane relative to hopane have been reported in oils from China and Indonesia (Grantham and others, 1984), most oils worldwide contain much lower amounts (Ekweozor and others, 1979a, 1979b; Benshan and others, 1984; Palmer, 1984; Schoell and others, 1984; Riva and others, 1986). In these previous studies, oleanane generally has been used as an indicator of organic matter derived from terrigenous sources, although Palmer (1984) noted that in Eocene rocks of the Elko Formation (Nevada, U.S.A.) oleanane occurred both in lignitic organic matter and oil shale even though the latter contained organic matter derived predominantly from aquatic organisms rather than land-plants.

In the Békés basin oils, oleanane occurs in some oils containing features otherwise interpreted to indicate predominantly marine organic precursors (low pristane/phytane ratios, even-carbon preference among *n*-alkanes) (samples 38, 44 and 49, Table 2). Because no obvious relationship exists between oleanane content and other indicators of the organic matter source of the oils, oleanane content in Békés basin oils cannot be used as an indicator of terrestrial versus marine organic matter. A possible explanation for this observation is that oleanane may have more than one source. While land plants may be a major source of oleanane, it is possible that a heretofore unknown source may exist, including possibly a bacterial and/or diagenetic source. Ten Haven and Rullkötter (1988) presented a diagenetic scheme wherein taraxer-14-ene is converted ultimately to oleanane. They point out that Howard (cited in Venkatesan and Kaplan, 1982, p. 2143) found an unknown terpene in a bacterium with a mass spectrum similar to taraxer-14-ene. However, no direct evidence is available in the current study to support the hypothesis of a bacterial source of oleanane precursors.

Although oleanane content may not provide clear evidence of the specific biological source of the organic matter contained in the Békés basin oils, it does seem to be a useful empirical correlation parameter. The chemical structure of oleanane is not sufficiently different from the chemical structure of hopane to explain the observed range of oleanane/hopane ratios by thermal maturity differences among the oils. The absence of oleanane and the high norhopane content in the Szeghalom-28 oil (Tables 1 and 7) are the main criteria for listing the sample as a subgroup of genetic group 2.

Oil migration

Ambiguities regarding correlation parameters (i.e., some inconsistencies among source-dependent correlation indicators) may be caused not only by maturation processes as discussed before, but also by mixing of oils from different sources during secondary migration. As noted by Clayton and others (1990), hydrocarbon gas and carbon dioxide have undergone substantial vertical and lateral migration. Similarly, the finding of oils with wide ranges of densities, boiling point distributions (condensates and medium-to-heavy oils), and chemical compositions in shallow fields (1000-2000 m depth) on the same structural highs (Battonya and Komádi fields, Fig. 1) is also evidence for substantial vertical and lateral secondary migration of oil.

The depth to a given maturation level (vitrinite reflectance value) (Fig. 8) varies throughout the Békés basin depending on local heat flow. The average geothermal gradient in the basin is high (5°C/100 m) (Dövényi and Horváth, 1988), but is somewhat lower (4°C/100 m or less) in the deep part of the basin compared to structurally high positions on the margins of the basin.

This variation in thermal-maturity-depth trend in the basin is small enough to allow some generalizations about depth and conditions of oil generation in the basin. Rocks of the upper Pannonian Törtel Formation and younger (Quaternary) are thermally quite immature with respect to liquid hydrocarbon generation (R_0 values much less than 0.6%). As noted above, parts of this rock sequence (delta-plain deposits penetrated in the Fábiánsebestyén-4 well) contain good potential source rocks. Equally organic-rich rocks representing delta-plain deposition have been reported at one location within the Békés basin proper (Békés-2 well) (I. Koncz, personal commun., 1989). These rocks are speculated to be sources of oil and gas in the basin, but no oil or gas occurrences have yet been directly correlated with upper Pannonian rocks.

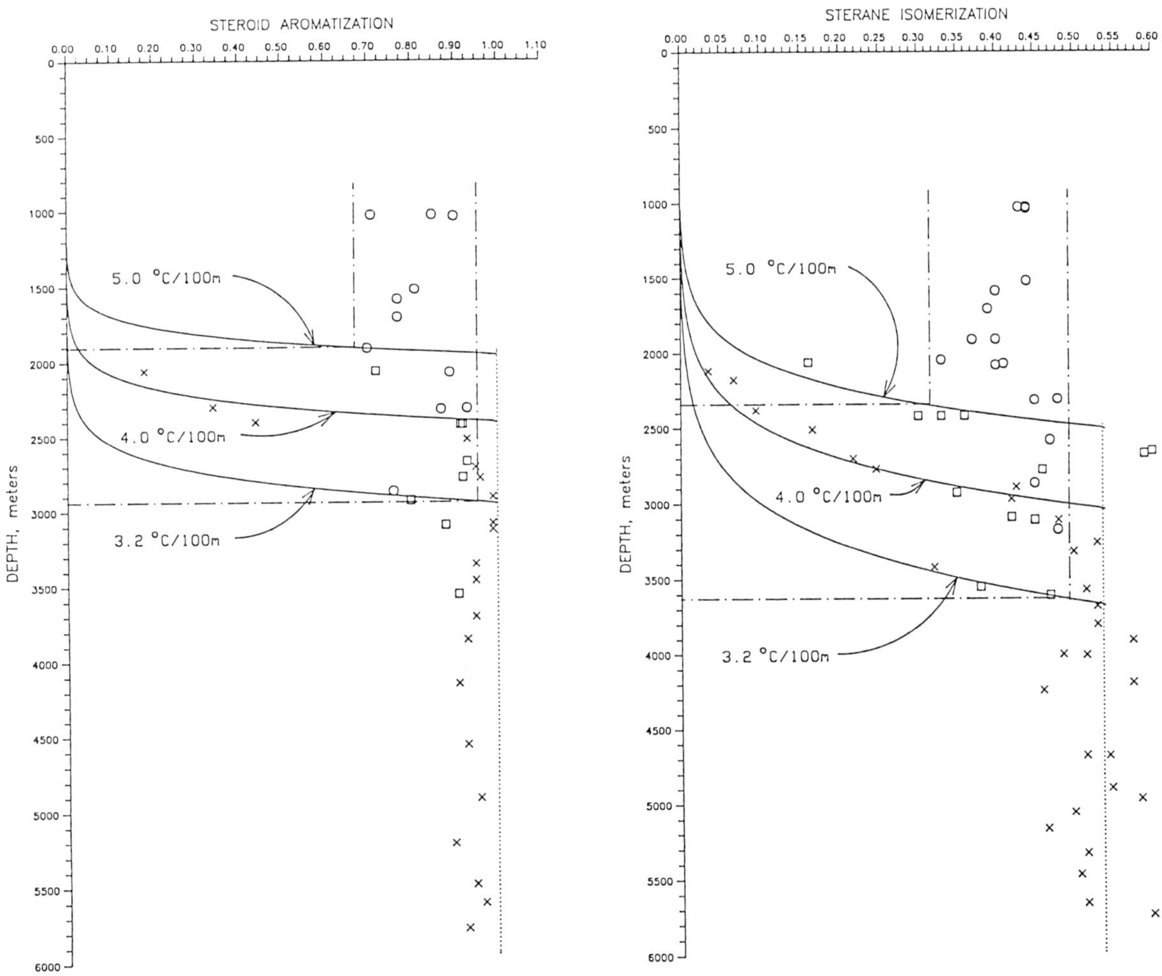

Figure 13. Isomerization of 24-ethylcholestane (20S/20S + 20R) and steroid aromatization (monoaromatic + mono- + triaromatic) in core sample extracts from Hódmezövásárhely (Hód)-1 well, located in Makó depression (Fig. 1), compared with Békés basin oils and core extracts. Circles represent Békés basin oil data; X's represent Hódmezövásárhely (Hód)-1 core extract data from Sajgó and others (1988); squares represent core extracts from several locations in Békés basin. Trend lines for various geothermal gradients were generated by computer for the two reactions using Mackenzie's (1984) kinetic parameters and the kinetic program of J.D. King (U.S.G.S., unpublished data).

Table 1. List of oil samples, locations, producing zones, and reservoir data. Location of fields and wells listed in column 2 are shown in Figure 1. Bsmt: basement; Pa_1: lower Pannonian; PC: Precambrian; Pz: Paleozoic; Mz: Mesozoic; Tr: Triassic; Mio: Miocene; cm: clayey marl; ss: sandstone; cgl: conglemerate; lm: limestone; sh: shale; dol: dolomite; m: marl; nd: not determined; Szoln : Szolnok Formation.

SAM-PLE No.	FIELD or WELL NAME	WELL No.	PRODUCING DEPTH (m)	API GRAV.	RESERVOIR AGE	FM. NAME	RESERVOIR LITHOLOGY	RES. TEMP. (^oC)
1	Battonya	1	1027-1029	42	Pa_1	nd	cm, ss, cgl	76
2	Battonya	70	1028-1030	43	Pa_1	nd	cm, ss, cgl	74
3	Battonya-East	63	1036-1041	43	Pa_1	nd	cm	65
4	Békés (well)	1	2995-3002	42	Pa_1	Szoln	ss	155
5	Biharkeresztes	1	1584-1588	32	Bsmt	nd	m	nd
6	Biharkeresztes	1	1865-1895	44	nd	nd	nd	nd
7	Biharugra	3	2295-2303	44	Mz	nd	lm	nd
8	Csanádapáca	3	1911-1930	35	Pa_1	nd	cm	111
9	Dévaványa	1	2440-2450	44	Mio+PC	nd	nd	144
10	Dévaványa	11	2573-2586	35	Pa_1	Szoln	ss	nd
11	Endröd	2	2394-2400	34	Pz	nd	m	141
12	Endröd	4	2450-2456	57	nd	nd	nd	nd
13	Endröd	5	2595-2603	29	nd	nd	nd	nd
14	Endröd-North	5	2795-2825	29	Pa_1	nd	cm	133
15	Endröd-North	5	2901-2909	30	PC	nd	m	171
16	Gyoma (well)	2	2918-2924	35	Pa_1	nd	m	nd
17	Hunya (well)	1	3169-3174	46	Pa_1	Szoln	ss, cgl	nd
18	Kaszapér-South	2	1570-1605	33	Pa_1	nd	cm	nd
19	Kaszapér-South	8	1628-1631	30	Pa_1	nd	cm	nd
20	Füzesgyarmat	11	1826-1836	46	Mio	nd	nd	nd
21	Komádi	1	1867-1873	nd	Pa_1+Mio	nd	sh	nd
22	Komádi	3	2527-2535	38	Mio	nd	ss	nd
23	Komádi	4	1920-1926	47	Pa_1	Algyö	ss	105
24	Komádi	6	2204-2212	39	Mio+Bsmt	nd	ss, m	117
25	Komádi	10	2134-2140	44	Pa_1	Algyö	ss	nd
26	Komádi	12	2553-2564	26	Pa_1	nd	cm	139
27	Komádi	15	2520-2537	28	Pa_1+Mio	nd	cm, sh	nd
28	Komádi	19	3115-3126	30	Mio	nd	ss, cgl	nd
29	Komádi-East	2	2275-2280	36	Pa_1	Szoln	ss	nd
30	Körösladány	1	2879-2887	48	Bsmt	nd	m	158
31	Köröstarcsa	1	3248-3260	42	Mio	nd	lm	nd
32	Köröstarcsa	1	3295-3302	nd	Mio	nd	lm	180
33	Medgyesbodzás (well)	1	2530-2545	39	Mio	nd	ss, cgl	nd
34	Mezögyán-well	1	2240-2276	32	Mio	nd	cgl, ss	133
35	Mezöhegyes	14	1188-1190	35	Pa_1	nd	cm	nd
36	Mezöpeterd	1	2917-2921	41	Mio	nd	cgl, ss	152
37	Mezösas	3	2568-2575	40	PC	nd	m	147
38	Mezötúr	1	2840-2890	32	Pa_1+Mio	nd	sh, cgl	nd
39	Örménykút-well	1	2983-2988	46	Mio	Szoln	ss	nd
40	Pusztaföldvár	46	1709-1716	nd	Pa_1	Földvár Zone	ss	nd
41	Pusztaföldvár	86	1607-1615	nd	Pa_1	Földvár Zone	ss	109
42	Pusztaföldvár	114	1776-1777	27	Pa_1	Békés Zone	cgl	nd
43	Pusztaföldvár	134	1766-1769	nd	Pa_1	nd	cm	nd
44	Pusztaföldvár	154	1698-1717	30	Pa_1	Földvár Zone	ss	nd

Table 1. Continued from the previous page.

SAMPLE No.	FIELD or WELL NAME	WELL No.	PRODUCING DEPTH (m)	API GRAV.	RESERVOIR AGE	FM. NAME	RESERVOIR LITHOLOGY	RES. TEMP. (°C)
45	Pusztaföldvár	177	1703-1706	30	Pa_1	Földvár Zone	ss	nd
46	Pusztaföldvár	185	1761-1763	28	Pa_1	Békés Zone	cm	nd
47	Pusztaszöllös	1	1808-1810	nd	Mz	nd	dol	121
48	Pusztaszöllös	35/2	1230-1235	nd	Pa_1	Komlós Zone	nd	84
49	Pusztaszöllös	29	1748-1750	30	Mz	nd	dol, lm	nd
50	Sarkadkeresztúr	1	2723-2729	50	Pz	nd	m	142
51	Sarkadkeresztúr	4	2948-2954	42	PC	nd	m	124
52	Sarkadkeresztúr	9	2551-2556	49	Pa_1	Szoln	ss	124
53	Sarkadkeresztúr	16	2861-2865	51	Pa_1	Szoln	ss	141
54	Sarkadkeresztúr	32	2916-2921	42	PC	nd	m	152
55	Szarvas	5	2204-2206	55	Pa_1	Algyö	ss	131
56	Szarvas	7	2256-2264	46	Pa_1	Szoln.	ss	nd
57	Szarvas	11	2309-2314	34	Pa_1	Szoln.	ss	145
58	Szeghalom	3	2101-2105	43	Mio	nd	cgl	nd
59	Szeghalom	11	2071-2076	44	Bsmt	nd	m	119
60	Szeghalom	12	2039-2075	46	PC	nd	m	119
61	Szeghalom	28	1097-1914	31	nd	nd	nd	nd
62	Szeghalom	38	2081-2086	39	nd	nd	nd	122
63	Szeghalom-N	1	1908-1914	44	Bsmt	nd	m	116
64	Szeghalom-N	5	2312-2323	45	Mz	nd	m	141
65	Tótlomlós	I	1755-1765	31	Pa_1	nd	cm	123
66	Tótkomlós	26	1898-1900	36	Tr	nd	dol	nd
67	Tótkomlós-East	1	1520-1525	44	Pa_1	nd	cm	118
68	Hód (well)	1	4152-4156	52	Pa_1	nd	m	nd
69	Hód (well)	2	4807-4815	38	Mio	nd	nd	nd

Figure 13 shows isomerization of 24-ethylcholestane (20S/20S + 20R) and aromatization (tri-/mono- + triaromatics) of steroids from Békés basin oils and rock extracts compared with data obtained for extracts from cores collected from the Hódmezövásárhely (Hód)-1 well (Fig. 1) (Sajgó and others, 1988). These reactions are complete at about 3000 and 2300 m, respectively, in the Hód-1 well (geothermal gradient approximately 4°C/100 m). The Hód-1 well is located in the Makó depression just to the west of the Pusztaföldvár-Battonya structure which forms the southwestern boundary of the Békés basin (Fig. 1). Although the Hód-1 well is located outside the Békés basin proper, the basin fill of the Makó depression was deposited in the same transgressive episode as the Békés basin fill (Mattick and others, this volume). Moreover, the heat flow in the Makó depression is similar to that of the Békés basin (Sajgó and others, 1988). Therefore, the geochemistry (and particularly the thermal history) of the Neogene section in the Hód-1 well is assumed to be representative of the average Neogene section in the deep parts of the Békés basin.

Isomerization data of Békés basin core extracts (squares in Fig. 13) from the Hód-1 well support this assumption. The reaction trend lines of Figure 13 were calculated for three geothermal gradients (5.0°C/100 m, 4.0°C/100 m, 3.2°C/100 m) using the kinetic parameters of Mackenzie (1984). The gradient of 4.0°C/100 m is the current-day gradient in the area of the Hód-1 well, and the 5.0°C/100 m and 3.2°C/100 m curves represent the extent of each reaction (isomerization or aromatization) for the range of present-day geothermal gradients in the Békés basin. The curves show the extent of each of the two reactions as a function of depth of burial for the geothermal gradients shown. The reaction curves are calculated using the method of J. D. King (U.S. Geol. Survey, unpublished data). Completion of the reactions (equilibrium in the case of isomerization) is indicated by vertical dotted lines. Circles show the reservoir depths and reaction extent for the least-mature Békés basin oils. These data are bracketed by horizontal, alternating dash-dot lines which show the depths corresponding to the maturity of these oils for the high and low geothermal gradients.

Comparison of the extent of steroid isomerization and aromatization of Békés basin oils with core extracts from the Hód-1 well and Békés basin wells (Fig. 13) provides an estimate of the approximate depths (maturities) at which oils were generated. Of course, this approach can be used only for oils which have incomplete reactions. Based on comparison of the degree of sterane isomerization and aromatization in the immature oils with the extent of these same reactions in the rock extracts versus depth, we conclude that some of the oils were expelled from source rocks at depths less than about 3300 m (R_o << 0.60%), perhaps as shallow as about 2000 m (R_o of about 0.35%). The odd-even carbon preferences noted above in some of the oils are, in some cases, coincident with fairly strong even- (or odd-) carbon preferences in the rocks. This coincidence of incomplete isomerization and carbon preferences supports the interpretation that some of the oils were expelled at fairly shallow depths, corresponding to low levels of thermal maturity, as indicated by vitrinite reflectance values (R_o of about 0.60% or less).

Not all of the oils with incomplete sterane isomerization contain *n*-alkanes with carbon preferences. This is not surprising because not all young sedimentary organic matter has a pronounced carbon preference. The intensity of an odd-even carbon preference depends on the relative contribution of lipid precursors from various sources. For example, variable proportions of higher-land-plant waxes would be reflected in the degree of

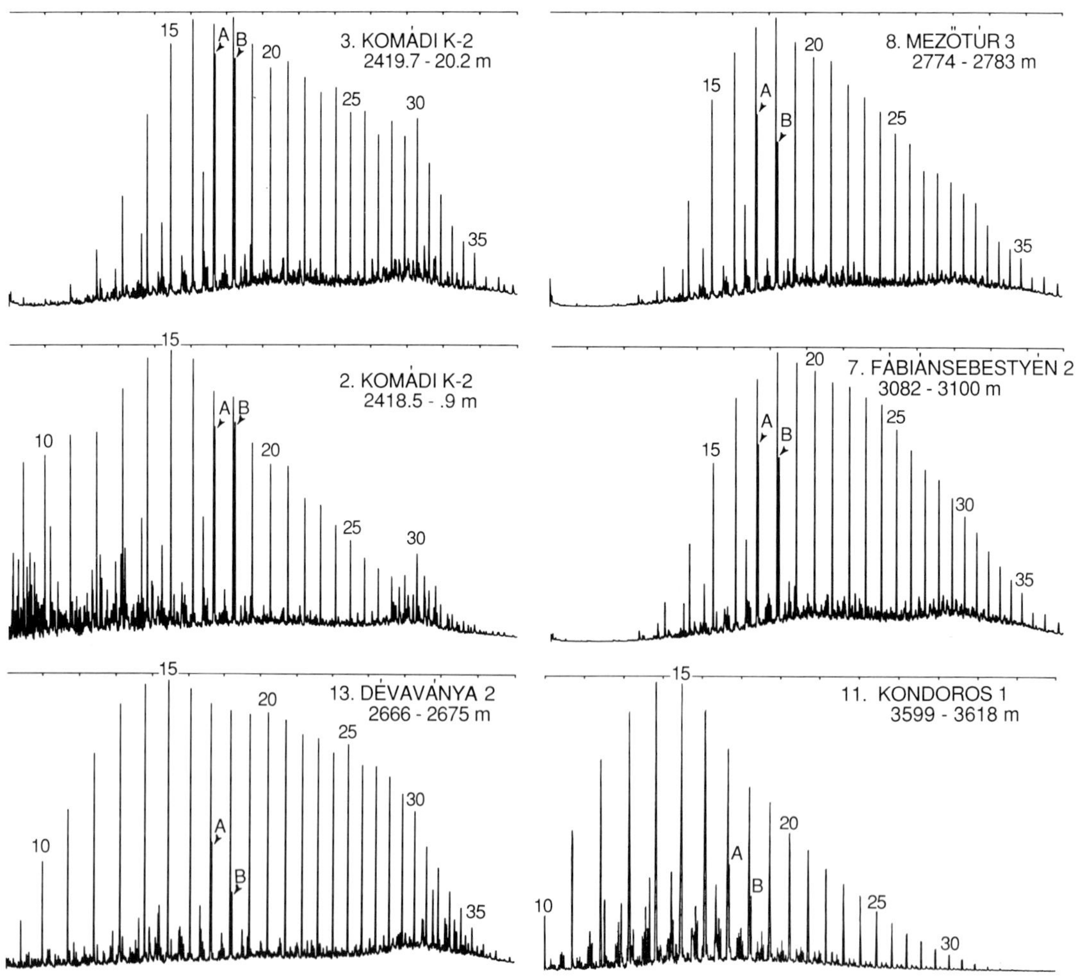

Figure 14. Gas chromatograms of C_{15}+ saturated hydrocarbons from rock extracts. Numbers preceeding well names refer to sample numbers in Tables 1 and 2.

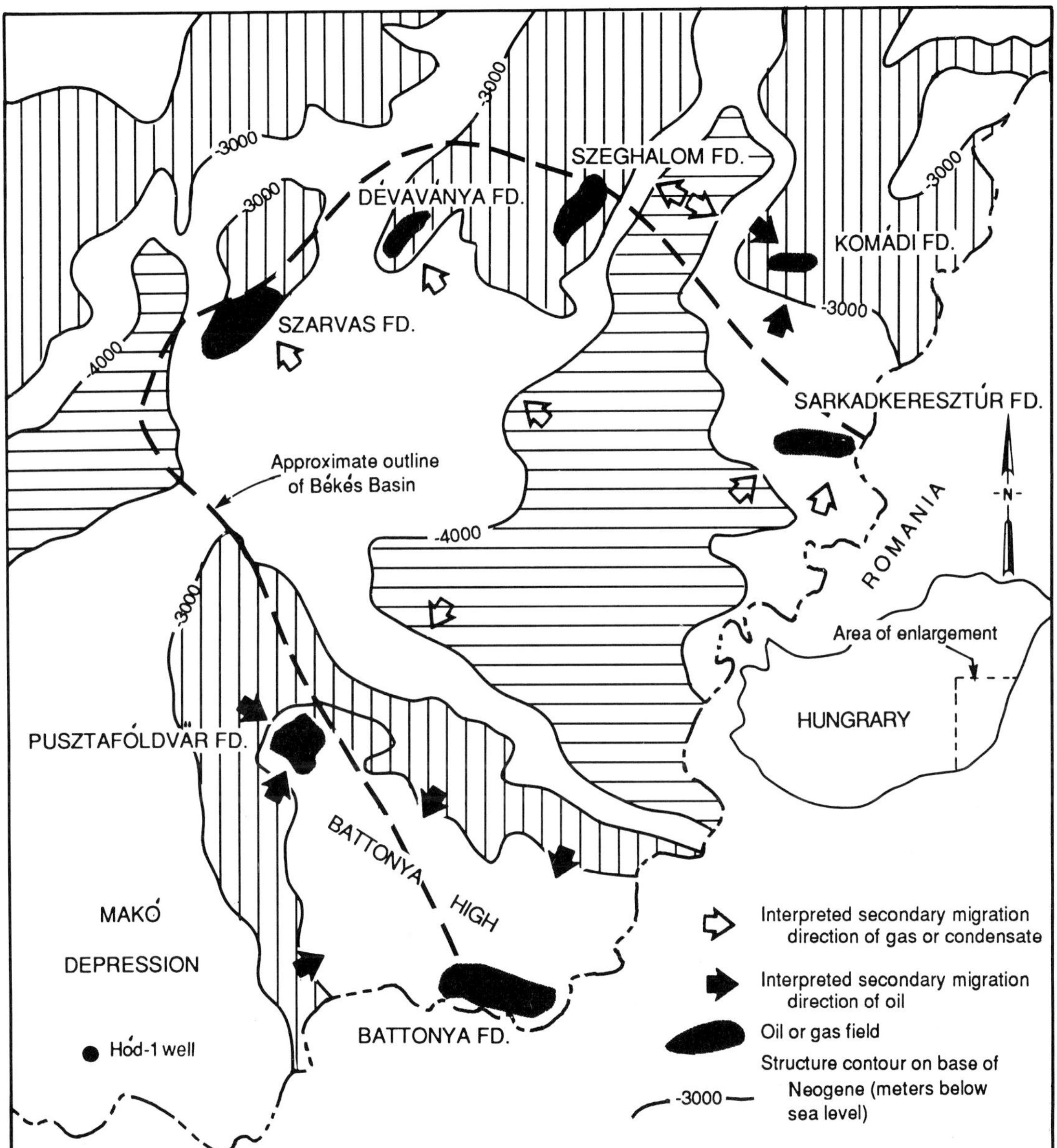

Figure 15. Generalized structure contour map on base of Neogene rocks. Datum is sea level. Vertically hatched area shows interpreted generation zone of least mature oils from Miocene source rocks (R_0 of about 0.35-0.60%). Horizontal hachures indicate generation depths of more mature oils. Unhachured area represents a transition between these two maturity zones. Solid arrows indicate directions of migration for oil. Open arrows indicate migration of condensate and gas.

Table 2. Summary of interpreted gechemical data for oils. Location of fields and wells listed in column 2 are shown in Figure 1. nd: not determined

Genetic Group 1

Sample No.	Field Name or Well name	A	B	C	D	E	F	G	H	I	J	K	L	M
1	Battonya	10	89	6	0.99	0.12	0.15	0.44	44	0.71	1.20	1.02	-26.9	-25.0
2	Battonya-East	70		4	1.21	0.19	0.21	0.52	43	0.85	1.19	1.00	-26.0	-24.5
3	Battonya	63		5	1.50	0.17	0.17	0.43	44	0.90	1.30	1.03	-26.3	-24.9
4	Békés (well)	1	94	4	nd	nd	nd	nd	nd	nd	2.16	1.02	-26.6	-26.2
5	Biharkeresztes	1	89	1	2.42	0.34	0.19	0.43	43	nd	1.07		-25.7	-25.0
6	Biharkeresztes	1	95	3	nd	nd	nd	nd	nd	nd	3.35	1.00	-25.9	-26.2
7	Biharugra	3	93	2	2.84	0.25	0.16	0.57	35	0.55	2.28	1.00	-28.6	-25.1
8	Csanádapáca	3	78	10	0.26	0.29	0.26	0.56	47	nd	2.17	1.03	-28.1	-26.5
9	Dévaványa	1	93	3	nd	nd	nd	nd	nd	nd	2.22	1.03	-27.5	-27.3
10	Dévaványa	11	88	7	0.90	0.35	0.21	0.46	47	nd	1.25	1.00	-25.9	-24.8
11	Endröd	2	90	5	1.24	0.22	0.21	0.49	55	nd	1.12	1.01	-26.8	-24.7
12	Endröd	4	92	5	nd	nd	nd	nd	nd	nd	1.92	1.05	-27.6	-25.9
13	Endröd	2	68	16	0.71	0.13	0.12	0.45	55	0.82	1.47	1.02	-26.4	-24.9
14	Endröd-North	5	92	4	0.66	0.51	0.74	0.60	nd	nd	1.87	1.02	-28.6	-26.9
15	Endröd-North	5	82	9	0.45	0.12	0.14	0.38	55	0.74	0.35	1.03	-26.8	-25.3
16	Gyoma (well)	2	88	6	0.27	0.20	0.33	0.39	55	nd	1.02	0.93	-28.2	-26.1
17	Hunya (well)	1	88	9	nd	nd	nd	nd	48	nd	1.73	1.02	-26.6	-25.8
18	Kaszaper-South	2	80	9	1.64	0.19	0.18	0.41	40	0.77	1.11	1.00	-26.9	-24.8
19	Kaszaper-D	8	74	8	1.24	0.09	0.13	0.36	56	0.85	1.77	1.01	-26.5	-24.7
20	Füzesgyarmat	11	94	4	nd	nd	nd	nd	nd	nd	1.55	1.04	-28.4	-26.6
21	Komádi	1	94	1	nd	nd	nd	nd	nd	nd	2.24	>1.0	-26.4	-26.3
22	Komádi	3	84	5	1.61	0.23	0.20	0.45	nd	0.83	2.07	nd	nd	
23	Komádi	4	91	6	1.13	0.25	0.24	0.49	53	nd	1.28	1.02	-25.3	-25.6
24	Komádi	6	63	27	0.68	0.04	0.16	0.35	50	0.94	1.53	nd	nd	
25	Komádi	10	93	3	nd	nd	nd	nd	nd	nd	1.75	>1.0	-25.6	-25.3
26	Komádi	12	62	17	1.33	0.05	0.15	0.49	53	0.79	1.40	0.99	-28.0	-26.2
27	Komádi	15	72	12	1.88	0.10	0.15	0.51	51	0.82	1.10	0.99	-27.1	-25.4
28	Komádi	19	88	6	0.30	0.45	0.34	0.22	nd	nd	1.41	1.05	-27.3	-26.2
29	Komádi-East	2	83	9	1.65	0.25	0.22	0.41	57	0.92	1.19	0.99	-26.3	-24.8
30	Körösladány	1	95	4	nd	nd	nd	nd	nd	nd	1.49	1.00	-23.7	-23.1
31	Köröstarcsa	1	88	8	1.90	1.63	0.15	0.83	nd	nd	1.15	0.98	-25.3	-23.6
32	Köröstarcsa	1	92	5	nd	nd	nd	nd	nd	nd	1.36	1.00	-26.6	-24.7
33	Medgyesbodzás (well)	1	87	8	0.41	0.22	0.32	0.46	55	nd	1.81	1.05	-26.8	-24.5
34	Mezögyán (well)	1	82	7	1.26	0.24	0.18	0.37	50	0.94	1.20	1.01	-26.8	-24.5
35	Mezöhegyes	14	87	5	nd	nd	nd	nd	nd	nd	1.12	1.02	-26.9	-24.7
36	Mezöpeterd	1	92	6	nd	nd	nd	nd	nd	nd	1.11	1.01	-26.8	-24.7
37	Mezösas	3	77	10	0.33	0.24	0.13	0.46	53	nd	1.68	1.04	-27.6	-27.2
38	Mezötúr	1	76	13	0.80	1.26	0.24	0.34	45	0.76	1.26	0.93	-24.5	-23.2
39	Örménykút -well	1	88	9	nd	nd	nd	nd	nd	nd	1.34	1.01	-25.7	-25.6
41	Pusztaföldvár	86	93	1	1.79	0.37	0.18	0.36	37	0.86	9	nd	-27.8	-26.8
43	Pusztaföldvár	134	87	7	0.87	0.07	0.11	0.39	57	0.95	1.40	1.00	-26.4	-24.8
46	Pusztaföldvár	185	91	4	0.32	0.16	0.31	0.31	nd	nd	2.04	1.04	-28.2	-25.8
48	Pusztaszöllös	3512	87	3	nd	nd	nd	nd	nd	nd	nd	nd	-27.9	-26.9
50	Sarkadkeresztúr	1	93	4	nd	nd	nd	nd	32?	nd	2.00	1.03	-27.6	-25.5
51	Sarkadkeresztúr	4	91	6	nd	nd	nd	nd	55	nd	1.76	1.02	-27.8	-25.6
52	Sarkadkeresztúr	9	91	8	0.69	0.71	0.26	0.57	nd	nd	1.57	1.03	-26.4	-25.3
53	Sarkadkeresztúr	16	88	5	nd	nd	nd	nd	nd	nd	1.72	1.03	-25.4	-24.8
54	Sarkadkeresztúr	32	92	5	nd	nd	nd	nd	nd	nd	1.71	1.03	-27.8	-25.6

Table 2. Continued from previous page.

Sample No.	Field Name or Well name	A	B	C	D	E	F	G	H	I	J	K	L	M
55	Szarvas	5	96	2	nd	nd	nd	nd	nd	nd	6.45	nd	-27.6	-27.3
56	Szarvas	7	91	5	nd	nd	nd	nd	nd	nd	1.55	nd	-26.2	-28.2
57	Szarvas	11	85	10	1.52	0.61	0.16	0.53	48	0.93	1.42	1.03	-26.1	-26.4
59	Szeghalom	11	89	5	1.35	0.18	0.18	0.41	41	0.89	1.27	1.03	-24.9	-24.7
60	Szeghalom	12	91	5	1.30	0.24	0.21	0.66	33	nd	1.25	1.04	-24.8	-24.7
61	Szeghalom	28	58	25	3.26	0.00	0.08	1.19	40	0.70	0.69	0.99	-27.2	-27.2
62	Szeghalom	38	92	5	1.49	0.25	0.22	0.54	40	nd	1.25	1.04	-25.2	-24.6
63	Szeghalom-North	1	92	5	1.16	0.27	0.23	0.49	37	nd	1.20	1.01	-25.4	-24.7
64	Szeghalom-North	5	88	7	2.90	0.32	0.21	0.45	45	0.87	1.29	1.04	-25.0	-23.9
66	Tótkomlós	26	82	3	0.81	0.09	0.11	0.35	53	nd	1.40	0.98	-26.4	-24.7
67	Tótkomlós-East	1	86	9	1.54	0.33	0.16	0.59	44	0.81	1.89	1.02	-26.1	-24.9

Genetic Group 2

Sample No.	Field name or Well name	A	B	C	D	E	F	G	H	I	J	K	L	M
40	Pusztaföldvár	46	78	12	nd	nd	nd	nd	nd	nd	0.76	0.91	-24.6	-24.4
44	Pusztaföldvár	154	78	12	1.63	0.35	0.15	0.41	39	0.77	0.76	0.94	-25.6	-24.2
47	Pusztaszöllös	1	76	13	nd	nd	nd	nd	nd	nd	0.95	0.98	-26.7	-24.9
61	Szeghalom	28	58	25	0.00	0.00	0.08	1.19	40	0.70	0.69	0.99	-27.2	-27.2

Genetic Group 3

Sample No.	Field name or Well name	A	B	C	D	E	F	G	H	I	J	K	L	M
42	Pusztaföldvár	114	64	10	2.11	0.34	0.15	0.32	39	0.81	0.74	0.89	-25.6	-24.2
45	Pusztaföldvár	177	78	6	1.20	0.19	0.14	0.34	37	0.79	0.72	0.90	-25.8	-24.9
49	Pusztaszöllös	29	76	6	1.77	0.27	0.18	0.39	nd	0.77	0.94	0.96	-26.6	-24.7
58	Szeghalom	3	66	11	3.38	0.40	0.19	0.61	43	0.85	1.08	1.00	-25.0	-24.4
65	Tótkomlós	1	78	8	2,16	0.24	0.18	0.46	43	0.79	0.89	0.96	-26.4	-24.8

Columns are as follows:
A: well number;
B: saturated hydrocarbons (ppm);
C: aromatic hydrocarbons;
D: 22,29,30-17a (H) trisnorhopane/22,29,30-18a 0(H) trisnorneohopane;
E: oleanane/hopane;
F: moretane/hopane;
G: C_{29}/C_{30} (norhopane/hopane);
H: 24-ethylcholestane (20S/20S + 20R);
I: triaromatic/triaromatic + monoaromatic steroid (Mackenzie, 1984);
J: pristane/phytane;
K: carbon preference index (n-C_{25}-C_{31})
L: δ^{13} C sats (‰ vs PDB);
M: δ^{13} C Aromatics (‰ vs PDB)

Table 3. Distribution of total organic carbon (TOC) for samples of rock cuttings from the Fábiánsebestyén-4 well. Determination by Rock-Eval. $M_{4,5}$: Miocene (Badenian and Sarmatian): K_3: Upper Cretaceous: T_2: Middle Triassic: T_1: Lower Triassic.

CLASSIFICATION[a]	1	2	3	4	5	6
TOC Carbonates[b]	<1.5	1.5-2.5	2.5-5	5-10	10-20	>20
TOC Shales	<1	5-10	10-20	20-40	40-80	>80
Törtel Formation (delta plain)	0%	0%	3%	20%	23%	54%
Algyö Formation (delta slope)	25%	43%	15%	15%	2%	0%
$M_{4,5}$	100%	0%	0%	0%	0%	0%
K_3	92%	8%	0%	0%	0%	0%
T_2	6%	19%	44%	25%	6%	0%
T_1	60%	30%	10%	0%	0%	0%

a) Classification according to total organic carbon content
b) Carbonate lithologies include limestone, dolomite and marl

odd-carbon preference. The cause of even-carbon predominance is unknown, but the occurrence and intensity of even-carbon preference undoubtedly depends upon the proportional contribution of lipid precursors from various sources or diagenetic processes leading to formation of even-carbon numbered molecules. While the presence of carbon preferences indicates thermal immaturity, absence of a carbon preference does not, in itself, prove that the oil has had a higher temperature history.

The extent of aromatization of steroids in the oils was calculated according to the method of Mackenzie (1984). In the oils which have incomplete sterane isomerization, the aromatization is also incomplete (Table 2). According to the kinetic models of Mackenzie (1984), the aromatization reaction has a higher activation energy (more temperature dependent) than the isomerization reaction. Therefore, in young, hot basins like the Békés basin, the aromatization reaction would be expected to proceed more rapidly than the isomerization reaction. An alternative explanation for the occurrence of "immature" biological marker configurations among the steroids is that the oils could have acquired biological markers from the host rocks during secondary migration (Mackenzie, 1984). This interpretation requires that the host rocks (carrier bed) contain sufficient amounts of biological

Table 4. Distribution of total organic carbon (TOC) for core samples from the Fábiánsebestyén-1,-3,-4, Kondoros-1, Gyoma-1, Hunya-1, and Doboz-1 wells. $M_{4,5}$, Miocene (Badenian and Sarmatian).

CLASSIFICATION[a]	1	2	3	3 & 4	4	FREQUENCY FOR CARBONATES
TOC Carbonates[b]	<1.5	1.5-2.5	2.5-5.0		5-10	
TOC Shales	<5	5-10	10-20		20-40	
Algyö Formation (delta slope)	63%	15%	22%		0%	19%
Szolnok Formation (basinal facies)	65%	14%	14%		7%	14%
Tótkomlós Formation (basal marl)	26%	8%	33%		33%	75%
$M_{4,5}$	0%	33%		67%		33%
TOTALS FOR NEOGENE	52%	14%	25%		9%	
FREQUENCY FOR CARBONATES	0%	13%	79%		100%	

a) Classification according to organic carbon content
b) Carbonate lithologies include limestone, dolomite and marl

Table 5. Summary of extractable organic matter yields (bitumen) for core samples from the Fábiánsebestyén-1,-3,-4, Kondoros-1, Gyoma-1, Hunya-1, and Doboz-1 wells in the Békés basin.

CLASSIFICATION	1	2	2 & 3	3
Bitumen (mg/g rock)	<0.25	0.25-0.50		0.5-1+
Algyö Formation (delta slope)	11%	67%		22%
Szolnok Formation (delta slope)[a]	7%	72%		21%
Tótkomlós Formation (basal marl)	0%	73%		27%
$M_{4,5}$ (Badenian, Sarmatian)	0%		100%	

a) basinal facies of delta slope sequences

marker compounds that are available for solution by an oil and/or gas phase moving through the rock. A major problem with this explanation is that the secondary migration carrier beds are not organic-rich rocks, or even marginally rich "source-quality" rocks, but are sandstones, or in many cases, are fractured igneous and metamorphic basement rocks (Clayton and others, 1990; Spencer and others, this volume). Further, the "immature" features occur only in the more dense oils which contain other features suggestive of fairly low-temperature origin. None of the low density oils or condensates exhibit any of the low-temperature features. While "stripping" of biological markers during secondary migration may occur in some basins, no evidence is available to support such an interpretation as a significant process in the Békés basin.

Most of the Békés basin oils do not exhibit features suggestive of low-temperature (low "maturity") origin. Instead, all of the biological marker reactions commonly used as maturation indicators have reached completion. CPI or OEP values are near 1.00, and in many cases, the alkane distributions are skewed toward low molecular weight components ($<C_{15}$). According to the maturation-depth relationships for rocks discussed above (biological marker reactions, vitrinite reflectance, etc.) these "mature" oils were probably generated and expelled from rocks at present-day depths greater than about 3000 m. This conclusion is supported by the gas chromatographic results from rock extracts (Fig. 14). The more deeply buried Miocene and lower Pannonian samples have little or no carbon preference in the C_{25+} range (samples at 3082 and 3599 m depth in the Fábiánsebestyén-2 and

Table 6. Organic carbon, extraction, and liquid chromatography results for core samples of early Pannonian and Miocene age core samples. nd, not determined.

SAMPLE	WELL NAME & NUMBER	AGE	DEPTH (m)	TOC[a]	EOM[b]	HS[c]	AR[d]
1	Komádi-2	$M_{4,5}$	2418.5-18.9	5.2	1706	499	169
2	Komádi-2	$M_{4,5}$	2418.5-9	5.2	3523	1941	424
3	Komádi-2	$M_{4,5}$	2419.7-20.2	1.1	2715	1056	200
4	Komádi-2	$M_{4,5}$	2420.7-0.9	3.3	5798	2401	748
5	Komádi-4/8	$M_{4,5}$	2053.5-72	3.5	5194	2325	646
6	Fábiánsebestyén-2	$M_{4,5}$	3100-18	3.1	2935	1784	374
7	Fábiánsebestyén-2	$M_{4,5}$	3028-3100	1.2	3258	2438	282
8	Mezötúr-3	$M_{4,5}$	2774-83	2.6	8852	6015	823
9	Gyoma-2	$M_{4,5}$	2923-30	nd	15000[e]	12600	1500
10	Kondoros-1	$M_{4,5}$	3580-99	0.7	980	nd	nd
11	Gyoma	Pa_1	3350-59	0.9	1213	945	58
12	Dévaványa	Pa_1	2666-75	1.8	4254	2801	529
13	Derecske	Pa_1	2646-57	0.7	666	386	62

a) Total organic carbon; weight percent
b) Extrable organic matter; ppm
c) Saturated hydrocarbons (ppm)
d) Aromatic hydrocarbons (ppm)
e) Possible oil-stained sample

Kondoros-1 wells, respectively, Fig. 14). The Miocene and lower Pannonian rocks from the Kondoros-1 well still contain about 1000 ppm extractable organic matter (Table 6) even though the R_o value is 1.2%. If the rock has lost a substantial proportion of its initial hydrocarbon mass, either through migration or thermal cracking, then the 1000 ppm extractable organic matter and residual organic carbon content (0.7%) could represent a small fraction of the original mass of organic material present (Cooles and others, 1986; Daly and Edman, 1987).

in quantity and quality (generation capacity) of organic matter on a regional scale is needed for improved interpretation of specific oil migration pathways in specific structural or stratigraphic trends and to further assess the potential of these trends.

Migration distances of the "low-maturity" oils are inferred to be relatively short. Production depths in most fields are generally in the range of 1000-2000 m. This suggests that future exploration for relatively dense, low-maturity oil in stratigraphic traps downdip from currently producing structures may be successful. In addition, oil migrating from more deeply buried source rocks (Fig. 15, horizontal hachures) could have charged shallow stratigraphic traps with light oil or condensate (or gas). The vertical migration distance for the light oil would be longer than the vertical migration distance for the heavy oil. The oil in these types of traps would be analogous to the oil being produced from fields located on basin-margin highs. Traps associated with these highs contain a wide variety of oils of varying maturity from at least two, and probably three, types of source rock as discussed previously.

Table 7. Summary of geochemical data showing ranges for genetic oil-types

Genetic Group	Sample[a] nos.	A	B	C	D	E	F	G	H	I
1	1-39, 41, 43, 46, 48, 50-57, 59, 60, 62-64, 66, 67	75-96% (2<75%)	>1	>1	0.22 to 0.83	0.11 to 0.74	0.30 to 2.84	0.05 to 1.63	mostly -25.0 to -28.6 (3>-25.0)	-24.5 to 28.2
2	40, 44, 47	76-78%	<1	<1	0.39 to 0.41	0.15 to 0.18	1.63 to 1.77	0.27 to 0.35	-24.6 to -26.7	-24.2 to -24.9
2	61	58%	<1	<1	1.19	0.08	3.26	0	-27.2	-27.2
3	42, 45, 49, 58, 65	64-78%	<1 (except no. 58)	<1	0.32 to 0.39	0.11 to 0.19	0.81 to 3.38	0.19 to 0.40	-25.0 to -26.4	-24.2 to -24.7

a) see Tables 1 & 2
Columns are as follows:
A: normalized prcent of saturated hydrocarbons + aromatic hydrocarbons + nonhydrocarbons
B: pristane/ phytane
C: carbon preference index (C_{25}--C_{31})
D: norhopane/hopane
E:moretane/hopane
F: 22, 29, 30-17a(H) trisnorhopane/ 22, 29, 30-18a(H) trisnorneohopane
G:oleanane/hopane
H: $^{13}C_{Sat}$ (‰ vs PDB); and I: $^{13}C_{Arom}$ (‰ vs PDB).

Figure 15 shows a modified structure contour map at the base of the Neogene with the shallow generation zone (R_o = 0.35-0.60%) vertically hachured to indicate the source rock interval from which the oils of lowest maturity were derived. As mentioned previously, the pre-Pannonian Miocene rocks (at the base of the Neogene section) are the most organic-rich source rocks in the basin. Because of the small number of wells at off-structure locations in the basin, the distribution of organic-rich facies within this shallow oil-generation zone (less than 3000 m; R_o = 0.35-0.60%) is not well-known on a basinwide scale. A better understanding of the variation

In contrast, stratigraphic traps at depths of about 3000 m or greater are likely to contain low density oil (including condensate) and/or gas. These traps would be charged with oil and/or gas generated from more mature source rocks, assuming that downward migration of oil is not a significant process in the Békés basin. In fact, the stratigraphic interval containing the effective source rocks (pre-Pannonian Miocene and lower Pannonian) for these deeper traps is the same as for the shallow traps but occurs in deeper parts of the basin and, therefore, is more deeply buried. As noted by Molenaar and others (this volume) and Mattick and others (this volume), the Neogene rocks were deposited on top of the existing pre-Tertiary topographic surface essentially filling basement lows and onlapping basement highs. Therefore, source rock intervals may not be time-equivalent across the basin, but may be physically continuous with considerable structural relief and attendant variation in thermal maturity. At the time of this writing, very few wells have been drilled in deep (central) parts of the basin and little is known regarding the distribution of organic matter and maturation levels in the deep zones.

Reservoir Temperatures

Available reservoir temperature data are given in Table 1. Many of the reservoirs have unusually high present-day temperatures, mostly over 100°C, with a few between 150 and 180°C. In some cases, relatively high-density oils (API gravities 30°-35°) occur at temperatures in excess of 100-150°C. This observation suggests that in-reservoir maturation is not a major process in the basin, notwithstanding the high reservoir temperatures. One explanation for the inferred lack of significant in-reservoir maturation is that the duration of heating has been very short owing to fairly recent emplacement of the oil (oil generation and migration is probably occurring today). Hence, kinetic limitations could have impeded the maturation process despite the high temperatures. A second explanation is that petroleum hydrocarbons are more thermally stable than generally recognized. The data of the present study are insufficient to fully explain these observations, although generation of oil under conditions of rapid heating may lead to the liberation of a high proportion of asphaltic components from kerogen as is typically observed in retort processes or rapid laboratory heating experiments. This process could result in initially fairly heavy oils which may have had insufficient time to undergo advanced in-reservoir maturation even at these high temperatures. This may be a tenable explanation for the more asphaltic oils which have low gravity values and occur at high reservoir temperatures (Komádi-12 well, sample 26, Tables 1 and 2) but is inadequate to explain the occurrence of low gravity, high aliphatic oils in high-temperature reservoirs (Szarvas-11 well, sample 57, Tables 1 and 2).

SUMMARY AND CONCLUSIONS

Oil generation has occurred over a range of thermal maturation levels, corresponding to a range of present-day source rock burial depths. Based solely on frequency of occurrence, without consideration of volumes of oil, the bulk of the oil seems to have been generated from source rocks buried to present-day depths of about 3000 m (R_o » 0.6%) or greater. Oil has also been expelled from source rocks at shallow depths (2000-3000 m; R_o = 0.4-0.6%) under conditions of fairly low thermal maturity, but is less common than the more mature type.

Based on consideration of a number of geochemical correlation parameters, three genetic oil-types are clearly indicated. At least one additional genetic oil-type may exist as indicated by complete absence of oleanane in samples analyzed in this study and the previous study of Sajgó (1984). Additional oil-types may be present; this question cannot be fully resolved with the data of the current study, particularly since mixing of oils from different sources may have occurred.

Source-rock evaluation and correlation of source rock extracts with oils indicate that the pre-Pannonian and lower Pannonian Miocene rocks are the probable source rocks of the oil produced in the Békés basin. Excellent source rock potential is present in younger, delta-plain deposits of the upper Pannonian Törtel Formation, but these rocks may not have reached sufficient maturation levels to generate commercial quantities of oil. Further, the geographic distribution of rocks of the Törtel Formation may be limited. To date, the organic-rich upper Pannonian rocks have been observed in the Makó depression to the west and in the area of the Fábiánsebestyén-4 well north of the Békés basin and at one location within the Békés basin (Békés-2 well).

Because of limited drilling in deep parts of the basin, the basinwide distribution of effective source rocks is not well-known. The occurrence of genetically-related oils at several fields scattered along the margin of the basin leads to the inference that the effective source rock interval is fairly continuous. The geology of the basin, especially the depositional setting of the Tertiary rocks, is consistent with the hypothesis that the source rocks are laterally continuous although diachronous.

Geological and geochemical evidence in support of vertical and lateral migration of oil in the basin is substantial. From an organic geochemical perspective, future exploration for stratigraphically-trapped hydrocarbons located downdip from current producing structures is most likely to be successful at shallow depths (about 3000 m or less). At depths of 3000 m or less, hydrocarbons in the form of light oil, condensate, and gas, as well as denser oils (25°-30° API), may exist. The light hydrocarbons would be derived from deeply buried source rocks, whereas, the dense oils would have been generated at low maturation levels.

ACKNOWLEDGMENTS

The authors would like to express their appreciation to the Hungarian Oil and Gas Co. (MOL Ltd.) for permission to publish the data. We gratefully acknowledge T. Kostick, U.S.G.S., for drafting the figures and D. Malone, U.S.G.S., for typing the manuscript. We appreciate constructive reviews by J. Palacas and C. Spencer.

REFERENCES

Benshan, W., Shanfa, F., Fenfang, X., Shanchun, J., and Jiamo, F., 1984, A preliminary organic geochemical study of the Fushan depression - a Tertiary basin of Eastern China; *in* Schenck, P.A., de Leeuw, J.W. and Lijmbach, G.W.M. (eds.), Advances in Organic Geochemistry 1983; Org. Geochem., 6, 108-113.

Clayton, J.L. and Koncz, I., Geochemistry of natural gas and carbon dioxide in the Békés Basin - Implications for Exploration, this volume, 187-199.

Clayton, J.L., Spencer, C.W., Koncz, I., and Szalay, Á., 1990, Generation and migration of hydrocarbon gas and carbon dioxide in the Békés basin, southeastern Hungary: Org. Geochem , 15 (3), 233-247.

Cooles, G.P., Mackenzie, A.S., and Quigley, T.M., 1986, Calculation of petroleum masses generated and expelled from source rocks; *in* Leythaeuser, D. and Rullkötter, J. (eds.), Org. Geochem., 10, 235-246.

Daly, A. and Edman, J.D., 1987, Loss of organic carbon from source rocks during thermal maturation: Amer. Assoc. Petrol. Geol. Bull., 71, 546.

Dövényi, P. and Horváth, F., 1988, A review of temperature, thermal conductivity, and heat flow data for the Pannonian Basin; *in* Royden, L.H. and Horváth, F.(eds.), The Pannonian Basin, a study in basin evolution; Amer. Assoc. Petrol. Geol. Mem. 45, Tulsa, 195-233.

Ekweozor, C.M., Okogun, J.I., Ekong, D.E.U., and Maxwell, J.R., 1979a, Preliminary organic geochemical studies of samples from the Niger Delta (Nigeria) I: Analyses of crude oils for triterpanes; Chem. Geol., 27, 11-28.

Ekweozor, C.M., Okogon, J.I., Ekong, D.E.U., and Maxwell, J.R., 1979b, Preliminary organic geochemical studies of samples from the Niger Delta (Nigeria) II: Analysis of shale for triterpenoid derivatives; Chem. Geol., 27, 29-37.

Espitalié, J., Madec, N., Tissot, P., Merrwig, J., and Leplat, P., 1977, Source rock characterization method for petroleum exploration; Proc., Offshore Technology Conference, Houston, Texas, 439-444.

Grantham, P.J., Posthuma, J., and Baak, A., 1984, Triterpanes in a number of Far-Eastern Crude oils; *in* Schenck, P.A., de Leeuw, J.W. and Lijmbach, G.W.M. (eds.), Advances in Organic Geochemistry 1983; Org. Geochem., 6, 675-683.

ten Haven, H.L., de Leeuw, J.W., Rullkötter, J., and Sinninghe Damasté, J. S., 1987, Restricted utility of the pristane/phytane ratio as a paleo-environmental indicator; Nature, 330, 641-643.

ten Haven, H.L. and Rullkötter, J., 1988, The diagenetic fate of taraxer-14-ene and oleanene isomers; Geochim. et Cosmochim. Acta, 52, 2543-2548.

Horváth, F., Dövényi, P., Szalay, Á., and Royden, L. H., 1988, Subsidence, thermal, and maturation history of the Great Hungarian plain; *in* L. H. Royden and F. Horváth, (eds.), The Pannonian Basin, A Study in Basin Evolution; Am. Assoc. Petrol. Geol. Mem. 45, 355-372.

Mackenzie, A.S., 1984, Applications of biological markers in petroleum geochemistry; *in* Brooks, J. and Welte, D. (eds.), Advances in Petroleum Geochemistry, 1, 115-214.

Mattick, R.E., Rumpler, J., Ujfalusy, A., Szanyi, B., Nagy, I., Sequence Stratigraphy of the Békés Basin, this volume, 39-65.

Molenaar, C.M., Révész, I., Bérczi, I., Kovács, A., Juhász, Gy.K., Gajdos, I. and Szanyi, B., Stratigraphic framework and sandstone facies distribution of the Pannonian sequence in the Békés Basin, this volume, 99-110.

Palmer, S.E., 1984, Hydrocarbon source potential of organic facies of the lacustrine Elks Formation (Eocene/Oligocene), northeast Nevada; *in* Woodward,J., Meissner, F.F. and Clayton, J. (eds.), Hydrocarbon source rocks of the Greater Rocky Mountain Region; 491-512.

Powell, T.G. and McKirdy, D.M., 1973, Relationship between ratio of pristane to phytane, crude oil composition and geological environment in Australia; Nature, 243, 37-39.

Riva, A., Salvatori, T., Cavaliere, R., Richuito, T., and Novelli, L., 1986, Origin of oils in Po Basin, northern Italy; *in* Leythaeuser, D. and Rullkötter, J. (eds.), Advances in Organic Geochemistry 1985; Org. Geochem., 10, 391-400.

Sajgó, Cs., 1984, Organic geochemistry of crude oils from south-east Hungary; *in* Schenck, P.A., De Leew, J.W. and Lijmbach, G.W.M. (eds.), Advances in Org. Geochem., 6, 569-578.

Sajgó, Cs., Horváth, Z.A., and Lefler, J., 1988, An organic maturation study of the Hód-1 borehole (Pannonian Basin); *in* Royden, L.H. and Horváth, F. (eds.), The Pannonian Basin - a study in basin evolution, Amer. Assoc. Petrol. Geol. Mem. 45, 297-309.

Scalan, R.S. and Smith, J.E., 1970, An improved measure of the odd-even predominance in the normal alkanes of sediment extracts and petroleum; Geochim. et Cosmochim. Acta, 34, 611-620.

Schoell, M., Teschner, M. and Wehner, H., 1984, Maturity related biomarker and stable isotope variations and their application to oil/source rock correlation in the Mahakam Delta Kalimantan; *in* Schenk, P.A., de Leeuw, J.W. and Lijmbach, G.W.M. (eds.), Advances in Organic Geochemistry 1983; Org. Geochem., 6, 156-163.

Spencer, C.W., Szalay, Á. and Tatár, É., Abnormal pressure and hydrocarbon migration in the Békés basin, this volume, 201-219.

Szalay, Á., 1988, Maturation and migration of hydrocarbons in the southeastern Pannonian Basin; *in* Royden, L.H. and Horváth, F. (eds.), The Pannonian Basin - a study in basin evolution; Amer. Assoc. Petrol. Geol. Mem. 45, 347-354.

Tissot, B., Durand, B., Espitalié, J., and Combaz, A., 1974, Influence of nature and diagenesis of organic matter in formation of petroleum; Amer. Assoc. Petrol. Geol. Bull., 58, 499-506.

Venkatesan, M.I. and Kapkan, I.R., 1982, Distribution and transport of hydrocarbons in surface sediments of the Alaskan Outer Continental Shelf; Geochim. Cosmochim. Acta, 46, 2135-2149.

9 Geochemistry of Natural Gas and Carbon Dioxide in the Békés Basin - Implications for Exploration

J. L. Clayton[1] *and I. Koncz*[2]

ABSTRACT

Natural gases are produced from reservoirs (Precambrian to Tertiary in age) located on structural highs around the margins of the Békés basin. Gas composition and stable carbon isotope data indicate that most of the flammable gases were derived from humic kerogen contained in source rocks located in deep parts of the basin. Depths of gas generation and vertical migration distances were estimated by comparing rock maturity and carbon isotopic composition of methane with Neogene source rock maturity-depth relationships. These calculations indicate that as much as 3500 m of vertical migration has occurred. Isotopically heavy (> -7 ‰) CO_2 is the predominant species present in some shallow reservoirs located on basin-margin structural highs and has probably been derived via long-distance vertical and lateral migration from thermal decomposition of carbonate minerals in Mesozoic and older rocks in the deepest parts of the basin. A few shallow reservoirs (<2000 m) contain isotopically light (-50 to -60 ‰) methane with only minor amounts of C_{2+} homologs (<3% v/v). This methane is probably mostly microbial in origin.

Porous horizons in Neogene rocks and fractured basement rocks have provided long-distance secondary migration routes for thermal gas and CO_2.

Little migration has occurred across formational boundaries. An understanding of the migration distances at certain oil and gas fields provides a guide which, when integrated with the geology of specific plays, can help predict occurrences of undiscovered gas accumulations.

INTRODUCTION

The stable carbon isotope composition of methane in natural gases varies according to the level of thermal maturation and the chemical composition of the parent organic matter (Galimov, 1973; Stahl, 1977; Rice, 1983; Schoell, 1983a). Stahl (1977) investigated the relationship between the $\delta^{13}C$ of methane and maturity of the source for sapropelic and coaly or humic kerogen sources. Schoell (1983a) published equations relating vitrinite reflectance to $\delta^{13}C$ of methane, although the equations may not be applicable in all cases worldwide (Rigby and Smith, 1981; Rice, 1983).

[1] U.S. Geological Survey, Denver, Colorado 80225, USA
[2] MOL Rt (Hungarian Oil and Gas Co., Ltd.), H-2443, Százhalombatta, Hungary

P. G. Teleki et al. (eds.), Basin Analysis in Petroleum Exploration, 187–199.

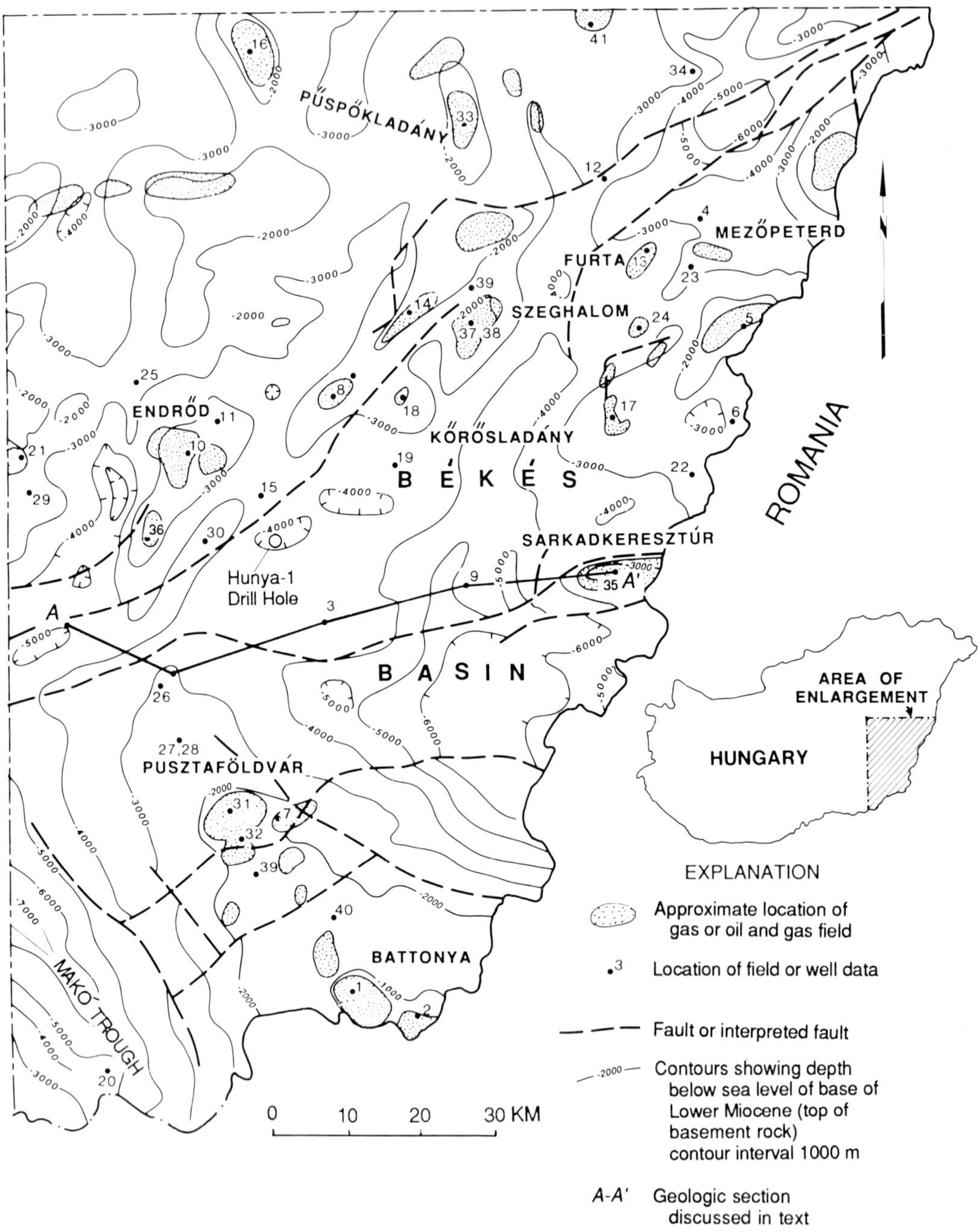

Figure 1. Map of Békés basin and vicinity showing approximate location of gas and oil fields. Numbers refer to location of gas samples. Structure contours are on top of pre-Tertiary (basement) rocks. Selected field names are shown.

In this paper, we present stable carbon isotope ratios for methane and CO_2, and compositional data for gases produced in the Békés basin. The location of oil and gas fields are shown in Figure 1. Geochemical data for the gases are given in our earlier paper (Clayton and others, 1990). The purpose of the present study is to apply quantitative maturity-isotopic ratio relationships of methane, gas composition, and carbon isotope ratio of carbon dioxide to questions of generation, migration, and accumulation of gas in the Békés basin, with the specific objective of predicting the occurrence of possible undiscovered accumulations.

Geologic setting

The Békés basin oil and gas reservoirs range in age from Precambrian through late Pannonian (*sensu lato*; i.e., Pliocene). Most of the gas production is from basement rocks or rocks of Miocene through lower Pannonian age in normal to slightly over-pressured basin-margin structural and combination structural-stratigraphic traps. A few reservoirs contain predominantly or exclusively CO_2. Chemical composition and carbon isotope ratios of the methane and/or CO_2 were determined for 203 gas samples, representing all of the major producing fields in the basin.

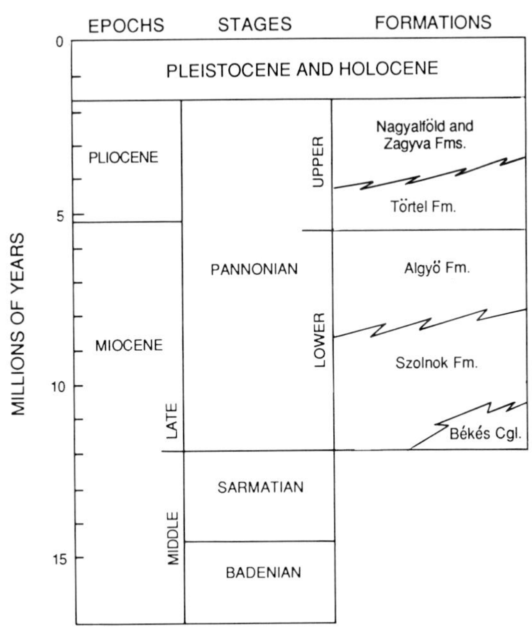

Figure 2. Generalized stratigraphic section for middle Miocene and younger rocks. Adapted from Bérczi and Phillips (1985) and Molenaar and others (this volume). Oil and gas is produced from the Pannonian-Badenian section as well as from underlying basement rocks (Precambrian through Mesozoic). Békés sandstone or conglomerate occurs only on the Battonya-Pusztaföldvár structure.

The geology of the Békés basin is discussed throughout this volume and, therefore, is not treated in the present paper. For reference, a generalized stratigraphic section showing epochs, stages and formations discussed is given in Figure 2.

Previous studies

Holczhacker (1981) reported the occurrence in Hungary of shallow accumulations (950 m or less) of ^{13}C-depleted methane of probable microbial origin and suggested that the beginning of the thermogenic generation of methane begins at about 800 m in the Pannonian Basin. Koncz (1983) examined the relationship of carbon isotopic ratio of methane to depth (or thermal maturity) for 79 natural gas samples from Hungary, and concluded that most of the hydrocarbon gas was derived from Neogene source rocks containing mainly humic kerogen. However, he also reported the occurrence of ^{13}C-depleted gases ($\delta^{13}C < -40$ %) at relatively great depths (≥3 km), which he attributed to derivation from kerogen of a more sapropelic source.

Kertai (1972) suggested that CO_2 found in Hungarian natural gas is derived predominantly from thermal metamorphism of carbonates in the basement rocks below the Tertiary fill. In the Békés basin, with few exceptions, CO_2 occurs in significant amounts only in fractured basement rocks. This fact is consistent with Kertai's hypothesis.

ANALYTICAL PROCEDURES

CO_2 was extracted from gas samples with a $Ba(OH)_2$ solution. The $BaCO_3$ precipitate was filtered, washed, dried and converted to CO_2 by reaction with orthophosphoric acid. methane was separated from the gas samples by gas chromatography. Isotope ratio measurements were performed on a Varian MAT 86 mass spectrometer. The standard deviation for replicate isotope ratio measurements was 0.01-0.02 ‰. For replicate gas geochemical, including sample preparation, a standard deviation of 0.2-0.4 ‰ was obtained.

RESULTS AND DISCUSSION

Source rock data

The geochemistry of possible source rocks in the basin is discussed in detail in our companion paper (Clayton and others, this volume). According to the pyrolysis assay, the Neogene sequence contains predominantly type III kerogen with lesser amounts of type II. Hydrogen indexes are generally less than 300.

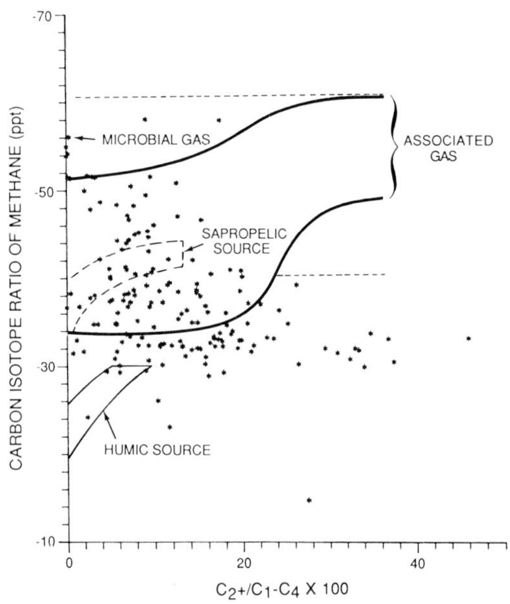

Figure 3. Carbon isotope ratio of methane (^{13}C, PDB) versus C_{2+} hydrocarbons. Regions for genetic classification from Schoell (1983a).

This finding of low hydrogen indexes is not surprising considering that the rocks were deposited in a deltaic setting where preservation of large amounts of hydrogen-rich kerogen is not favored. It is noteworthy that the Békés basin produces mostly gas and only secondary amounts of oil as would be expected in view of the predominance of type III kerogen.

Habitat and source of the gases

Carbon isotope composition of methane is plotted against percent of C_{2+} homologues in Figure 3. According to the isotope-maturity curves of Stahl (1974) and the genetic classification scheme of Schoell (1983a), the Békés basin gases are derived from a "mixed" source, i.e., in part from terrestrial ("humic") kerogen and partly from marine ("sapropelic") kerogen. The maximum gas wetness (C_{2+}) lies in the region defined by Stahl (1974) for gases derived from humic sources, and near the boundary for mixed and humic sources according to Schoell (1983a). As noted above, most of the Neogene rocks in the basin contain a humic type of kerogen. The prodelta Miocene rocks containing sapropelic kerogen could have contributed some of the gas. However, the limited occurrence of these Miocene rocks containing hydrogen-rich kerogen and the predominance of rocks containing humic kerogen suggest that most of the gas has been

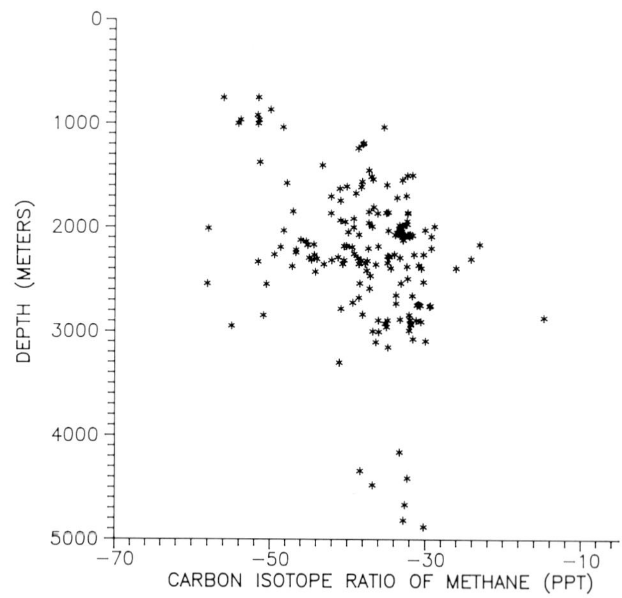

Figure 4. Carbon isotope ratio of methane (^{13}C, PDB) versus present-day reservoir depth. Lack of correlation is the result of extensive vertical gas migration.

derived from humic kerogen with secondary amounts derived from sapropelic kerogen.

The occurrence of isotopically heavy methane (about -30 to -33 ‰) associated with high amounts of C_{2+} hydrocarbons is unusual and unexplained, although it may represent mixing of methane generated at high temperatures with C_{2+} hydrocarbons during migration to cooler reservoirs (Schoell, 1983a,b). Schoell (1983b) reported ^{13}C-enriched gases (more positive than -40 ‰) containing about 10-20% C_{2+} in southern Germany, which he attributed to a gain of C_{2+} compounds during migration.

In general, there is little or no relationship between reservoir depth and the carbon isotope ratio of the methane (Fig. 4). Some shallow, isotopically light methane ($\delta^{13}C$ <- 50‰) is present at reservoir depths between about 500 and 1000 m. These gases are composed predominantly of methane (Fig. 3), and most likely are microbial in origin. Some ^{13}C-depleted methane also occurs at depths between 2000 and 3000 m. This gas could also be microbial gas generated earlier in the burial history of the basin and preserved during subsequent basin subsidence. A few of the ^{13}C-depleted methane samples occurring between 2000-3000 m contain 10 percent or more C_{2+} hydrocarbons. These gases probably represent mixing of microbial and thermally-generated gases.

It is noteworthy that some fields contain methane with a wide range of carbon isotope ratios. For example, at Battonya field the methane produced from deeper horizons (about 1000-1200 m) is enriched in ^{13}C (-35 to -38 ‰); whereas, shallow reservoirs (750 m) in the same field are depleted in ^{13}C (-51 ‰; Fig. 5). This suggests that very little migration has occurred vertically within the field. The shallow gas is probably microbial and the deeper, thermal gas has likely migrated from deeper parts of the basin through fractured basement rocks and/or Miocene clastic rocks.

The situation is similar at Pusztaföldvár field (Fig. 6). Here, a range of 18 ‰ for methane is observed between shallow (microbial) and deep (thermal) gas reservoirs. Further, the amount of CO_2 (v/v %) is substantially different for the separate reservoir zones. For example, in the Pusztaföldvár field, the gas produced from the basal conglomerate contains about 62% CO_2. In contrast, the lower Pannonian sandstones above (upper Földvár zone) contain only 2-3% CO_2. The same trend is present at several other fields in the Békés basin area (e.g., Endröd, Endröd-É (North), Tótkomlós, and Hajdúszoboszló). In all of these cases, no communication exists between the shallow and deep reservoirs to allow vertical migration of gas or oil across formation boundaries.

Gas migration within the basin appears to have been controlled mainly by the distribution of laterally continuous porous horizons. The most important of these secondary migration routes are the fractured basement (pre-Tertiary) rocks and lower Neogene sequence (pre-Pannonian Miocene and lower Pannonian). The Neogene rocks were deposited on top of the existing pre-Tertiary topographic surface (which is composed of Precambrian through Mesozoic rocks) during a cycle of deltaic progradation that filled basement lows and onlapped basement highs (Mattick and others, 1985). Therefore, porous units within the deltaic sequence are not time-equivalent across the entire basin, but are likely physically continuous and exhibit considerable structural relief (from about 1000 m to 6000 m below sea level). Certainly, the porous fractured basement rocks represent a laterally continuous migration conduit exhibiting the maximum structural relief present in the basin. Few major fault systems of Miocene or younger age are present (Grow and others, this volume) to allow significant vertical migration across stratigraphic boundaries. The Sarkadkeresztúr and Szarvas fields (Figs. 1 and 11) are examples of the few fields that contain faulting sufficient to allow significant vertical migration across formation boundaries. Isotopic and pressure data (discussed later) suggest that all of the producing zones in these fields are in hydrodynamic communication.

These geologic considerations (i.e., general lack of migration along faults) are consistent with the interpretation that the lowermost Miocene; hydrogen-rich (sapropelic) source rocks are probably not the major source of gas in the basin. Instead, source rocks containing more humic kerogen that occur within the Neogene sequence (predominantly lower Pannonian and pre-Pannonian Miocene) stratigraphically adjacent to the secondary migration carrier beds are probable sources for much of the gas.

Carbon Dioxide

As described by Koncz (1983), carbon dioxide can have several different sources. Primary sources are the atmosphere and the mantle (juvenile carbon dioxide). Atmospheric carbon dioxide may be incorporated into carbonate rocks or may be assimilated via photosynthesis into living tissue (0-25 %). At high temperatures, carbonate minerals may undergo hydrolysis or metamorphism to generate carbon dioxide with a generally heavier isotope composition than that derived by reaction of organic carbon. The carbon fixed by photosynthesis (organic carbon) may undergo various diagenetic or higher temperature reactions to form carbon dioxide with variable carbon isotope ratios. CO_2 of metamorphic origin resulting from decarbonation reactions is typically depleted in ^{13}C by about 5 ‰. However, metamorphic reactions may include processes other than simple decarbonation and may preclude assignment of a specific narrow range of $\delta^{13}C$ with metamorphic CO_2. Most CO_2 in geothermal areas has $\delta^{13}C$ values between about -3 ‰ and -4 ‰ (Hoefs, 1980). Even though more ^{13}C-depleted CO_2 is

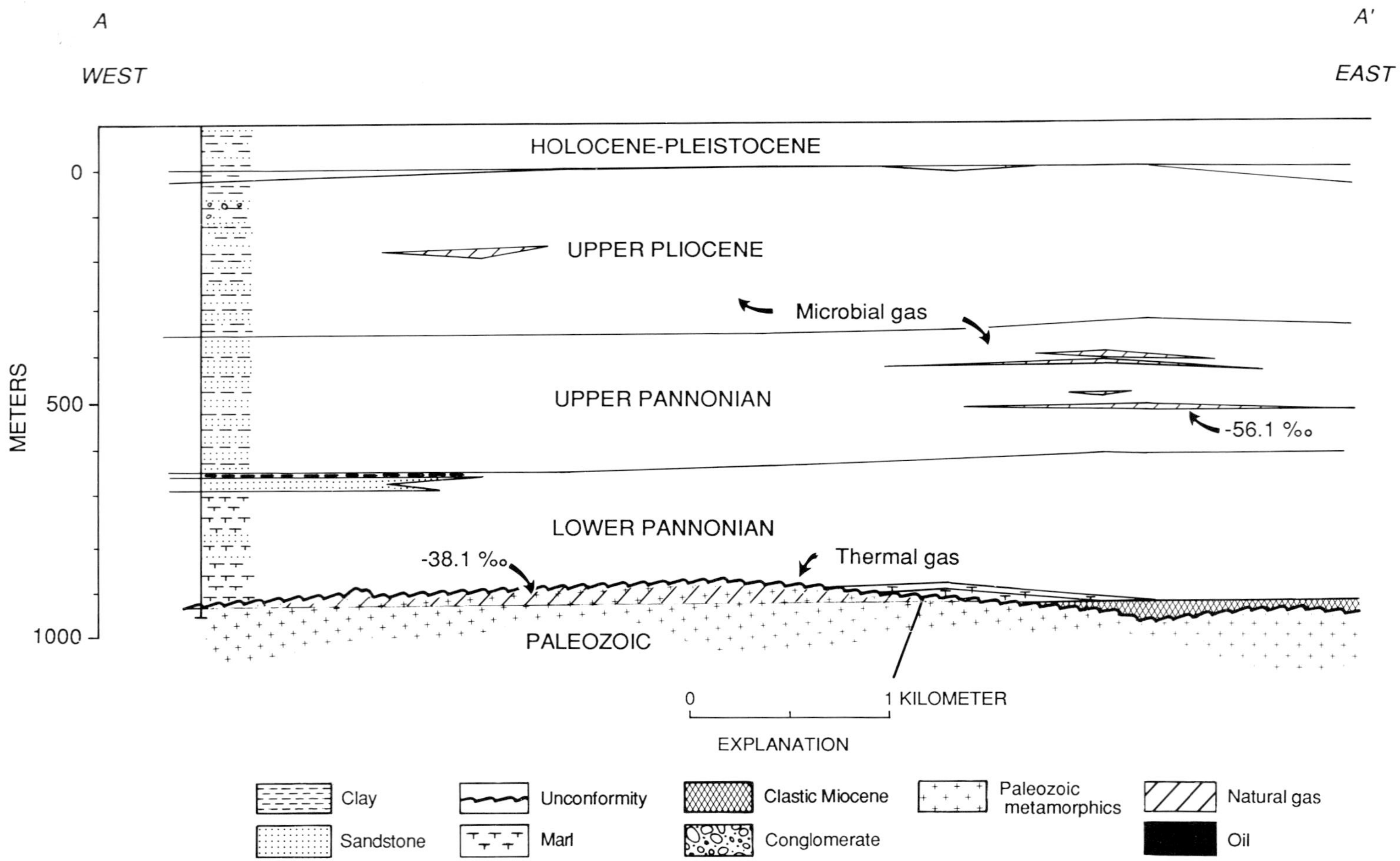

Figure 5. Generalized geologic west to east section A-A' across the Battonya field. See Figure 1 for location. Gases enriched in ^{13}C occur in deeper reservoirs (lower Pannonian, Miocene and basement), but only microbial gas is present in upper Pannonian (and younger Pliocene) reservoirs.

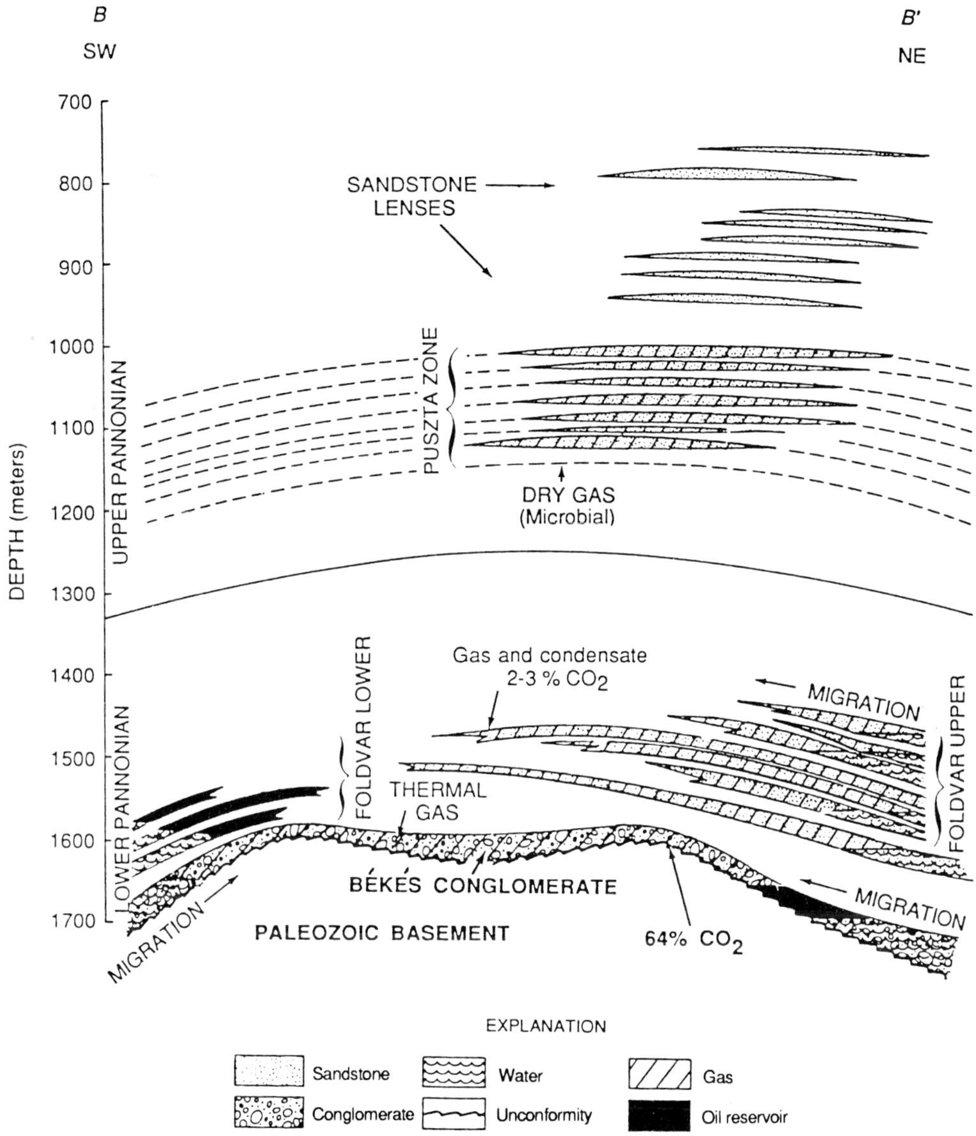

Figure 6. Generalized geologic west to east section B-B' across the Pusztaföldvár field. See Figure 1 for location.

sometimes present in metamorphic terranes, the $\delta^{13}C$ values are readily distinguishable from CO_2 derived from organic materials. The carbon isotope ratios for carbon dioxide derived from organic carbon is generally depleted in ^{13}C relative to carbon dioxide of inorganic origin.

However, determination of the origin of carbon dioxide in a gas reservoir is difficult because of possible mixing of carbon dioxide from several sources. Although carbon dioxide derived from carbonate minerals is enriched in ^{13}C relative to that derived from the oxidation of organic matter, mixing or simultaneous generation of gas from different sources may obscure the isotopic signature of the accumulated gas. Additional problems can be introduced during vertical migration of carbon dioxide because dissolution of carbonate minerals may occur. Therefore, interpretation of the origin of carbon dioxide gas is possible only by careful consideration of the geochemistry within the geologic framework.

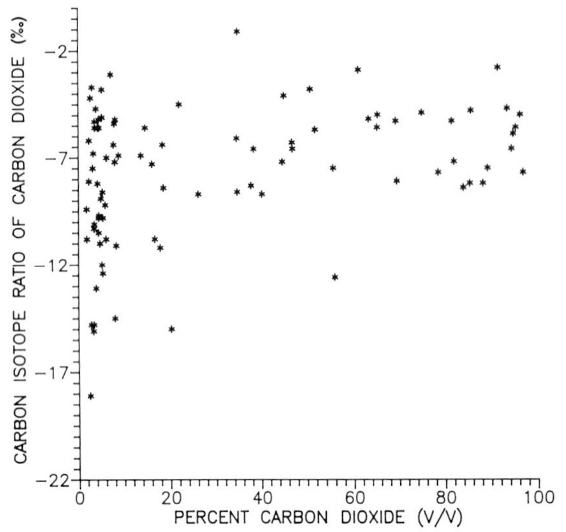

Figure 7. Carbon isotopic composition of carbon dioxide versus percent carbon dioxide (v/v) in the reservoir.

In the Békés basin, occurrences of high concentrations of CO_2 are restricted almost entirely to reservoirs in basement (pre-Tertiary) rocks. Within the basement reservoirs, CO_2 occurs in a wide range of concentrations (Koncz, 1983). The carbon isotope ratio of the CO_2 ranges from -1 ‰ to -21 ‰, with most samples between -3 ‰ and -11 ‰ (Fig. 7). CO_2 is isotopically heavier than about -10 ‰ in reservoirs containing more than about 25 percent (v/v) CO_2. One interpretation of these data is that the bulk of the carbon dioxide is derived from regional metamorphism of carbonate basement rocks in deep zones. As stated previously, the pre-Tertiary basement rocks include various metamorphic lithologies (Dank and Kókai, 1969). This interpretation of metamorphic CO_2 is consistent with the geology of the basin, because the basement rocks are highly fractured and probably serve as migration conduits for oil and gas from source rocks deeper in the basin. Both the carbon isotope ratio of the CO_2 and its occurrence virtually exclusively in basement rocks are consistent with this interpretation. Abnormally high reservoir pressures observed in the deeper parts of the basin would cause migration into shallower (lower-pressure) reservoirs (Spencer and others, this volume). Another possibility is that juvenile (mantle) CO_2 is present in some of the Békés basin reservoirs. High $^3He/^4He$ ratios have been reported in Hungary from reservoirs that contain predominantly CO_2 (Cornides and others, 1986). Moreover, Koncz (1983) found isotopically heavy methane (-14 to -17 ‰) in western Hungary which he attributed to derivation from the mantle. No helium isotope data are available from the Békés basin. However, the common occurrence of metamorphosed carbonate rocks in the basement and the absence of ^{13}C-enriched methane suggest that mantle-derived gases are probably not major components of the Békés basin gases.

Evidence for hydrocarbon gas migration

No relationship exists between the carbon isotope composition of the methane and present-day reservoir depths or temperatures (Fig. 4). Methane with carbon isotope ratios of about -25 ‰ to -42 ‰ occurs at depths from 5000 m to as shallow as 1000 to 1500 m, although most of the gas samples are from reservoirs at 3200 to 1200 m present-day burial depth. In other words, methane in reservoirs at a given depth has a wide range of isotope ratios except, at depths greater than 4000 m where methane with $\delta^{13}C$ = -30 to -40 ‰ is present.

The Békés basin has been continuously subsiding since sedimentary filling began in early Miocene time. Avail-

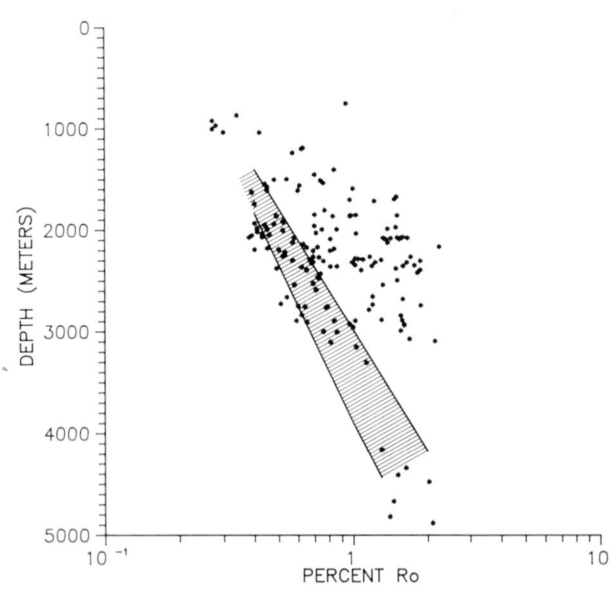

Figure 8. Gas reservoir depth compared to calculated gas maturity (percent R_o). R_o values, or "R_o equivalent" correspond to maturities at which the gas was generated according to the equations of Schoell (1983a). The two lines and cross-hatched area show range of depth-reflectance relationship for Neogene source rocks in the Békés basin.

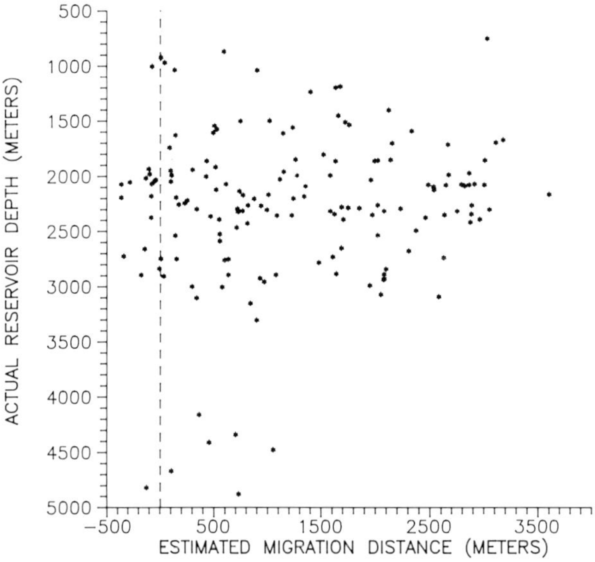

Figure 9. Comparison of calculated depth of gas generation (from gas maturity and R_0-source rock depth of burial curves) to present-day depth of gas accumulation. Horizontal axis represents inferred distance of vertical migration of the gases.

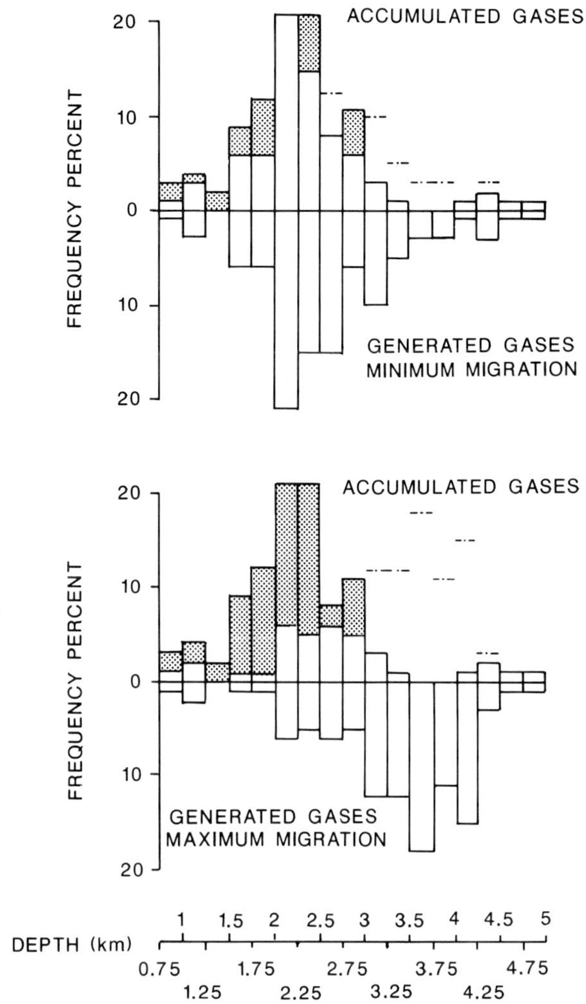

Figure 10. Histogram of vertical component of migration of methane. Gas maturities (and, hence, vertical migration distances) were calculated using the equations of Schoell (1984) for both humic and sapropelic sources to yield minimum and maximum possible migration distances.

able geological and geochemical evidence indicate that the Pannonian Basin's Tertiary rocks are at or near their maximum paleotemperature (Sajgó and Lefler, 1986). Therefore, the absence of any relationship between reservoir depth (temperature) and isotope composition suggests substantial redistribution (i.e. vertical and lateral migration) of the gas subsequent to generation and expulsion from the source rock. The most probable source rocks are the lowest part of the Pannonian and pre-Pannonian Miocene shales and marls. The lower part of the section (Miocene) is missing on structurally high positions where gas production occurs. Mesozoic and older basement rocks are unlikely sources for significant quantities of hydrocarbon gas. Studies of these rocks in the Békés basin (Clayton, unpublished data) and in outcrops in Romania (D. Hajdú, personal communication, 1987) have not revealed the existence of any rocks with sufficient amounts of organic carbon to be potential source rocks. Therefore, gas accumulations in structurally-high positions along the margins of the basin indicate that significant secondary migration of gas occurred.

Based on the quantitative relationship between the carbon isotope ratio of methane and thermal maturity of the source recognized by Stahl (1974) and the equations published by Schoell (1983a), we estimated the maturity (the vitrinite reflectance (R_O) value, or "R_O equivalent") of the methane in the Békés basin (Fig. 8). maturation levels (R_O) were estimated for all methane samples using both humic and sapropelic kerogen equations (Schoell, 1983a). This calculation is based on the assumptions that (1) the equations of Schoell (1983a) are applicable to the Pannonian Basin or, more particularly, the Békés basin, and that (2) secondary migration of the gas occurs as bulk phase transfer. Based on the relationships between R_O values and range of depths obtained from analyzing core samples of source rocks, migration distances were calculated for each methane sample using both the humic and sapropelic kerogen model. Most, but not all, of the gases are best described by the humic kerogen equation because the sapropelic equation requires mainly downward migration (discussed further below) and in some cases unrealistically high R_0 values for generation of the gas.

Figure 8 shows the relationship between actual reservoir depth and the R_0 equivalent, or R_0 values, at which the methane was generated according to the equations of Schoell (1983a). The cross-hatched area between the two subparallel lines on Figure 8 shows the relationship between depth and maturity (R_0) for the presumed Neogene source rocks. These data include the observed maximum and minimum slopes for R_0 versus depth for Neogene rocks resulting from variation in burial history of the Neogene sediment at different geologic settings within the basin. Most of the data of Figure 8 lie above the Neogene source-rock maturity lines or within the boundaries of the two source-rock lines. The gases that lie between these lines are interpreted to have undergone very little secondary migration. Data that plot above the source-rock maturity lines correspond to gases that have undergone variable degrees of vertical migration. Data below the lines probably represent gases derived from more sapropelic sources (hydrogen-rich) because significant downward migration seems unlikely. As noted above, the gas composition data also indicate that some mixing of gases from different sources may have occurred. Calculated R_0 values that are low probably include some gases of microbial or mixed thermogenic-microbial origin rather than strictly thermogenic gas. Isotopically light, dry gases of obvious microbial origin were not included in the calculation.

IMPLICATIONS FOR EXPLORATION

From a practical standpoint, petroleum geochemists and exploration geologists need to evaluate the possibility that gas accumulations are trapped downdip along the migration route between the deep gas-source rock and shallower producing reservoirs. Figure 9 illustrates the relationship between the distance of the vertical component of migration (calculated minus actual reservoir depth) and depth of actual gas accumulation determined from the quantitative carbon isotope-maturity relationship. According to this comparison, most of the gas occurring in reservoirs at present-day depths shallower than about 3500 m has undergone significant vertical secondary migration. Although as much as 3000 m or more of vertical migration is indicated in some cases,

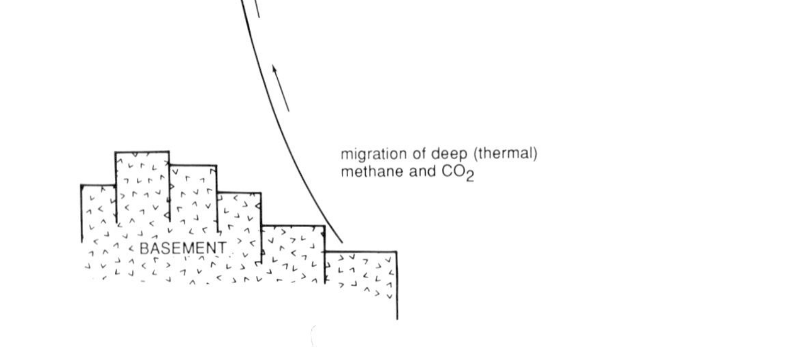

Figure 11. Schematic section through the Szarvas field showing growth faulting in the lower part of the field. Location of the fault from Pogácsás (written communication, 1988). Unfaulted reservoirs contain predominantly indigenous gas, whereas faulted reservoirs contain migrated gas from deep sources.

most of the gases appear to have undergone vertical migration distances of less than 2000-2500 m. Where traps are present (especially in lower Pannonian and older rocks), gas should be trapped on the downdip flanks of current producing structures (see Fig. 1) at depths less than about 4000 m.

Vitrinite reflectance equivalent values calculated for each of the methane samples using the equations published by Schoell (1983a) yields two vertical migration distances of the gases depending on which equation (sapropelic or humic) is used. Figure 10 shows the frequency distribution of non-migrated (indigenous) and migrated gas for the two end-member cases. The shaded area in Figure 10 represents the frequency distribution of gas accumulations at various depths derived from vertical migration (i.e., non-indigenous gas), and the unshaded areas below the zero line represent the frequency distribution of gases generated at various depths (maturities). The unshaded areas above the zero line represent gases in reservoirs which are indigenous and are interpreted to have not undergone any significant component of vertical migration. Both the maximum and minimum migration models indicate that the migration of gas has occurred from source rocks between the depth interval 3000-4000 m, except the maximum migration model suggests more migrated gas is present in the shallow zones (<3000 m). The overall distribution of migration distances is not significantly different in the two cases; only the proportion of gas contained in shallow reservoirs derived from various generation zones varies. In both cases, gases occurring in deep reservoirs (approximately 4000 m and greater) are interpreted to be mostly indigenous.

Oil and gas are produced from reservoirs ranging in age from Precambrian through upper Pannonian. To better evaluate secondary migration patterns in the basin, we examined the carbon isotopic ratio of methane and gas composition for the five main producing zones in the Békés basin for each producing structure or field. For every field in the basin, we evaluated the migration distances for gas within each of the five producing zones based on the quantitative relationship between maturity and the isotopic composition of the methane and the gas chemical composition. This procedure was carried out for each individual well in all of the fields for which data were available. All of the data were published in another paper (Clayton and others, 1990). Geographically proximal fields are grouped together for the purposes of discussing migration of gas and evaluation of possibilities for undiscovered accumulations.

A critical consideration in this assessment of migration is whether vertical migration occurs solely along basinward dipping, laterally continuous porous horizons, or whether some vertical migration occurs across formation boundaries, along fractures or fault zones. The two alternative migration routes could result in significantly different distributions of gas, depending upon the stratigraphic relationships between the various secondary migration carrier beds and effective source rocks in the basin. Basinwide, migration appears to have occurred predominantly along laterally continuous porous rock intervals which exhibit substantial structural relief from the deep basin onto the structurally high positions around the basin margins. As noted earlier, little post-Cretaceous faulting has been recognized in the basin (Grow and others, this volume). The data of Spencer and others (this volume) demonstrate substantial vertical pressure differential between most of the producing zones in the basin, indicating limited or no communication of fluids across formation boundaries in most fields. The geochemical data support this interpretation. Several fields discussed previously contain gas of vastly different isotopic and chemical composition in producing zones of different ages even though the vertical distance between reservoirs is short (as little as about 500 m). These observations suggest that migration is largely confined to formational, or more specifically, lithological boundaries and has not occurred to any large degree vertically across lithologies. An exception to this is the Sarkadkeresztúr field, where faulting has lead to a common pressure regime among the various producing zones and vertical migration of gas along the fault zones.

Szarvas--Endröd

Gas is present in both upper and lower Pannonian strata at the Endröd field, but only in lower Pannonian rocks (lowermost Szolnok Fm.) at the Szarvas field. The gas at the Endröd field is mostly indigenous (early diagenetic) with some mixed early diagenetic and catagenetic gas. These gases have undergone only short migration distances. In contrast, some of the gas at the Szarvas field has apparently migrated from generation depths of about 3000-3800. This gas contains substantial quantities of CO_2 derived from the deep parts of the basin. Figure 11 shows a schematic cross section of the Szarvas field. The lowermost Neogene rocks contain gas with high amounts of CO_2 (>40%) and isotopically heavy methane.

As Figure 11 shows, the deeper Szarvas field reservoirs are offset by growth faults which provide migration routes for gas generated at depth (high maturities). In contrast, gas with low amounts of CO_2 and isotopically lighter methane (<about -40 ‰) occurs in shallower Neogene reservoirs which are not faulted. This isotopically light gas was generated at shallow depths according to the maturity-isotope ratio relationship and has undergone only limited vertical migration. No significant mixing of gas from the faulted lower Neogene reservoirs with gas from shallower reservoirs is indicated. Traps in shallow reservoir rocks located on the basinward side of the Szarvas field are prospective exploration targets for low maturity gas and gas-condensate according to this model of generation. Identification of additional faults is a critical element for predicting the occurrence of further accumulation of the CO_2-rich, high maturity gas.

Sarkadkeresztúr

Gas occurs in Pannonian rocks in the Sarkadkeresztúr field, and some low density (i.e. thermally very mature) oil occurs in lower Pannonian sandstones. Gas and gas-condensates are produced from Miocene and basement rocks that form a common reservoir. The gas composition and carbon isotope ratio of the methane indicates that the gas was generated from source rocks adjacent to the reservoirs or, in some cases, downdip from the reservoirs at present-day depths of about 3300 to 3800 m. Therefore, the area to the southwest of the field, towards the deep part of the Békés basin, is prospective for additional accumulations of condensate and gas if suitable trapping mechanisms are present.

Pusztaföldvár-Battonya

Gas is produced from upper and lower Pannonian rocks, as well as basement rocks. The upper Pannonian gas is most likely indigenous, of diagenetic and biogenic origin, although only one sample was analyzed from this horizon. Supporting evidence for this interpretation is the occurrence of indigenous, diagenetic and possibly microbial gas in lower Pannonian sandstones (depths of about 900 to 1500 m). The more deeply buried lower Pannonian sandstones contain some catagenetic, migrated gas which has undergone moderate migration distances (about 500-1000 m). The basement and lower Pannonian rocks contain only gas which has migrated from estimated depths of about 2000-3600 m. These gases contain high concentrations of CO_2, and presumably were derived from source rocks in the deep basin. The CO_2 may facilitate migration of the gas and/or oil, at least in basement rocks or near the basement-Tertiary contact.

Martfü, Mezötúr, Öcsöd

Gas production occurs in nearly all of the pre-upper Pannonian rocks in this group of fields. In general, interpreted gas migration distances are short. Production occurs from depths of about 2000-2800 m. The chemical isotopic composition indicates that the gas was generated under thermal conditions (maturation levels) close to those associated with these production depths, and significant vertical migration is not indicated. More deeply buried traps in or adjacent to those fields could be expected to contain gas of similar composition.

SUMMARY AND CONCLUSIONS

This study represents a case history where studies of gas composition and carbon isotope ratios have been correlated with stratigraphy, structure, and reservoir pressures to understand the origin, migration, and accumulation of gas in the Békés basin. Although the quantitative relationship between source rock thermal maturity (R_o) and methane isotope ratio is imperfectly known because of variation in the maturity-depth gradient of Neogene source rocks and variation in the type of parent organic matter, some limitations can be recognized regarding the generation and migration history of the gas. The isotope ratios of the methane and gas composition data indicate that the parent source organic matter is of mixed composition, but mainly humic. Further, consideration of the isotope ratios of methane and R_o data for Neogene source rocks indicate that gases have undergone vertical migration of 2000-2500 m and, in some cases, possibly as much as, but not more than 3500 m. The occurrence of ^{13}C enriched methane (about -30 ‰) associated with significant amounts of C_2-C_4 components (20 to 40+ %) is unusual and cannot be explained by commonly known processes of thermal generation of gas. Therefore, mixing of gases from two or more sources seems likely.

CO_2 is common in basement rocks in structurally high positions on the basin's margin. The carbon isotopic composition of the CO_2 and the occurrence of carbonate basement rocks in the basin is consistent with an interpretation of metamorphic origin, although some contribution from mantle sources cannot be discounted. Updip migration of CO_2 has occurred by pressure differential and buoyancy through fractured basement rocks. Some isotopically light (as negative as -21 ‰) CO_2 is present and probably represents contributions from thermal decomposition of organic matter. The occurrence of CO_2 of metamorphic origin in shallow reservoirs is convincing evidence for long distance lateral and vertical migration of gas in the basin. Some of the CO_2 could have been derived from the mantle, but this question cannot be resolved with the data of the present study.

ACKNOWLEDGMENTS

The authors would like to express their appreciation to the Hungarian Oil and Gas Company, Ltd. (previously the Hungarian Oil and Gas Trust) for permission to publish the data. T. Kostick, U.S. Geological Survey, drafted the figures. D. Malone, U.S. Geological Survey typed the manuscript.

REFERENCES

Bérczi, I. and Phillips, R. L., 1985, Processes and depositional environments within Neogene deltaic-lacustrine sediments, Pannonian Basin, southeast Hungary; Eötvös Lóránd Geophys. Inst., Geophys. Transactions, 31, 55-74.

Cornides, I., Takaoka, N., Nagao, K. and Matsuo, S., 1986, Contribution of mantle-derived gases to subsurface gases in a tectonically quiescent area, the Carpathian Basin, Hungary, revealed by noble gas measurements; Geochem. Jour., 20, 119-125.

Clayton, J. L., Spencer, C. W., Koncz, I., and Szalay, A., 1990, Generation and migration of hydrocarbon gas and carbon dioxide in the Békés basin, southern Hungary; Org. Geochem., 15, (3), 233-247.

Clayton, J. L., Koncz, I., King, J. D., and Tatár, E., Organic Geochemistry of crude oils and source rocks, Békés basin; this volume......

Dank, V. E. and Kókai, J., 1969, Oil and gas exploration in Hungary; *in* Hepple, P. (ed.), The exploration for petroleum in Europe and North Africa; Inst. Petroleum, London, 131-145.

Galimov, E. M., 1973, Carbon isotopes in oil and gas geology (In Russian); Moscow: Nedra, English Translation NASA TTF, 682 p.

Grow, J. A., Mattick, R.E., Bérczi-Makk, A., Péró, C., Hajdú, D., Pogácsás, Gy., Várnai, P., and Varga, E., Structure of the Békés basin inferred from seismic reflection, well, and gravity data; this volume, 1-38.

Hoefs, J., 1980, Stable isotope geochemistry; Springer-Verlag, New York, 208 p.

Holczhacker, K., 1981, A stabil szénizotóparány-adatok felhasználási lehetöségei, (The possibilities of utilizing carbon-isotope ratio data; in Hungarian); Köolaj és Földgáz, 14, 178-186.

Kertai, Gy., 1972, A köolaj és földgáz vegyi összetétele és keletkezése (Chemical composition and generation of oil and gas; in Hungarian); Akadémiai Kiadó, Budapest, Hungary, 112 p.

Koncz, I., 1983, The stable carbon isotope composition of the hydrocarbon and carbon dioxide components of Hungarian natural gases; Acta Mineral.-Petrog., Szeged, Hungary, 26,(1), 33-49.

Mattick, R. E., Rumpler, J. and Phillips, R. L., 1985, Seismic stratigraphy of the Pannonian Basin in southeastern Hungary; Eötvös Lóránd Geophys. Inst., Geophys. Trans., 31, 13-54.

Molenaar, C.M., Révész, I., Bérczi, I., Kovács, A., Juhász, G.K., Gajdos, I., and Szanyi, B., Stratigraphic framework and sandstone facies distribution of the Pannonian sequence in the Békés basin; this volume, 99-110.

Rice, D. D., 1983, Relation of natural gas composition to thermal maturity and source rock type in San Juan Basin, northwestern New Mexico and southwestern Colorado; Amer. Assoc. Petrol. Geol. Bull., 67, 1199-1218.

Rigby, D. and Smith, J. W., 1981, An isotopic study of gases and hydrocarbons in the Cooper basin; APEA Journal 21, 222-229.

Sajgó, Cs. and Lefler, J., 1986, A reaction kinetic approach to the temperature-time history of sedimentary basins; *in* Buntebarth, G. and Stegena, L. (eds.), Lecture Notes in Earth Sciences, v. 5: Paleogeothermics, Springer-Verlag, Berlin, 119-172.

Schoell, M., 1983a, Genetic characterization of natural gases; Amer. Assoc. Petrol. Geol. Bull., 67, 2225-2238.

Schoell, M., 1983b, Isotope techniques for tracing migration of gases in sedimentary basins; Jour. Geol. Soc. London, 140, 415-422.

Schoell, M., 1984, Stable isotopes in petroleum research; *in* Brooks J. and Welte D., (eds.), Advances in Petroleum Geochemistry, v. 1: Academic Press, New York, 215-247.

Spencer, C. W., Szalay, Á., and Tatár, É., Abnormal pressure and hydrocarbon migration in the Békés basin, this volume, 201-219.

Stahl, W., 1974, Carbon isotope fractionations in natural gases; Nature, 251, 134-135.

Stahl, W., 1977, Carbon and nitrogen isotopes in hydrocarbon research and exploration; Chem. Geol., 20, 121-147.

10 Abnormal Pressure and Hydrocarbon Migration in the Békés Basin

Charles W. Spencer[1], *Árpád Szalay*[2] *and Éva Tatár*[2],

ABSTRACT

The Békés basin is a relatively young, hot basin. Oil, gas, and condensate are produced from many relatively shallow (<3000 m) structures around the margins of the basin. The producing rocks range in age from Precambrian to Pliocene. Exploration for stratigraphic traps should result in the discovery of many new oil and gas fields. A study of reservoir pressures and migration has been made to help evaluate the hydrocarbon potential of various stratigraphic-trap play areas.

Overpressuring commonly occurs in reservoirs at depths greater than 1800 m. Several mechanisms could cause this abnormally high pressure. Undercompaction (incomplete dewatering) and active hydrocarbon generation are the most likely mechanisms in sedimentary rocks. High pressure in basement rocks may be caused by thermally generated CO_2 (metamorphism of carbonate rocks), downward and lateral expulsion of hydrocarbons from actively generating source beds, and aquathermal heating (expansion of water) in fractures in very low-permeability basement rocks.

High pressures, deep in the basin, are an important factor in the migration of oil and gas laterally and vertically to shallower stratigraphic traps. Migration occurs in carrier beds (sandstones) and fractured basement rocks. Many low-temperature producing fields yield hydrocarbons that were originally generated much deeper in the basin. Stratigraphic traps and structural-stratigraphic traps flanking the basement highs should have trapped large amounts of oil and gas along the paths of migration updip to shallower, structurally controlled producing areas.

INTRODUCTION

Structural prospects at depths <3000 m in the Békés basin in the southeast part of the Pannonian Basin have been relatively well explored by drilling. A few deep (>3000 m) on-structure wildcat wells have also been drilled in the basin; however, very few off-structure, stratigraphic tests have been drilled to date. Shallow drilling on basement topographic and structural highs has been relatively successful in finding and developing oil and gas fields (Fig. 1). This study attempts to evaluate these shallow hydrocarbon occurrences in order to predict the potential of stratigraphic and deep structural prospects. In the Békés basin, hydrocarbons are pro-

[1] U.S. Geological Survey, Denver, Colorado 80225, USA
[2] MOL Rt. (Hungarian Oil and Gas Co., Ltd.), H-5001, Szolnok, Hungary

P. G. Teleki et al. (eds.), Basin Analysis in Petroleum Exploration, 201–219.

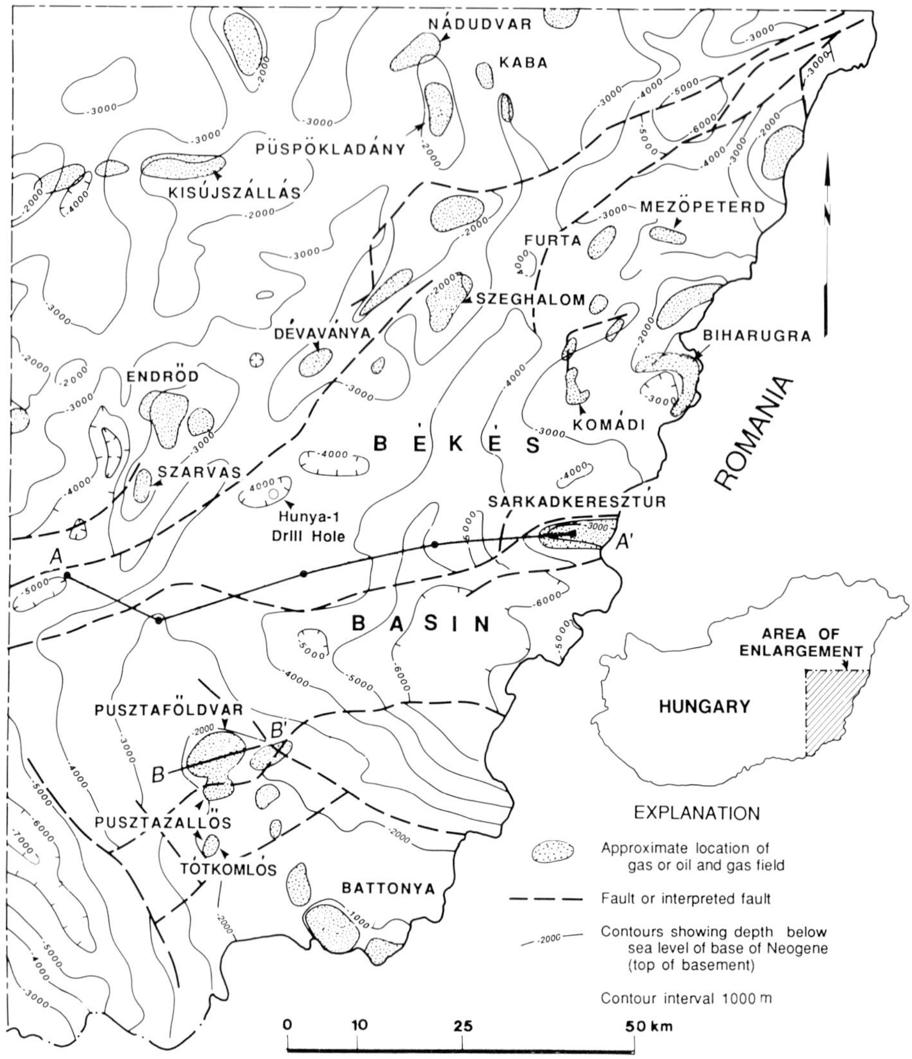

Figure 1. Index map of Békés basin area showing depth to the top of the pre-Neogene basement, oil and gas fields, and locations of cross sections (Figs. 3, 4, and 20). Modified from Clayton and others (1990).

duced from Neogene sedimentary rocks and from fractured basement rocks of Mesozoic, Paleozoic, and Precambrian age. Where present, early to middle Miocene-age rocks often form a common reservoir with the basement rocks. Following the Hungarian usage, Miocene refers to rocks of Badenian and Sarmatian age in the Békés basin.

Available data (Clayton and others, this volume) indicate that the most likely primary source rocks are Miocene marls and shales. The lower part of the Pannonian section contains sufficient organic matter to be a potential source rock where it is thermally mature. Clayton and others (this volume) have determined that at least some of the oils in Békés basin fields migrated from Miocene and Pannonian source beds that have been heated only to the very early stages of hydrocarbon generation (i.e. "immature" or low temperature oils). They also noted that oils in some fields migrated from different depths (maturity) and originated from different source beds.

The pressure and migration studies initially involved analysis of data on reservoir pressure, source-bed richness, subsurface temperature, thermal maturation, and selected producing field and core information. The later phases of the evaluation included a determination of the timing of oil and gas generation and interpretation of subsequent secondary migration routes in carrier beds and fracture systems. It was known that the deep, hotter parts of the basin are overpressured. Consequently, the distribution, magnitude, and origin of the overpressuring was studied in order to better predict possible oil and gas paleomigration routes and entrapment areas (accumulations). In order to further analyze secondary migration, an attempt was made to correlate oils back to their source beds. Sixty-nine oils and more than 200 source-bed samples were studied by Clayton and others (this volume) to determine where the oils originated. Also, Clayton and Koncz (this volume) and Clayton and others (1990) analyzed 203 methane and CO_2 samples for their composition and isotope ratios. These data were used to estimate approximate depths of gas formation versus present-day depths of occurrence and other attributes. The results of these studies show that there are three major genetic oil types. Low maturity oils were expelled from Pannonian source rocks at depths of 2000-3000 m. More mature oils and condensates were expelled from lower Pannonian and Miocene rocks at depths of 3000 to 4000 m. Oils of mixed origin and maturities are found in shallower structures around the margins of the basin.

The studies by Clayton and Koncz (this volume) and Clayton and others (1990) show that the gases in the basin originated over a wide range of depths and were mostly sourced by humic organic matter. Most of the gases occurring at present-day depths of 3500 m have undergone some secondary vertical (upward) migration from deeper source beds. Most of the gases in shallower (<3500 m deep) reservoirs have experienced less than 2000 to 2500 m of vertical migration.

OVERPRESSURING

Normal (hydrostatic) pressures are present in most of the Békés basin to a depth of about 2500 m. However, abnormally high-reservoir pressure can also occur at depths as shallow as 1800 m in various sub-basins of the Pannonian Basin, including the Békés basin. Szalay (1983 and 1988) used well logs and drill-stem test (DST) data to determine the depths and magnitude of overpressuring in the southeast part of the Great Hungarian Plain region (which includes the Békés and adjacent basins). His investigations were based on the concept that rocks compact as they are buried, but compact less when part of the overburden pressure is partially supported by high-pore pressure. In most such studies, porosity logs (acoustic, density) are used, but in the Békés basin few porosity logs are available. Therefore, the lateral log curves were mostly used to estimate porosity (Szalay, 1983). This method has been widely used to study pressure variations in marine deposits in other basins (MacGregor, 1965).

Szalay and Szentgyörgyi (1979) and Szalay (1983) describe a method, using well logs, to subdivide the Pannonian strata into generalized lithogenetic units, comprising non-reservoir rocks (shales and marls) and reservoir rocks (sandstones). Parts of the sequences containing shales generate hydrocarbons and the sandstone intervals often serve as carrier beds for fluids migrating updip and for fluids migrating away from areas of abnormally high pressure. For purposes of later discussion, Figure 2 shows the approximate correlation of the time-transgressive, Pannonian formations as used by Molenaar and others (this volume) and lithogenetic units used by Szalay. For example, the sandy delta-plain deposits of the Törtel Formation are approximately equivalent to the Pa_2 unit of Szalay (1988).

Figure 3 shows interpreted hydrocarbon migration directions along cross section A-A' based on pressure data and geochemistry of oils and gases. Also shown are estimated and measured vitrinite reflectance (R_o) levels at the top of the main oil generation ($R_o = 0.6\%$) zone and the top of the main gas-condensate generation zone ($R_o = 1.3\%$). A few samples of low-temperature oils have also been recovered, and these are believed to have been expelled from source rocks at maturation levels of less than $R_o = 0.6\%$ (J.L. Clayton, oral communication, 1986).

Figure 4 shows mud-weight pressure profiles and extrapolated DST pressure values along section A-A'. Several such cross-basin pressure profiles were constructed in order to better interpret the secondary migration of oil and gas.

Analysis of DST and mud-weight data indicate that normal pressures occur at shallow depths and low temperatures, but almost all deep (2500 m) formations are overpressured (Fig. 5). High quality DST pressure data are sparse in the lower part of the Szolnok Formation but, where the Szolnok Formation contains permeable carrier beds, these beds should serve as a pressure sink toward which hydrocarbons under high pressure should have migrated. This pressure-driven migration should cause lateral oil and gas movement through the lower Szolnok Formation and other carrier beds into the updip edges of

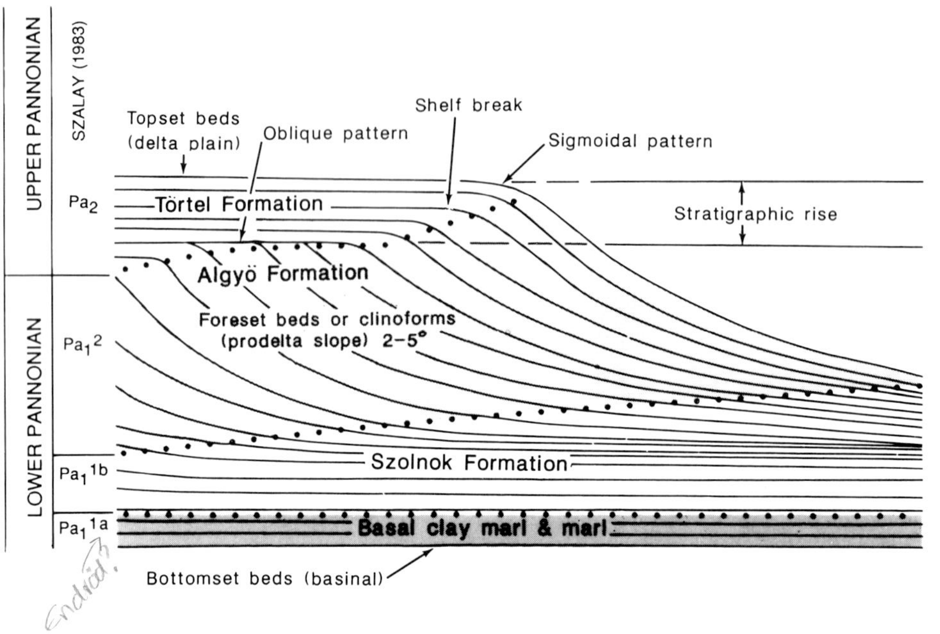

Figure 2. Schematic diagram showing depositional patterns of Pannonian sediments as indicated by seismic-stratigraphic interpretations made by Molenaar and others (this volume). At left are lithogenetic units of Szalay (1988) as determinated from resistivity and other logs. The lithogenetic units are identified by letters and numbers. For ages of formations see Phillips and others (this volume).

stratigraphic traps along the flanks of the basement highs (Mattick and others, this volume).

Minor amounts of downward migration may have occurred from the overpressured lower part of the Algyö Formation into the sandy turbidites in the Szolnok Formation. Migration from the basal clayey marl and Miocene source beds should also occur downward into fractured basement rocks and upward into the Szolnok carrier (reservoir) beds under the proper pressure gradients.

In some cases the highest measured pressure gradients occur in basement rocks or near the top of basement. In these cases, fluids will tend to be expelled upward and laterally to shallower lower pressure environments (Fig. 3). It is recognized that the pressures in the basal clayey marl may be higher than those measured in basement rocks; but, because the clayey rocks have very low permeability, the true pressures cannot be extrapolated.

CAUSES OF OVERPRESSURING

To predict migration paths of oil and gas it was important to evaluate the causes of abnormal pressure in the Békés basin. A variety of mechanisms are known to cause overpressuring (Fertl, 1976). Notwithstanding the origin of overpressuring, there is a general agreement that low-permeability "seals" must be present if abnormally high pressures are to be retained through time. In the Békés basin, core analyses, DST, and other subsurface data indicate thick sequences of seals are present in Pannonian sedimentary rocks deep in the basin.

Potential causes of overpressuring in sedimentary rocks in the Békés basin are: (1) dewatering of shales and siltstones, (2) hydrocarbon generation (3) aquathermal heating, and (4) generation of carbon dioxide (CO_2) as a result of thermal decomposition of carbonate rocks in the presence of silica. Clay mineral transformations, that release water, may also contribute to abnormally high pressures but data available to evaluate this mechanism were insufficient.

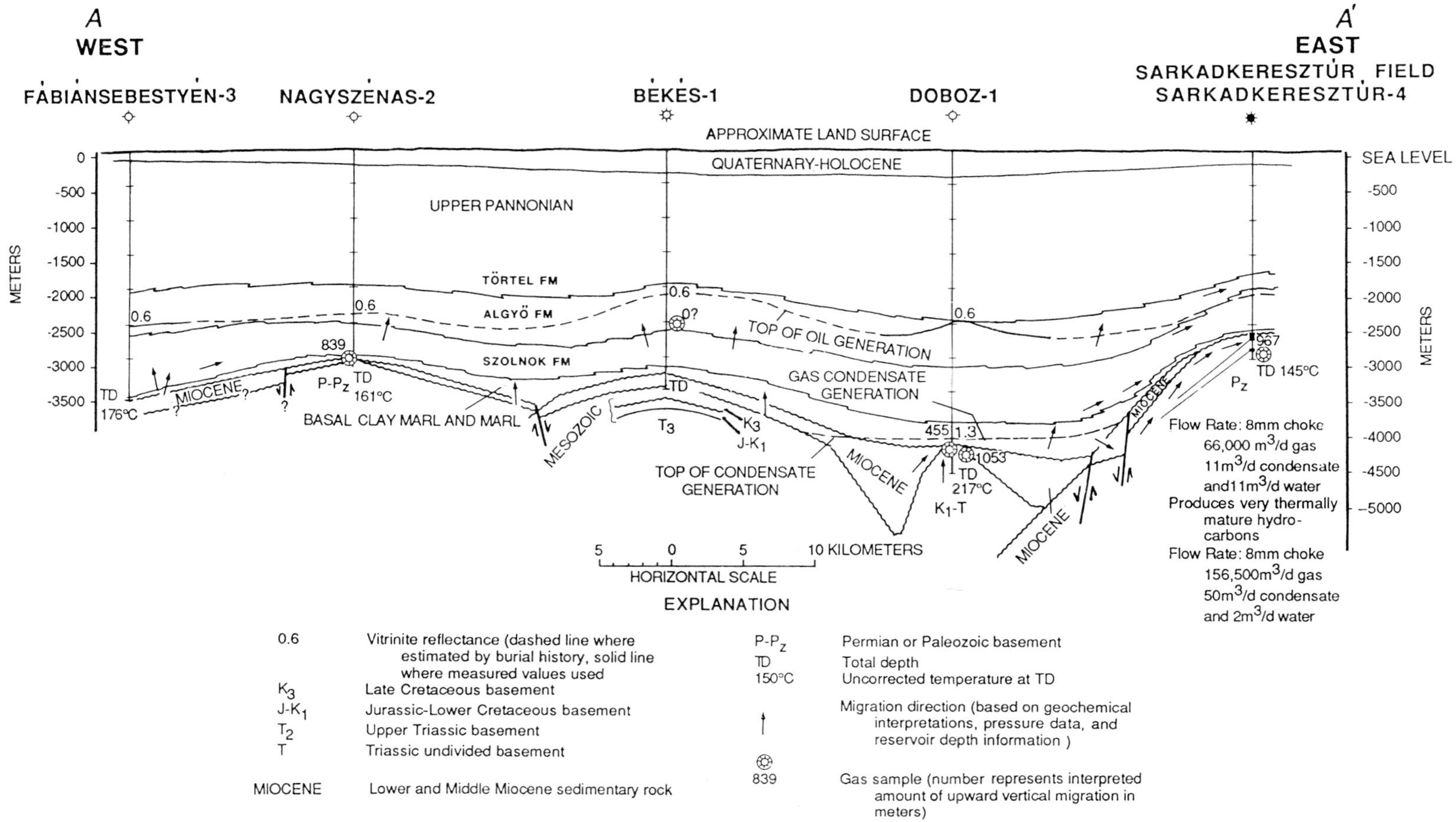

Figure 3. West to east cross section A-A' showing depths to various formations, the top of the oil window, and interpreted hydrocarbon migration directions (see Fig. 1 for location of section). Figure from Clayton and others (1989).

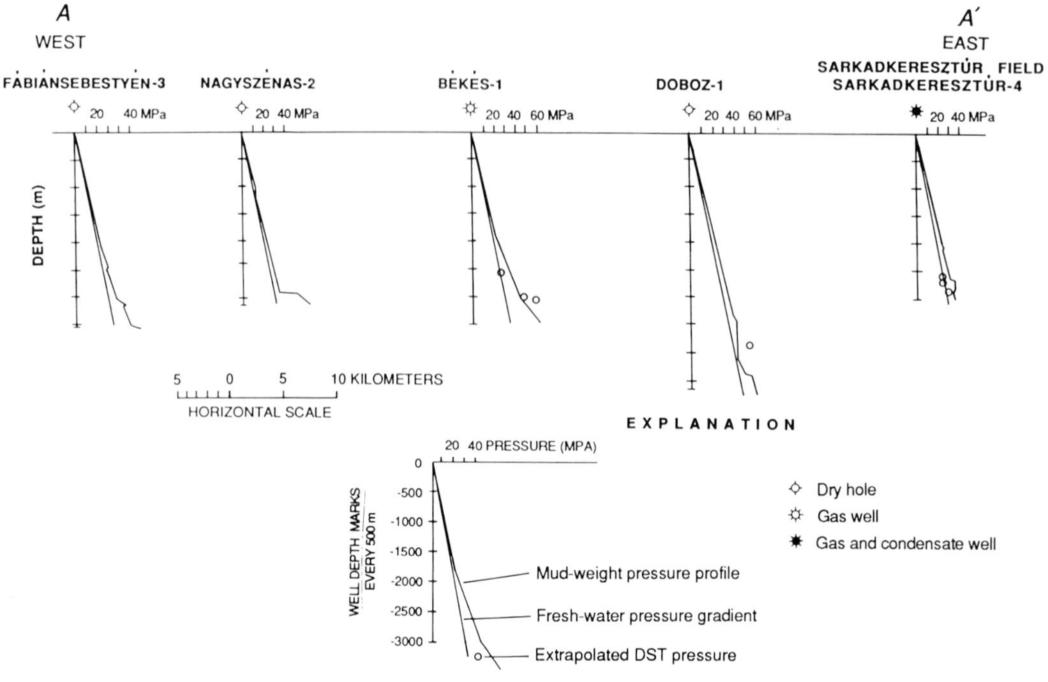

Figure 4. Mud-weight pressure profiles for wells shown in Figure 3. True reservoir pressures from available extrapolated drill stem tests are also shown. Pressures in the Sarkadkeresztúr-4 well are lower than original pressures because of production from the field. See Figure 19 for additional pressure data in this field.

Overpressuring also occurs in the basement rocks, most of which are fractured. Overpressuring in fractured basement rocks may be caused by; (1) aquathermal heating (Barker, 1972) and (2) thermally generated CO_2 in the basement rocks concurrent with downward migration of fluids (hydrocarbons and water) from overpressured Miocene and lower Pannonian basal clayey marl and marl.

Carbon isotope studies by Koncz (1983) of Hungarian gases with high concentrations of CO_2 indicate that most of the CO_2 may have resulted from thermal metamorphism of carbonate rocks at high temperature deep in the lithosphere. Carbon dioxide appears to occur in higher concentrations in those areas where deep basement faults are present. Pressures in the CO_2-bearing reservoirs range from normal to overpressured. An analysis of the distribution of CO_2 in gases in the Békés basin and vicinity (Clayton and Koncz, this volume) shows that high concentrations (20%) of CO_2 are also clearly associated with fractured basement rocks. Most Pannonian sandstones have low concentrations (<5%) of CO_2 or none at all. Much of the normally pressured oil and gas produced from shallow (2000-1000 m), relatively cool, structurally high basement rocks in the area of the Battonya-Pusztaföldvár fields (Fig. 1) contain gases that are composed of more than 40 percent CO_2. This CO_2 also is believed to have originated deep in the basement and, as will be discussed later, is mixed with high-temperature gas and condensate and low-temperature oil. These occurrences illustrate the complexity of fluid migration in this basin. The Sarkadkeresztúr field (Figs. 1 and 3) produces high-temperature gas and condensate with low CO_2 content from slightly overpressured basement rocks and Pannonian sandstones. The low CO_2 content in basement rocks here may be caused by dilution by large volumes of hydrocarbon gases charging the fractured basement.

Relatively few measurements of reservoir pressure are available for the Békés basin area. Therefore, it was necessary to examine data from all basins in the southeastern part of the Great Hungarian Plain. Figure 5 shows measured reservoir pressure versus depth for tests in the southeast part of the Great Hungarian Plain. The plot shows that pressures are mostly normal (hydrostatic) to below normal to a depth of about 1500 m. From depths of about 1700 m to 2500 m pressures are normal to well above normal. Probably these above-normal pressures at shallower depths are caused mostly by sediment dewa-

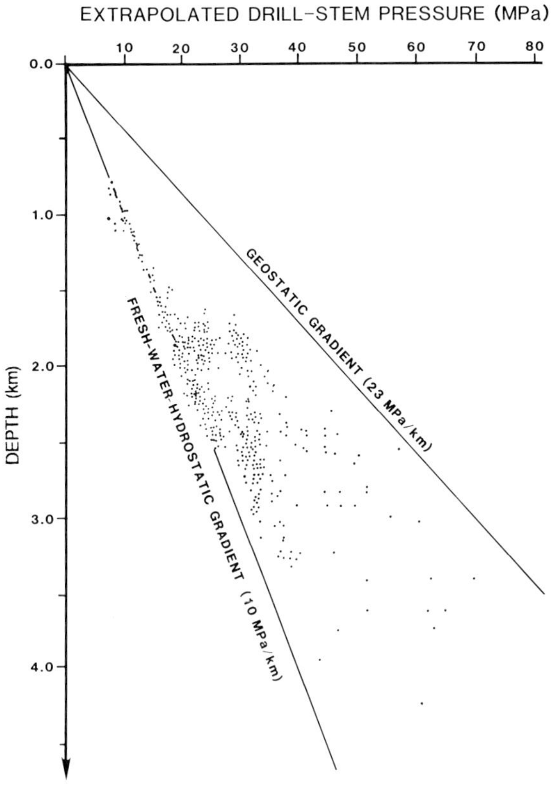

Figure 5. Reservoir pressure versus depth for the southeastern part of the Great Hungarian Plain (from Szalay, 1988).

tering with some pressure increase from temperature-related hydrocarbon generation (Meissner, 1978; Hedberg, 1979, 1980; Spencer, 1987).

Below depths of about 2500 m all pressures exceed hydrostatic values. Some low values of apparent overpressure can be attributed to the higher density of more saline waters in the deep Miocene rocks, but many pressures are clearly abnormally high. High pressures at these depths are interpreted to be caused mostly by active hydrocarbon generation related to high temperature with some contribution from undercompacted shales and marls, deep aquathermal heating, and CO_2 generation. Note that there is a fair degree of variation in the pressures at any given depth (Fig. 5).

The distribution of measured pressure versus temperature for the southeastern part of the Great Hungarian Plain shows a break in the visually-fitted trend (Fig. 6). The upper segment of the trend line, from the surface to about 120-125^{0}C, conforms to expected trends. The lower segment of the trend line, starting at about 125^{0}C, shows an anomalous increase in pressure with increasing temperature. This increase is interpreted to be caused by active hydrocarbon generation (Meissner, 1978; Hedberg, 1980, Law and others, 1980; Spencer, 1987), together with pressure contributed from CO_2 generation and some water yielded by undercompacted shales.

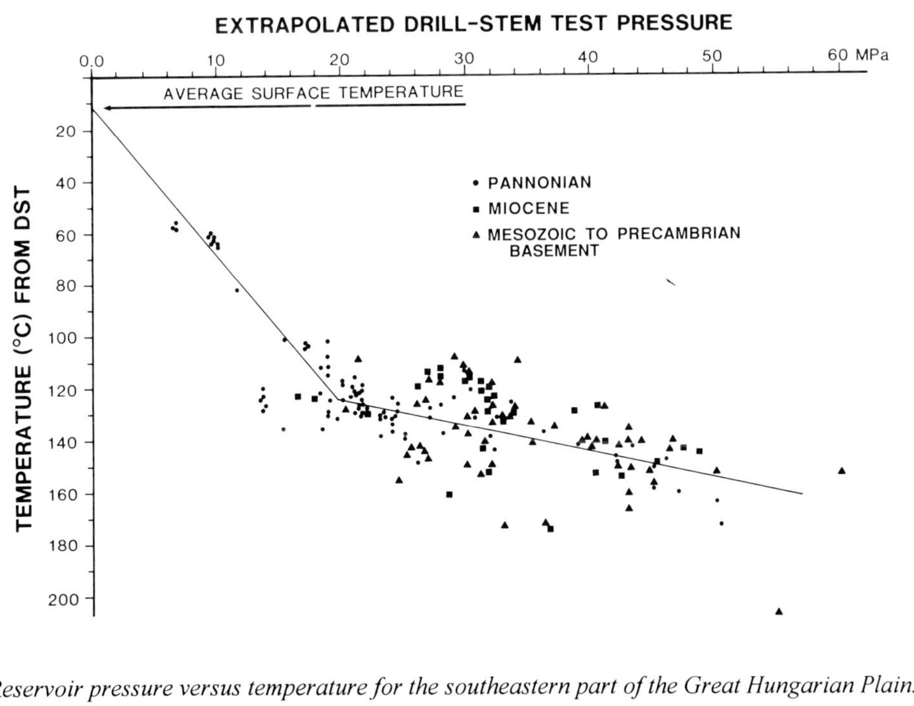

Figure 6. Reservoir pressure versus temperature for the southeastern part of the Great Hungarian Plain.

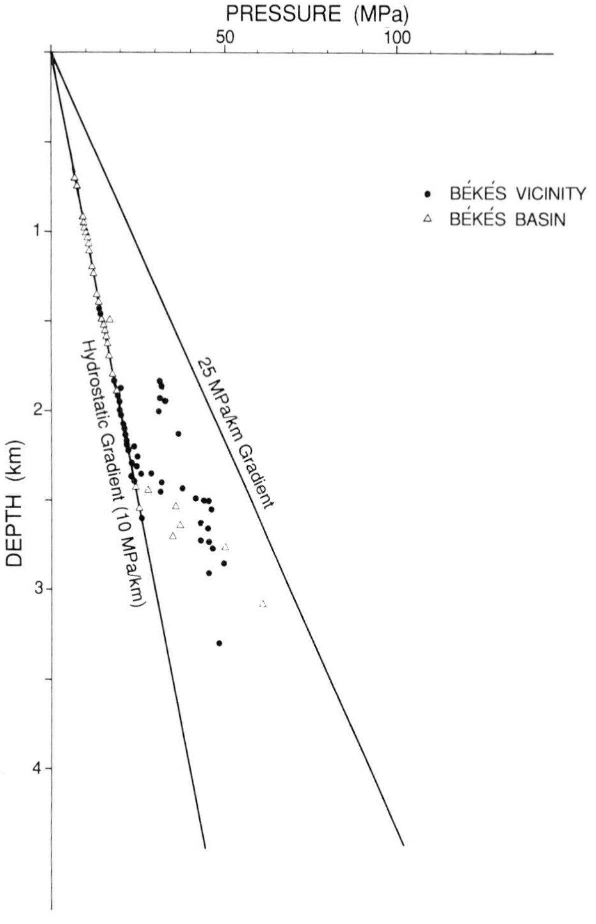

Figure 7. Lower Pannonian reservoir pressure versus depth for the Békés basin area and vicinity.

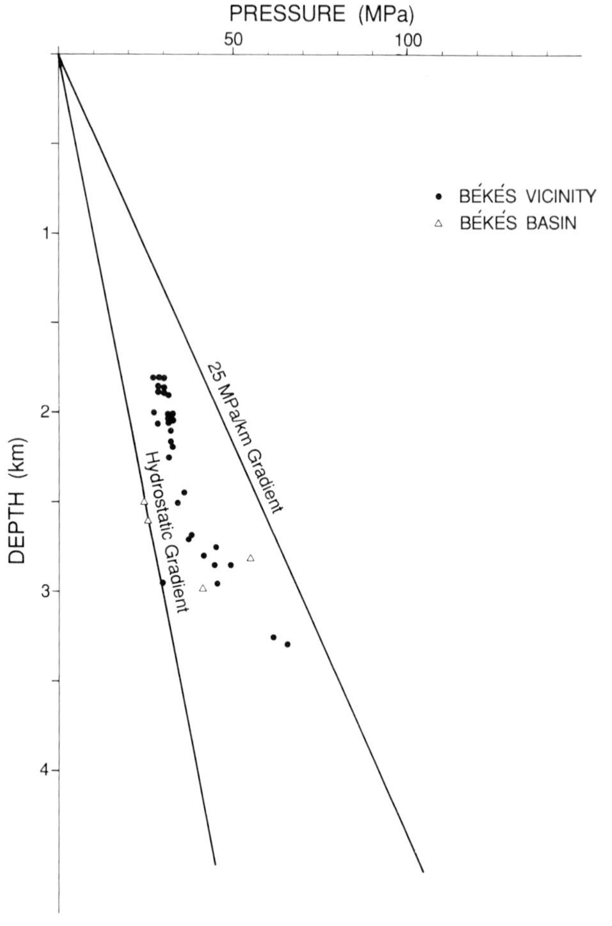

Figure 8. Miocene reservoir pressure versus depth for the Békés basin and vicinity.

Pressures in lower Pannonian rocks (Fig. 7) in the Békés basin and surrounding areas are normal to above normal from about 1800 m to about 2400 m. With a few exceptions, the pressures in Miocene rocks (Fig. 8) are abnormally high below a depth of about 1800 m. Paleozoic and Mesozoic reservoir pressures (Fig. 9) are mostly normal to a depth of 1800 m, and from 1800 m to 2800 m are normal to above normal.

Temperatures in the Békés basin (Fig. 10) have been compiled from well logs. These data exhibit some variability and are not corrected for the minor effects of cooling by drilling mud, but they show that temperatures deeper than 2500 m are mostly 120^{0}C or higher and are at levels sufficient to cause overpressuring by active hydrocarbon generation (Spencer, 1987). The effect of undercompaction (shale dewatering) has probably also contributed to overpressuring to a small degree. This interpretation is based on the recovery of water from overpressured sandstones during DSTs and undercompaction as suggested by interpretation of porosity logs (L. Kormos, oral communication, 1987).

The distribution of mean maximum and minimum core porosity versus depth for shale (clayey marl and marl) is shown in Figure 11. The decrease in porosity is mostly caused by loss of water resulting from compaction and clay mineral transformations such as a conversion of smectite to illite/smectite and illite. The trend of maximum porosity values suggests undercompaction in the depth intervals 2400-3000 m and 3400-3800 m. There is also an anomalous increase in maximum shale porosity from 4200 to 5000 m. The cause of this increase is probably undercompaction caused by incomplete shale dewatering (Szalay, 1983; L. Kormos, oral communication, 1987) and very high pore pressure caused by hydrocarbon generation.

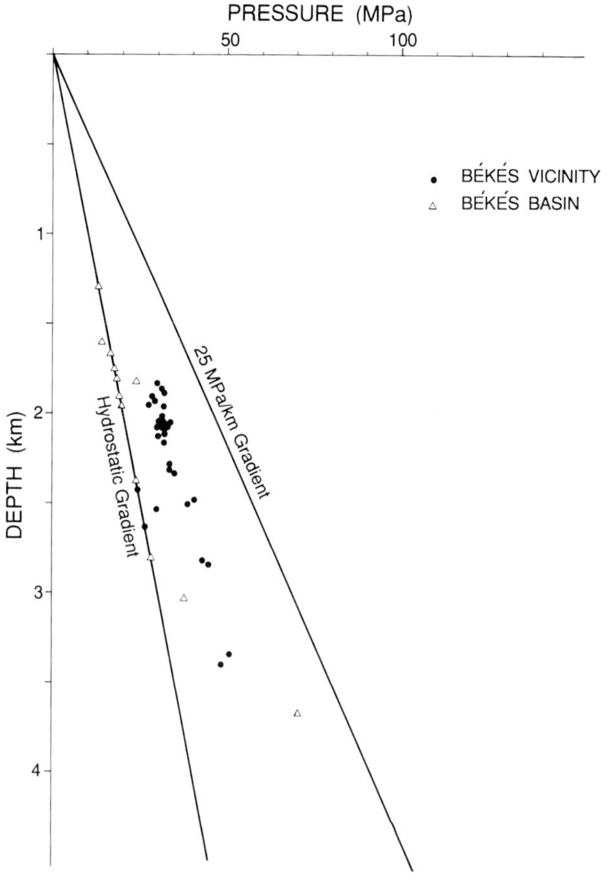

Figure 9. Mesozoic and Paleozoic reservoir pressure plotted against depth for the Békés basin and vicinity.

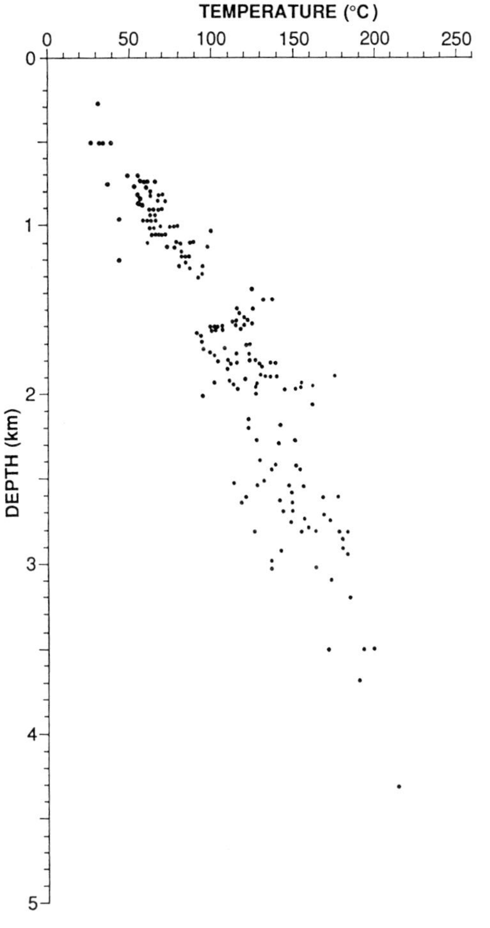

Figure 10. Subsurface temperatures plotted against depth, derived from uncorrected, well-log temperatures in the Békés basin.

High pressure caused by the slight increase of water volume, as a result of increased rate of heating (aquathermal heating) of sediments, requires almost perfect seals (Barker, 1972). Such aquathermally-produced pressure increases are difficult to generate in most rocks, except in evaporite deposits or in very rapidly heated rocks of very low permeability. Generally, high pressures caused by aquathermal heating would not be expected to occur over a wide range of depths and temperatures, but would be restricted to a few of the most perfectly sealed intervals. However, active oil and gas generation does occur over a wide range of temperatures (and depths).

Pressure maintenance by active hydrocarbon generation requires: (1) low permeability rocks, (2) kerogen still capable of generating oil or gas, and (3) temperature and time sufficient for generation at a rate that exceeds loss of pressure by fluid migration (Law and others, 1980). If any of these requirements are not met, then pressures may be normal even at high temperature. All these conditions are present in most of the overpressured parts of the Békés basin.

POROSITY AND PERMEABILITY

Secondary migration of oil and gas generally occurs in strata with the most permeable carrier beds, and along fractures in low-permeability rocks. Porous and permeable reservoirs are obviously critical to economic hydrocarbon production. It is important to be able to predict good reservoir quality with increasing depth of burial and diagenesis. Figure 12 shows average porosity versus depth for cored Pannonian sandstones from the southeastern part of the Great Hungarian Plain. Although the average sandstone porosity in Pannonian rocks shows a

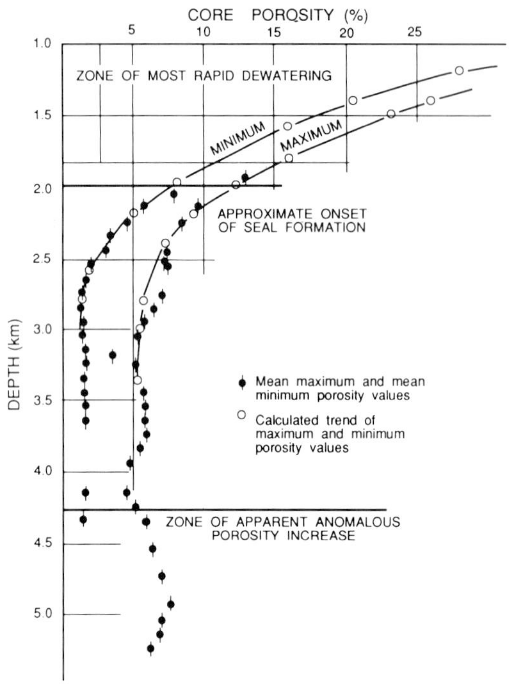

Figure 11. Average shale (clayey marl and marl) maximum and minimum core porosity values versus depth for the southeastern part of the Great Hungarian Plain.

substantial decrease with depth, there is a broad range of values at depths of 3000 m and deeper; therefore, some sandstones with fairly good porosity(10%) may be expected to occur at depths greater than 3000 m.

Horizontal and vertical permeability data for sandstone and shale (including pelitic marl and marl) for the southeastern part of the Great Hungarian Plain (Fig. 13) show that the average sandstone permeability decreases quite rapidly from depths of about 1500 to about 3000 m, and then more slowly down to depths of 4700 m and greater. A value of 0.1 mD is the lowest permeability that could be measured by the available laboratory methods. Because low-permeability cores are stress-sensitive (Thomas and Ward, 1972; Jones and Owens, 1980), the permeability measurements of about 1.0 mD or less are much higher than true in-situ permeabilities. It should be noted that, in the United States, many low-permeability ("tight") reservoirs produce gas from sandstones with permeabilities even less than 0.001 mD (Spencer, 1989). In fact, 0.1 mD in-situ permeability to gas is the upper limit generally used to define tight-gas reservoirs. The United States produces more than 28 x 10^9 m^3 of gas per year from tight gas reservoirs (Spencer, 1989).

As expected, the lateral permeabilities of shale are much higher than their vertical values (Fig. 13). The vertical permeability of shale is very low at a depth of about 1900 m and the horizontal permeability is very low at 2800 m. On the basis of core analysis, the vertical seals formed by shales appear to be fairly well developed at shallow depths (2000 m). This development of seals is generally consistent with the shallowest occurrences of overpressuring (Fig. 5). Seals are necessary to retard pressure loss from rocks with pressures higher than hydrostatic. Horizontal permeability at about 2800 m is very low, but laboratory-measured permeabilities are short-term measurements. Through geologic time, even shales have the ability to allow water to migrate (and gas to diffuse).

Usually oil can only migrate through shales at high-differential pressures because the shale pores are mostly water-wet, or nearly water-wet, and have very high capillary pressures (5-15 MPa). Therefore, seals for oil reservoirs can be effective over longer periods of time than seals for water and gas. High pressure in water and gas reservoirs may be substantially decreased over relatively short periods of time (500,000 years to 1 million years). Some kind of mechanism, that maintains pressures, must be active to keep pore pressures at high levels. As mentioned earlier, continuous dewatering of shales and active hydrocarbon generation must be at least two of the factors causing pressure maintenance, and some of the occurrences of deep overpressures in hot basement rocks may be caused by aquathermal heating and active thermal generation of CO_2.

TIMING OF HYDROCARBON GENERATION

Oil, gas condensate, and most natural gas are generated as a result of the heating of kerogen. It is generally accepted that this generation is time- and temperature-dependent. The approximate time of generation of oil and gas condensate can be estimated after reconstructing the subsidence, compaction, and thermal history of the sediments. Figure 14 shows the subsidence and maturation history of the Neogene and Quaternary (Q) strata in the Hunya-1 well (Fig. 1). The timing of hydrocarbon generation in the Békés basin can be estimated by compiling the subsidence and thermal history of potential source rocks. This onset of generation varies from place-to-place in the basin because of the combined effect of

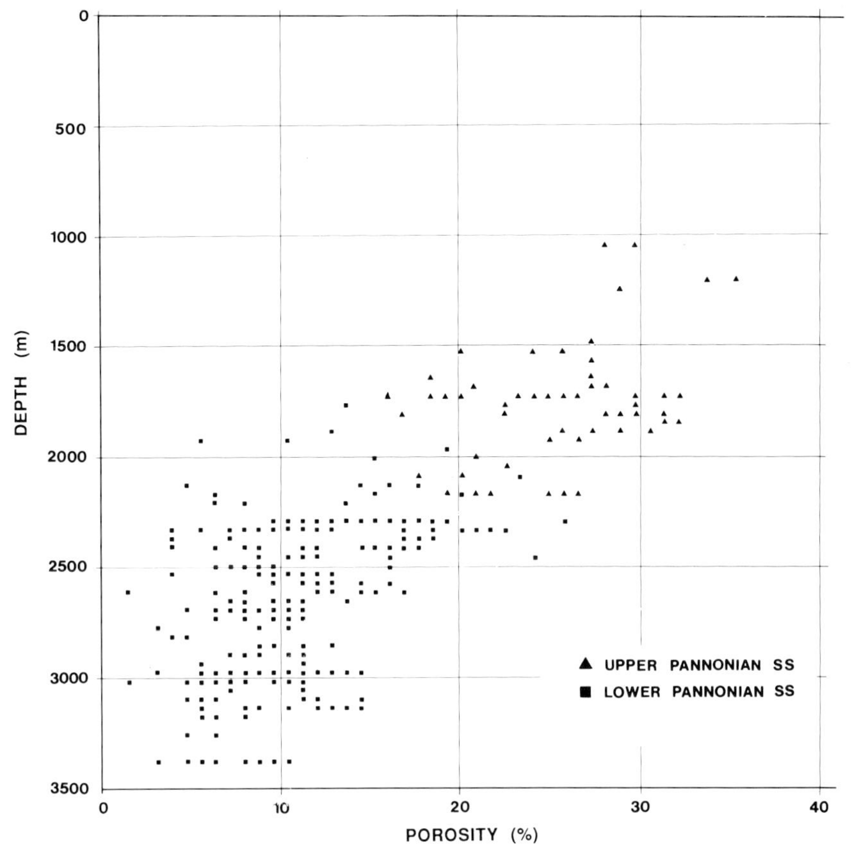

Figure 12. Average core porosity values versus depth for Pannonian sandstones of the Békés basin. Data compiled by B. Kiss.

slightly different subsidence histories and paleogeothermal conditions.

The base of the basal clayey marl and marl ($Pa_{1(1a)}$) unit in the Hunya-1 well reached a maturation level of 0.6% (R_o) at about 6.2 million years ago. The top of the $Pa_{1(1a)}$ source rock entered the "oil window" about 5 million years ago. Source rocks in the base of the Pa_2 unit are at a thermal maturation level of about $R_o = 0.5\%$ today, a value that appears to be near the minimum level in the Békés basin for the onset of generation of thermally-low-maturity, low-temperature oil (Clayton and others, this volume). Generally speaking, the Miocene source rocks in the deeper parts of the Békés basin are within the zone of wet or dry gas generation (R_o of 1.3%), source rocks in the basal clayey marl and marl ($Pa_{1(1a)}$) are within the oil to gas and condensate zone, the Algyö Formation (Pa_{12}) has reached the oil generation zone, and oil-prone source beds (Clayton and others, this volume) in the base of the Törtel Formation (Pa_2) may have just reached the onset of low-temperature oil generation. Rocks of the Algyö Formation occur mostly within the main phase of oil generation, but a limited number of analyses of these source beds indicates that they are relatively lean and contain more gas-prone than oil-prone kerogen (Clayton and others, this volume). Where oil-prone kerogen is present, the rocks should have good oil potential.

MIGRATION MECHANISMS

There are several ways that oil and gas migrate. The two most common mechanisms are: (1) buoyancy, and (2) pressure differential. Migration due to buoyancy is mostly restricted to conventional reservoirs (i.e., rocks with good permeability and porosity and low capillary pressure) or well-fractured tight reservoirs. For oil and gas to migrate by buoyancy, the continuous hydrocarbon phase (column) must be high (long) enough to exceed the capillary entry pressures of the carrier bed, but not exceed

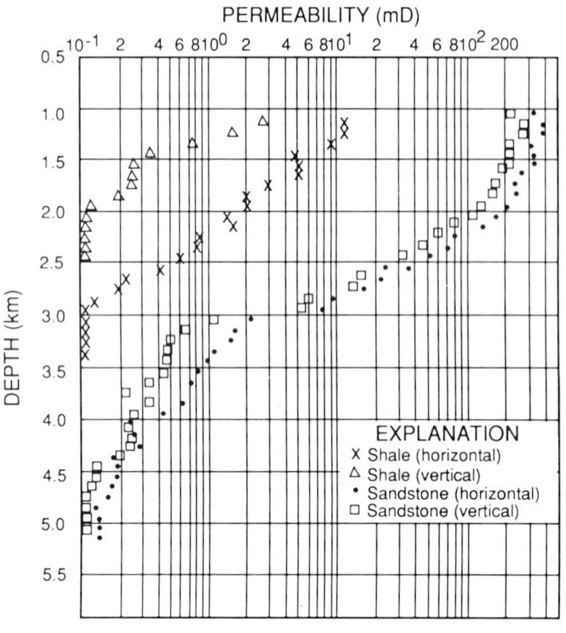

Figure 13. Average shale (including clayey marl and marl) and sandstone permeability values versus depth for the southeastern part of the Great Hungarian Plain.

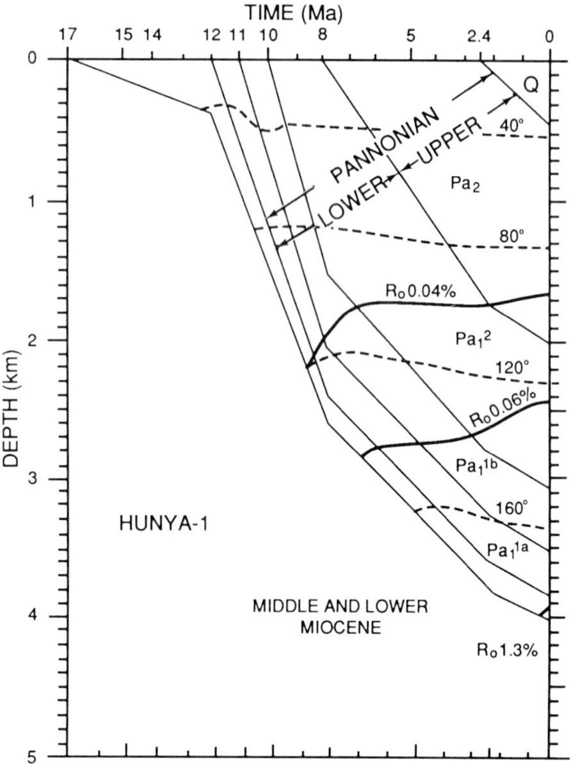

Figure 14. Burial curve for rocks penetrated by the Hunya-1 well modified from Horváth (1986). R_0 = 0.6% marks the beginning of main oil generation and 1.3% marks the beginning of condensate and gas generation. See Figure 2 for a correlation of Pannonian lithogenetic units with geologic formations. Location of well shown is in Figure 1.

the capillary pressure of the seals. Capillary pressure is an important controlling mechanism regulating the amount (column) of oil and gas trapped in stratigraphic traps (Schowalter, 1979).

In strata that have relatively high capillary pressures and low permeability, oil and gas usually do not migrate by buoyancy. Economic quantities of oil and gas in these rocks migrate mostly by high-pressure differential (or diffusion in the case of gas). However, once the oil and gas enters a well-fractured rock (e.g. basement rocks), migration updip to a lower pressure environment can take place in open fractures by buoyancy, perhaps aided by pressure differential.

Deep reservoirs will have higher pressures than shallow reservoirs even under hydrostatic conditions. Therefore, reservoir pressure alone is not useful to help understand fluid migration. One technique used to map fluid migration in rocks is to convert reservoir pressure to the height of a column of water that a given reservoir pressure will support (potentiometric surface). For example, 20 MPa will support a 2000 m column of fresh water. A map showing these values can be contoured to make a potentiometric-surface map of a particular hydrologic system. The data for such maps are always calculated in units of meters, relative to sea level, in order to convert pressures (migration energy) to a common datum. Usually, fluids will tend to migrate from a higher potentiometric value to a lower one. This can be vertical or lateral migration.

Calculated potentiometric surface data for various hydrologic units in the basin area are shown in Figures 15, 16, 17, and 18. In some areas there is considerable local variation in potentiometric values. Some of the anomalously low values were found to be caused by (1) poor pressure buildup resulting in an erroneous extrapolation, (2) pressure depletion of a limited (small) reservoir, or (3) pressure depletion caused by oil and gas production in the case of producing fields (example, Sarkadkeresztúr-4 well in Fig. 4). It was not possible to contour these maps, because it was difficult to determine which potentiometric values were affected by the three pre-

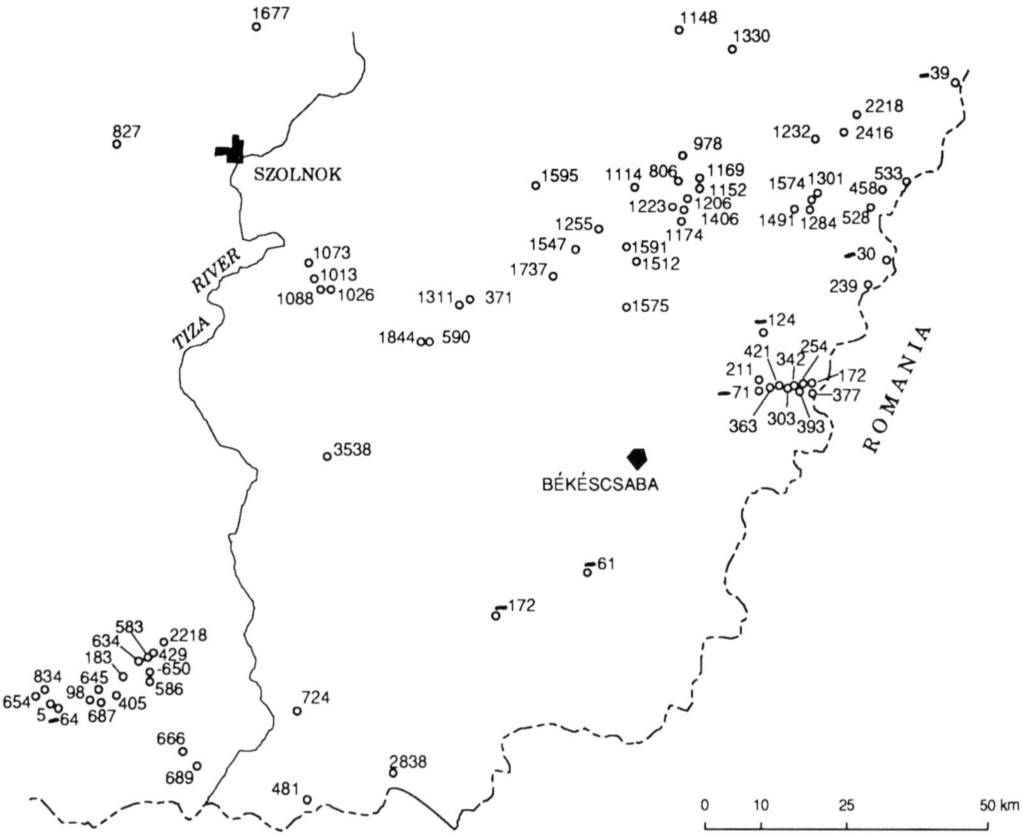

Figure 15. Map showing potentiometric values of basement rocks (Precambrian, Paleozoic, and Mesozoic rocks). Values are in meters relative to sea level. Minus values are in meters below sea level.

viously noted factors. Even without contours, these maps are useful to portray the overall conditions. The basic concept is that fluids will tend to move up and down and laterally from points of high calculated potentiometric-elevation surfaces to lower ones.

Very few potentiometric values from deep basement rocks are available. The Fábiánsebestyén-4 well (Fig. 15) had a basement potentiometric surface of 3538 m above sea level. This well encountered hot water (steam at the well head) in basement rocks. Possibly the sealing effect of the Miocene and basal clayey marls was sufficient to retain some of the pressure created by rapid aquathermal heating.

Because of low permeability, almost no usable DSTs were obtained from the basal clayey marl ($Pa_{1(1a)}$) interval. The potentiometric map of the Szolnok Formation ($Pa_{1(1b)}$) (Fig. 17) has very few data points. The Algyö ($Pa_{1(2)}$) map (Fig. 18) indicates there are many areas, and fields, with slightly above normal potentiometric surfaces. The shallow formations have normal (hydrostatic) to slightly below normal pressures. The Törtel (Pa_2) Fm. has source beds rich in oil-prone kerogen, where penetrated in wells outside the Békés basin, but cores from the lower part of the Törtel Fm. in the Békés basin have not been studied (Clayton and others, this volume). However, where the lower part of the Törtel Fm. is within the oil window (R_o= 0.6%), it could be a good source of oil and gas for stratigraphic traps flanking the basement highs. It may also be one of the sources of the interpreted low-temperature oils analyzed by Clayton and others (this volume).

The lower part of the Algyö Formation ($Pa_{1(2)}$) is thermally mature (Figs. 3, 4, 14) and appears to be

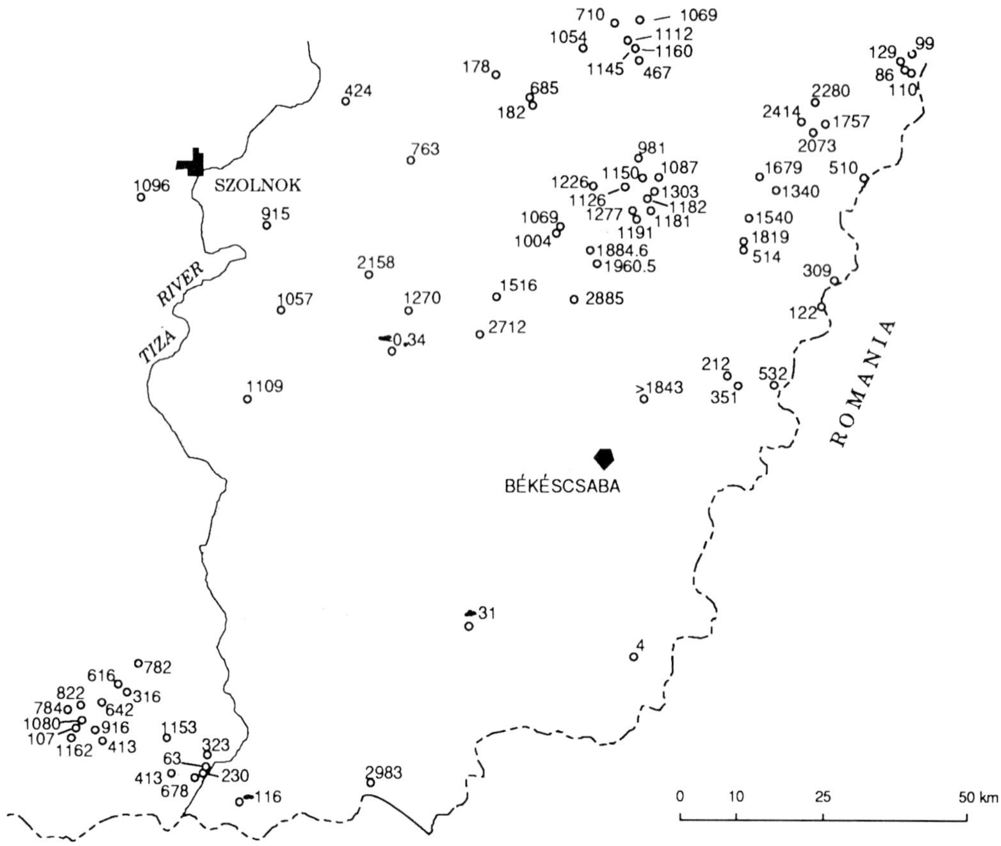

Figure 16. Map showing potentiometric values of middle and lower Miocene rocks.

overpressured where temperatures are greater than 120^{0}C (Fig. 6).

SELECTED EXAMPLES OF HYDROCARBON MIGRATION AND ENTRAPMENT

The deep, hot parts of the basin are overpressured. The pressuring fluids are oil with gas, gas with condensate, gas, CO_2 and water. High pressure conditions are most prevalent in lower Pannonian, Miocene, and basement rocks. Core analyses (Fig. 13) show that the horizontal permeabilities of sandstone, shales (including clayey marl and marl) are higher than the vertical permeabilities. Therefore, in the absence of extensive vertical faults and fractures, fluids should tend to migrate parallel to bedding. The seismic stratigraphic interpretations (Mattick and others, this volume) and lithostratigraphic studies (Molenaar and others, this volume) also show shale and clayey marl seals interbedded with sandstone reservoir (carrier) beds. On a large scale, such a reservoir-seal geometry would tend to focus migration along bedding planes in an updip direction to stratigraphic traps in areas where reservoir rocks onlap basement rocks.

Source beds overlying basement are interpreted to also have expelled oil and gas downward (Fig. 3) into fractured basement rocks. These expelled hydrocarbons could then migrate updip by pressure differential and buoyancy through fractured basement rocks to shallower basement structural (paleomorphologic) highs such as at the Sarkadkeresztúr field (Fig. 3).

Studies by Clayton and Koncz (this volume) show gas may have migrated vertically for distances of as much as 3500 m. Much of this vertical migration is interpreted to have taken place in fractured basement and Miocene rocks from deep source beds over a lateral distance as much as 5 to 10 km (Fig. 3).

Some examples of local vertical migration of gas and condensate can be observed using data from the Sarkadkeresztúr field (Figs. 1, 3, 4). This field, located on the northeastern edge of the Békés basin, is overpres-

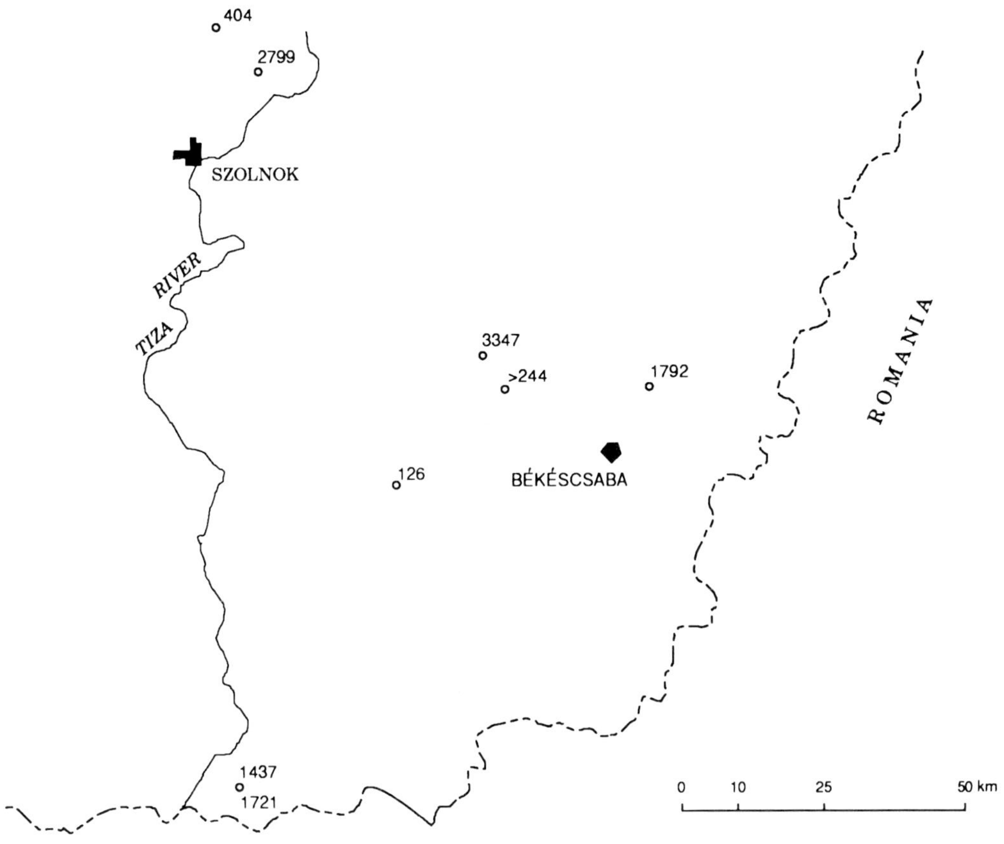

Figure 17. Map showing potentiometric values for turbidite sandstones in the Szolnok Formation ($Pa_{1(1b)}$).

sured in many reservoirs. A plot of reservoir pressures (Fig. 19) shows that the pressures in some pools align along a gas-condensate pressure gradient of a common reservoir. The interpretation is that the pressure communication is caused by the presence of faults and fractures. Slightly overpressured water occurs at the bottom of the gas-condensate column indicating pressure communication to deeper, overpressured aquifers, otherwise the pressure in the water beneath the gas-condensate would be closer to the hydrostatic gradient (shown to the left in Fig. 19).

Projection of the R_o = 0.6% line eastward into the Sarkadkeresztúr-4 well (Fig. 3), shows that the rocks in the producing interval should be barely thermally mature and not in the high-temperature, gas-condensate window. However, geochemical studies of the condensate and gas produced from this field indicate the hydrocarbons are very thermally mature and are approximately in the top of the gas condensate zone (I. Koncz, oral communication, 1986); therefore, the most likely source of this production is oil and gas that has migrated updip from thermally very mature, deep, overpressured Miocene rocks (and possibly, from the basal clayey marl) on the flanks of the structure. The flanks of this structure, and similar basin-margin structures, should, thus, be prospective for structural and stratigraphic entrapment of migrated oil and gas in fractured basement, in Miocene strata, and in the Szolnok and Algyö Formations.

The Endröd field (Fig. 1) is a good example of the lack of vertical hydrocarbon migration in the Pannonian sediments. This field does not produce oil and gas from basement rocks even though the basement rocks form a structurally high feature. Instead it produces gas, interpreted to be indigenous to the structure (Clayton and Koncz, this volume). The potentiometric surface of the lower Pannonian reservoirs at Endröd is about normal to slightly above normal.

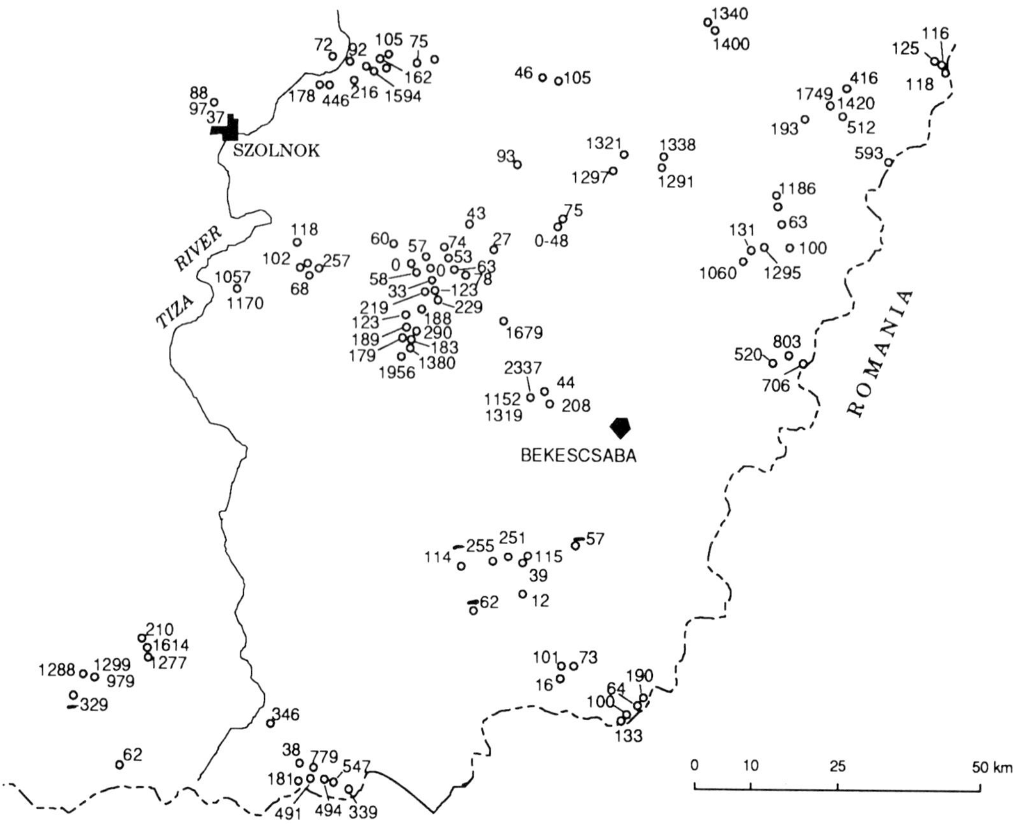

Figure 18. Map showing potentiometric values for delta-slope sandstones in the Algyö Formation (Pa1(2)).

In contrast, the Szarvas field (Fig. 1), southwest of Endröd, produces gas and condensate from basement rocks at a depth of about 3000 m and shallower, as well as from overlying lower Pannonian sandstones. The high potentiometric surface of fluids in the basement rocks at Szarvas shows that the basement is overpressured, as is part of the lower Pannonian section. Clayton and Koncz (this volume) noted that the gas at Szarvas has migrated from depths of about 3000 to 3800 m and some of this gas has a high CO_2 content. In addition, vertical migration may have occurred along growth faults.

The Szeghalom field (Fig. 1) is overpressured and produces gas and condensate that appear to have migrated laterally and updip from a structural low (trough) to the southeast. Stratigraphic traps flanking this structural trough should have high potential for production of gas and oil.

Pressure and geochemical data, and studies of fields and geological structure indicate that, substantial migration should have taken place in an updip direction out of the basin into stratigraphic traps on the updip edges of Algyö and Szolnok reservoirs flanking the highs. Migration should also have occurred into shallower structural traps through the Szolnok carrier beds and through the fractured Miocene and basement rocks.

SUMMARY AND CONCLUSIONS

The Békés basin is a relatively young, hot basin with shallow oil and gas production from structurally high basement and Neogene rocks. The deep, hot, overpressured parts of the basin have been tested by only a few wells. Geochemical studies of Clayton and others (this volume) indicate that some of the oils in the structurally high areas were generated at lower temperatures than the generally accepted top of the oil window. These same geochemical studies report very mature (high temperature) gas and condensate from other structurally high closures. The present authors concur with Clayton and

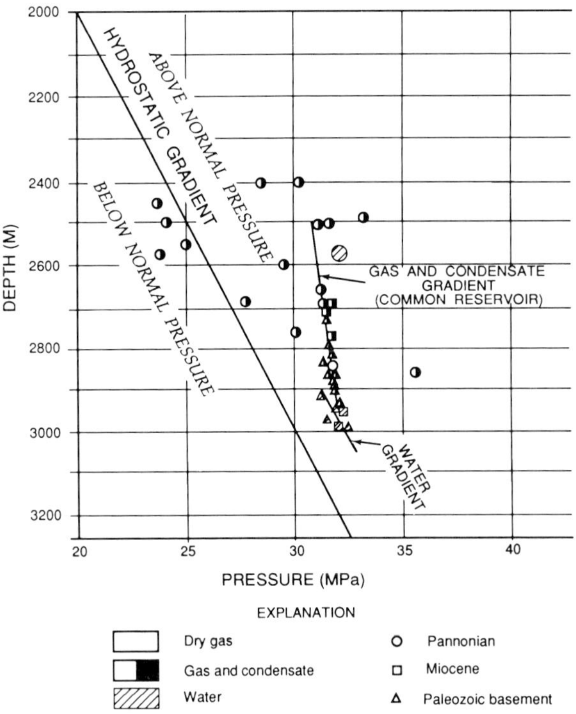

Figure 19. Pressure versus depth for water, dry gas, and gas and condensate reservoirs in the Sarkadkeresztúr field. The plot shows general overpressuring in the field and a common gas-condensate pressure gradient for many fault- and fracture-connected basement, Miocene, and Pannonian reservoirs.

others (this volume) that the immature oils migrated short distances through Neogene carrier beds and fractured basement rocks and that the high temperature hydrocarbons migrated upward and laterally from the deep hot parts of the basin.

These geochemical results should influence exploration approaches to be used in the search for stratigraphic traps and basin-flank structural traps. It was previously assumed that the oil and gas on the highs migrated from source beds deep in the basin. It now appears that some short-range migration of low-temperature oils has occurred by buoyancy in conventional reservoirs. It is also evident that thermally mature or post-mature gas has migrated (by diffusion and pressure differential) from deep in the basin into structural highs, where it has mixed with the low-temperature oils (Clayton and others, this volume). As noted earlier, gas is generally more mobile than oil. The fields with low thermal maturity oils and (or) mixed low and high temperature oils occur in structurally high areas such as the Pusztaföldvár, Kaszaper, Mezöhegyes, and Battonya fields on the southwest side of the Békés basin, and at the Komádi and Biharugra fields (Fig. 1), to the north of the Békés basin (Clayton and others, this volume). The Sarkadkeresztúr field produces mostly highly mature gas and condensate (I. Koncz, oral communication, 1986). On the basis of thermal maturity and burial history work by Szalay (1983), as well as data of Figures 3 and 4, these hydrocarbons are interpreted to have migrated updip from a depth of about 4000 m and laterally for a distance as much as 2 to 10 km.

The absence of large volumes of thermally mature oil on the Battonya-Pusztaföldvár high indicates that such oils, if generated, have been blocked from migrating updip by stratigraphic traps flanking the structures. The optimum exploration depths for stratigraphic traps ap-

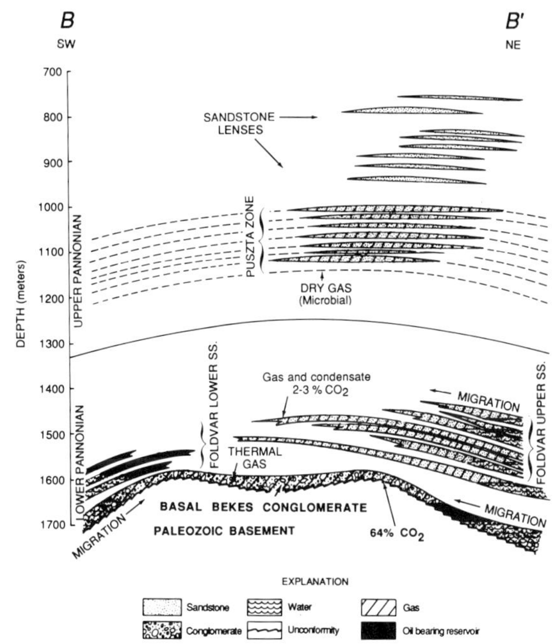

Figure 20. Generalized cross section B-B' in the Pusztaföldvár area. (See Fig. 1 for location). The basal Békés Conglomerate produces low-temperature oil and gas with a high CO_2 content, that is mixed with high-temperature gas and condensate (Clayton and others, this volume). The Földvár Lower sandstones produce low-temperature oil and high-temperature gas, together with small a quantity of CO_2. The Földvár Upper sandstones produce high-temperature gas and gas condensate; the CO_2 content ranges from low to none. The upper Pannonian Puszta zone sandstones contain dry gas of probable biogenic origin.

pear to be between 3500 m (updip from R_o = 1.3% at 4000 m) and 2400 m (updip from R_o = 0.6%). Oil and gas should also be trapped in local structures on the flanks of the major basement highs.

The complexity of migration and accumulation of oil and gas in shallow structures in the Békés basin is illustrated using the Pusztaföldvár area as the example (Fig. 20). The drape of the Pannonian rocks over this basement structure results from differential compaction of Neogene sediments. The basement rocks here contain large amounts of CO_2 gas and low-temperature oil. The low-temperature oil has probably migrated a short distance (Clayton and others, this volume). The highest concentrations of CO_2 occur in the basement rocks and in the basal Békés conglomerate. The amount of CO_2 in the lower and upper Pannonian sandstones ranges from low to none. The CO_2 had originated in basement rocks (Koncz, 1983) and has probably migrated into the lowest reservoirs of the structure through basement fractures. The gas and condensate, that have high-temperature origins (Clayton and Koncz, this volume; Clayton and others, this volume), have migrated through basement fractures and lower Pannonian sandstones from depths of 3500 m or deeper. The dry gases in the upper Pannonian beds are probably biogenic (Clayton and Koncz, this volume). Gas of biogenic (bacterial, low-temperature) origin is a major potential resource in the United States and the Russia (Rice and Claypool, 1981).

In the course of our migration studies we found that a potentially large amounts of biogenic gas appear to be present in the Békés basin in shallow water-bearing reservoirs. A specially designed exploration effort will be needed in order to evaluate this potential gas resource. We also noted that potentially large tight-gas (Spencer, 1989) resources are probably present in the deep part of the Békés basin.

ACKNOWLEDGMENTS

Many of our colleagues provided data and advice. We would particularly like to thank C. M. Molenaar, I. Révész, I. Bérczi, D. Hajdú, and G. K. Juhász, for providing stratigraphic and other information. J. L. Clayton and I. Koncz provided geochemical data and interpretations. Mrs. G. L. Horváth furnished data and interpretations from producing fields. J. Rumpler, G. Pogácsás, J. A. Grow, B. V. Szanyi, A. Kovács, E. Varga, and A. Ujfalusy provided geophysical information. L. Kormos and B. Kiss provided core and well log data and interpretations. R.R. Charpentier and C.W. Keighin provided constructive reviews of this paper.

REFERENCES

Barker, C., 1972, Aquathermal pressuring - role of temperature in development of abnormal pressure zones; Amer. Assoc. Petrol. Geol. Bull., 56, 2068-2071.

Clayton, J. L., and Koncz, I., Geochemistry of natural gas and carbon dioxide in the Békés basin - Implications for exploration; this volume, 187-199.

Clayton, J. L., Koncz, I., King, J. D., and Tatár, E., Organic geochemistry of crude oils and source rocks, Békés basin, Hungary; this volume, 161-185.

Clayton, J.L. Spencer, C.W., Koncz, I., and Szalay, Á., 1990, Generation and migration of hydrocarbon gas and carbon dioxide in the Békés basin, southeastern Hungary; Org. Geochem. 15, (3), 233-247.

Fertl, W. H., 1976, Abnormal formation pressure environments; *in* Fertl, W. H., (ed.), Abnormal formation pressures, Developments in petroleum science 2; Elsevier Scientific Publishing Company, New York, 1-48.

Hedberg, H. D., 1979, Methane generation and petroleum migration; Oil and Gas Journal, 77, 186-192.

Hedberg, H. D., 1980, Methane generation and petroleum migration; *in* Roberts, W. H. III, and Cordell, R. J. (eds.), Problems of petroleum migration; Amer. Assoc Petrol. Geol. Studies in Geol. 10, 179-206.

Horváth, F., 1986, Maturity prediction of disperse organic material in the Neogene sedimentary rocks of the Pannonian Basin on the basis of paleotemperature reconstruction. Internal report for Köolaj Vállalat (Oil Co., Szolnok, Hungary), 263 p.

Jones, F. O. and Owens, W. W., 1980, A laboratory study of low-permeability gas sands; Jour. Petrol. Technol., 32, 1631-1640.

Koncz, I., 1983, The stable carbon isotope composition of the hydrocarbon and carbon dioxide components of Hungarian natural gases; Acta Mineral-Petrog., Szeged, Hungary, 26, (1), 33-49.

Law, B. E., Spencer, C. W., and Bostick, N. H., 1980, Evaluation of organic matter, subsurface temperature and pressure with regard to gas generation in low-permeability Upper Cretaceous and lower Tertiary sandstones in Pacific Creek area, Sublette and Sweetwater Counties, Wyoming; The Mountain Geol., 17, 23-25.

MacGregor, J. R., 1965, Quantitative determination of reservoir pressures from conductivity log; Amer. Assoc. Petrol. Geol. Bull., 49, 1502-1511.

Mattick, R., Rumpler, J., Ujfalusy, A., Szanyi, B. and Nagy, I., Sequence stratigraphy of the Békés basin; this volume, 39-65.

Meissner, F. F., 1978, Petroleum geology of the Bakken Formation, Williston Basin, North Dakota and Montana, in 24th Annual Conference, Williston Basin Symposium; Montana Geol. Soc., 207-227.

Molenaar, C. M., Révész, I., Bérczi, I., Kovács, A., Juhász, G. K., Gajdos, I., and Szanyi, B., Stratigraphic framework and sandstone facies distribution of the Pannonian sequence in the Békés basin; this volume, 99-110.

Phillips, R.L., Révész, I., and Bérczi, I., Lower Pannonian deltaic-lacustrine processes and sedimentation, Békés basin; this volume, 67-82.

Rice, D. D., and Claypool, G. E., 1981, Generation, accumulation, and resource potential of biogenic gas; Amer. Assoc. Petrol. Geol. Bull., 65, 5-25.

Schowalter, T. T., 1979, Mechanics of secondary hydrocarbon migration and entrapment; Amer. Assoc. Petrol. Geol. Bull., 63, 723-760.

Spencer, C. W., 1987, Hydrocarbon generation as a mechanism for overpressuring in Rocky Mountain region; Amer. Assoc. Petrol. Geol. Bull., 71, 368-388.

Spencer, C.W., 1989, Review of characteristics of low-permeability gas reservoirs in western United States; Am. Assoc. Petrol. Geol. Bull., 73, 613-629.

Szalay, Á., 1983, The role of basin evolution models for predicting hydrocarbon potential in the area of the depressions of the southeast Great Hungarian Plain (in Hungarian); Candidate thesis, Hungarian Academy of Sciences, Budapest, 125 p.

Szalay, Á., 1988, Maturation and migration of hydrocarbons in the southeastern Pannonian Basin, *in* Royden, L.H., and Horváth, F., (eds.), The Pannonian Basin: A study in basin evolution; Amer. Assoc. Petrol. Geol. Memoir 45, 347-354.

Szalay, Á., and Szentgyörgyi, K., 1979, Contribution to the knowledge of lithologic subdivision of Pannonian Basin formations explored by hydrocarbon drilling - Reconstruction based on trend analysis; Hungarian Academy of Sciences, Bulletin of Dept. 10, Budapest, 12, 401-423.

Thomas, R. D. and Ward, D. C., 1972, Effect of overburden pressure and water saturation on gas permeability of tight sandstone cores; Jour. Petrol. Technol., 24, 120-124.

11 Structural Control on Hydrocarbon Accumulation in the Pannonian Basin, Hungary

György Pogácsás[1], *Robert E. Mattick*[2], *Gábor Tari*[3], *and Péter Várnai*[1]

ABSTRACT

The Pannonian Basin is a back-arc basin superimposed on the Alpine compressional megasuture that resulted from continental collision between Europe and smaller continental fragments following southward subduction of Tethyan ocean floor. A set of discrete basins, whose development was dominated by extensional listric and wrench faults, formed inside the Carpathian loop in middle Miocene time. The basin-forming tectonic activity culminated in middle to late Miocene time. The Pliocene to recent development of the region was characterized by widespread thermal subsidence and limited local tectonic deformation.

This geodynamic setting governs the characteristic styles of hydrocarbon accumulations in the Pannonian Basin. As revealed by the analysis of multichannel seismic-reflection and drill-hole data, the oil and natural gas fields are typically associated with basement highs. The most prospective of these basement highs are associated with: (1) wrench faulting, (2) extensional faulting, (3) intra-basement arches, (4) thrust faulting, and (5) structural inversion (uplifting of formerly depressed areas by a combination of compression and wrenching). In addition, oil and natural gas traps are associated with listric normal faults that resulted from differential compaction of sediments.

INTRODUCTION

The aim of this chapter is to provide a general overview of specific structural features that are associated with prominent oil and natural gas traps in the Pannonian Basin. The search for oil and natural gas in the area is shown to be dependent on the interrelations between the tectonic setting and the accumulation mode of hydrocarbons. By outlining these structural features, we hope to offer an insight into possible field extensions and new prospects in the Pannonian region.

Following the scheme of Harding and Lowell (1979), we distinguish structural styles based on the involvement or non-involvement of the basement complex in the

[1] MOL Rt. (Hungarian Oil and Gas Co., Ltd.) H-1068, Budapest, Hungary
[2] U.S. Geological Survey, Reston, Virginia 22092, USA
[3] Rice University, Houston, Texas 77251, USA

P. G. Teleki et al. (eds.), Basin Analysis in Petroleum Exploration, 221–235.

structural development. This is relevant to the Pannonian Basin because its major hydrocarbon accumulations appear to be dominantly associated with structures that involve the basement complex (pre-Tertiary rocks). Five types of structures that involve the basement complex are discussed with examples shown on seismic sections: (1) wrench faulting with an example of a flower structure in the Derecske basin; (2) extensional block faulting coeval with the deposition of synrift sediments with an example from the Nagylengyel area; (3) positive basement warps with an example from the Algyö area; (4) compressive thrusts with an example from the Földes area; and (5) structural inversion with an example from the Budafa anticline. In addition, an example from the Dévaványa area of oil and natural gas traps associated with a growth-fault structure that does not involve the basement complex is also presented.

The intra-Carpathian region, which consists of a number of subbasins separated by basement highs, is surrounded by the Carpathian arc (Fig. 1). The largest subbasin is the Pannonian basin (also called the Great Hungarian Plain). The pre-Tertiary basement complex of the region is composed of several plate fragments juxtaposed by Cretaceous to Eocene tectonic events (Channell and Horváth, 1976; Balla, 1987). The Cretaceous-Paleogene evolution of the area was marked by the convergence and collision of two major continental units: Europe and the rigid Moesian plate and a smaller, nonrigid and highly deformed unit consisting of a complex assemblage of small continental fragments (Horváth and

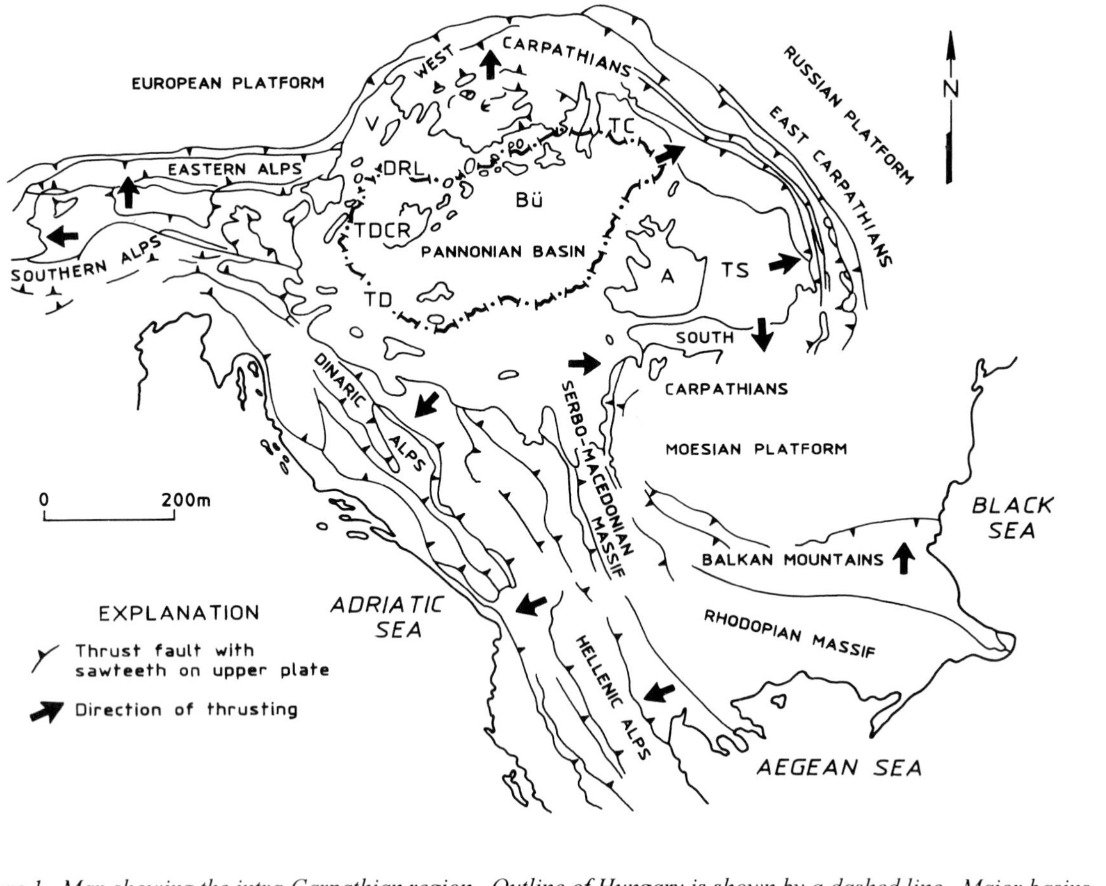

Figure 1. Map showing the intra-Carpathian region. Outline of Hungary is shown by a dashed line. Major basins within this region are the Pannonian, Vienna (V), Southwest Transdanubian (TD), Transcarpathian (TC), and Transylvanian (TS) Basins. Danube-Rába Lowland (DRL); Transdanubian Central Range (TDCR); Bükk Mountains (Bü); Apuseni Mountains (A). Figure modified from Burchfiel and Royden (1982).

Royden, 1981; Royden and others, 1982; Balla, 1985, 1987; Royden, 1988). Convergence was probably the result of southward and westward directed subduction of a Jurassic-Lower Cretaceous oceanic terrane.

According to Horváth and Royden (1981), Horváth (1986), and Royden (1988), the intra-Carpathian basins are typical back-arc extensional basins that formed during and after the main stages of thrusting in the outer Carpathians. The basins can be classified as either peripheral basins that lie adjacent to and superimposed on thrust belts (Vienna basin and Transcarpathian depression), or as central basins that lie in a more internal position (Pannonian and Danube-Rába basins). In the first case, the basins formed mainly in middle Miocene time (Karpatian and Badenian stages); whereas in the second case, the basins formed in a later phase of thermal subsidence during late Miocene-early Pliocene time (Sarmatian and Pannonian *sensu lato* stages).

The Miocene opening of the intra-Carpathian basins resulted from strike-slip motion along a set of roughly NE-SW-trending, left-lateral shears accompanied by a conjugate set of NW-trending, right-lateral shears (Horváth and Royden, 1981; Horváth and others, 1987), in which setting the right-lateral faults are dominant. The resulting extension and normal faulting produced discontinuous pull-apart basins and push-up swells along divergent strike-slip faults. Deep sedimentary troughs are usually bounded by a set of normal faults. Large amounts of displacement along listric normal faults resulted in tilting of the originally horizontal strata and a regional unconformity developed between the middle Miocene (synrift) and Pannonian (postrift) sedimentary rocks (Horváth and others, 1987). Seismic illustrations of these structural features have been published by Pogácsás (1984, 1985, 1986), Horváth and Rumpler (1984), Molnár and others (1987), and Rumpler and Horváth (1988).

The style of sedimentation within individual basins was influenced by the proximity of the basin to a thrust front (Royden, 1988). Basins located close to a thrust front contain a thick, syn-extensional, faulted sedimentary rock section overlain by a thin, post-tectonic, sedimentary rock section. Basins located in a more internal position within the basin system, such as the Pannonian Basin, can be characterized by the dominance of a post-extensional, unfaulted, flat-lying sedimentary sequence. The thickness of the Neogene-Quaternary sedimentary fill may reach 7000 m in some basins.

WRENCH FAULTING, DERECSKE BASIN

On the northern side of the Derecske basin near the eastern border of Hungary (Fig. 2, profile I), irregular seismic reflection patterns occur in the Pannonian section, which were attributed by Albu and others (1976) to the presence of a "mobile zone" in the crust. According to Pozsgay and others (1981), this mobile zone is related to a deeper fault zone and lamellation near the Mohorovicic discontinuity. Such lamellar structures may be characteristic of wrench faults (Fuchs and Schultz, 1976). Numerous other authors, based on interpretation of seismic profiles in this region (Varga and Pogácsás, 1981; Hajdú and others, 1982; Gajdos and others, 1983; Balla, 1985; Pogácsás, 1985, 1986; Samu, 1985; Rumpler and Horváth, 1988), have concluded that the mobile zone represents significant strike-slip movement. Seismic stratigraphic investigations in the Derecske basin (Pogácsás and Völgyi, 1982; Berkes and others, 1983; Pogácsás, 1984, 1985) suggested that different seismic facies are in contact with one another in the upper Miocene-Quaternary section along the inferred strike-slip fault zone .

A seismic profile (Fig. 2, profile I) that crosses the suspected strike-slip fault in the Derecske basin is shown in Figure 3. The interpretation of this profile shows a flower structure. High-amplitude reflections in the upper Pannonian (Pa_2), lacustrine-fluviatile, marshy facies are offset by about 50 m in an extensional collapse structure believed to be the result of strike-slip movement. The discontinuous reflections from lower Pannonian (Pa_1) delta plain facies also appear to be disturbed by faulting, and the Precambrian basement surface has been offset by about 400 m.

At the Földes-2 well (Fig. 3), natural gas reservoirs occur in the Pannonian section at depths of 2032-2041 m and 2669-2675 m with production, respectively, of 5000 m^3/day and 11,000 m^3/day. The upper natural gas pool occurs in southeastward dipping delta slope facies and the lower one in prodelta turbidities. The updip seal on these reservoirs appears to be a Riedel shear fault related to lateral displacement. Lower in the section, natural gas production comes from Miocene tuffaceous sandstone (246,600 m^3/day) and fractured Precambrian metamorphic rocks (129,000 m^3/day) at depths of 3247-3255 m and 3315-3540 m, respectively. These combination structural-stratigraphic traps are inferred to be related to vertical displacement of the basement that resulted from lateral movement.

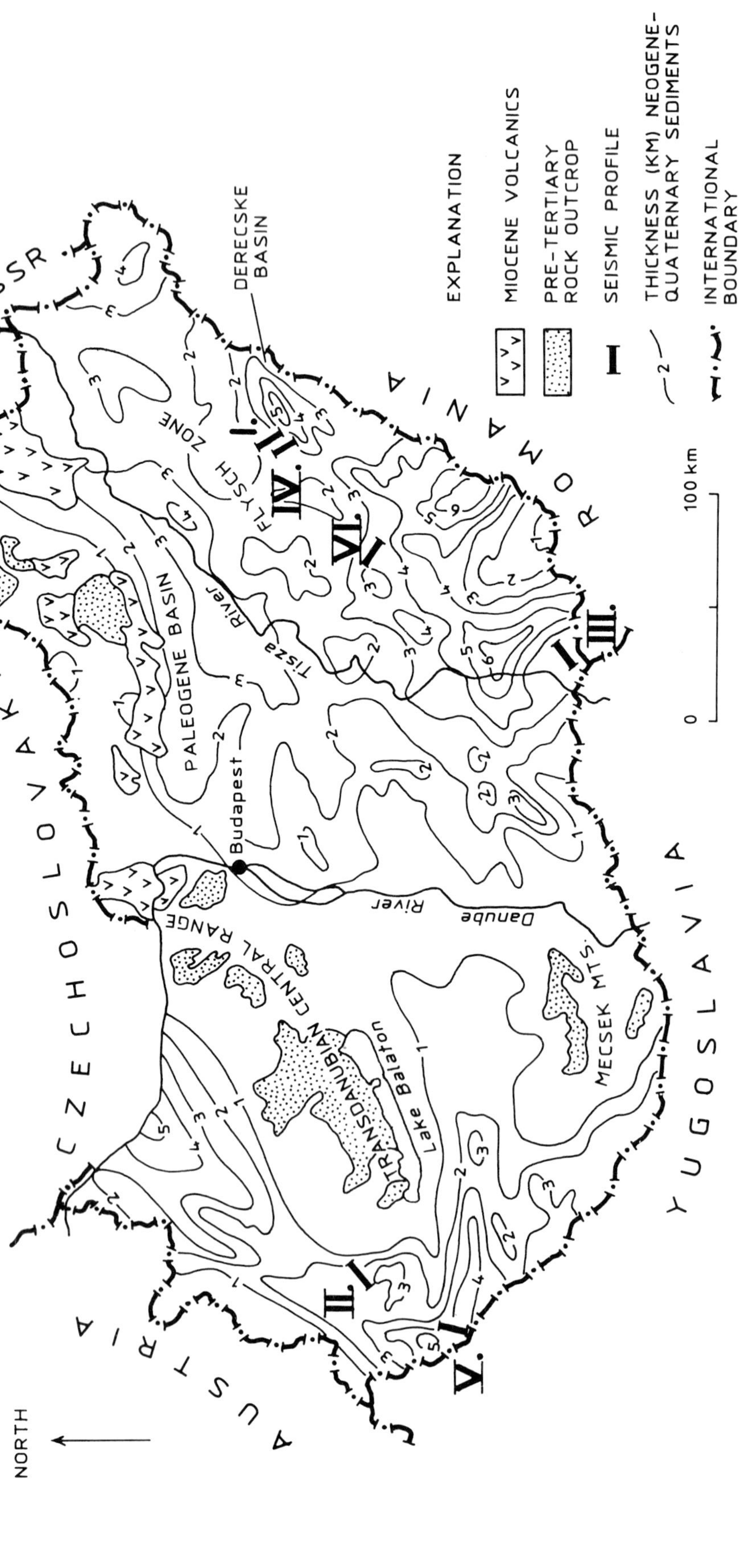

Figure 2. Isopach map of Neogene-Quarternary sedimentary rocks in Hungary and location of seismic profiles discussed in text (modified after Royden and Sandulescu, 1988). Contours in kilometers.

About 13 km northeast of the Földes-2 well, along the same strike-slip fault, hydrocarbon pools have been discovered in delta front as well as delta plain facies. There is a good likelihood, therefore, that along such relatively unexplored, lateral displacement zones more hydrocarbon accumulations will be discovered.

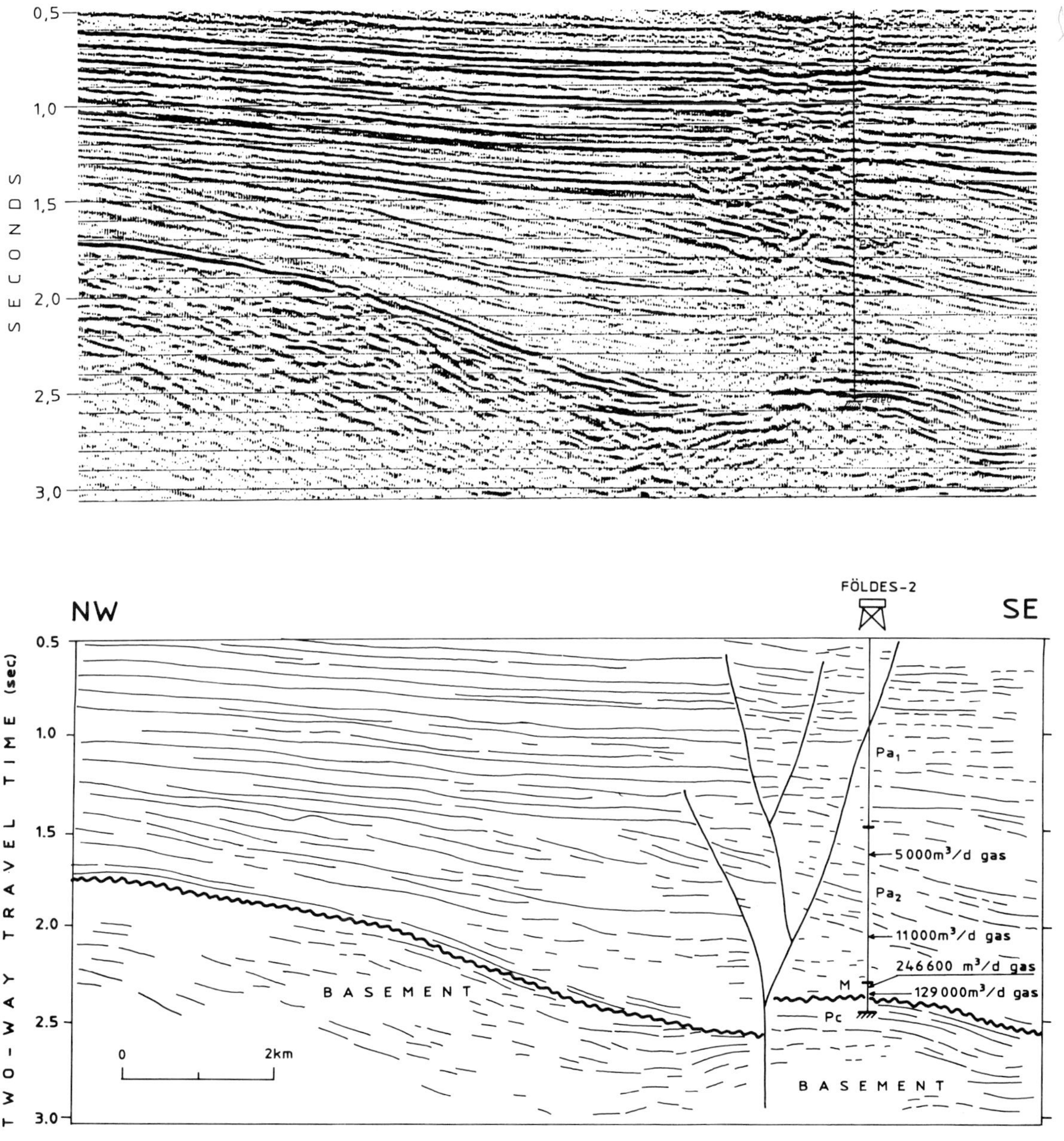

Figure 3. Derecske basin: (top), seismic profile I (uninterpreted, migrated 24-fold section); (bottom), interpretation of profile I. A well-developed flower structure can be observed on the right-hand side of the profile above 2.5 sec. Location of profile is shown in Figure 2. Wavy line represents unconformity between Neogene sedimentary rocks and the Mesozoic-Precambrian basement complex. Pa_1: lower Pannonian; Pa_2: upper Pannonian; M: Miocene; PC: Precambrian.

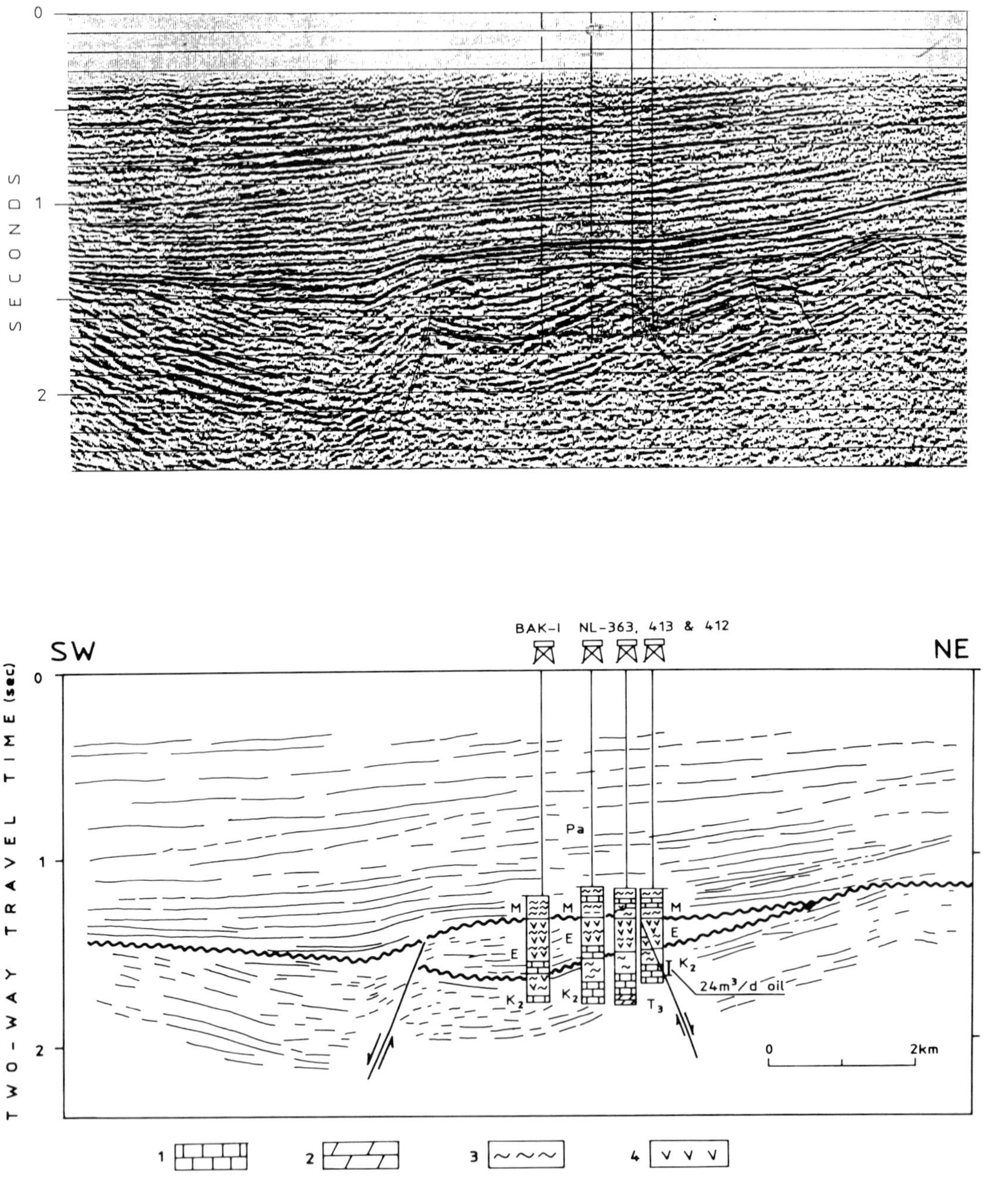

Figure 4. Nagylengyel field: (top), seismic profile II (uninterpreted, migrated 24-fold section); (bottom) interpretation of profile II. The most significant reservoirs are in Cretaceous rudistid limestone and Upper Triassic dolomite. Reservoirs in Cretaceous rudistid limestone are sealed above by marl beds. Location of profile is shown in Figure 2. (1): limestone; (2): dolomite; (3): marl; (4): volcanic rocks; Pa: Pannonian; M: Miocene; E: Eocene; K_2: Upper Cretaceous; T_3: Upper Triassic.

EXTENSIONAL BLOCK FAULTING IN THE NAGYLENGYEL AREA

During the 1950's and early 1960's (prior to production from the Algyö field), the Nagylengyel field, in terms of oil production, was one of the largest in Hungary. The original, in-place reserves are estimated to have been 45 million metric tons of oil, including a small percentage of natural gas expressed in BTU equivalents of oil; but, for the most part, these reserves have been nearly exhausted since the field was discovered by gravity and seismic-refraction methods in 1951. The main hydrocarbon reservoirs are in Upper Cretaceous rudistid limestones and Triassic dolomites. traps are associated primarily with uplifted fault blocks, and hydrocarbon reservoirs are sealed above by arenites and calcarenites of Late Cretaceous age. Geochemical analyses (Clayton and others, this volume) indicate that the oil in the Nagylengyel field was generated from Triassic source rocks that reached maturation in Neogene time.

Seismic profile II (Figs. 2 and 4) crosses the Nagylengyel field in a SW-NE direction. Based on the interpretation of seismic reflection patterns, cores and well logs, the postrift Pannonian sedimentary section represents a deltaic sequence that prograded from the northeast. Beneath, middle Miocene sedimentary rocks overlie a strongly eroded synrift Eocene section composed of sandstone, marl, limestone, and volcanic rock unconformably. The base of the synrift Eocene section is marked by another strong unconformity. The postrift (Miocene-Quaternary) section is relatively undeformed in contrast to the synrift Eocene and the prerift Mesozoic sections which are highly faulted and deformed. Faulting appears to be limited to the pre-Miocene section where Triassic-Eocene strata are offset by as much as 400-500 m (Fig. 4).

Dip directions of Upper Cretaceous-Eocene strata in the Nagylengyel area are concordant to the surface forms of the Triassic (Körössy, 1971; Bodzay, 1971). As previously stated, the Triassic-Eocene section is cut by numerous normal faults that strike in a NW-SE direction with as much as 400-500 m of throw. Because there appears to be no evidence of lateral displacement or reverse faulting, the structure in the Nagylengyel area is inferred to be a result of extensional block faulting of a large ancient dome. The strike directions of individual blocks point off the axis of the ancient dome (Bodzay, 1971).

In the Nagylengyel field, 15 oil pools are known. The pools are under normal hydrostatic pressure and the reservoirs are mostly hydrodynamically isolated. The thicknesses of Cretaceous reservoirs vary from 7-90 m, and Triassic reservoirs from 20-160 m (Dank, 1988). The permeability of Cretaceous reservoir rocks is relatively low, a result of the presence of vertically and horizontally oriented zones of compact marl. The oil, generated in Triassic dolomites, probably migrated along faults into Cretaceous and Triassic reservoirs. According to Bodzay (1971), oil migration probably took place during the Miocene because the highest parts of the basement structures do not contain oil, and these basement highs were breached by erosion during Miocene time.

POSITIVE BASEMENT WARPS, ALGYÖ AREA

The Algyö field was discovered using seismic methods in the early 1960's and the first production well was completed in 1965. Today the field covers an area of about 80 km^2. The original, in-place reserves of Algyö are estimated to be 75 million metric tons of oil and 100 billion m^3 of natural gas. Presently, this is Hungary's largest oil and gas field and it yields 52% of the total hydrocarbons produced in Hungary (Dank, 1988).

Seismic profile III (Figs. 2 and 5) crosses the southern part of the Algyö field. The large elevated basement block is inferred to have formed by differential subsidence, commencing in middle Miocene time, whereby the flanks of the high subsided faster than the central core area. Prior to subsidence, the basement complex, composed of highly metamorphosed Precambrian rocks, was deeply eroded. Middle Miocene-age synrift sediments, which comprise the basal part of the Neogene sedimentary section, are thin on the top of the structure and thicken on the flanks where they were deposited in an onlap pattern (Fig. 5). Postrift Pannonian sediments were later draped over the structure.

The thickness of individual reservoirs in the Algyö field vary from 5-70 m (Dank, 1988). The deepest hydrocarbon reservoirs occur in the upper few meters of the basement complex where porosity and permeability have been enhanced by fractures and fissures. Reservoirs in the basement complex contain relatively small amounts of hydrocarbons. Major production is from upper Pannonian sandstones associated with delta plain and delta front facies. Völgyi and others (1970) list 7 horizons of late Pannonian age that contain oil and natural gas reservoirs. The lower Pannonian section contains natural gas reservoirs in turbidite sandstone bodies and oil reservoirs in transgressive, basal formations.

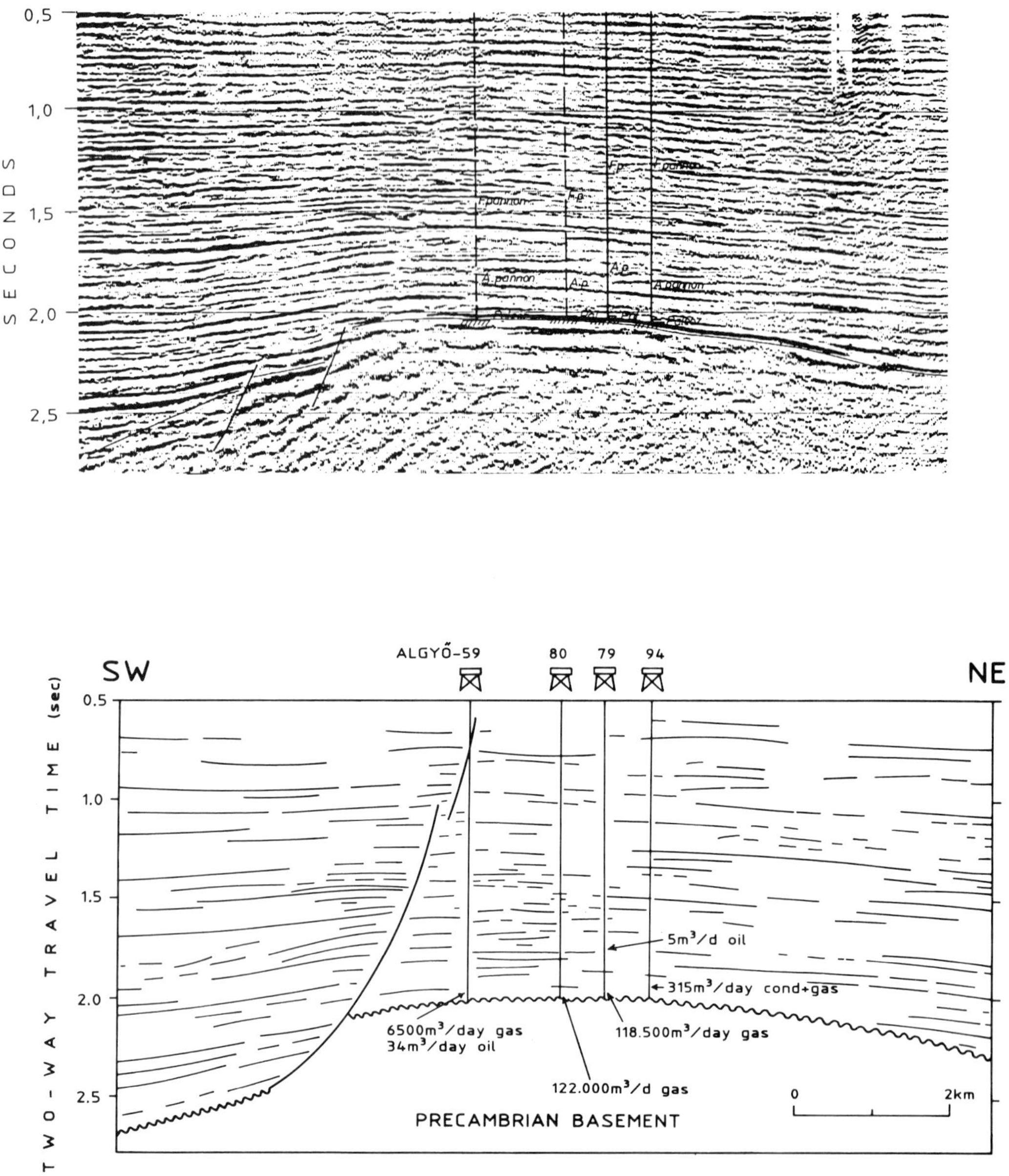

Figure 5. Algyö field: (top), seismic profile III (uninterpreted, migrated 24-fold section); (bottom), interpretation of profile III. Hydrocarbon pools are associated with turbidite and basal transgressive sands in the lower part of the Pannonian section and with weathered and fractured zones in the upper part of the metamorphic basement complex. The Algyö field, in relation to hydrocarbon reserves, is the largest in Hungary. Location of profile is shown in Figure 2.

The primary source rocks in the Algyö area are believed to be in the lower part of the Neogene section (I. Koncz, written communication, 1988; Clayton and others, this volume).

BASEMENT HIGHS ASSOCIATED WITH THRUSTS, FÖLDES AREA

In the Földes area (Fig. 2, profile IV), recent discoveries of oil and natural gas appear to be associated with a basement high that developed along a major thrust fault.

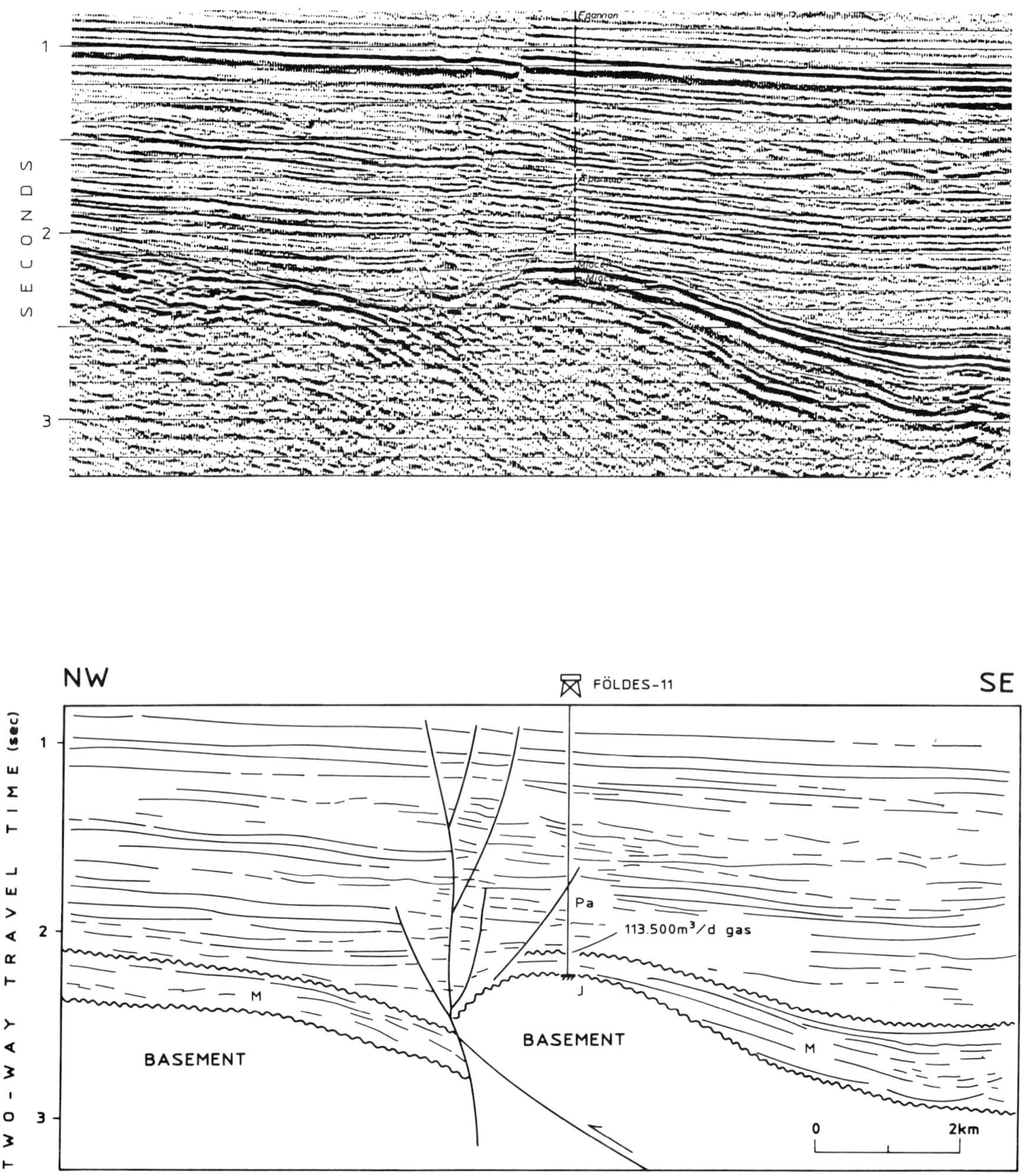

Figure 6. Földes area: (top), seismic profile IV (uninterpreted, migrated 24-fold section); (bottom), interpretation of profile IV. Thrust faulting in the basement occurred in Cretaceous or Miocene time. See text for explanation. The inferred strike-slip related flower structure is younger than the thrust fault. In the Földes-11 well, a natural gas pool was discovered in sandstone near the top of the Miocene section, as well as an oil pool in basal Miocene transgressive conglomerate. Location of profile is shown in Figure 2. M: Miocene; Pa: Pannonian; J: Jurassic.

Seismic profile IV crosses the Földes-11 well which bottomed in Jurassic rocks (Fig. 6). The position of the thrust fault was inferred based on a change in attitude of basement reflections. Additional evidence of thrusting in the area comes from the Sáránd-I well which penetrated Cretaceous carbonate rocks beneath a thick section of Precambrian metamorphic rocks several kilometers to the north (Bérczi-Makk, 1985; Haas, 1987; Pap, 1987).

As interpreted in Figure 6, Miocene strata as well as older strata are cut by the thrust fault. The fault, therefore, is inferred to have been active in Miocene time. However, on other seismic profiles from the same area, thrust faulting appears to cut only pre-Miocene strata. Therefore, what appears to be a fault in Miocene strata, in reality, may be a result of improper migration of the seismic data - that is to say, the fault may be a result of a "bow tie" effect related to the narrow syncline located 2 km northwest of the Földes-11 well at 2.3 sec.

The flower structure, which is interpreted to cut basement strata as well as some of the youngest sedimentary sequences (Fig. 6), is inferred to represent a zone of strike-slip movement. This zone appears to be a southward continuation of a strike-slip zone that was mapped about 5 km north in the Derecske basin (Fig. 3).

The Földes-11 well, drilled in 1987, produces 113,500 m^3/day of natural gas from a Pannonian sandstone reservoir immediately above the pre-Pannonian Miocene section and from a Miocene basal transgressive conglomerate and sand unit (Fig. 6). Exploration of the Földes area continues; preliminary estimates suggest that the structure could contain large recoverable reserves of oil and natural gas.

STRUCTURAL INVERSION, BUDAFA ANTICLINE

Positive structural inversion is a reversal of deformational processes; a depressed region is uplifted, usually by a combination of compression and wrenching (Lowell, 1985; Harding, 1985). This type of structure is represented by the Budafa anticline located in the western part of the Pannonian Basin (Fig. 2, profile V). The Budafa anticline is part of a larger EW-trending fold belt that may extend into Yugoslavia where surface expression of folds is referred to as the Sava folds (Dimitrijevic, 1982; Molnár and others, 1987; Rumpler and Horváth, 1988). Oil and natural gas is presently being produced from the Sava fold-belt in Yugoslavia.

The Budafa anticline was discovered in the early part of the 20th century (about 1919) using dip needle, torsion balance, seismic, and surface geologic mapping techniques (Dank, 1988). The first discovery well on the structure was drilled in 1937. Seismic profile V (Figs. 2 and 7) crosses the structure perpendicular to the fold axis.

Wells drilled on the crest and on the flanks of the Budafa anticline have penetrated Karpatian (middle Miocene) to Quaternary sedimentary rocks (Fig. 7). Core data from these wells indicate that the Karpatian section (M_{34}) consists of conglomerate and breccia (derived from Triassic carbonate rocks) that grade upward to limy marls and limestone and clayey marls and sandstone. The Badenian section (M_4, middle to upper Miocene) consists of marls and siltstones, sandy marls and clayey marls containing thin beds of andesite tuff. The Sarmatian section (M_5, upper Miocene) chiefly consists of marls and sandstone. The lower Pannonian (Pa_1) and upper Pannonian (Pa_2) sections are composed of marls, fine-grained turbidites and sandstones. The sandstones in the Pannonian section represent delta front and river-mouth-bar depositional environments.

Structurally, the Budafa feature is an asymmetrical anticline. Below the crest, subsurface horizons have more than 500 m (400 milliseconds) of relief (Fig. 7). Karpatian (M_{34}) and older strata are cut by a major thrust fault and a subsidiary reverse fault. The Badenian (M_4) through Pannonian (Pa) sequences are asymmetrically folded with the northwestern limb of the fold steeper than the southeastern limb.

Analysis of well and seismic data (Fig. 7) indicate that the Budafa anticline represents an inverted structure. On the southwestern flank of the structure, at the Letenye-I (Le-I) well, the Badenian (M_4) section is about 400 m (320 milliseconds) thick; whereas, below the crest of the structure, at the Budafa-V (B-V) well, the same age section is about 760 m (610 milliseconds) thick. In addition, the seismic data indicate that the Karpatian section thickens in the same direction. This lateral thickening of Badenian and pre-Badenian beds towards the axis of the anticline is interpreted as evidence that the area beneath the crest of the structure was a depression during Badenian and Karpatian time. The major phase of inversion caused by thrust faulting probably occurred subsequent to Badenian time and prior to Sarmatian time. This timing is evidenced by the onlap and thinning of Sarmatian (M_5) beds from the southwest flank towards the axis of the Budafa structure. By Sarmatian time, the structure had been transformed from a depression (perhaps a half graben) to an anticline. The marked doming

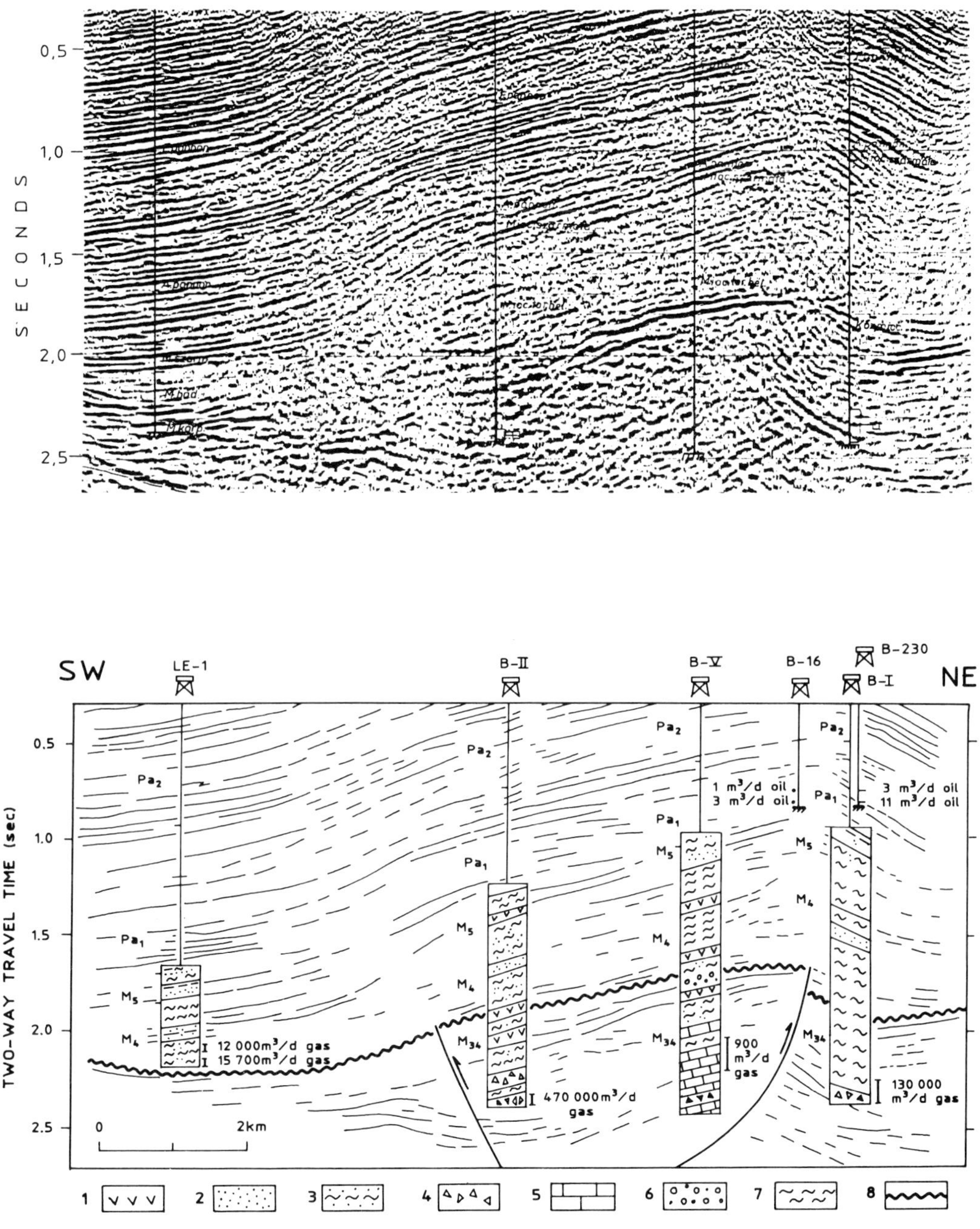

Figure 7. Budafa anticline: (top), seismic profile V (uninterpreted, migrated 24-fold section); (bottom), interpretation of profile V. Location of profile is shown in Figure 2. Positive structural inversion resulted in an anticline prolific in oil and natural gas. Eighty-one oil pools have been discovered in Pannonian prodelta turbidites and river-mouth-bar deposits. Low-permeability, Miocene carbonate breccias and conglomerates contain natural gas below a depth of 4000 m. (1): volcanic rocks; (2): sandstone; (3): marl and sandstone; (4): breccia; (5): limestone; (6): conglomerate; (7): marl; (8): unconformity; Pa_2: upper Pannonian; Pa_1: lower Pannonian; M_5: Sarmatian; M_4: Badenian; M_{34},: Karpatian.

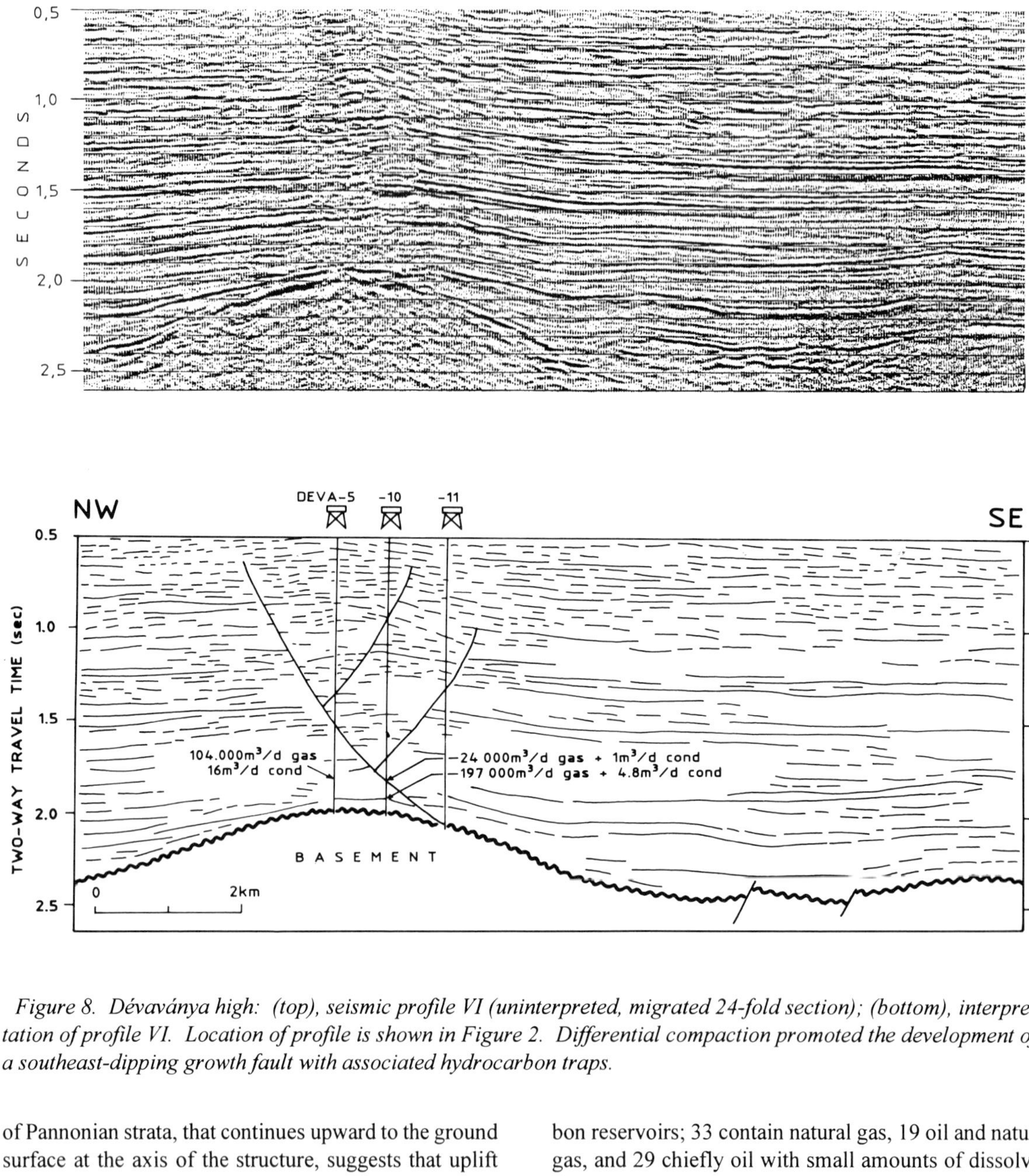

Figure 8. Dévaványa high: (top), seismic profile VI (uninterpreted, migrated 24-fold section); (bottom), interpretation of profile VI. Location of profile is shown in Figure 2. Differential compaction promoted the development of a southeast-dipping growth fault with associated hydrocarbon traps.

of Pannonian strata, that continues upward to the ground surface at the axis of the structure, suggests that uplift (compression) continued throughout much of Pannonian time. According to Horváth and Rumpler (1984), the Budafa anticline is a very young compressional structure that is probably related to convergent and/or discontinuous transcurrent faults.

Hydrocarbon accumulations in the Budafa anticline occur in a number of zones, chiefly in the lower Pannonian, Badenian and Karpatian sections (Fig. 7). The lower Pannonian section contains 81 known hydrocarbon reservoirs; 33 contain natural gas, 19 oil and natural gas, and 29 chiefly oil with small amounts of dissolved methane gas. The Pannonian reservoirs are associated mainly with turbidite bodies in the lower part of the lower Pannonian section and with river-mouth-bar sandstone bodies in the upper part of the lower Pannonian section. The oil from the lower Pannonian section is intermediary-paraffin with a density of 0.79-0.86 g/cm^3 at 20^0 C. Fifteen million metric tons of oil, 5 billion m^3 of natural gas, and 10 billion m^3 of CO_2 gas have been produced from the Pannonian section. In addition, significant amounts of natural gas have been produced from Badenian as well as Karpatian reservoirs.

Probable source rocks in the Budafa area are thick (80-120 m) beds of lower Pannonian bituminous marl.

GROWTH FAULTS, DÉVAVÁNYA AREA

In the Dévaványa area (Fig. 2, profile VI), growth faults and associated rollover structures play a role in the trapping of oil and natural gas. Seismic profile VI, which crosses the Dévaványa basement high, is shown in Figure 8. In this area, the basement complex is composed of Paleozoic metamorphic rocks. The sedimentary rocks that overlie the basement complex with marked unconformity are chiefly of Pannonian age and represent deposition in prodelta, delta slope and delta plain environments. Above the basement high, a prominent listric normal fault, that dips southeastward, cuts much of the sedimentary section. At the base of the sedimentary section, the dip of the fault appears to parallel the basement surface.

The fault above the Dévaványa high is thought to be associated with a paleo-slope. According to Mattick and others (1988), deltas began to prograde from the north to the south in early Pannonian time in this area. The basement high, which was oriented approximately perpendicular to the direction of progradation, initially acted as a dam to southward sedimentation. That is to say, coarse-grained sediments (delta plain and delta front) were deposited on the northeastern flank of the basement high while finer-grained sediments (delta slope and prodelta) were deposited on the southwestern flank. Later, as a result of differential compaction during sediment dewatering, the sedimentary section on the southwestern flank of the high compacted more than the sedimentary section on the northeastern flank. This mechanism promoted the development of major, southwestward-dipping, growth faults and antithetic faults above the basement high (Fig. 8). Similar growth faults that parallel basement highs have been described from other areas of the Pannonian Basin where the trends of basement highs are oriented normal to the direction of delta progradation, as in the Sarkadkeresztúr and Szarvas areas (Pogácsás, 1987; Mattick and others, 1988).

Some of the hydrocarbon traps discovered on the Dévaványa high appear to be related to rollover structures that developed in association with growth faulting (Fig. 8). Production of 24,000 m^3/day of natural gas and 1 m^3/day of condensate appears to be related to a rollover structure on the downthrown side of the fault, and production of 19,700 m^3/day of natural gas and 4.8 m^3/day of condensate appears to be related to a trap on the upthrown side of the fault. In addition to fault related traps in the Dévaványa area, hydrocarbons are produced from fracture zones in the basement complex, from turbidite and delta front sandstone bodies, and from the lower part of the basal transgressive Miocene sandstones. Recoverable hydrocarbon reserves discovered to date in the Dévaványa area are about one million metric tons of oil, including natural gas expressed in BTU equivalents of oil.

CONCLUSIONS

As a result of extensive oil and gas exploration in Hungary during the past 60 years, the largest hydrocarbon traps most likely have been discovered, especially those related to Neogene tectonics. An exception to this statement is that potential traps associated with Neogene strike-slip zones have not been explored adequately; however, little is known about the fractured and fissured traces of these zones in the pre-Tertiary basement complex.

The Paleogene basin (Fig. 2) and the Upper Cretaceous-Paleogene flysch zone (Fig. 2) of northeast Hungary are relatively unexplored in comparison to most other parts of the country. The Paleogene basin locally contains thick sedimentary rock sections and a better understanding of how tectonics was involved in the development of this basin could result in new discoveries of oil and gas. Similarly, the internal structure of the Upper Cretaceous-Paleogene flysch zone is poorly known.

Finally, the Mesozoic section throughout Hungary remains relatively untested and unexplored. Recent studies in northwest Hungary (unpublished) indicate that the basement complex contains many nappe structures, which were emplaced during pre-Neogene Eoalpine and Mesoalpine deformational episodes. Preliminary examination of seismic records indicates that nappe structures involving Mesozoic rocks may be more extensive in Hungary than previously thought. Intrabasement nappe structures of similar age in the Vienna Basin in Austria are associated with promising oil and gas prospects (Wessely, 1988).

REFERENCES

Albu, I., Bodoky, T., Jánváry. J., Pintér, A., Szeidovitz, Gy., and Varga, G., 1976, Geofizikai kutatás as Alföldön (Geophysical exploration on the Great Hungarian Plain); Eötvös Lóránd Geophys. Inst. 1976 Yearbook, 53-55.

Balla, Z., 1985, The Carpathian loop and the Pannonian Basin: A kinematic analysis; Eötvös Lóránd Geophys. Inst., Budapest, Geophys. Transactions, 30, (4), 313-354.

Balla, Z., 1987, Tertiary paleomagnetic data for the Carpatho-Pannonian region in the light of Miocene rotation kinematics; Tectonophysics, 139, 67-98.

Bérczi-Makk, A., 1985, Typical Mesozoic formations of the Great Hungarian Plain (in Hungarian); Általános Földtani Szemle, 21, 3-47.

Berkes, Z., Pogácsás, Gy., and Szanyi, B., 1983, Seismic stratigraphic interpretation of the Neogene sediments in the Derecske depression of Eastern Hungary; Proc. of the 28th Int. Geophys. Symposium, Balatonszemes, Hungary, 158-172.

Bodzay, T., 1971, Fluid geology of the Nagylengyel oil field (in Hungarian); Candidate thesis, Hungarian Academy of Sciences, Budapest, Hungary, 1-110.

Channell, J.E.T. and Horváth, F., 1976, The African/Adriatic promontory as a paleogeographical premise for alpine orogeny and plate movements in the Carpatho-Balkan region; Tectonophysics 35, 71-101.

Clayton, J.L., Koncz, I., King, J.D., and Tatár, E., Organic geochemistry of crude oils and source rocks, Békés basin; this volume, 161-185.

Dank, V., 1988, Petroleum geology of the Pannonian Basin, Hungary, an overview; *in* Royden, L.H. and Horváth, F. (eds.), The Pannonian Basin a study in basin evolution; Amer. Assoc. Petrol. Geol. Memoir 45, 319-332.

Dimitrijevic, M.D., 1982, Dinarides: an outline of the tectonics; Earth Evol. Sci., 2, 4-23.

Fuchs, K. and Schultz, K., 1976, Tunneling of low-frequency waves through the substructural lithosphere; Jour. Geophys., 42, 175-190.

Gajdos, I., Pap, S., Somfai, A., and Völgyi, L., 1983, Lithostratigraphic units of the Pannonian (s.l.) of the Great Hungarian Plain (in Hungarian); Hung. Geol. Survey Special Publ., Budapest, 1-70.

Haas, J., 1987, The main structural geological features of Hungary (in Hungarian); *in* Szederkényi, T. (ed.), Az Alföld medencealjzatának szerkezetfejlödése (Structural evolution of the basement of the Alföld); Hung. Acad. Sci., Szeged, Special Publ., 9-18.

Hajdú, D., Pap S., and Völgyi, L., 1982, New tectonic results of exploration for basement on the Great Hungarian Plain (in Hungarian); Földtani Kutatás, 25, 39-49.

Harding, T.P. and Lowell, J.D., 1979, Structural styles, their plate-tectonic habitats and hydrocarbon traps in petroleum provinces; Amer. Assoc. Petrol. Geol. Bull. 63, 1016-1058.

Harding, T.P., 1985, Seismic characteristics and identification of negative flower structures, positive flower structures and positive structural inversion; Amer. Assoc. Petrol. Geol. Bull. 69, 582-600.

Horváth, F., 1986, A Pannon medence kialakulásának geofizikai modellje (Evolution of the Pannonian Basin based on geophysical modeling); Candidate thesis, Hungarian Academy of Sciences, 1-148.

Horváth, F. and Royden, L., 1981, Mechanism for the formation of the intra-Carpathian basins - a review; Earth Evol. Sci., 1, 307-316.

Horváth, F. and Rumpler, J., 1984, The Pannonian basement - extension and subsidence of an alpine orogene; Acta Geol. Hung., 27, 229-235.

Horváth, F., Dövényi P., Pap, S., Szalay, A., and Hargitay, M., 1987, Tectonic control on migration and trapping of hydrocarbons in the Pannonian Basin; *in*, Doligez, B. (ed.), Migration of hydrocarbons in sedimentary basins; Editions Technip, Paris, 667-681.

Körössy, L., 1971, Geological and evolutionary studies of the Pannonian rocks of Hungary (in Hungarian), *in* Bartha, F. and others (eds.), A magyarországi pannonkori képzödmények kutatásai (Research on the Pannonian age deposits of Hungary), Akadémiai Kiadó, Budapest, 199-221.

Lowell, J.D., 1985, Structural styles in petroleum exploration; OGCI Publ., Tulsa, 1-458.

Mattick, R.E., Phillips, R.L., and Rumpler, J., 1988, Seismic stratigraphy and depositional framework of sedimentary rocks in the Pannonian Basin in southeastern Hungary; *in* Royden, L.H. and Horváth, F. (eds.), The Pannonian Basin - a study in basin evolution; Amer. Assoc. Petrol. Geol. Memoir 45, 153-170.

Molnár, K., Pogácsás, Gy., and Rumpler, J., 1987, Seismic reflection investigations in the Hungarian part of the Pannonian Basin: Application to hydrocarbon exploration; Ann. Inst. Geol. Pub. Hung., 70, 593-600.

Pap, S., 1987, Geological results from boreholes drilled deeper than 4.5 km in East Hungary (in Hungarian); Alföldi Tanulmányok, 11, 7-43.

Pogácsás, Gy., 1984, Results of seismic stratigraphy in Hungary; Acta Geol. Hung., 27, (1-2), 91-108.

Pogácsás, Gy., 1985, Seismic stratigraphic features of Neogene sediments in the Pannonian Basin; Eötvös Lóránd Geophys. Inst., Geophys. Trans., 30, 373-410.

Pogácsás, Gy., 1986, Hydrocarbon geology of neotectonic deformation based on seismic, paleomagnetic and radiometric data; Proc., 31st Int. Geophys. Symp., Gdansk, 1, 221-231.

Pogácsás, Gy., 1987, Limits of the application of seismic data for stratigraphic studies within the Pannonian Basin with special emphasis on Neogene rocks (in Hungarian); Öslénytani Viták (Paleontological Discussions), 34, 31-47.

Pogácsás, Gy. and Völgyi, L, 1982, Correlation of the East Hungarian Pannonian sedimentary facies on the basis of hydrocarbon prospecting and seismic and well-log sections; Proc., 27th Int. Geophys. Symp., Bratislava, A-I, 322-336.

Pozsgay, K., Albu, I., Petrovics, I. and Ráner, G., 1981, Character of the Earth's crust and upper mantle on the basis of seismic reflection measurements in Hungary; Earth Evol. Sci., 1, (3-4), 272-279.

Royden, L.H., 1988, Late Cenozoic tectonics of the Pannonian Basin system; *in* Royden, L.H. and Horváth, F. (eds.), The Pannonian Basin - a study in basin evolution; Amer. Assoc. Petrol. Geol. Memoir 45, 27-48.

Royden, L.H., Horváth, F. and Burchfiel, B.C., 1982, Transform faulting, extension and subduction in the Carpathian Pannonian region; Geol. Soc. Amer. Bull. 93, 717-725.

Rumpler, J. and Horváth, F., 1988, Some representative seismic reflection lines and structural interpretations from the Pannonian basin; *in* Royden, L.H. and Horváth, F. (eds.), The Pannonian Basin - a study in basin evolution; Amer Assoc. Petrol. Geol. Memoir 45, 153-170.

Samu, L., 1985, Tectonics of the Derecske basin based on the interpretation of seismic profiles (in Hungarian); Magyar Geofizika, 26, (5-6), 182-199.

Varga, I. and Pogácsás, Gy., 1981, Reflection seismic investigations in the Hungarian part of the Pannonian Basin; Earth Evol. Sci., 1, (3-4), 232-239.

Völgyi, L., Suba, S., Balla, K., and Csalagovics, I. 1970, Hydrocarbon fields of Hungary; Müszaki Kiadó, Budapest, 1-423.

Wessely, G., 1988, Structure and development of the Vienna Basin; *in* Royden, L.H. and Horváth, F. (eds.), The Pannonian Basin - a study in basin evolution; Amer. Assoc. Petrol. Geol. Memoir 45, 333-346.

12 History of Oil and Natural Gas Production in the Békés Basin

András Kovács[1] *and Paul G. Teleki*[2]

ABSTRACT

Exploration for oil and natural gas in Hungary began in 1915 when the Eötvös torsion balance was used to measure the earth's gravitational field. Since that time, as a result of extensive exploration, 17 fields associated with basement highs on the periphery of the Békés basin have been discovered. In this trend, oil and gas is produced from compaction anticlines associated with basement highs. Weathered and fractured basement rocks and basal sedimentary rocks usually form a common reservoir, which is 38 m thick in the Battonya field, and more than 400 m thick in the Sarkadkeresztúr field. Higher in the section, natural gas is produced from lower Pannonian sandstone reservoirs that reach thicknesses of 56 m in the Pusztaföldvár field.

Source rocks are thought to be Miocene shales and Pannonian marls and clay marls deposited in the adjacent troughs.

Many of the fields are nearly depleted. Discovered recoverable hydrocarbons for the Békés basin as of 1985 were 3.215×10^6 metric tons of oil and 25.397×10^9 m^3 of natural gas. By the end of 1985, 63.40% of the gas and 83.01% of the discovered oil had been produced. Limited exploration in deep (central) parts of the basin was not encouraging. The best prospects for future exploration are probably stratigraphic traps on the flanks of the known fields at depths of 3000-4000 m.

INTRODUCTION

This chapter reviews the strategy of past oil and gas exploration in the Békés basin area and its production history, in order to provide some insight into future oil and gas prospects. To this end, the petroleum geology of three characteristic fields (Battonya, Pusztaföldvár and Sarkadkeresztúr) are discussed in detail.

The Hungarian part of the Békés basin (3930 km^2) is bounded on the east by the border between Hungary and Romania (Fig. 1). This political boundary cuts across the deepest part of the basin. In Hungary, the basin is bounded by the Battonya, Pusztaföldvár, Szarvas, Endröd, Szeghalom and Sarkadkeresztúr basement highs (see Phillips and others, this volume, Fig. 2). All known oil and natural gas fields are associated with these peripheral basement highs (Figs. 2 and 3). The thickness of the

[1] MOL Rt. (Hungarian Oil and Gas Co., Ltd.), H-5000, Szolnok, Hungary
[2] U. S. Geological Survey, Reston, Virginia, 22092, USA

P. G. Teleki et al. (eds.), Basin Analysis in Petroleum Exploration, 237–256.

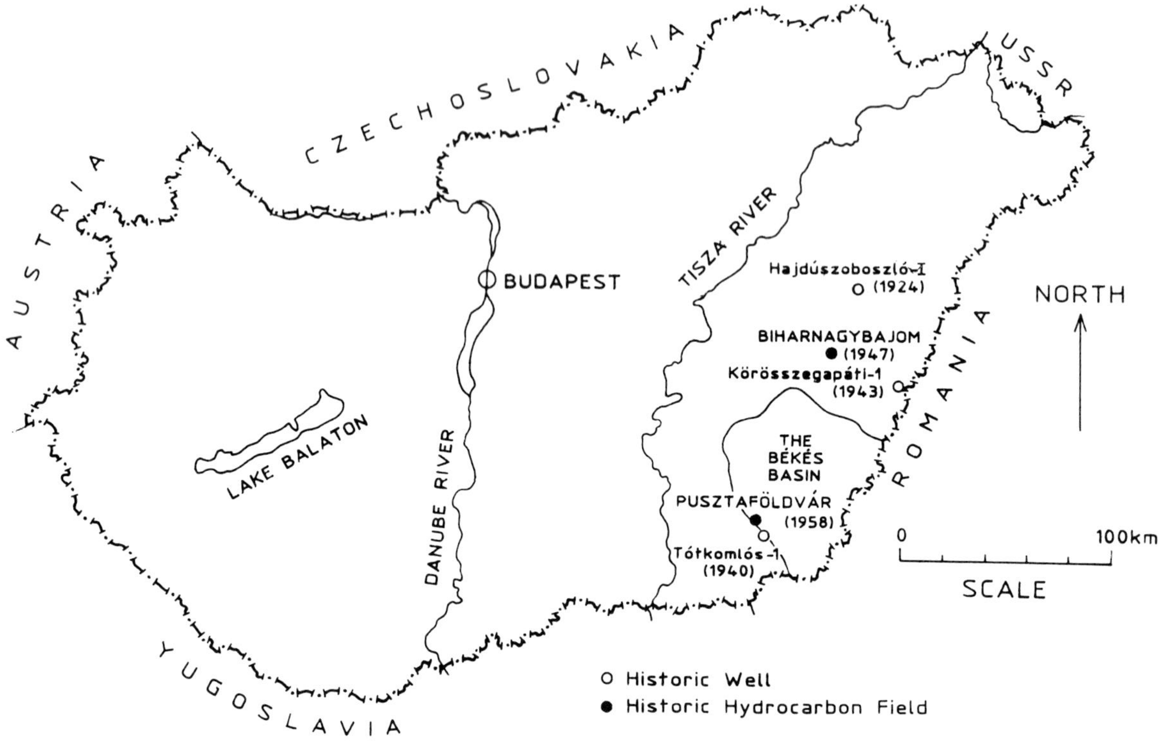

Figure 1. Map showing location of Békés basin in Hungary and some historic exploration localities. The year of discovery is shown in parenthesis.

Neogene and younger sedimentary deposits in the basin varies from 1000-2600 m along its periphery to as thick as 6500 m in its center (Fig. 2). Neogene and younger sediments thin to 1000 m over the Battonya-Pusztaföldvár basement high, to 2200 m over the Dévaványa high, 2500 m over the Szarvas high, and 2600 m over the Endröd and Sarkadkeresztúr highs. The average surface elevation of the basin is 85-90 m above sea level and the ground surface has only minor topographic relief.

Reservoir rocks range in age from Precambrian to late Pliocene in the basin. From several structures, oil and natural gas are produced from fractured basement composed of igneous and metamorphic rocks that range in age from Precambrian to Triassic. On some structural highs, hydrocarbons are produced from reservoirs in basal Miocene rocks commonly composed of breccia and/or conglomerates. These pre-Pannonian Miocene rocks, usually fractured and brecciated, rest unconformably on basement rocks and are thin to absent on the shallowest paleomorphologic highs such as the Battonya-Pusztaföldvár high (Szentgyörgyi and Teleki, this volume). The pre-Pannonian Miocene and basement rocks usually form a common oil and gas reservoir (Spencer and others, this volume). In many fields, oil and natural gas are produced from reservoirs composed of lower Pannonian conglomerate, calcareous marl, and sandstone. Small amounts of natural gas are produced from upper Pannonian lens-shaped sandstone bodies. Some of this gas is biogenic in origin (Clayton and Koncz, this volume).

Geochemical analyses indicate that pre-Pannonian (Badenian and Sarmatian) and lower Pannonian shales and marls in the deeper parts of the basin are the probable source of much of the natural gas and oil produced in the Békés basin (Szalay, 1988; Clayton and Koncz, this volume; Clayton and others, this volume). These rocks contain predominantly type III (gas-prone) kerogen and it is not surprising, therefore, that chiefly natural gas is produced from the basin, with only secondary amounts of oil (Clayton and Koncz, this volume). The bulk of the

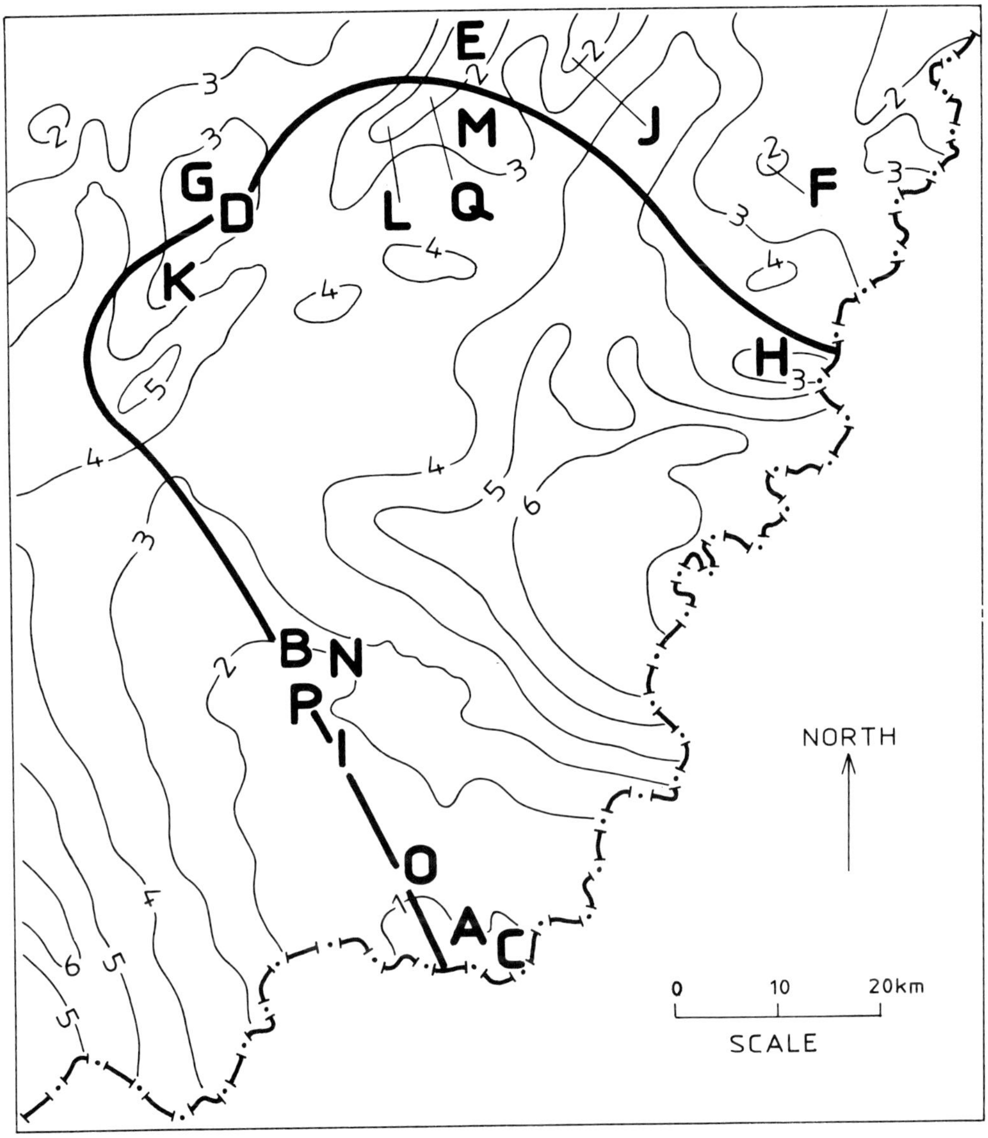

Figure 2. Isopach map of the Neogene sedimentary fill in the Békés basin. Contours in kilometers. The outline of the basin is shown by a heavy line. Letters show locations of hydrocarbon fields: A: Battonya; B: Pusztaföldvár; C: Battonya-East; D: Endröd-I; E: Füzesgyarmat; F: Komádi; G: Endröd-III; H: Sarkadkeresztúr; I: Kaszaper-South; J: Szeghalom; K: Szarvas; L: Dévaványa; M: Körösladány; N: Csanádapáca; O: Mezöhegyes; P: Pusztaszöllös; Q: Dévaványa-North.

oil was generated at present-day burial depths of about 3000 m or less; whereas, much of the natural gas was generated at somewhat deeper depths (3000-4000 m) (Clayton and Koncz, this volume; Clayton and others, this volume). The source rocks passed into the oil generation window at about 8 Ma, and in the deep troughs, fluid migration was downward and/or lateral due to overpressuring in middle Miocene and lower Pannonian shales (Szalay, 1988; Spencer and others, this volume). Upward migration appears to have occurred in areas of substantial structural relief, predominantly in areas of maximum basement relief around the margins of the basin. These observations have led to speculations that future exploration, especially for natural gas, should be

Figure 3. Index map of the Békés basin showing location of significant hydrocarbon fields and wells. Many wells inside field outlines are not shown.

targeted at stratigraphic traps on the downdip flanks of current producing structures, at depths of less than about 4000 m (Dank, 1988; Clayton and Koncz, this volume; Spencer and others, this volume).

EXPLORATION HISTORY

The exploration history of the Békés basin and the immediate surrounding areas is summarized from Kertai (1960), Papp (1963), Alliquander (1968), Csiky (1980), Dank (1985), Csontos and others (1987), and unpublished reports of the former Hungarian Central Office of Geology.

The first geophysical exploration work was undertaken by the Royal Hungarian Geological Institute in 1915, when an Eötvös torsion balance was used to measure the earth's gravitational field in western Hungary. The computed gravity anomalies suggested that large structures existed buried beneath Cenozoic sediments. This exploration technique was subsequently used in the Békés basin area in the 1920's.

The first partly successful wildcat well (Hajdúszoboszló-I, Fig. 1) was drilled in 1924-25, near the town of Hajdúszoboszló, about 60 km north of the Békés basin (Fig. 1). The well produced hot (73^0 C), iodine-enriched water at a rate of 1600 liters/minute and methane gas at a rate of 3700 m^3/day from a depth of 1090 m. The iodine-rich waters eventually turned Hajdúszoboszló into a health spa and the methane gas was used locally to generate electricity. Exploration for oil and natural gas in western Hungary, using gravity methods during the 1920's and 1930's, continued, but without success, except for occasional discoveries of small amounts (1000-2000 m^3/day) of methane gas and "medicinal" waters.

These unsuccessful exploration efforts gave way to joint ventures. The MANÁT (Hungarian-German Oil Co.) was formed in the late 1930's and received exploration rights in the Békés basin and surrounding areas. In 1940, this company reentered and deepened the Tótkomlós-1 well located on the site of a gravity high (Fig. 1). While deepening the well, drilling mud was lost and a blowout occurred at a depth of 1619 m. Methane and CO_2 gases, oil, and water flowed to the surface. The volume of gas measured during this blowout was 88,000 m^3/day, of which 51.4% was CO_2. The well could not be controlled because of casing problems, and the buildup of excess pressure caused the formation to fracture. The well-head equipment sank. The blowout eventually extinguished itself, leaving a water-filled hole which is now a lake. A second well (Tótkomlós-2) was drilled nearby; this well was also dry.

In 1943, the site of a gravity high at Körösszegapáti was drilled by MANÁT. The first well (Körösszegapáti-1, Fig. 1) was not completed, because of high pressure in a fractured zone, casing problems, and the presence of large volumes of CO_2 gas.

MASZOVOL (a Hungarian-Soviet enterprise) discovered the first oil field in the vicinity of the Békés basin, at Biharnagybajom, in 1947 (Fig. 1) on the site of a gravity high. This field produced 40,000 metric tons of oil and 55 million m^3 of natural gas before the field was depleted and abandoned in 1967.

In the early 1950's, after the Hungarian-Soviet MASZOLAJ Company was founded, seismic reflection and refraction methods began to replace gravity methods as the principal exploration tool. This proved to be a turning point in the exploration history of the Békés basin area, because gravity methods are not particularly amenable to oil and gas exploration in the basin. As discussed by Grow and others (this volume), the central part of the Békés basin, contrary to prevailing wisdom during early exploration, is marked by a large gravity high and some of the surrounding basement highs are marked by gravity lows. Drilling activity, however, was minimal until 1957.

In 1957, the Hungarian Oil and Gas Trust (OKGT) was established. This company completed the Tótkomlós-9 as a producing well in that year.

By 1958, the earlier investment in detailed seismic exploration began to pay off. The Békés basin's most important oil and natural gas field was discovered early that year. This discovery, at Pusztaföldvár (Fig. 1), was followed by other discoveries in the same year; one being the largest natural gas field in Hungary at Hajdúszoboszló (Fig. 1), the other the oil and natural gas pools of the Battonya field (Fig. 2). In the following year (1959), the Pusztaszöllös and Mezöhegyes fields were discovered, then, in 1961, the Szarvas field (Fig. 2). In the late 1960's, and during the following decades, discoveries were made in relatively rapid succession: Csanádapáca in 1967, Battonya-East in 1970, Endröd-I in 1971, and Füzesgyarmat and Komádi in 1974 (Fig. 2). Between 1976 and 1980, discoveries were made at Endröd-III, Sarkadkeresztúr and Kaszaper-South (Fig. 2). Since 1981, the Szeghalom, Dévaványa, Dévaványa-North and Körösladány fields have been discovered (Fig. 2).

Prior to the mid-1980's, exploration in the Békés basin area was concentrated on basement highs that fringe the basin. Starting in 1972, and continuing through the mid-1980's, 12 wells (Békés-1,2,3,4,5, Doboz-I, Örménykút (Örm)-I, Hunya-1, Köröstarcsa (Köt)-I, Köt-1, Kondoros (Kond)-1, and Gyoma-1) were drilled to explore for deep hydrocarbons reservoirs in the central (deep) part of the basin (Fig. 3). In general the deep drilling was unsuccessful. In the Doboz-I and Örm-I wells, CO_2 gas flowed from Mesozoic carbonate rocks. The deepest well, Békés-2, was drilled to a total depth of 5500 m. This well penetrated 2161 m of Mesozoic rocks but no hydrocarbons shows were reported. For the most part, only methane gas shows mixed with CO_2 and commonly accompanied by salt water were encountered at depths greater than 3000 m. Overpressuring is reported throughout most of the basin in reservoirs deeper than about 2500 m (Spencer and others, this volume).

OIL AND GAS FIELDS

Oil and natural gas is presently being produced from 17 fields in the Békés basin and the immediate surrounding area (Figs. 2 and 3). All of these fields are associated with basement highs located along the periphery of the basin. With the exception of the Sarkadkeresztúr field, the geological characteristics of these fields are somewhat similar. The Battonya and Pusztaföldvár fields, the oldest in the area (discovered in 1958), are considered to be characteristic of the fields in the basin. These two fields are described in detail in later sections of this chapter, as is the Sarkadkeresztúr field, presently the largest natural gas producing field in the basin.

The characteristics of fields outside of the Battonya-Pusztaföldvár trend differ somewhat from those within the trend. The most significant difference between fields in this trend and other fields in the Békés basin is reservoir pressure. reservoir in the Battonya-Pusztaföldvár trend are not significantly overpressured; whereas, in fields outside the trend, basal turbidites, pre-Pannonian Miocene rocks, and basement rocks are significantly overpressured. The Szarvas field differs from fields in the Battonya-Pusztaföldvár trend in that some of its turbidite reservoir contain mostly CO_2 gas, and pro-delta sandstones that overlie the turbidites are slightly overpressured and contain hydrocarbon gas mixed with CO_2. In fields such as Szarvas and Dévaványa, listric faults associated with differential sediment compaction were probably important in providing the migration path for natural gas into the traps. Whereas oil and natural gas occur throughout much of the geologic section in the Battonya-Pusztaföldvár trend, oil and natural gas pools in the Dévaványa field are limited mostly to the lowest part of the sedimentary section (pre-Pannonian Miocene section and basal Pannonian turbidite pinchouts). In comparison to the Battonya-Pusztaföldvár trend, large accumulations of hydrocarbons occur in fractured basement rocks in the Szeghalom, Sarkadkeresztúr, and Dévaványa fields.

Battonya

Drilling on the Battonya basement high was preceded by gravity surveys in the 1940's and by seismic surveys in 1951. In 1959, the Battonya (Bat)-1 well (Fig. 3) discovered the multi-pool Battonya field. The ultimate recoverable reserves were estimated to be 144×10^3 metric tons of oil and 3.1×10^9 m^3 of natural gas.

In the Battonya field, hydrocarbon production is chiefly from two intervals referred to as the Battonya Lower and Battonya Upper Zones (Fig. 4). Minor amounts of natural gas are also produced from the Upper Pliocene Zone (Fig. 4). The most significant production of hydrocarbons is from the Battonya Lower Zone at depths between 880 and 930 m (Figs. 5 and 6). This interval contains a large natural gas accumulation above a thin oil column. The reservoir rocks in the Battonya Lower Zone, in ascending order, are: (1) weathered and fractured Paleozoic granites and quartz-porphyries, (2) the transgressive Békés Conglomerate, and (3) calcareous marl beds of early Pannonian age (Fig. 4). The three rock units comprise a common, fractured reservoir system about 38 m thick. The main producing mechanism is artificial water drive. The Békés Conglomerate and calcareous marl beds pinch out updip toward the west (Figs. 5 and 6). Basement rocks contain about 6% of the hydrocarbons and their porosity is also 6%. Conglomerates contain about 73% of the hydrocarbons and have an average porosity of 19%. Calcareous marl, with an average porosity of 27%, contains about 21% of the hydrocarbons. The top seal is a consolidated clayey marl about 250 m thick.

Above the Lower Zone, hydrocarbon gas is found in 5 relatively thin reservoir beds (2-11 m thick, Table 1) in the Battonya Upper Zone between depths of 403-504 m below sea level (Fig. 5). These reservoir beds, in the upper Pannonian section, are composed of alternating sandstones and marls, associated with delta plain facies of the basin's sedimentary fill (Molenaar and others, this volume). The 5 reservoirs, in descending order, are referred to as the Upper-1 through Upper-5 reservoirs (Fig. 5). The reservoir beds mostly consist of lens-

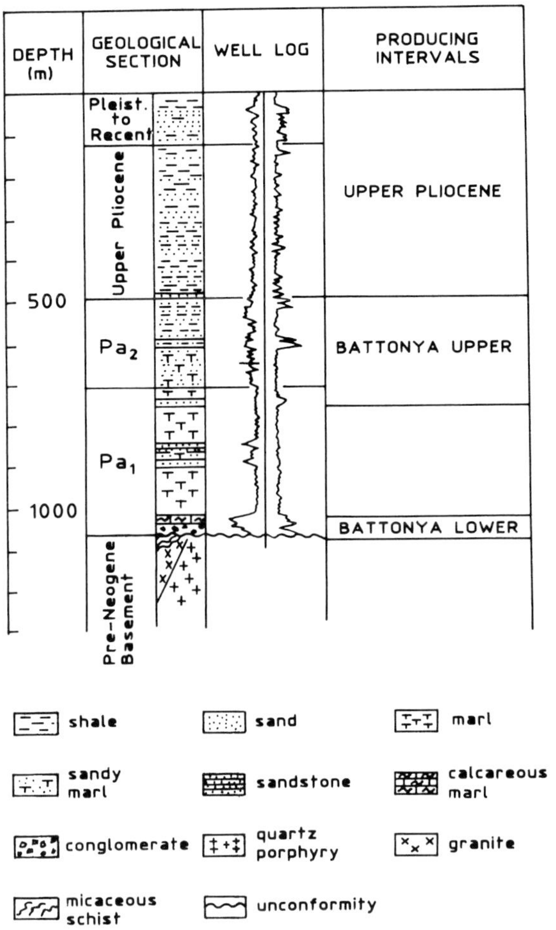

Figure 4. Typical stratigraphic column at the Battonya field area supplemented by lithologic and well-log information and names of producing intervals. Pa1: lower Pannonian; Pa2: upper Pannonian. Datum is ground surface.

shaped bodies of clayey sandstones of lacustrine-alluvial origin (Table 1). The average porosity of the sandstones is 20% (Table 1). The lithology of the reservoir beds is not uniform, some contain more clay than others. These reservoirs are stacked above the eastern part of the Battonya high (Fig. 7). Except for the Battonya Upper-2 and Upper-5 shut-in pools, these reservoirs have been exhausted and abandoned.

The largest hydrocarbon gas reservoir is the Battonya Upper-6, located on the north-northwestern part of the Battonya high (Fig. 7). The reservoir beds, at the lower-upper Pannonian boundary, consist of delta-plain clayey sandstone deposits that pinch out southward (Fig. 7). The highly porous (30%) sandstones are covered by a thin marl unit that seals the top of the 25-m-thick reservoir.

The physical attributes, as well as the cumulative hydrocarbon gas and oil production, for all of the reservoirs in the Battonya field are summarized in Table 1. As seen in this table, 100% of the oil and about 50% of the hydrocarbon gas produced from the Battonya field has been recovered from the Battonya Lower Zone. The cumulative production of hydrocarbon gas and oil for this zone are shown graphically in Figures 8 and 9, respectively. These graphs show that the yearly production of oil and gas declined steadily since the early 1970's. In the early 1970's, gas production averaged about 52×10^6 m^3/year; however, in the period 1980-1985, gas production decreased to about 40×10^6 m^3/year. The decline in oil production was much more dramatic. From yearly production rates of about 9×10^3 metric tons in the early 1970's, the average production rate for oil declined to less than 0.5×10^3 metric tons/year for the period 1982-85.

Pusztaföldvár

In the 1950's, following exploration of the nearby Tótkomlós area, detailed seismic surveys were conducted in the Pusztaföldvár area (Figs. 2 and 3). In 1958, the Pusztaföldvár (Pf)-1 well was drilled on a buried anticline (Fig. 10). This well penetrated medium-size, stacked oil and natural gas pools. Exploration of the central part of the Pusztaföldvár area was completed by 1963. By 1970, step-out wells toward the southwest were drilled (Fig. 10) and the ultimate recoverable reserves were estimated to be about 2.0×10^6 metric tons of oil and 12.3×10^9 m^3 of natural gas. The total area of production is 28 km^2 at depths between 1600-1680 m below sea level (Fig. 10). According to Dank (1988), the Pusztaföldvár field contributed about 2.5% of the total hydrocarbon production of Hungary in the early to mid 1980's.

The basement rocks in the Pusztaföldvár area consist largely of Paleozoic metamorphic rocks (mica schist), but Mesozoic dolomites and metamorphosed limestones and marls are also present (Fig. 11) in the southwestern part of the area explored by step-out wells. Above the basement, thin Badenian beds, spottily distributed on this high, are composed of conglomerates and breccias of continental origin in a red-clay matrix. Lower Pannonian marine beds unconformably overlie the basement complex. The lowest unit is an 80-m-thick, coarse-grained, transgressive conglomerate, known as the Békés Conglomerate (Phillips and others, this volume), that is over-

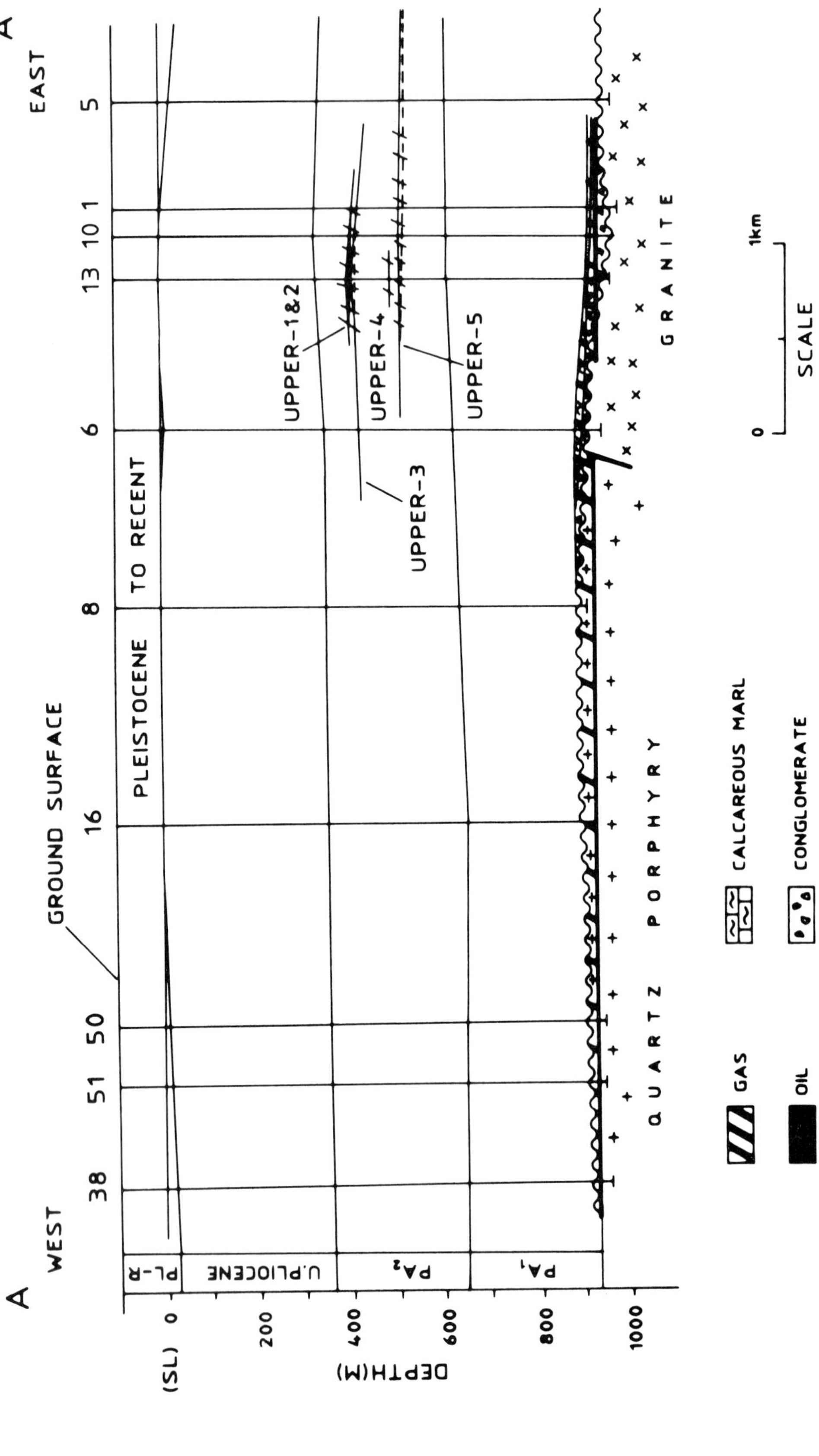

Figure 5. Geologic section across the Battonya structure. Location of section is shown in Figure 6. Depths are referenced to sea level. Numbers on top of section refer to wells shown in Figure 6. Pl-R: Pleistocene-Recent; Pa2: upper Pannonian; Pa1: lower Pannonian; SL: sea level. Age of basement rocks is thought to be Paleozoic.

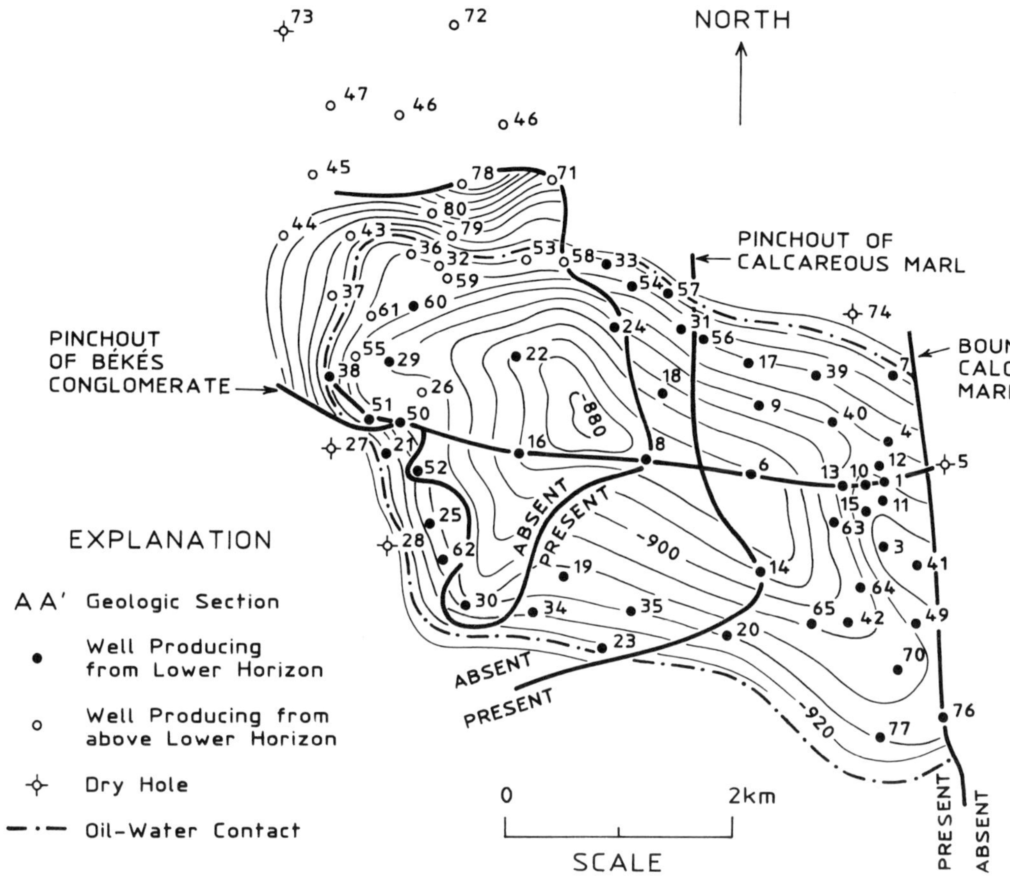

Figure 6. Map showing structure on top of the Battonya Lower Zone in part of the Battonya field and location of geologic section shown in Figure 5. Contour interval is 5 m. Datum is sea level.

lain by marl (Fig. 11). The calcareous content of the marl is variable. The marl is interbedded with sandstone beds of varying thicknesses. The interbedded marls and sandstones are interpreted to represent delta-slope and prodelta depositional environments (Phillips and others, this volume). The sandstone units comprise 15-20% of the total lower Pannonian sequence. Alternating clays, clayey sands, and sandstones comprise the upper Pannonian delta-plain alluvial and lacustrine facies (Fig. 11). These facies are predominantly sandy in the 650-700 m depth interval.

In the Pusztaföldvár field, hydrocarbons occur in 4 zones referred to, in ascending order, as the Békés, Lower Földvár, Upper Földvár, and Puszta Zones (Figs. 11 and 12). These are described below.

Békés Zone

This interval contains the field's largest and deepest (1600-1680 m below sea level) pool (Fig. 11). It consists of a gas cap of hydrocarbon gas and CO_2 above an oil column. The reservoir rocks, which directly overlie basement, are lower Pannonian basal, marginal-marine, transgressive deposits (Békés Conglomerate). The Békés Conglomerate is widespread along the Battonya-Pusztaföldvár trend and forms productive reservoirs in the Pusztaszöllös, Tótkomlós, Csanádapáca, Mezöhegyes, and Battonya fields. The reservoir in the Puszta-földvár field is sealed above by calcareous marl and clayey marl.

Approximately 83% of the oil and 38% of the gas produced from the Pusztaföldvár field has come from the

Table 1. Summary of reservoirs (in descending order) of the Battonya field and their physical attributes. T: thickness of reservoir; Pa_1: lower Pannonian; Pa_2: upper Pannonian; Pz: Paleozoic; ss: sandstone; cgl: conglomerate; c/ss: clayey sandstone; Por: porosity; H_2O Sat: water saturation; Perm: permeability; u: unknown; P: pressure; mPa: megaPascals; Temp: temperature; CPO: cumulative production of oil through 1985; MT: metric tons; CPG: cumulative production of natural gas through 1985 (associated and nonassociated); PS: present status; PR: producing; SI: shut in; A: abandoned. Depths shown in column 3 are referenced to sea level.

Reservoir Name	**Area** (km^2)	**Depth** (m)	**T**(m)	**Age**	**Rock**	**Por** (%)	**H_2O Sat** (%)	**Perm** (mD)	**P** (mPa)	**Temp** (°C)	**CPO** (10^3 MT)	**CPG** (10^6 m^3)	**PS**
Upper-1	1.4	403	11.0	Pa_2	ss&cgl	20	30	u	4.45	40	0	23	A
Upper-2	0.3	408	8.5	Pa_2	c/ss	20	30	u	5.40	40	0	0	SI
Upper-3	0.2	475	2.0	Pa_2	c/ss	20	30	u	5.40	40	0	0.9	A
Upper-4	1.5	488	6.0	Pa_2	c/ss	20	30	u	5.78	48	0	15	A
Upper-5	0.6	504	4.5	Pa_2	c/ss	20	30	u	5.40	50	0	0	SI
Upper-6	8.7	645	25	Pa_2	c/ss	30	30	u	7.26	59	0	1078	PR
Lower	16.7	927	38	Pa_1&Pz	ss&cong	20	50	u	10.59	75	174	974	PR

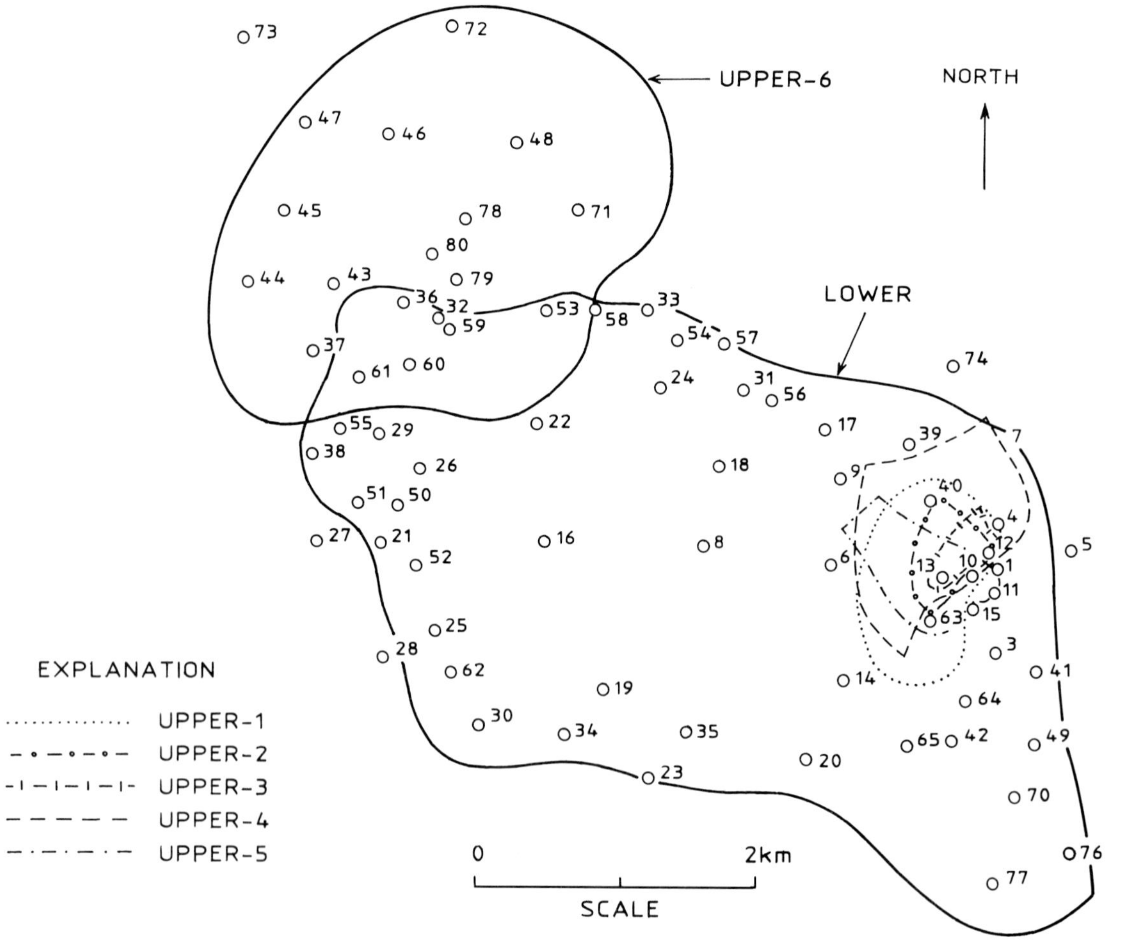

Figure 7. Map showing pool boundaries in the Battonya field.

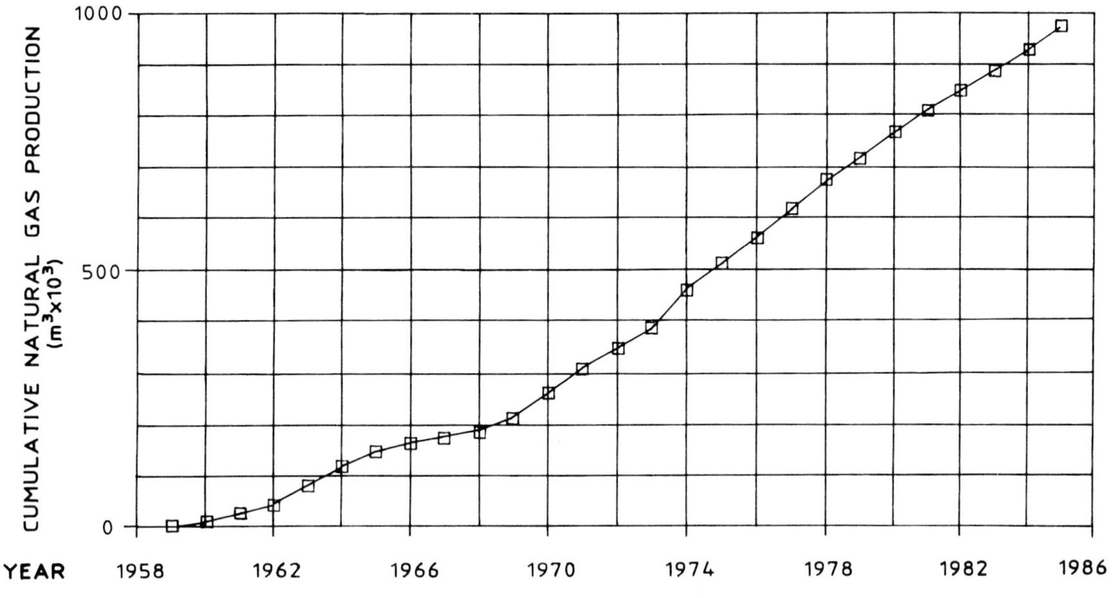

Figure 8. Cumulative natural gas production from the Battonya Lower Zone in the Battonya field.

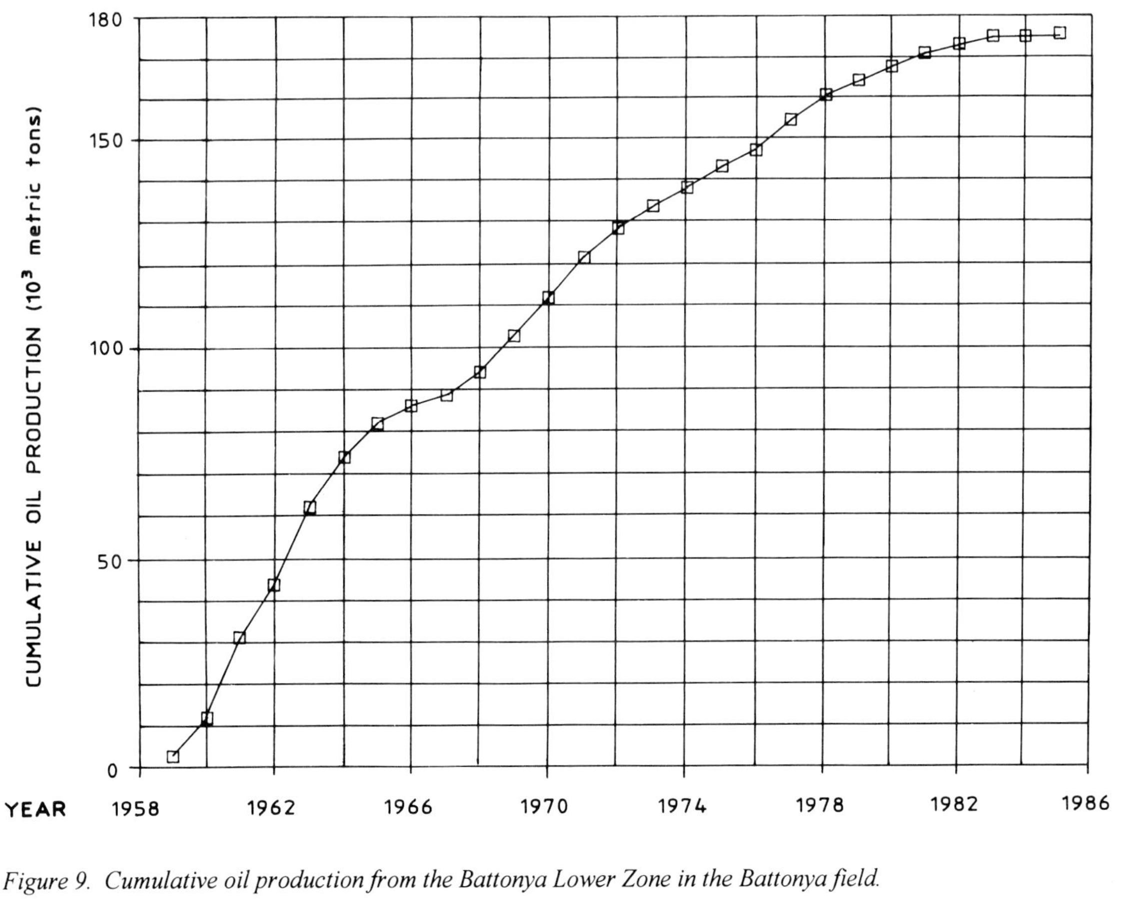

Figure 9. Cumulative oil production from the Battonya Lower Zone in the Battonya field.

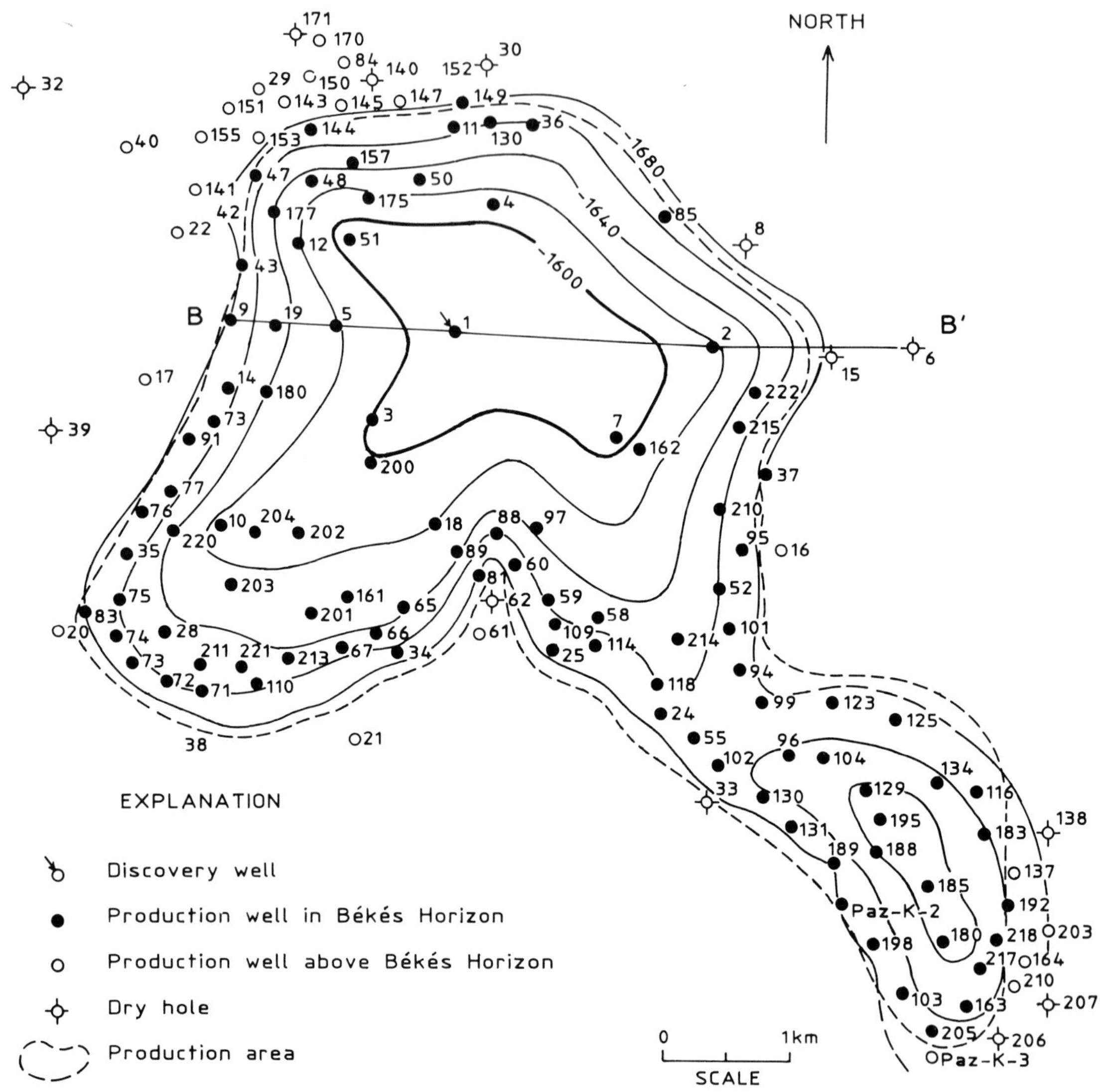

Figure 10. Structure map on the top of the Békés Zone, Pusztaföldvár field. Contour interval is 20 m. Datum is sea level.

Békés Zone (Table 2). The cumulative production of natural gas and oil from this zone are shown graphically in Figures 13 and 14, respectively. These graphs illustrate that oil and gas production has been steadily declining since the early 1970's. In the early 1970's, oil production averaged about 71×10^3 metric tons/year; however, the average production rate was about 26×10^3 metric tons/year in the 1982-85 period. During the same time period, the average gas production declined from about 185×10^6 m^3/year to about 77×10^6 m^3/year.

Lower Földvár Zone

This zone contains 3 undersaturated oil pools and one natural gas pool located on the western flank of the Battonya high north of section BB' in the vicinity of the Pusztaföldvár (Pf)-153 well (Fig. 10). The reservoir beds range in thickness from 23.5-90.0 m (average thickness 58 m) and consist of lower Pannonian sandstones (Table 2). The reservoir units are usually sealed above and at their updip termination by clayey marl. Along section BB', lower Pannonian sandstones (labeled L in

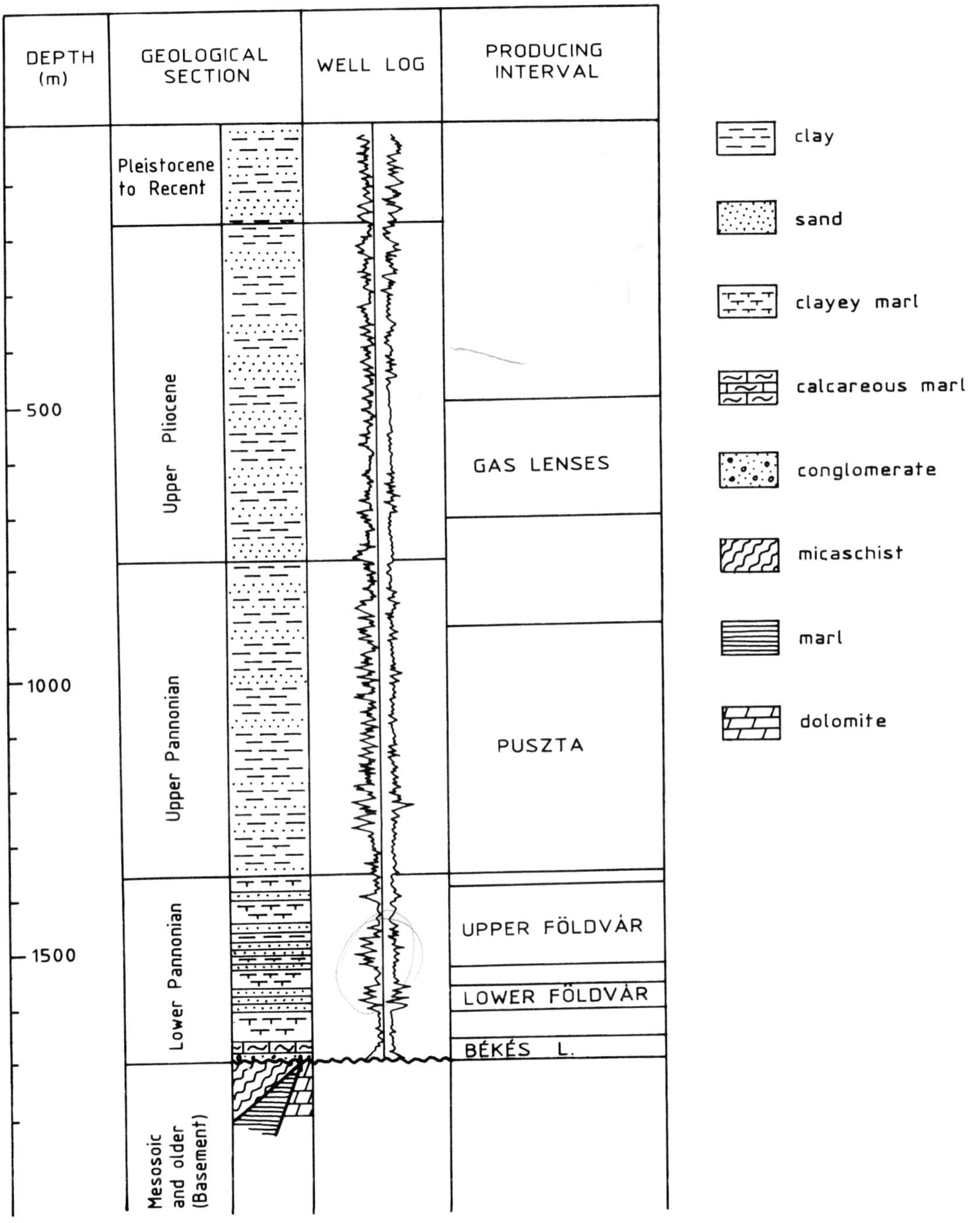

Figure 11. Typical stratigraphic column at the Pusztaföldvár field supplemented by lithologic and well-log information and names of producing intervals. "Gas lens" refers to upper Pannonian sands that have been charged with natural gas that escaped from deeper reservoirs during completion of the Pusztaföldvár (Pf)-50 well. Locally, thin Miocene(?) continental conglomerate and breccia deposits (not shown) directly overlie the basement complex. Datum is sea level.

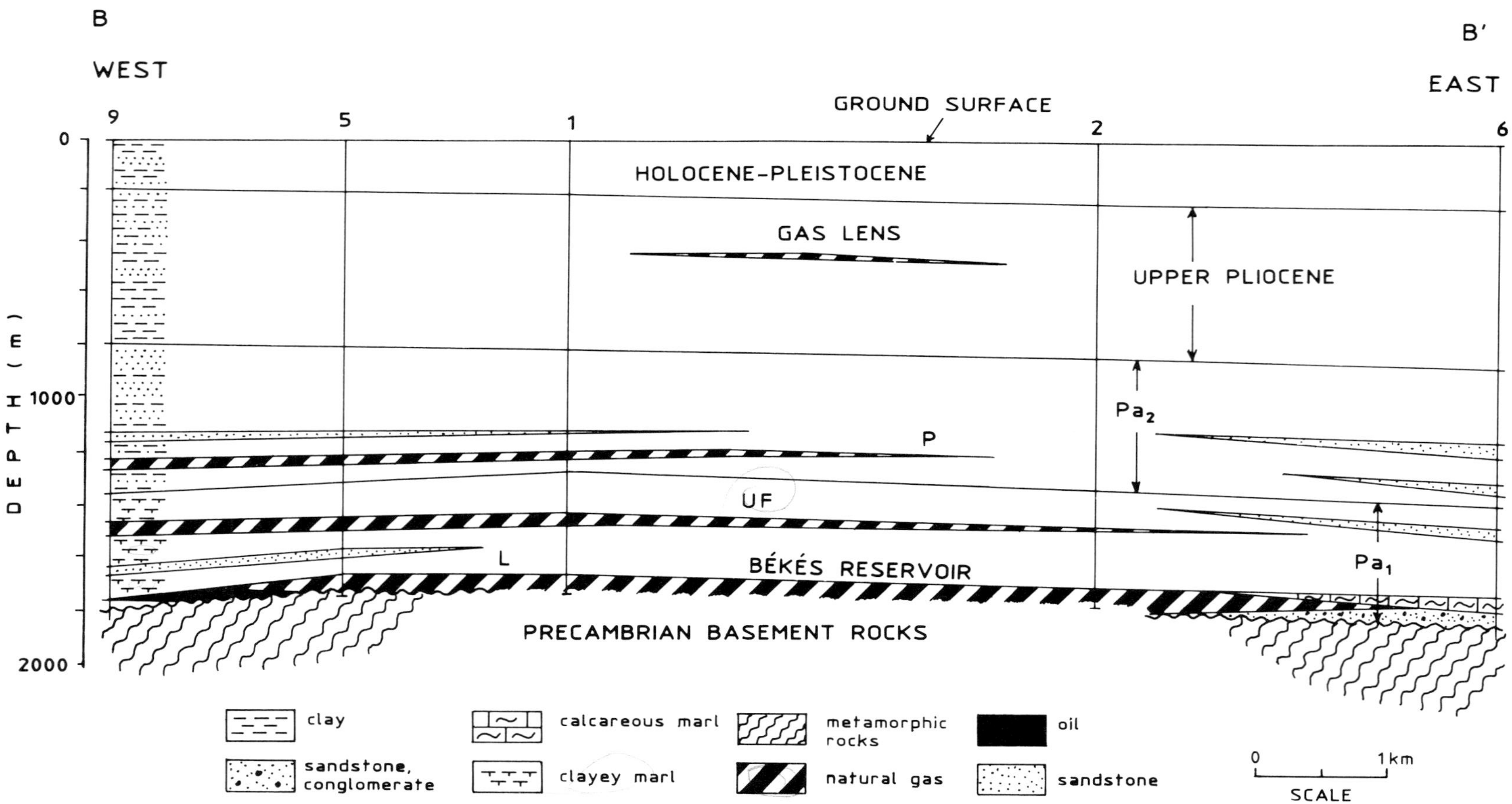

Figure 12. *Generalized geologic section across the Pusztaföldvár field. Location of section is shown in Figure 10. Numbers on top of section refer to wells shown in Figure 10. L: lower Pannonian sandstone lens (similar lenses comprise the 4 Lower Földvár reservoirs to the north); UF: one of 6 Upper Földvár reservoirs; Pa1: lower Pannonian; Pa2: upper Pannonian.*

Table 2. Summary of producing reservoirs (in ascending order) in the Pusztaföldvár field and their physical attributes. T: thickness of reservoir; Pa_1: lower Pannonian; Pa_2: upper Pannonian; ss: sandstone; cong: conglomerate; Por: porosity; H_2O Sat: water saturation; Perm: permeability; P: pressure; mPa: megapascals; Temp: temperature; CPO: cumulative production of oil through 1985; MT: metric tons; CPG: cumulative production of natural gas through 1985 (associated and nonassociated); PS: present status; PR: producing; SI: shut in; A: abandoned. Depths shown in column 3 are referenced to sea level.

Reservoir Name	Area (km^2)	Depth (m)	T (m)	Age	Rock	Por (%)	H_2O Sat (%)	Perm (mD)	P (mPa)	Temp (°C)	CPO (10^3 MT)	CPG (10^6 m^3)	PS
Békés	25.0	1688	99	Pa_1	ss&cong	19.7	30	360	17.72	125	1782	3296	PR
Lower Földvár[1]	3.2	1627	58	Pa_1	ss	19.5	27	115	17.35	120	371	677	A&PR
Upper Földvár[2]	7.8	1507	39	Pa_1	ss	26.1	20	181	16.00	113	0	3575	PR&SI
Puszta[3]	1.0	925	6	Pa_2	ss	27.5	20	987	6.50	75	0	1144	A&SI

1 Values in row are an average for 4 reservoirs, except CPO and CPG which represent sums.
2 Values in row are an average for 6 reservoirs, except CPO and CPG which represent sums.
3 Values in row are an average for 36 reservoirs, except CPO and CPG which represent sums.

Figure 12) are not productive, but their updip termination can be observed between depths of 1600-1700 m between the Pf-5 and Pf-1 wells.

Upper Földvár Zone

This zone is the second largest producing interval in the field. It contains 3 large and 3 small natural gas pools, some of which produce high-BTU gas. The reservoirs are lens-shaped, marly sandstone bodies between depths of 1375-1525 m below sea level (Fig. 11). The reservoir beds range in thickness from 5.0-56.5 m (average thickness 39 m) and consist of sandstone (Table 2). One of the Upper Földvár reservoirs (labeled UF) is shown in Figure 12.

Puszta Zone

Thirty-six, small-to-large, unconnected natural gas pools occur in upper Pannonian sandstones between depths of 900-1350 m below sea level (Fig. 11). The composition of the gas in these reservoirs is mostly methane. The reservoir beds range in thickness from 1-17.5 m (average thickness 6 m) and consist of sandstone (Table 2). One of the Puszta reservoirs (labeled P) is shown in Figure 12.

Above the Puszta Zone, between depths of 320-540 m, upper Pannonian sands and sandstones have been charged with natural gas that escaped from deeper reservoirs during completion of the Pf-50 well. These sands and sandstones are labeled "Gas Lenses" in Figures 11. One of the lenses is shown in Figure 12.

Pressure in the Pusztaföldvár field is hydrostatic. Natural fractures in lower Pannonian marls often induce loss of drilling mud; thus, it is often difficult to maintain circulation while drilling.

Sarkadkeresztúr

The Sarkadkeresztúr field was discovered in 1976 and in the mid 1980's it contributed about 5% of Hungary's total production of natural gas (Dank, 1988). The field is related to a narrow basement high (possibly a horst) consisting of Precambrian (?) metamorphic rocks (Fig. 15). Middle Miocene (Sarmatian) conglomerates pinchout on the flanks of the basement high and the structure is overlain by Pannonian-Holocene sedimentary rocks.

All of the present hydrocarbon production (chiefly gas) is from the Sarkad reservoir locally in excess of 400 m thick (Fig. 16). The reservoir rocks consist of fractured and fissured mica schist and overlying Miocene conglomerates. The average porosity of Sarkad reservoir

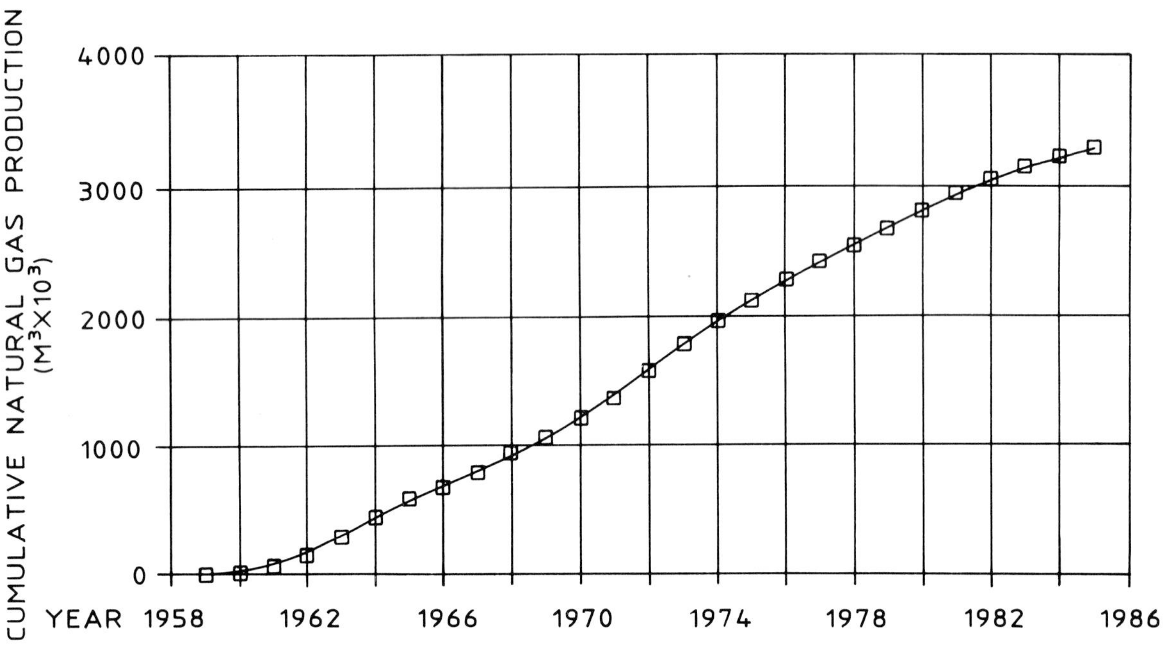

Figure 13. Cumulative natural gas production from the Békés Zone in the Pusztaföldvár field.

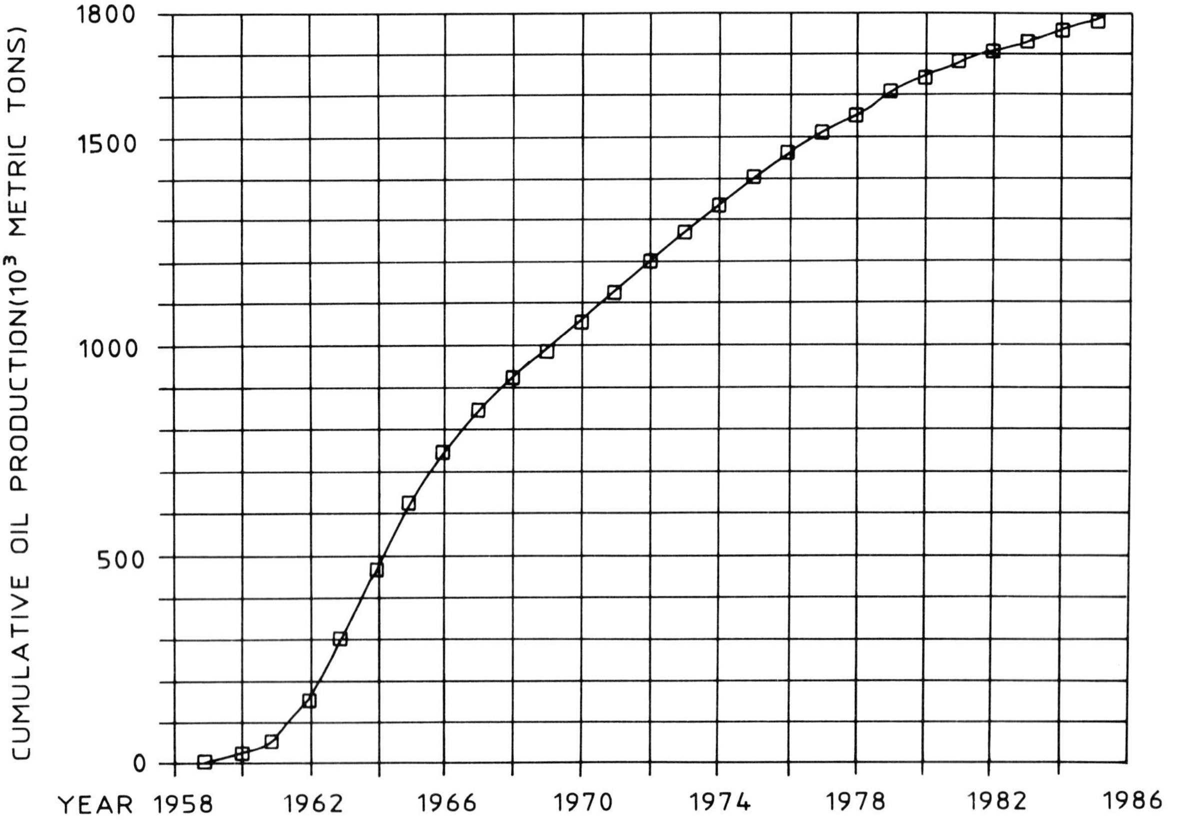

Figure 14. Cumulative oil production from the Békés Zone in the Pusztaföldvár field.

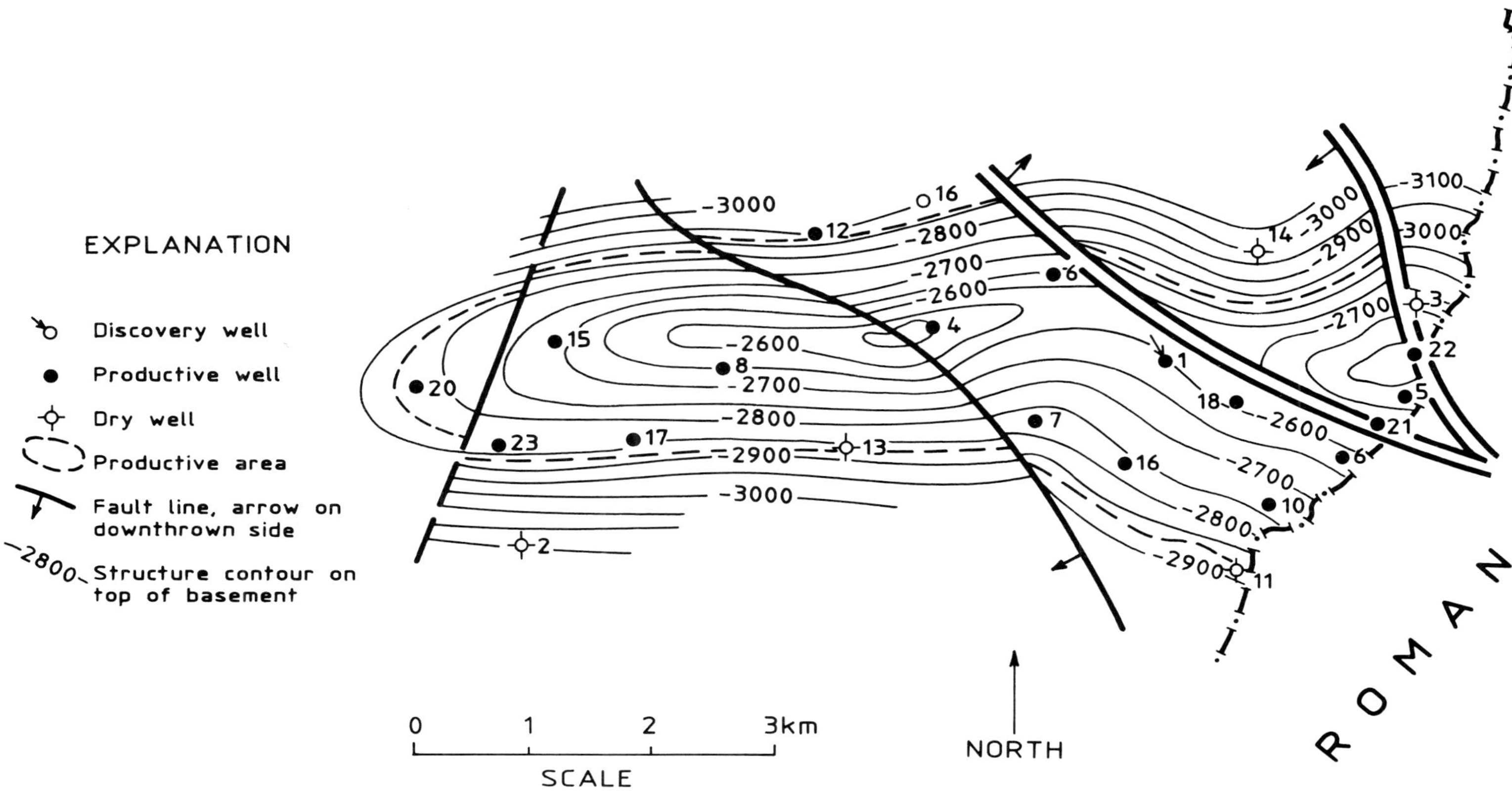

Figure 15. Structure map on the top of basement (Sarkad reservoir) in the Sarkadkeresztúr field. Contour interval is 50 m. Datum is sea level.

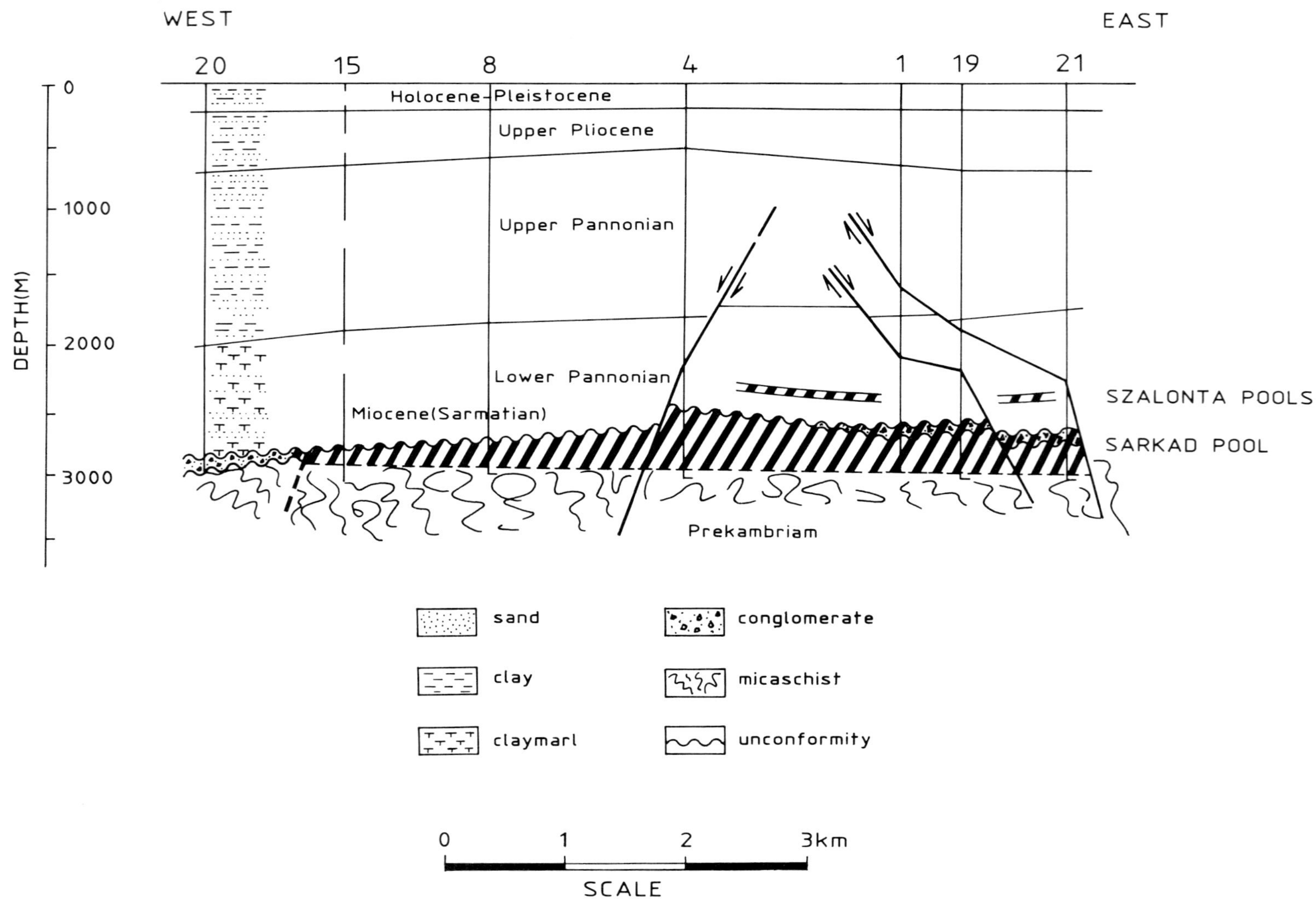

Figure 16. Geologic section across the Sarkadkeresztúr field. Numbers at the top of section refer to wells shown in Figure 15. Depths relative to sea level (SL).

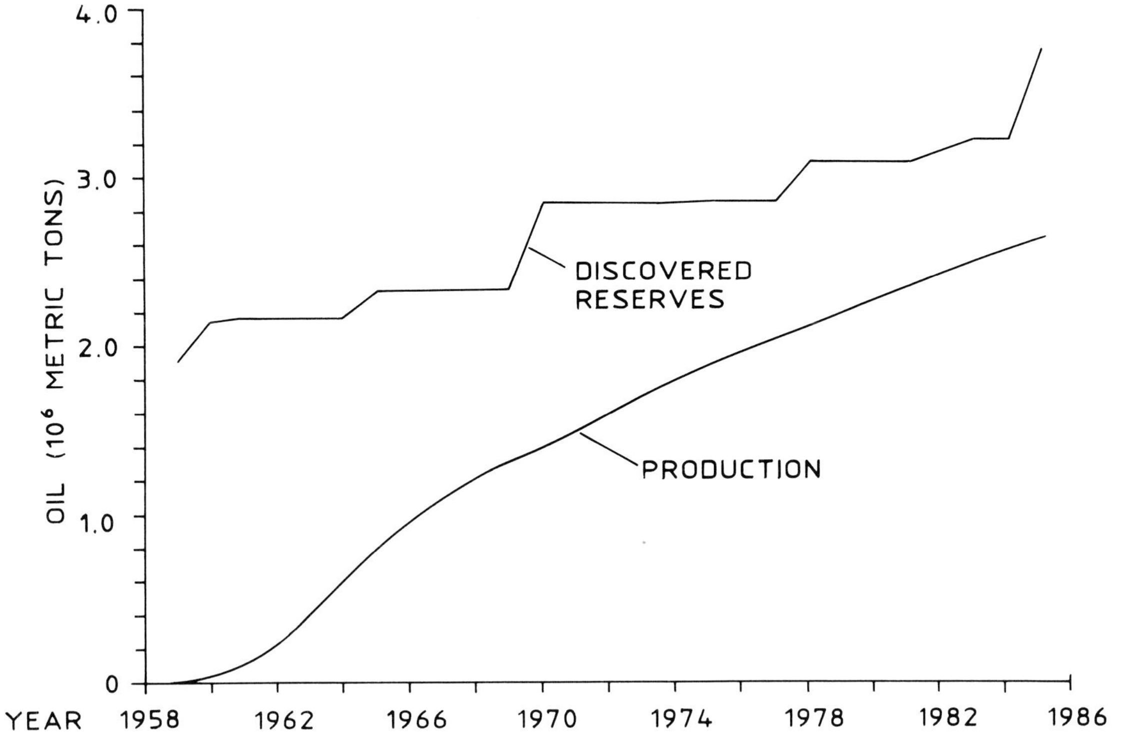

Figure 17. Cumulative production and known recoverable reserves of oil in the Békés basin starting from 1958 when production began.

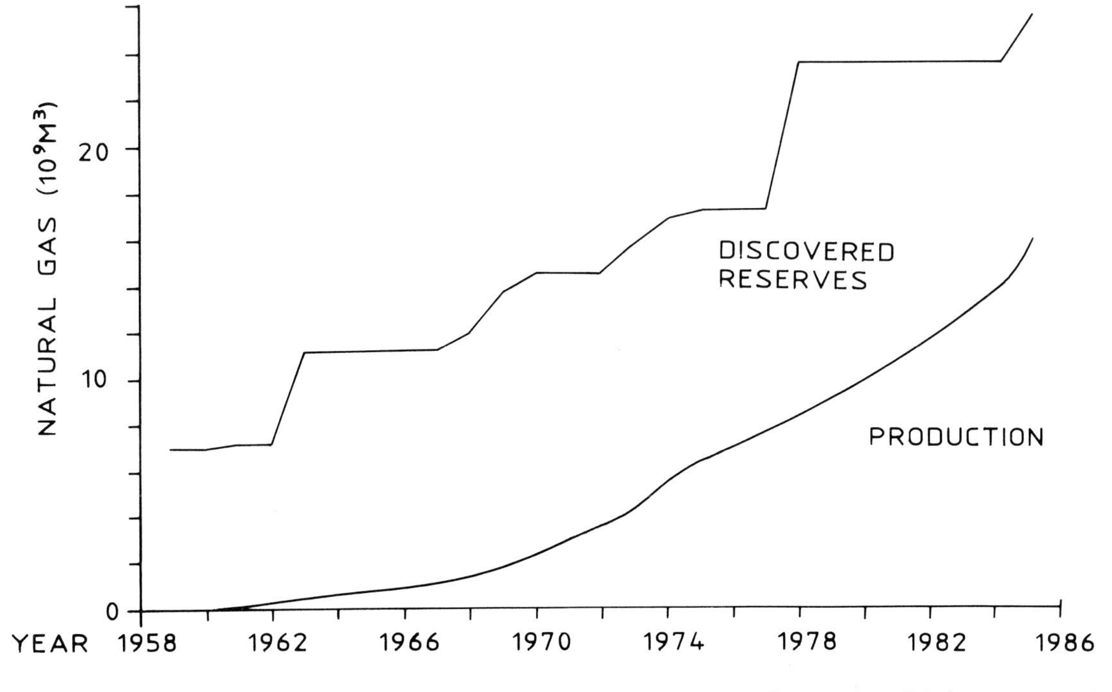

Figure 18. Cumulative production and known recoverable reserves of natural gas in the Békés basin starting from 1958 when production began. Depths relative to sea level.

rocks is about 5% and the average permeability about 94 millidarcies. Small amounts of hydrocarbon gas were also produced from lower Pannonian turbidite sandstones called the Szalonta reservoirs. Peak production of natural gas was reached in 1981 when 603×10^6 m^3 were produced. By 1985, the production rate had declined to about 353×10^6 m^3/year.

According to Dank (1988), discovery of the Sarkadkeresztúr field is important because it shows that the metamorphic basement rocks are locally permeable and can contain good hydrocarbon reservoirs. This enhances the exploration prospects on the flanks of known structures.

PRODUCTION AND RESERVES, BÉKÉS BASIN

In the mid-1980's approximately 3% of the oil and 11.5% of the natural gas produced in Hungary came from the Békés basin. The cumulative production and recoverable reserves of oil and natural gas of the basin for the period 1958-85 are summarized in Figures 17 and 18, respectively. The ultimately recoverable hydrocarbons consist of 3.215×10^6 metric tons of oil and 25.397×10^9 m^3 of natural gas. By the end of 1985, 63.40% of the natural gas (16.10×10^9 m^3) and 83.01% of the oil reserves (2.669×10^6 metric tons) had been produced.

CONCLUSIONS

The Békés basin and the immediate surrounding area have been extensively explored since exploration began in the early part of the 20th century. All the known oil and gas fields are associated with basement highs located along the periphery of the basin. Production of oil and gas from these fields is declining rapidly and, by the end of 1985, 63.40% of the known recoverable oil reserves had been produced. Limited exploration in deep (central) parts of the basin during the mid-1980's was not encouraging. The best prospects for future exploration are stratigraphic traps on the flanks of known fields at depths between 3000-4000 m.

REFERENCES

Alliquander, Ö., 1968, Adalékok a magyarországi mélyfúrás történetéhez, tekintettel a szénhidrogén-kutató és feltáró fúrásokra (Addenda to the history of deep drilling in regard to petroleum exploratory and development wells) Bányászati és Kohászati Lapok, Köolaj és Földgáz, 1, 8-18.

Clayton, J.L., and Koncz, I., Geochemistry of natural gas and carbon dioxide in the Békés basin - implications for exploration; this volume 187-199.

Clayton, J.L., Koncz, I., King, J.D., and Tatár, É., Organic geochemistry of crude oils and source rocks, Békés basin; this volume, 161-185.

Csiky, G., 1980, History of petroleum and natural gas exploration in Hungary from the beginning till 1920; Földtani Közlöny, 110, (1), 15-18.

Csontos, Sz. Á., Szili, J., and Vadász, Gy., 1987, Az olaj tükrében (In the oil mirror); Köolajkutató Vállalat (Hungarian Oil Exploration Co.), Szolnok, Hungary, Internal report, 1-192.

Dank, V., 1985, Hydrocarbon exploration in Hungary; *in* Hála, János (ed.), Neogene mineral resources in the Carpathian Basin: Historical studies on their utilization; Hung. Geol. Survey, Spec. Pub., 107-213.

Dank, V., 1988, Petroleum geology of the Pannonian Basin, Hungary: an overview; *in* Royden, L.H., and Horváth, F., (eds.), The Pannonian Basin a study in basin evolution: Amer. Assoc. Petrol. Geol. Memoir 45, 319-331.

Grow, J.A., Mattick, R.E., Bérczi-Makk, A., Péró, C., Hajdú, D., Pogácsás, Gy., Várnai, P., and Varga, E., Structure of the Békés basin inferred from seismic reflection, well and gravity data; this volume, 1-38.

Kertai, György, 1960, A magyarországi szénhidrogén-kutatás eredményei 1945-1960-ig (The results of Hungarian hydrocarbon exploration from 1945 to 1960); Földtani Közlöny, 90, 406-418.

Molenaar, C.M., Révész, I., Bérczi, I., Kovács, A., Juhász, G.K., Gajdos, I., and Szanyi, B., Stratigraphic framework and sandstone facies distribution of the Pannonian sequence in the Békés basin,; this volume, 99-110.

Papp, Simon, 1963, A magyarországi köolaj-és földgázkutatás az 1780-tól 1945-ig terjedö idöszakban (Hungarian oil and gas exploration during the 1780-1945 time period); Hung. Acad. Sci., Pub. of Dept. 10, 32 (1-4), 450-465.

Phillips, R.L., Révész, I., and Bérczi, I., Lower Pannonian deltaic-lacustrine processes and sedimentation, Békés basin; this volume, 67-82.

Spencer, C.W., Szalay, Á., and Tatár, É., Abnormal pressure and hydrocarbon migration in the Békés basin; this volume, 201-219.

Szalay, Á., 1988, Maturation and migration of hydrocarbons in the southeastern Pannonian basin; *in* Royden, L.H., and Horváth, F., (eds.), The Pannonian Basin a study in basin evolution; Amer. Assoc. Petrol. Geol. Memoir 45, 347-354.

Szentgyörgyi, K., and Teleki, P.G., Facies and depositional environments of Miocene sedimentary rocks; this volume, 83-97.

13 Vertical Seismic Profile Experiments at the Békés-2 Well, Békés Basin

M.W. Lee[1] *and Gábor Göncz*[2]

ABSTRACT

Vertical seismic profile (VSP) experiments were conducted during 1986-1987 at the Békés-2 well site, where the well penetrated 2161 m of Mesozoic carbonate rocks beneath basin fill (TD = 5500 m). For these experiments a vertical-component, wall-locking geophone was used. Energy was generated by an explosive source for one near-offset and two far-offset VSP experiments, and by a Vibroseis source for a walk-away VSP experiment. These sets of VSP data were analyzed in conjunction with a seismic reflection profile recorded on the ground surface.

The aim of this work was to improve understanding of the structure of complexly deformed, buried Mesozoic sedimentary rocks at the well site. The geometry of deformed Mesozoic strata was modeled by means of ray-tracing techniques. The near-offset VSP data were used to identify prominent reflections below 3500 m in the Mesozoic section. These were interpreted as reflections from low-impedance layers within decollement folds. Although the far-offset and walk-away VSP data did not define the geometry of the deformed strata definitively, these data were useful in identifying truncations of strata. In contrast to the surface-recorded reflection data, the VSP data contained a lower signal-to-noise ratio for the deep part of the stratigraphic section. Because of complex deformation, individual Mesozoic stratum cannot be mapped as coherent seismic units over distances of more than a few kilometers in the vicinity of the Békés-2 well.

INTRODUCTION

Recently, vertical seismic profiling (VSP) techniques have proven to be important in hydrocarbon exploration. Oristaglio (1985) summarized many of the important applications of VSP, such as stratigraphic investigation, mapping of structurally complex strata, and theoretical investigation of wave fields. Caldwell (1984) successfully applied offset-VSP techniques in the structurally complex Western Overthrust Belt of the United States. In the North Sea area, Kennet and Ireson (1981) applied VSP techniques to identify truncated beds in fault zones.

This study was initiated primarily to understand the source of low-frequency reflections that appear on seismic reflection profile Gyula (Gyu)-30 below 2.5 sec.

[1] U.S. Geological Survey, Denver, Colorado 80225, USA
[2] MOL Rt. (Hungarian Oil and Gas Co., Ltd.), H-1068, Budapest, Hungary

P. G. Teleki et al. (eds.), Basin Analysis in Petroleum Exploration, 257–275.

(Fig. 1, left). The first VSP data (near-offset) were acquired in March of 1986 at the Békés-2 well. This well was drilled to a depth of 5500 m and penetrated 2161 m of Mesozoic carbonate rocks beneath basin fill. As correlation of the 1986 VSP data with the seismic events of seismic profile Gyu-30 (Fig. 1, left) was not satisfactory, the VSP data were reprocessed and additional VSP data were acquired.

The field acquisition phase for recording the new data was based on a preliminary interpretation of profile Gyu-30 (Fig. 1, right). This interpretation involved thrusting of Mesozoic rocks during Late Cretaceous time. These thrust sheets, or nappes, were thought to have been later offset during a phase of Miocene extensional faulting (Fig. 1, right). The stippled zone shown in Figure 1 was the main target of subsequent investigations.

New VSP data (4 sets) were acquired in March of 1987. Two far-offset sets of VSP data were recorded, one with the energy source (dynamite) located 2 km south of the Békés-2 well and the other with the source 2 km north of the well. Dynamite was also used as an energy source to record one set of near-offset data (offset distance from the well was less than 5 m). The far-offset and near-offset data sets were recorded between well depths of 1000-3500 m at intervals of 20 m. In addition, a set of walk-away VSP data were recorded using a wall-locking geophone located at a depth of 2000 m in the well. The energy sources (Vibroseis) for the walk-away VSP experiment were located along a line extending from 4 km south of the well to 4 km north of the well at 100 m intervals. Because a three-component geophone was not available, the quality of the acquired offset and walk-away VSP data was judged to be less than optimal. During the walk-away VSP experiment, an additional surface-recorded, seismic reflection profile, referred to as "New CDP (common depth point) profile", was simultaneously acquired. The New CDP profile crossed the Békés-2 well site. All 4 sets of VSP data were analyzed in conjunction with the New CDP profile and the geological interpretations of Grow and others (this volume).

DATA PROCESSING AND ANALYSIS

The data were processed as outlined in Figure 2. A detailed discussion of the processing procedures can be found in Balch and Lee (1984), Lee (1984), and Hardage (1983). A brief discussion of the processing follows under 3 headings: (1) near-offset VSP data, (2) far-offset VSP data, and (3) walk-away VSP data.

Near-offset VSP Data

Wavelet-shaping filters were derived based on the results from the first 150 milliseconds (ms) of the monitor-geophone recording. Although the wavelet-shaping filters stabilized the initial signals received in the well, long-period multiple mismatches between recordings continued to persist. To remove the variations of the long-period multiples, wavelet deconvolution was applied to the data set using the variable-norm deconvolution technique of Gray (1978). The deconvolution operator was designed using a 2.5-sec. window after onset time. Based on the variation of the long-period multiples, six different deconvolution operators were derived and applied to the corresponding data sets.

One example of wavelet deconvolution is shown in Figure 3. Comparison of Figures 3a and 3b indicates that after deconvolution and flattening of the frequency content (up to 60 Hz), the waves traveling upward are more clearly defined, especially near 2.5 sec. (shown by arrow in Fig. 3b).

Based on seismic character, the reflections in the vicinity of the Békés-2 well can be divided into four zones (Fig. 4). The upper zone (S_1), between depths of 920-2000 m, contains numerous, continuous high-frequency reflections. The average interval velocity of S_1 is 2.82 km/sec. S_2, between depths of 2000-2800 m, has an average velocity of 3.56 km/sec. and exhibits very strong continuous reflections, as well as reflections that appear to terminate abruptly. S_3, between depths of 2800-3550 m, has an interval velocity of 4.14 km/sec., and discontinuous, irregular reflections dominate the zone. In S_3, strong reflections appear near the well, but their amplitudes rapidly diminish at short distances from the well. S_4, below a depth of 3550 m, has an average velocity of 5.81 km/sec. and is characterized by two strong reflections (one at about 1.3 sec. and the other at about 1.6 sec.) and numerous discontinuous reflections. In S_4 some calculated interval velocities were abnormally high - an indication that strata in this zone dip steeply.

In order to define an "average reflection character" near the well and investigate relative changes in reflection character, cumulative summation and lateral stacking of data (Lee, 1984) with a spatial sampling rate of 1 m were performed. These results are shown in Figure 5. Key reflections are labeled with subscripts that indicate the calculated depth of origination. In S_1 most of the cumulatively-summed reflections are similar to corresponding reflections that appear in the laterally-stacked data. This implies that the strata in S_1 near the borehole are rela-

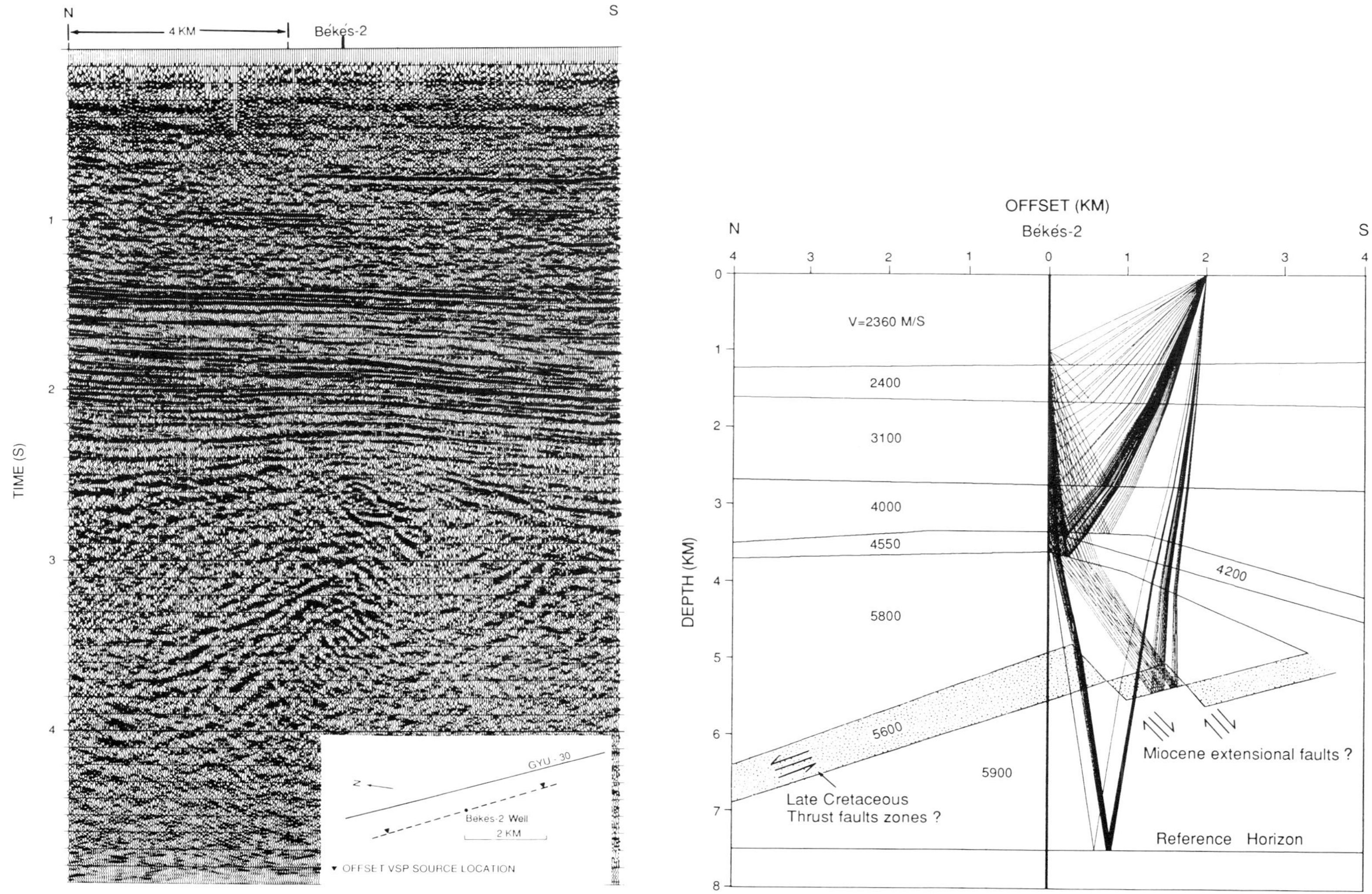

Figure 1. (Left) Migrated seismic profile Gyu-30 located near the Békés-2 well. Location of the profile with respect to the well is shown in the inset. (Right) Far-offset, ray-trace model constructed from an initial interpretation of seismic profile shown on left. The stippled area was the main zone of interest. Figure provided by John Grow (1986, written communication).

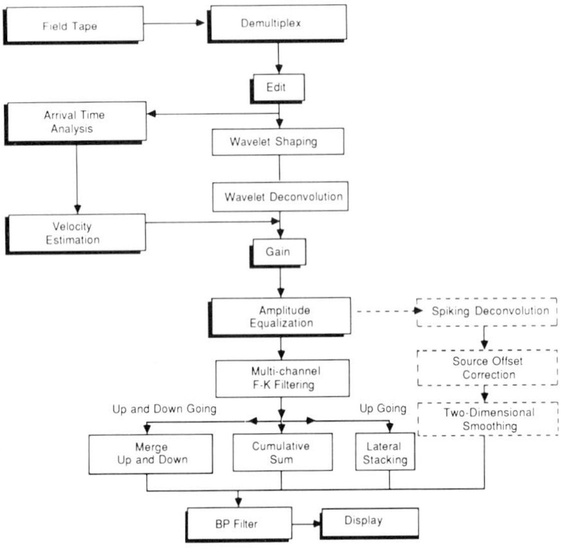

Figure 2. Flow diagram showing processing procedures applied to VSP data. Boxes outlined with heavy lines denote processing steps common to both near- and far-offset VSP data; boxes outlined with light lines denote steps used in processing of near-offset data only. Dotted boxes represent steps used in processing of far-offset data.

tively horizontal and continuous. In S_2 some reflections are seen to terminate abruptly at short distances from the well. For example, at about 2.0 sec. (Fig. 5b), the reflection labeled with an arrow terminates at a distance of about 20 m from the well. The loss of amplitude apparent on the corresponding cumulative-summed trace is believed to be caused by the termination of strata, perhaps by faulting, near the well. In zone S_3, the character of seismic reflections near the borehole is different from the reflection character appearing on the cumulatively-summed trace. Therefore, this zone is interpreted to consist of discontinuous beds. In zone S_4, two previously-mentioned strong reflections dominate the zone, and these are the only clearly identifiable reflections in the lower part of the section.

Far-Offset VSP Data

Two sets of far-offset VSP data, one recorded with an energy source located north of the well and the other with the source south of the well, are shown in Figures 6a and 6b after two-dimensional smoothing (McDonnell, 1981). Comparison of the two data sets indicates that they are quite different, despite the fact that the source offset distances were the same and the deep strata dip gently. For example, on the data set recorded with the source south of the well (Fig. 6b), four well-defined reflections (R_{3650}, R_a, R_{4500}, and R_b) are apparent below 2.2 sec. However, on the data set recorded with the source north of the well (Fig. 6a), only one strong reflection, R_{4500}, can be identified in the same depth interval. Above 2.2 sec., R_{3400} can be observed clearly on the data set recorded with the source on the north side of the well, but is difficult to identify on the data set with the source on the south side of the well. These observations indicate that near-surface geologic conditions are different at the north and south shotpoints, which are 4 km apart, and that the subsurface rocks near the well are not laterally homogeneous. This conclusion is further documented by the fact that the reflections recorded with the source on the north side of the well are associated with a complex reverberation pattern that is not seen on the corresponding data set. Although the complex reverberations could affect the ability to identify primary reflections, the observed differences in reflection character between the two sets of offset VSP data also could be caused chiefly by lateral heterogeneity of subsurface rocks near the well.

The far-offset VSP data sets are believed to contain complex multiples, as well as converted waves. For example, conversions from P-waves to S-waves are apparent at boundaries near depths of 2000 m and 2700 m (Fig. 6) as characterized by waves traveling downward at low velocities. The high-amplitude seismic event labeled "?" at a depth of 3.4 sec. in Figure 6 may be associated with a critically refracted wave that was converted from a wide-angle reflected wave. This explanation, however, could not be verified.

In Figure 6c, first arrival times for data recorded with the source located north of the well and with the source located south of the well are plotted with different symbols so that differences in first arrival times from the two directions can be observed. The arrival times associated with the source located south of the well are always greater than those associated with the source located north of the well. A maximum time difference of 38 ms occurs at a depth of 2040 m; at a depth of 3500 m, the difference is 26 ms. These time differences are attributed partly to the dip of the strata and partly to a difference in near-surface geologic conditions on the north versus the south side of the well.

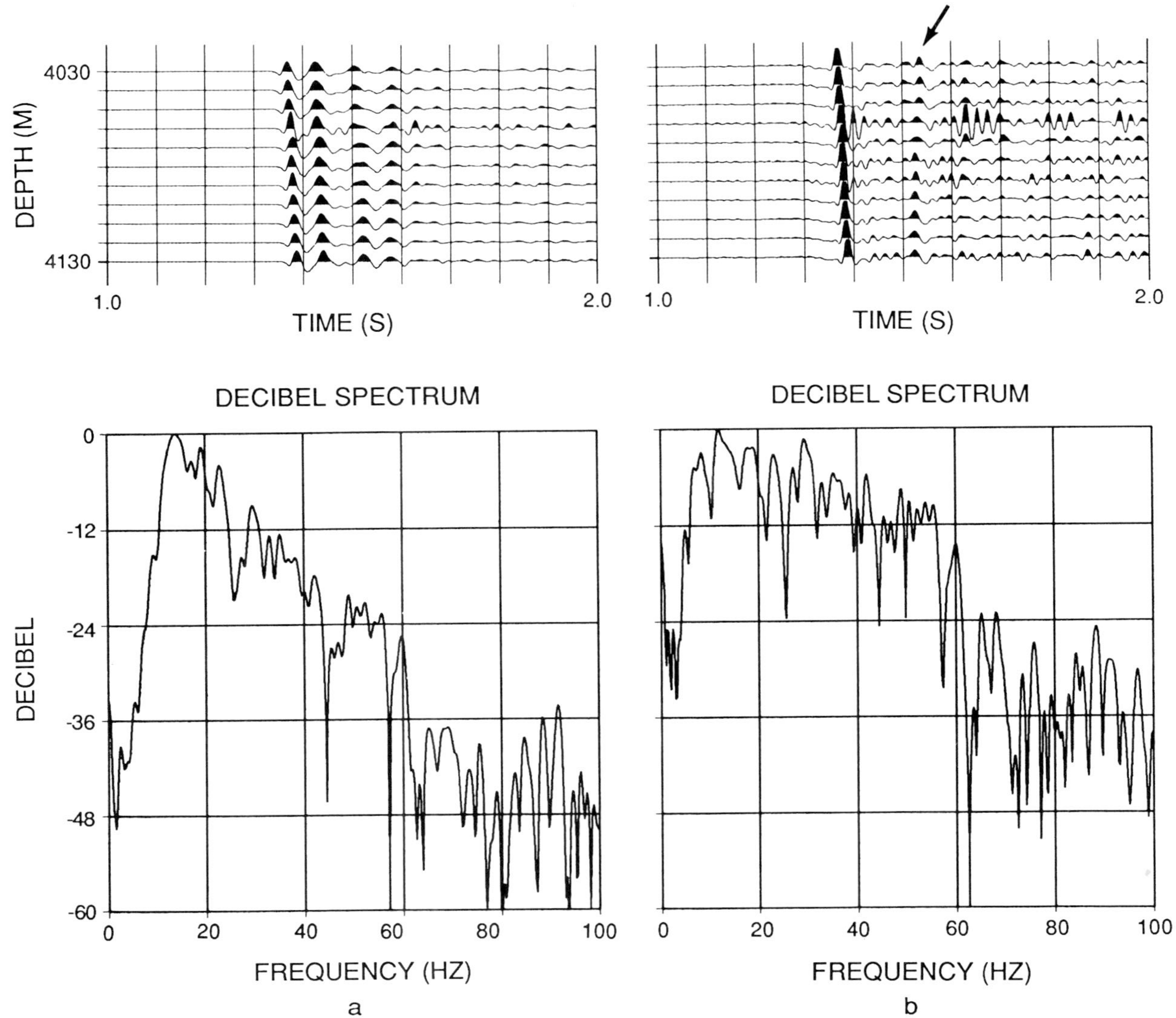

Figure 3. Examples of wavelet deconvolution and amplitude spectrum analysis for near-offset VSP data. (a) Plots showing data in the 4030-4130-m depth interval before deconvolution (top), and the amplitude spectrum of the data at a depth of 4030 m with a time gate of 1-2 sec. (bottom). (b) corresponding plots after deconvolution.

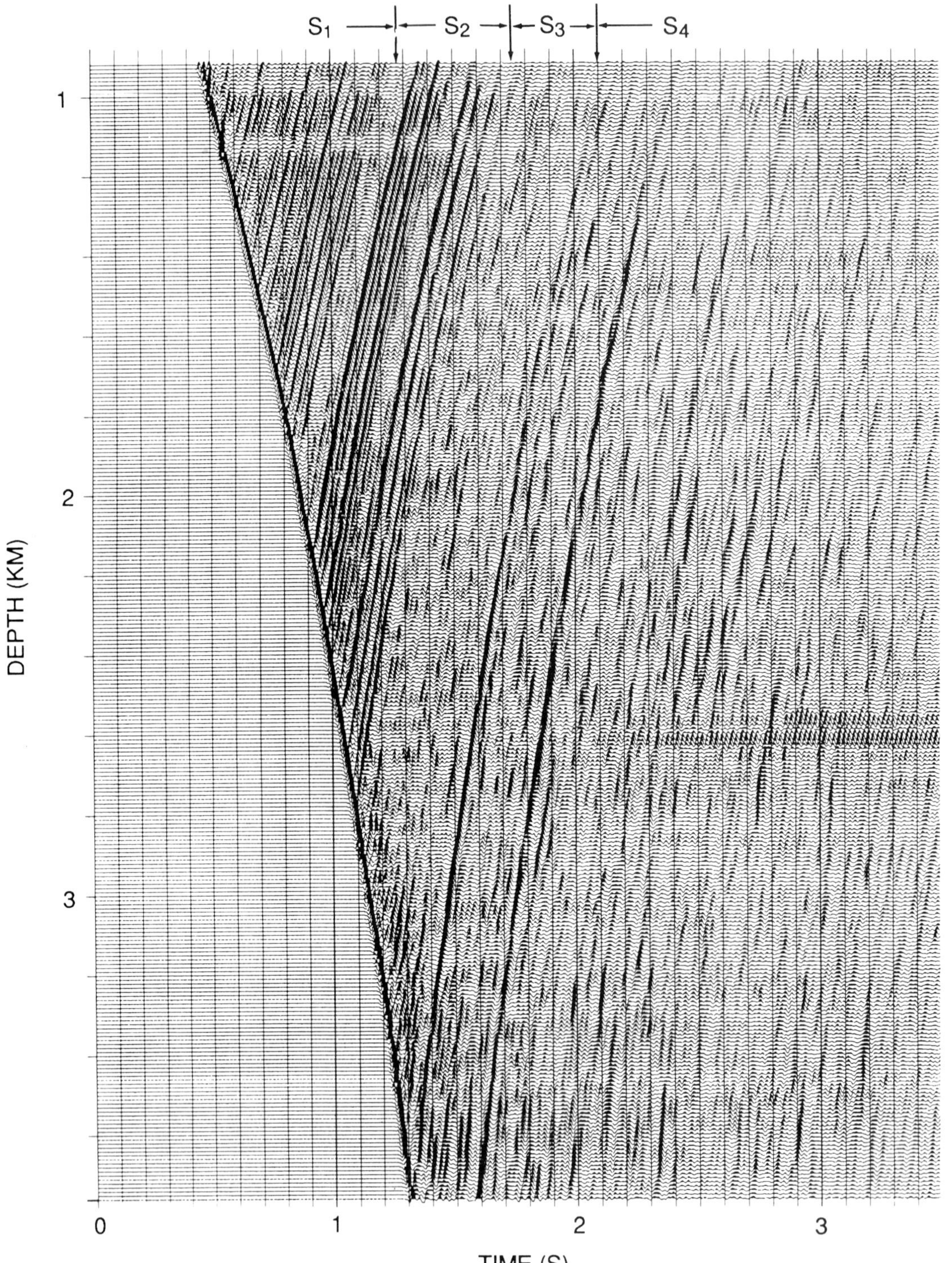

Figure 4. Processed near-offset VSP data (920-3760 m depth) with magnification of upward-traveling waves by a factor of 4. Division into 4 zones (S_1-S_4) is based on differences in seismic character as discussed in the text.

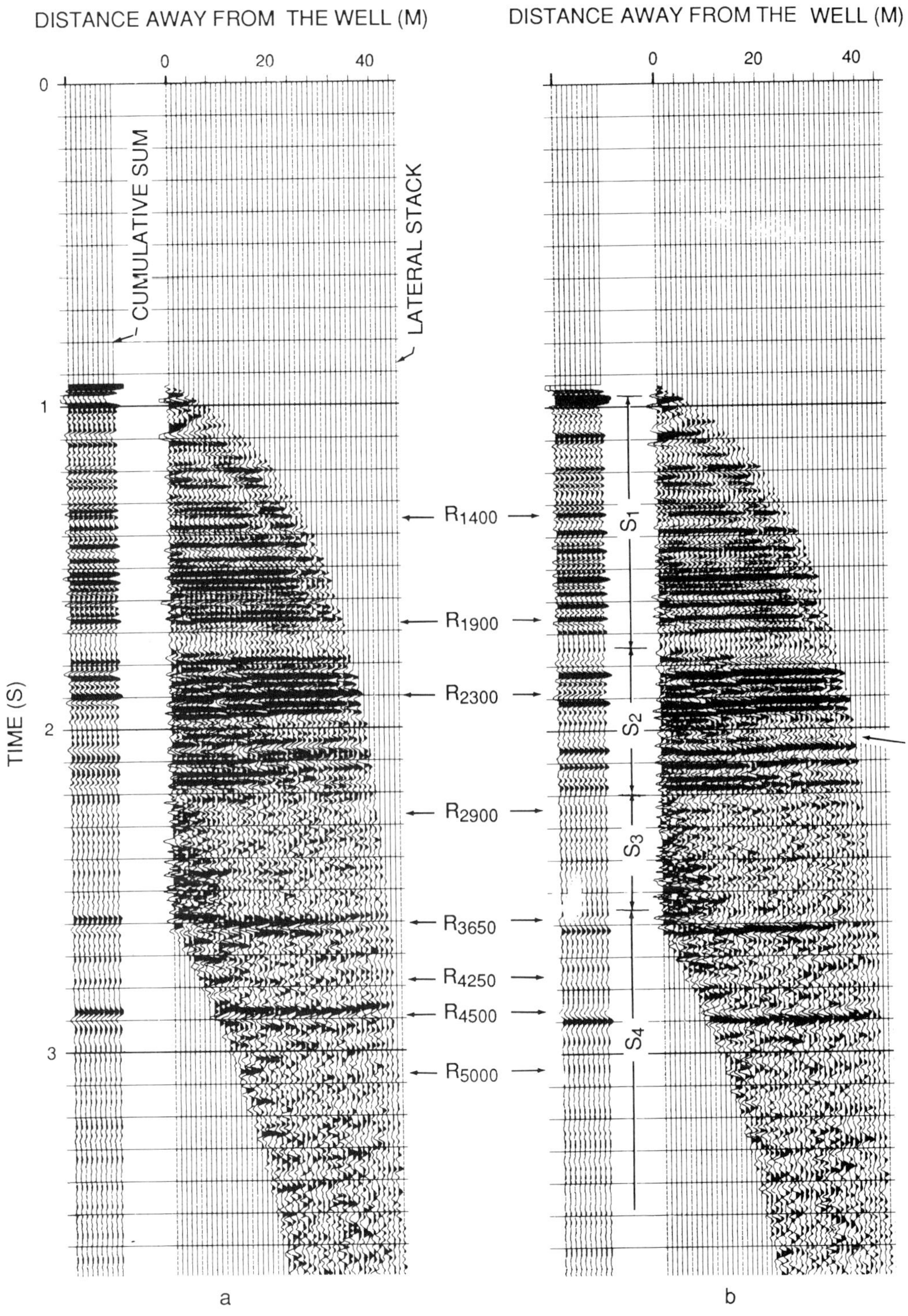

Figure 5. Cumulatively-summed and laterally-stacked near-offset VSP data with several identified reflections: (a) reverse polarity; (b) normal polarity. S_1 through S_4 are seismic zones discussed in text. R with a subscript indicates an identified reflection. The subscript refers to depth (in m) at which the reflection is inferred to have been generated. An arrow at about 2 sec. on Figure 6b points to a reflection that terminates abruptly.

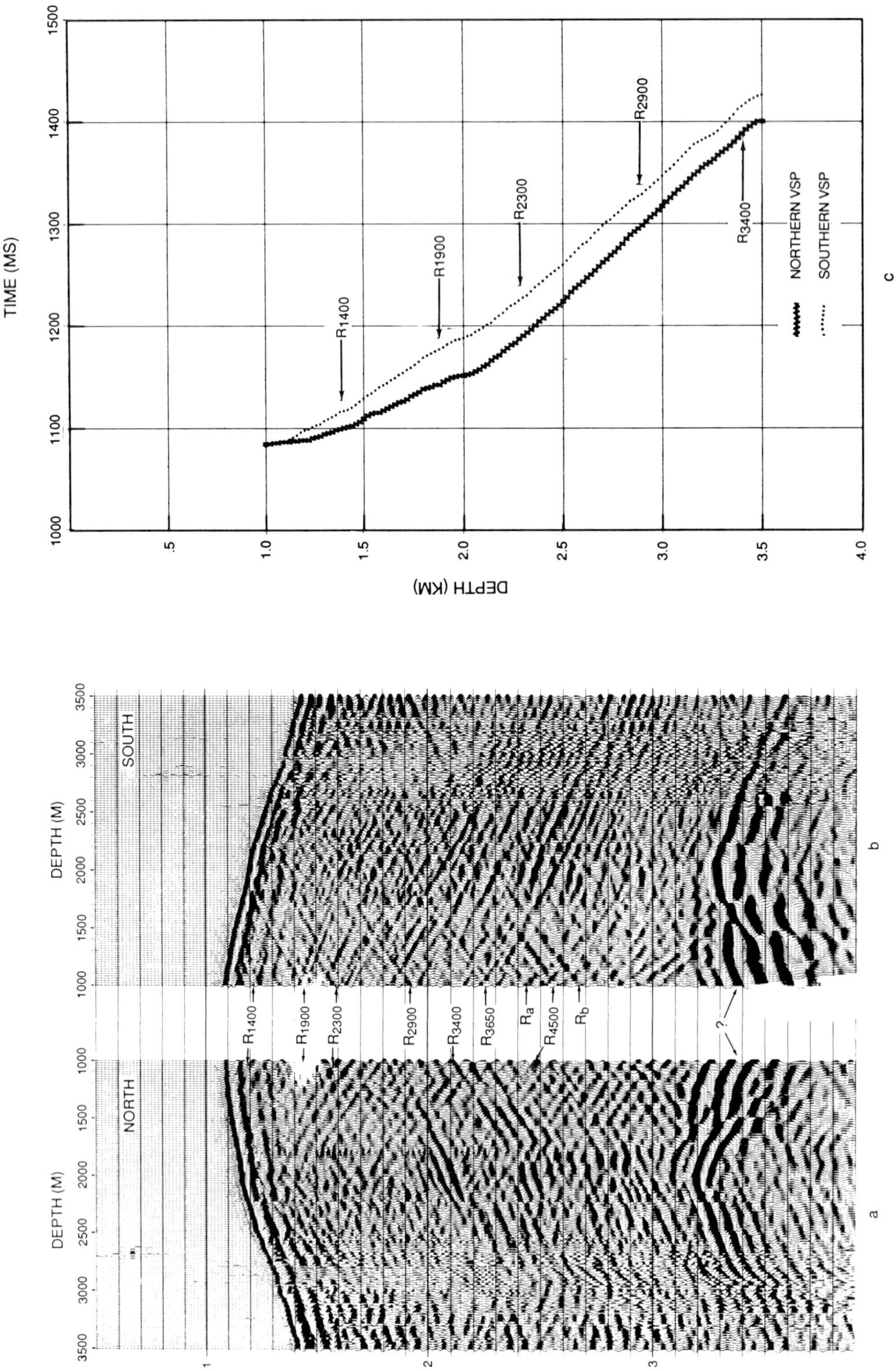

Figure 6. Far-offset VSP data showing selected identified reflections: (a) with the energy source located on the north side of the well; (b) with the energy source located on the south side of the well; (c) comparison between the first arrival times for far-offset data recorded with the source on the north side of the well (crosses) and with the source on the south side of the well (dots).

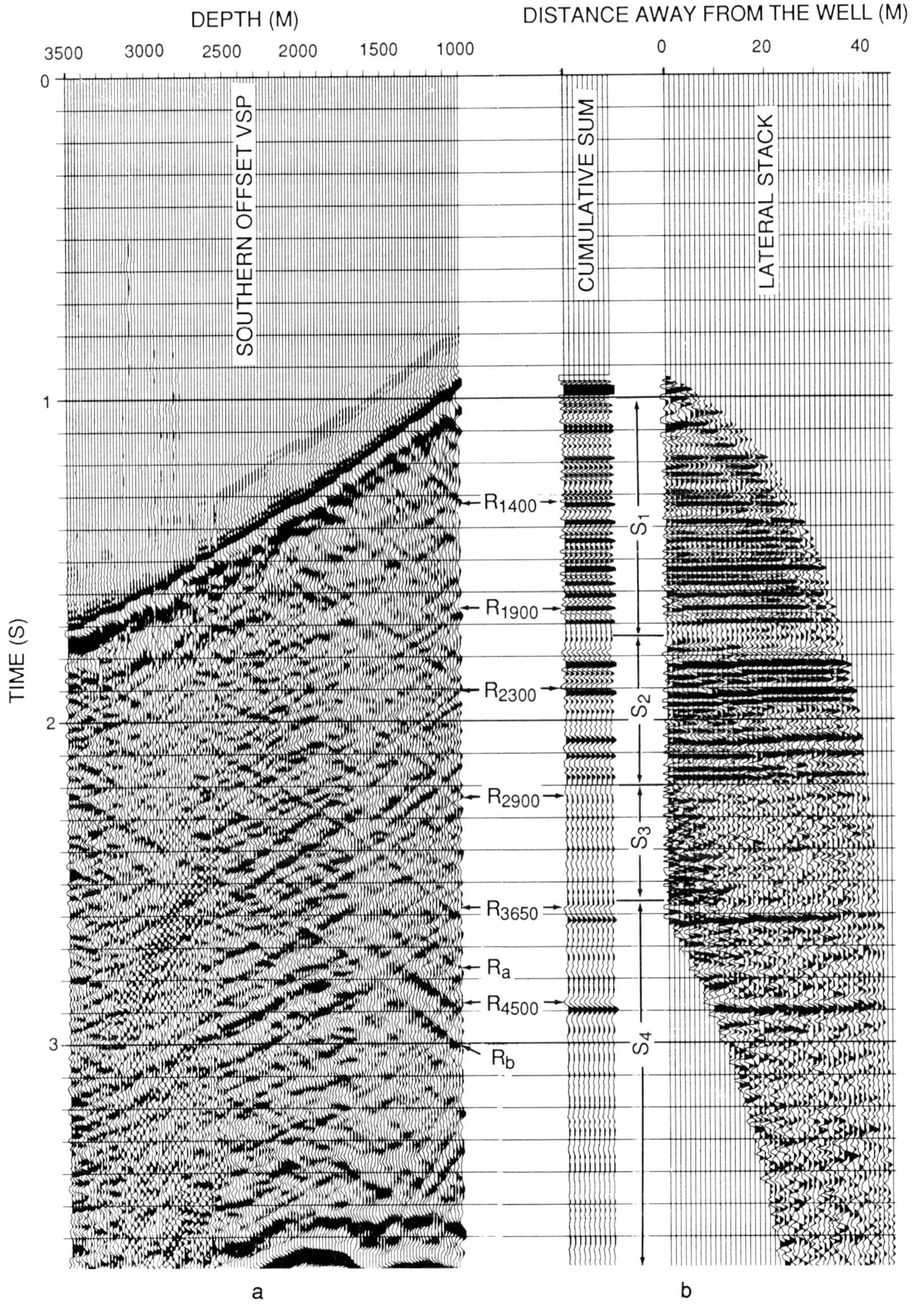

Figure 7. Processed offset VSP data: (a) far-offset VSP data recorded with energy source located south of well (after application of source-offset corrections); (b) cummulatively-summed and laterally-stacked near-offset VSP data.

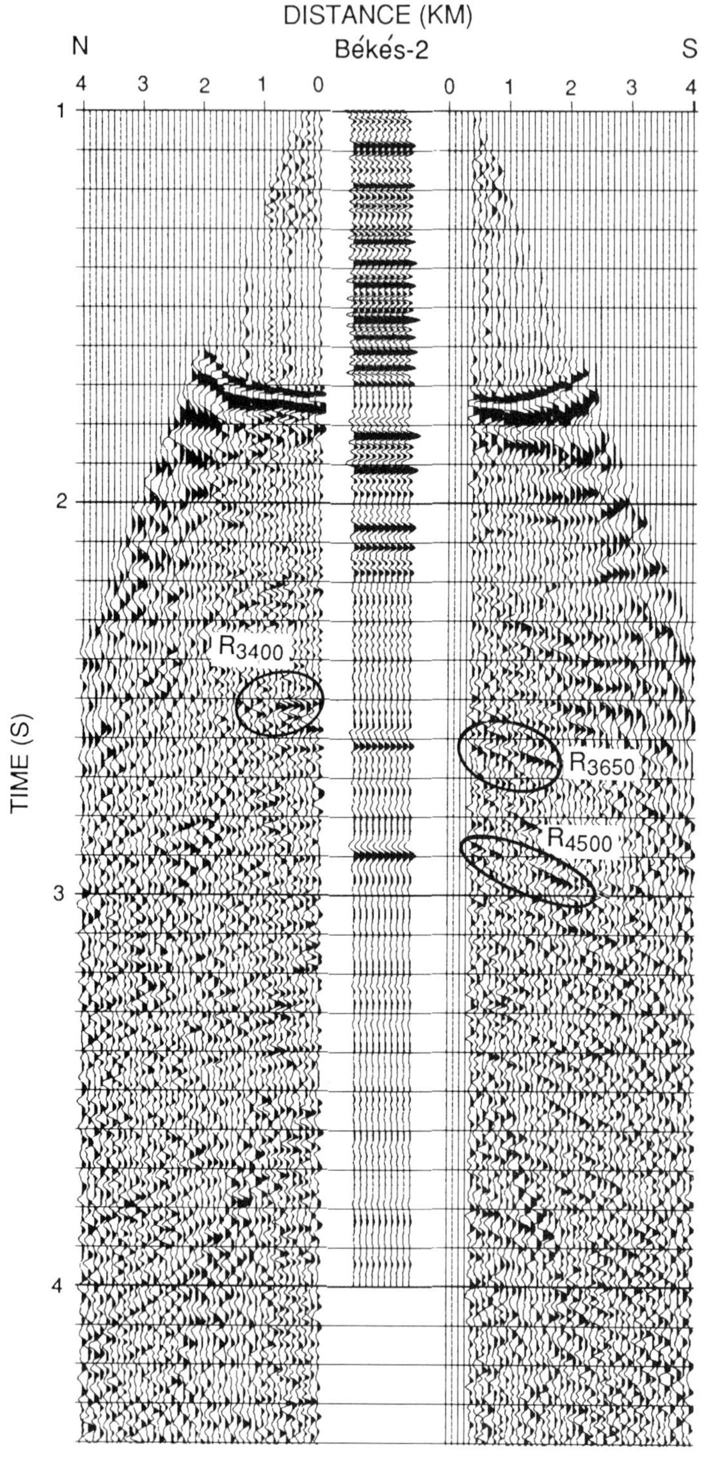

Figure 8. Walk-away VSP data after application of normal moveout and static shift correlations. Cumulatively-summed near-offset data are inserted in the middle of the section. Several reflections, discussed in the text, are labeled.

Owing to the presence of complex multiplies, converted waves, and other types of noise, the wavelet deconvolution technique that improved the near-offset data failed to improve the far-offset VSP data. Thus, robust spiking deconvolution was applied to the far-offset data set in order to enhance temporal resolution. After source-offset correction (Lee, 1984) and two-dimensional smoothing, the resulting far-offset VSP data with the source located south of the well are shown together with the cumulatively-summed and laterally-stacked near-offset VSP data in Figure 7. Comparison of the two data sets indicates that arrival times and seismic signatures are in good agreement, although the near-offset VSP data contain much higher temporal frequencies (Fig. 7).

Walk-away VSP Data

The walk-away VSP data corrected for normal moveout (NMO) are shown in Figure 8 together with the cumulatively-summed, near-offset VSP data. After application of the NMO correction, the walk-away data were converted to corresponding two-way travel times from the ground surface and shifted in phase by 90° in order to improve correlation with the near-offset VSP data. Comparison of the two data sets indicates several differences that suggest that the subsurface geology on the north side of the Békés-2 well is somewhat different from that on the south side of the well. For example, R_{3400} on the left side of Figure 8 is strong; whereas, on the right side of the figure it is much weaker.

In summary, the data set consisting of the near-offset, the far-offset, and the walk-away VSP data is internally consistent, even though some ambiguities among the multiple sets of VSP data remain owing primarily to the fact that only single-component measurements were made. The observed high variability in reflection character among the VSP data sets below seismic zone S_3 suggests that the Mesozoic rocks below 2800 m depth near the Békés-2 well site are complexly deformed as will be discussed below.

INTERPRETATION AND MODELING

The near-offset VSP data are summarized in Figure 9 (a-e). Figure 9a shows a corridor stacked trace (Hardage, 1983) which emphasizes reflected events near the well with minimal interference from multiples. A computed impedance log is shown in Figure 9b and a synthetic seismogram in Figure 9c. In general, the synthetic seismogram correlates better with the corridor-stacked trace than with the cumulatively-summed data (Fig. 9d). This might be expected because the cumulatively-summed data, in contrast to the sonic or corridor-stacked data, show the seismic character averaged over a distance of 40 m near the well site.

The tops of major geologic units identified in the Békés-2 well are shown in Figure 9. A major unconformity between postrift (late Miocene to Recent) and synrift (middle Miocene) sedimentary rocks (Grow and others, this volume) occurs at a two-way travel time of 2376 ms (3176 m depth). This unconformity is characterized seismically by a quiet zone. Another unconformity between synrift and prerift (Mesozoic) sedimentary rocks occurs at a two-way travel time of about 2450 ms (3339 m depth). Thrust faulting is inferred from the repetition of strata in the Mesozoic section. Grow and others (this volume) infer the existence of a major thrust fault just below 2700 ms where Middle Triassic rocks overlie Upper Cretaceous rocks. The only strong reflections in the lower part of the section, however, are at 2610 and 2900 ms. These reflections correspond to R_{3650} and R_{4500} respectively (Figs. 7 and 8), and do not appear to correlate with major geologic boundaries or faults.

A comparison between surface-recorded seismic data (new CDP line) and the cumulatively-summed, near-offset VSP data is shown in Figure 10. Within seismic zones S_1 and S_2, the seismic signatures of the two data sets are similar, and many reflection horizons can be correlated. However, within zone S_3, the high-amplitude, low-frequency seismic events which are observable on the surface-recorded profile do not appear (or are reduced in amplitude) on the near-offset VSP data set. The high-amplitude, low-frequency seismic events can also be observed on seismic profile Gyu-30 (Fig. 1a). In S_3, the strongest seismic event that appears on the near-offset data set is labeled R_{2900} (Fig. 10); its amplitude is low compared to the amplitude of the same event on the CDP line. The differences in seismic character between the near-offset VSP data and the surface-recorded reflection profile in S_3 support our previous analysis of zone S_3, namely that in this zone, discontinuous strata are present in the vicinity of the well.

In seismic zone S_4, R_{3650} and R_{4500} on the VSP data set match the corresponding reflections on the surface-recorded seismic profile quite well in terms of amplitude and character (Fig. 10). With the exception of these two events, however, there is little correlation between the VSP and surface-recorded data.

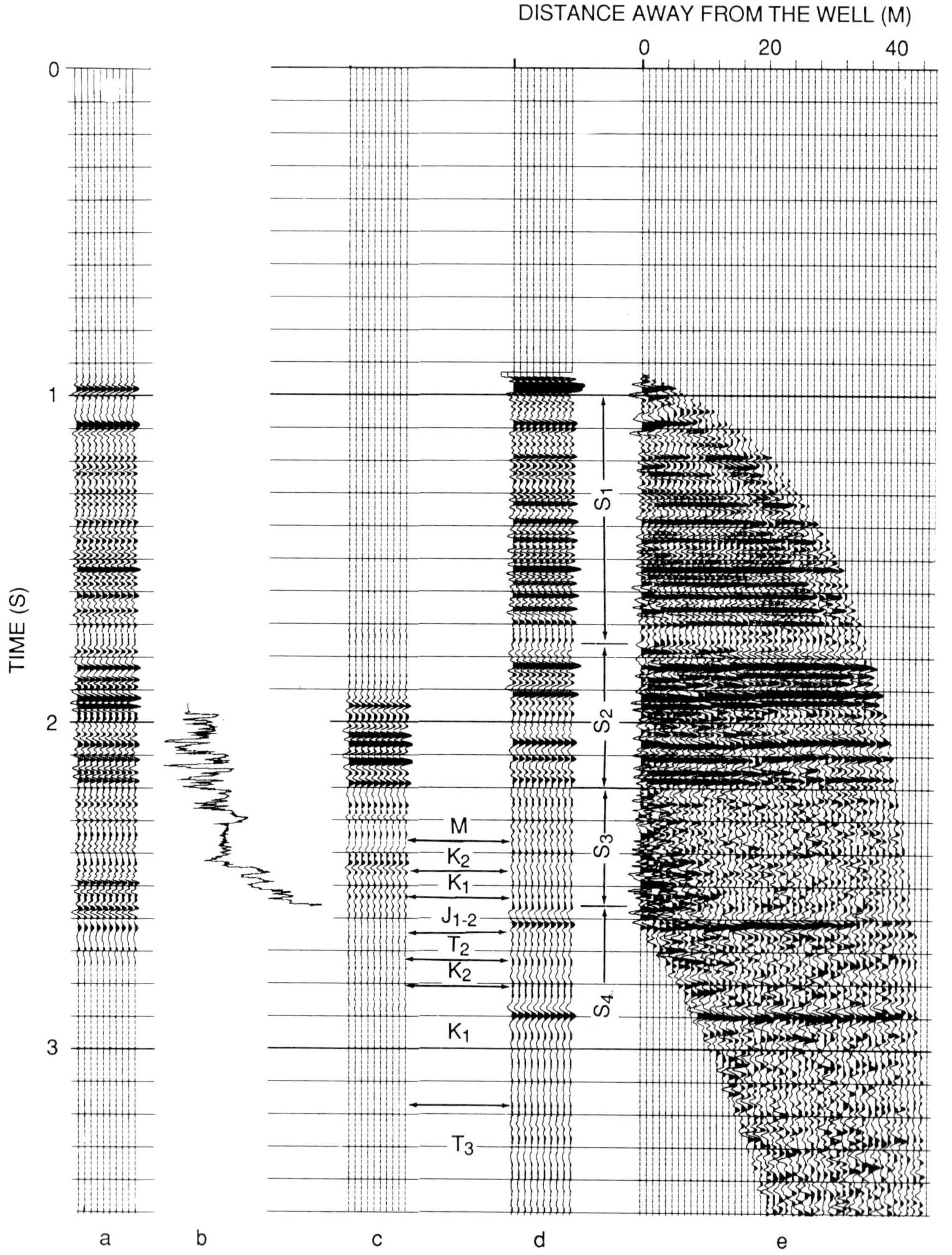

Figure 9. Summary of near-offset VSP data: (a) corridor-stacked data; (b) impedance log in time calculated from sonic and density logs between depths of 2347-3550 m in the Békés-2 well; (c) synthetic seismogram; (d) cumulatively-summed data; (e) laterally-stacked data. The tops of geologic units penetrated in the well are indicated by: M, Miocene; K_2, Upper Cretaceous; K_1, Lower Cretaceous; $J_{1\text{-}2}$, Upper and Middle Jurassic; T_2, Middle Triassic; and T_3, Upper Triassic.

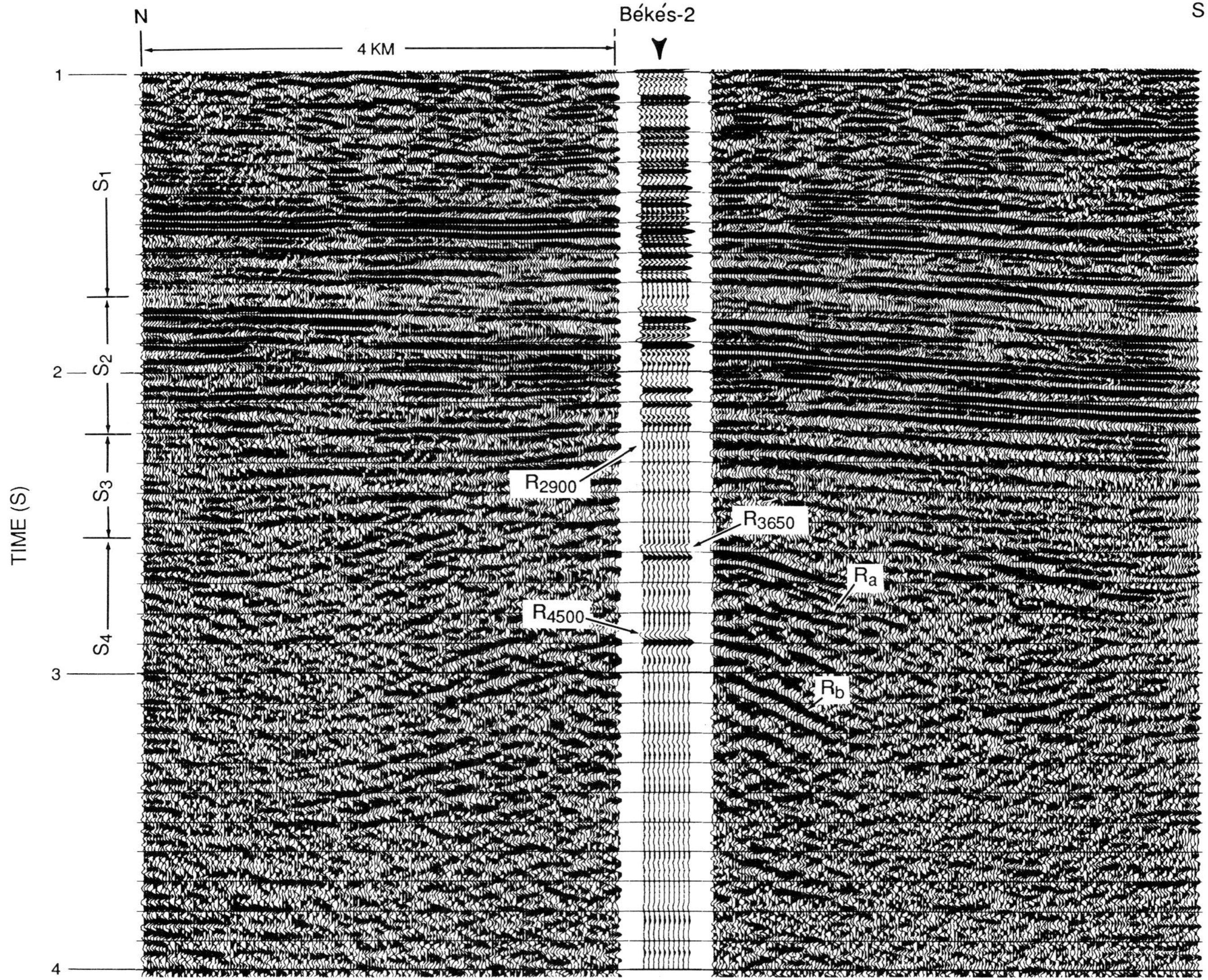

Figure 10. New CDP profile obtained along the same line of profile as the walk-away VSP data. Cumulatively-summed, near-offset VSP data are inserted at Békés-2 well. Labeled reflections and seismic zones S1-S4 are discussed in text.

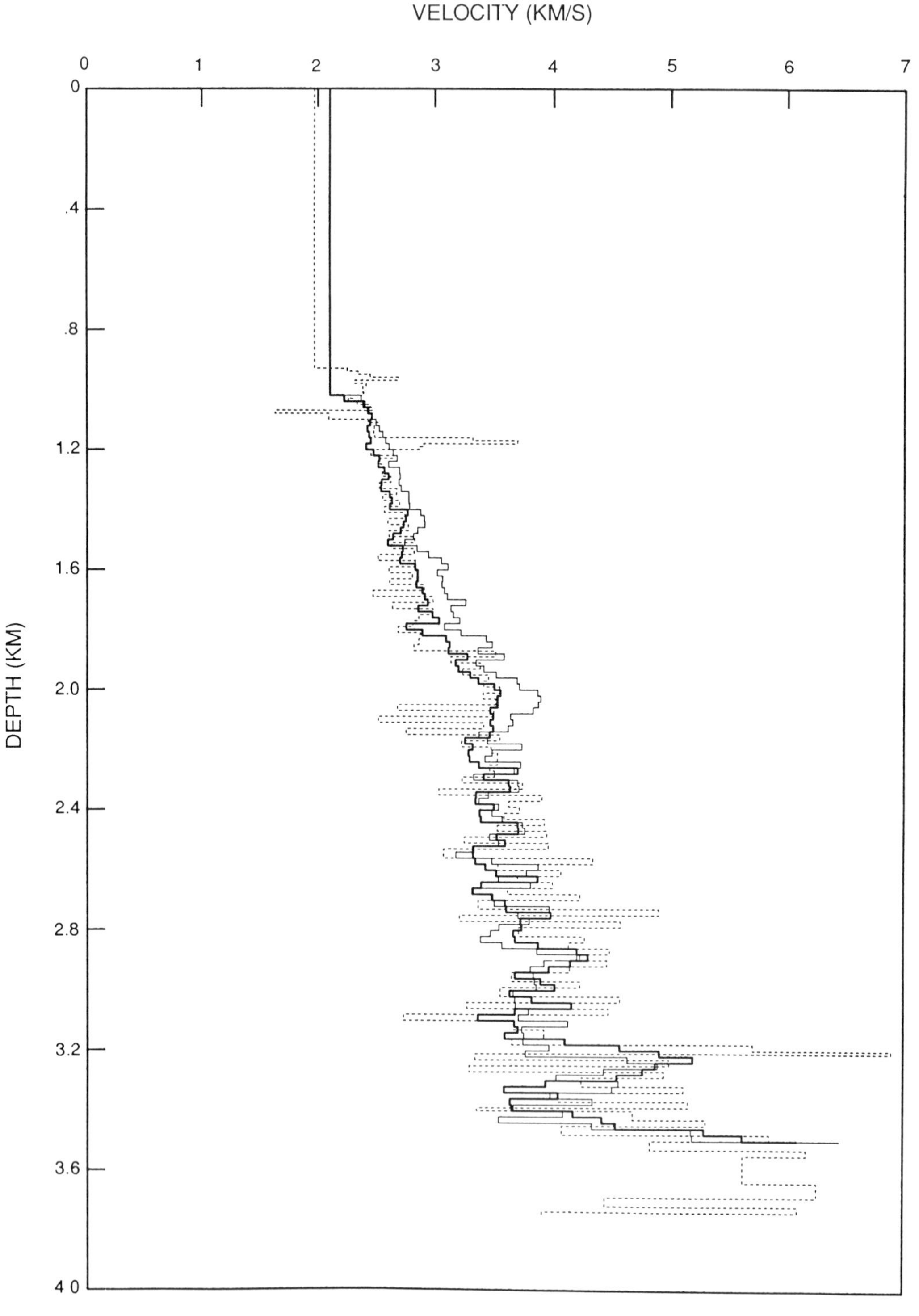

Figure 11. Calculated interval velocities. Dashed line represents interval velocities computed from near-offset VSP data; light line represents those calculated from offset VSP data with the energy source located north of the well; heavy line represents those calculated from offset VSP data with the energy source located south of the well.

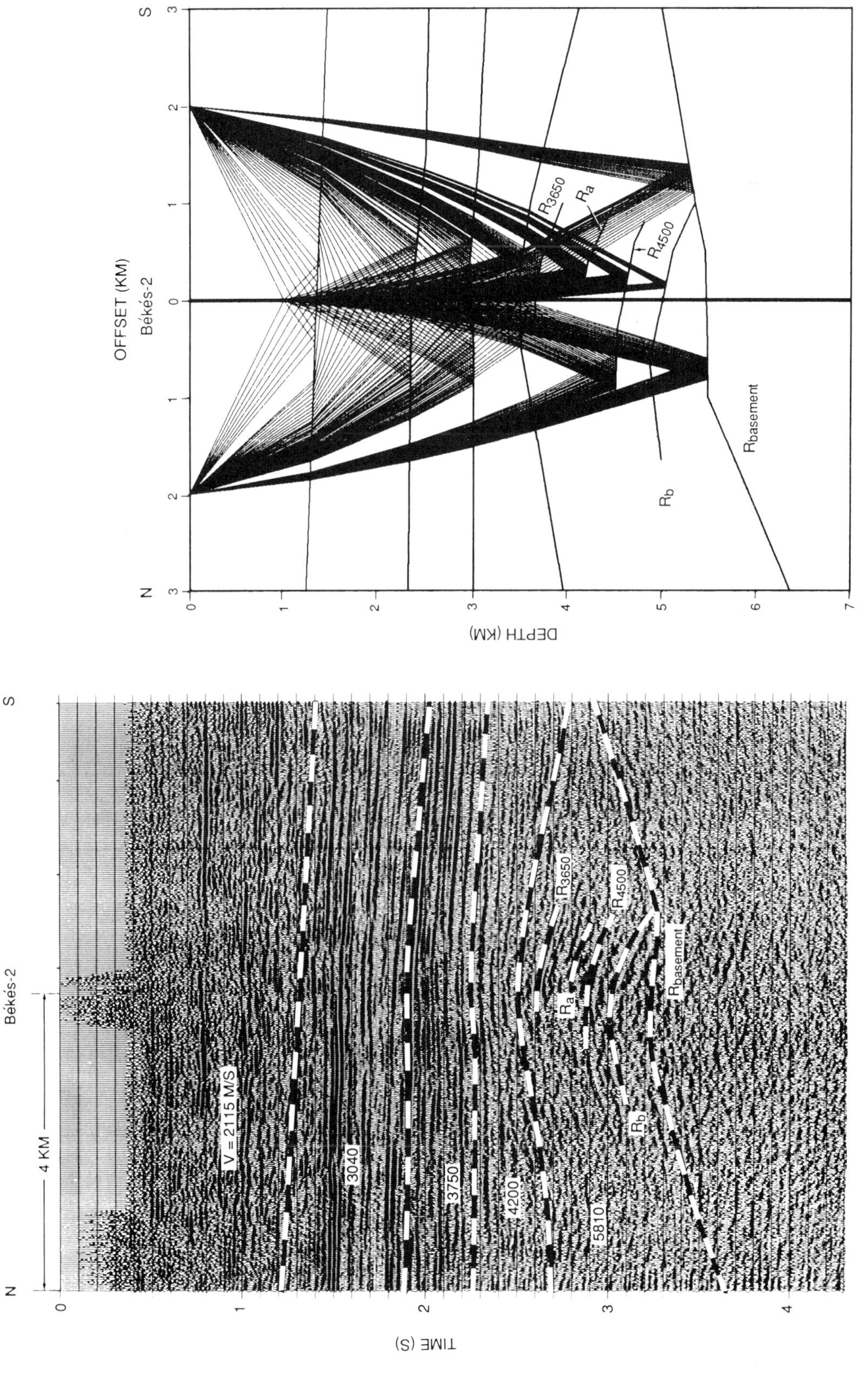

Figure 12. (Left) Interpretation of surface-recorded seismic profile shown in Figure 10. Interval velocities were calculated between prominent reflections. (Right) Results of applying offset VSP ray-tracing techniques to model shown on left. Assumed geophone spacing in the well was 100 m between depths of 1000-3500 m.

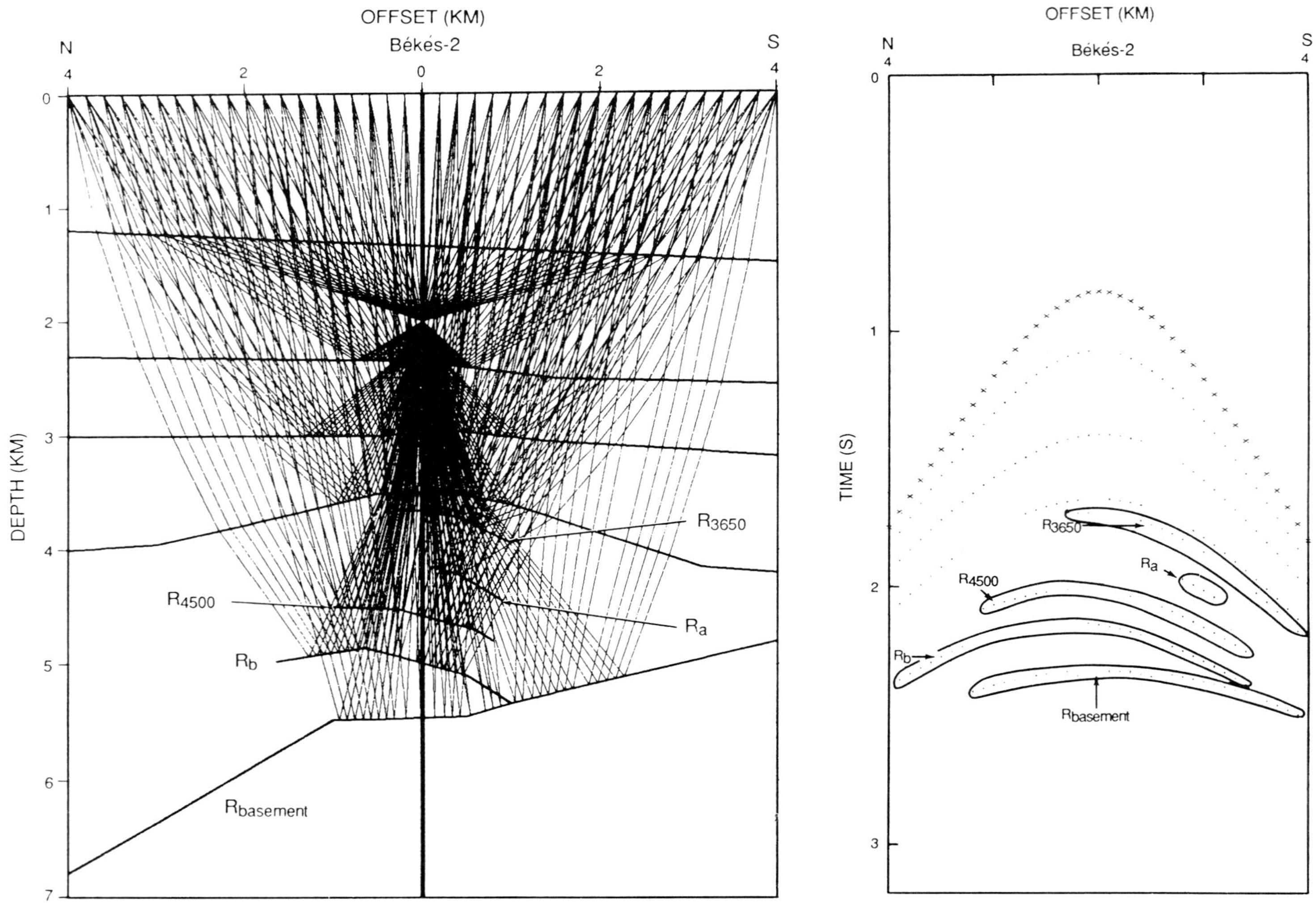

Figure 13. (Left) Results of walk-away, ray-trace model with a source interval of 200 m and the geophone located at a depth of 2000 m in the well. Note that the illuminated subsurface area is larger in comparison with the illuminated area of Figure 12b. (Right) Calculated arrival times for model shown on left. Crosses indicate arrival times for direct waves and dots indicate arrival times for reflected waves. Reflections which originated from the Mesozoic section are shown by dots enclosed by a line.

Table 1. Comparison between observed and computed first arrival times (in milliseconds (ms)) for offset VSP data. Observed (obs.) and calculated (cal.) arrival times are shown for an energy source located north of the well (columns 2 and 3) and for an energy source located south of the well (columns 4 and 5).

Depth (m)	1st Arrivals (ms) From North		1st Arrivals (ms) From South	
	Obs.	Cal.	Obs.	Cal.
1000	1080	1057	1078	1057
1100	1082	1080	1083	1080
1200	1084	1103	1094	1103
1300	1090	1128	1103	1128
1400	1096	1108	1113	1124
1500	1105	1113	1125	1129
1600	1113	1122	1138	1137
1700	1122	1132	1150	1141
1800	1134	1147	1165	1162
1900	1141	1162	1175	1174
2000	1147	1178	1184	1194
2100	1156	1197	1194	1213
2200	1171	1216	1209	1232
2300	1186	1235	1224	1258
2400	1203	1252	1240	1270
2500	1220	1264	1256	1283
2600	1239	1278	1276	1299
2700	1258	1293	1294	1315
2800	1276	1309	1311	1332
2900	1294	1327	1325	1350
3000	1313	1344	1343	1367
3100	1332	1359	1362	1382
3200	1351	1374	1378	1397
3300	1366	1391	1391	1394
3400	1384	1408	1411	1412
3500	1396	1424	1422	1448

Besides the R_{3650} and R_{4500} reflections, which are the two most persistent reflection events in the Mesozoic section, two other strong reflections, R_a and R_b, appear in the lower part of the Mesozoic section (Fig. 10). In order to further define reflections R_{3650} and R_{4500} and to confirm whether R_a and R_b are real (related to geology), ray-tracing techniques were used to construct a model. The first step in computing a model was to calculate interval velocities.

Interval velocities were first estimated from three VSP data sets assuming that all strata are horizontal (Fig. 11). In general, above a depth of 2200 m, the interval velocities computed from the VSP data with the energy source located north of the well are consistently higher than the velocities values computed from the other two sets of data (Fig. 11). Below 2200 m, the velocity values computed from the 3 data sets are similar. The average interval velocities calculated from the near-offset VSP survey were used in the model study.

Next, a simple model (in time) with discrete velocity layers was constructed from an interpretation of the surface-recorded seismic profile (Fig. 12, left). In this model, $R_{basement}$ is assumed to represent a decollement surface or the top of a deeply buried thrust sheet. This surface is assumed to be subparallel to the low-frequency reflectors that dip northward from 3.2 sec. beneath the Békés-2 well (Fig. 1, left). South of the well, the existence and location of this surface are speculative.

Ray-tracing techniques were applied to the model; results for a far-offset VSP configuration, assuming a 100-m geophone spacing in the well, are shown in Figure 12 (right). Two sets of observed and computed travel times for downward-traveling waves are given in Table 1 (one for an energy source located 2 km south of the well, and the other for an energy source located 2 km north of the well). The difference in computed first arrival times for downward-traveling waves from south and north source locations, at a depth of 3500 m, is 24 ms (1448 ms versus 1424 ms, respectively). The corresponding observed arrival times are 1422 and 1396 ms, a difference of 26 ms. The similarity between computed time differences (24 ms) and observed time differences (26 ms) indicates that the observed time difference is caused by dipping strata, rather than lateral velocity variation. To better fit the calculated arrival times to observed arrival times, an iterative least-squares method (Lines and others, 1984) could have been applied to the data. However, the amount of error involved is considered tolerable; therefore, such a method was not applied.

The results of modeling predict that reflections will be generated by R_{3650}, R_a, R_{4500}, R_b and $R_{basement}$ when the energy source is located south of the well; however, when the energy source is located north of the well, reflections are generated only by horizons R_{4500} and $R_{basement}$ (Fig. 12, right). With the exception of $R_{basement}$, the observed reflection pattern seen in Figure 6 correlates well with the reflections generated from ray-tracing techniques. On the observed VSP data set (Fig. 6), R_{4500} from the northern source is about 100 ms earlier than the corresponding reflection from the southern source for a geophone depth of 1000 m. On the model (Fig. 12, right), the arrival time of R_{4500} from the southern energy source is about 120 ms later than that calculated from the northern energy source. This is interpreted to mean that the model is consistent with the observed data.

The results of modeling (Fig. 12, right) demonstrate that reflections from Mesozoic strata for a given source location are generated only from a very small subsurface area. Thus, in order to map details over a large area, multi-offset VSP measurements should be made.

In contrast to the offset VSP method, the walk-away VSP method illuminates a larger subsurface area as illustrated in Figure 13, left. The general pattern of the computed arrival times of R_{3650}, R_a, R_{4500}, and R_b for the walk-away VSP model (Fig. 13, right) correlates well with the observed arrival times; but, the arrival time from the basement is very difficult to identify on the observed data.

In general, results of ray-trace modeling for the Mesozoic section correlate well with the measured VSP data except for $R_{basement}$. It is possible that three-component measurements would have provided better data in this area of complexly deformed rocks. R_{3650}, R_a, R_{4500}, and R_b, all with similar seismic expression, are interpreted to be reflections from relatively low-impedance lenses embedded in a relatively high impedance media. Because these data were recorded with a wall-locking geophone, the polarity of reflections generated by a change from high to low impedance is the same as that of the downward-traveling wave. R_{4500} (labeled with an arrow in Fig. 3) is characterized by a peak-trough combination separated by about 28 ms, and its peak amplitude is about 20% of the input of a downward-traveling wavelet. The polarity of the reflected event together with the sharpness of the downward-traveling wave indicates that the peak-trough reflection comes from a high-low-high impedance contrast - that is, a low-impedance layer embedded in a high-impedance media. The same is true for R_{3650}.

CONCLUSIONS

Based on the analysis of 3 sets of VSP data recorded at the Békés-2 well together with surface-recorded data from the same area, we conclude that high amplitude reflections are not generated from thrust-fault surfaces in the Mesozoic section. However, high-amplitude, low-frequency reflections from the Mesozoic section appear to be associated with low-impedance layers that do not extend very far from the well site. The trend of these low-impedance layers appears to parallel the trend of thrust faults and probably represent fracture zones associated with complex Mesozoic decollement folds. Reflections from the Mesozoic section cannot be mapped as coherent seismic events over distances of more than a few kilometers from the well site.

REFERENCES

Balch, A. H., and Lee, M. W., 1984, Vertical seismic profiling - Techniques, applications and case histories: International Human Resources Development Corp., Boston, 488 p.

Caldwell, J. G., 1984, Vertical seismic profiling in the Western U.S. Overthrust Belt; *in* Balch, A.H., and Lee, M. W., (eds.), Vertical seismic profiling - techniques, applications, and case histories; International Human Resources Development Corp., Boston, 425-457.

Grow, J. A., Mattick, R.E., Bérczi-Makk, A., Péró, C., Hajdú, D., Pogácsás, Gy., Várnai, P., and Varga, E., Structure of the Békés basin inferred from seismic reflection, well and gravity data, this volume, 1-38.

Gray, N.C., 1978, Variable norm deconvolution: Ph. D. thesis, Stanford University, Palo Alto, California, 88 p.

Hardage, B.A., 1983, Vertical seismic profiling, Part A: Principles, *in* Handbook of Geophysical Exploration Series; Geophysical Press, 14A, 450.

Kennet, P., and Ireson, R.L., 1981, The VSP as an interpretation tool for structural and stratigraphic analysis (abs.); Proc., 43rd Annual EAEG Meeting, Venice, Italy.

Lee, M.W., 1984, Processing of vertical seismic profile data, *in* Simaan, M., (ed.), Advances in geophysical data processing; JAI Press Inc., Greenwich, Connecticut, 1, 129-160.

Lines, L.R., Bourgeois, A., and Covey, J.D., 1984, Travel-time inversion of offset vertical seismic profiles - a feasibility study; Geophysics, 49, 250-264.

McDonnell, M.J., 1981, Box filtering technique; Computer Graphics and Image Processing, 23, 258-272.

Oristaglio, M. L., 1985, A guide to current uses of vertical seismic profiles; Geophysics, 50, 2473-2479.

14 Modeling Seismic Reflection Data in the Vicinity of the Békés-2 Well

John J. Miller[1] *and István Véges*[2]

ABSTRACT

In the vicinity of the Békés-2 well, located near the center of the basin, seismic reflection data show a complex geologic structure in the Mesozoic-age basement rocks below a depth of 3400 m. To assist in the interpretation of the geometry of the structure, three nearby seismic profiles were reprocessed, vertical seismic profiles (VSPs) were recorded in the well, and a new surface seismic line was recorded concurrently with one of the VSPs. Interpretations of the surface seismic data were integrated with those of the VSP data and a structural model of the Mesozoic-age rocks was developed. Two-dimensional seismic modeling aided in modifying the interpretation and the synthetic seismic response of the final interpretation gave a reasonable match to the observed data. This final interpretation is a good approximation of the geometric relationships between seismic reflection boundaries in the Mesozoic-age rocks in the area of the Békés-2 well site.

INTRODUCTION

As part of a comprehensive investigation of the geologic evolution and the oil and gas potential of the Békés basin, seismic reflection and well data were correlated and interpreted by Mattick and others (this volume), Molenaar and others (this volume) and Grow and others (this volume). Investigations of Grow and others (this volume) were aimed at deciphering the tectonic history of the basin. Although the reflection data were easy to interpret in the undeformed Miocene-Holocene sedimentary section (to a depth of about 3400 m; Grow and others, this volume), the complex geologic structure of the basement rocks beneath the sedimentary section could not be interpreted uniquely. Thus, as a corollary study, a structure involving Mesozoic rock sequences was chosen for evaluation in the center of the basin near the Békés-1 and Békés-2 well sites (Fig. 1). The test involved reprocessing and interpreting three seismic lines (Gyula(Gyu)-29, Gyu-30, and Békés(Bé)-43; Fig. 1). These three lines were chosen because they appeared to be in a critical location for developing a tectonic model of the basin.

The results of a study by Lee and Göncz (this volume), who recorded 4 vertical seismic profiles (VSPs) at the Békés-2 well site: a near-offset VSP, two far-offset VSPs (offset = 2 km), and a walk-away VSP, were used to aid in the interpretation. The VSPs were supplemented by a surface seismic line (line Bé-2; Fig. 1) recorded concur-

[1] U.S. Geological Survey, Denver, Colorado 80225, USA
[2] MOL Rt. (Hungarian Oil and Gas Co., Ltd.), H-1062, Budapest, Hungary

P. G. Teleki et al. (eds.), Basin Analysis in Petroleum Exploration, 277–294.

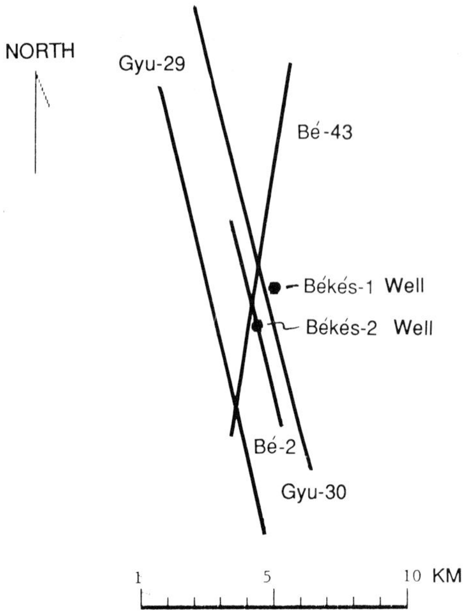

Figure 1. Map showing relative locations of Békés-1 and Békés-2 wells and seismic lines Gyu-29, Gyu-30, Be-43, and Békés-2 (see also: Grow and others (this volume), for a generalized map of the Békés basin).

rently with the walk-away VSP. To test the validity of the interpretation, two-dimensional seismic modeling techniques were applied, similar to those commonly used in petroleum exploration to test geologic interpretations (Ryder and others, 1981; Davis and Jackson, 1988; Anderson and others, 1989) and to aid in thc selection of processing parameters (May and Covey, 1983). This chapter describes the reprocessing of the surface data and the two-dimensional modeling of line Gyu-30.

DATA ACQUISITION

Seismic reflection lines Gyu-29, Gyu-30 and Bé-43 were conventionally recorded using an impulsive energy source, 48 receiver channels per shot and a nominal 24-fold subsurface coverage. Table 1 shows details of the recording parameters used for these lines. Note that the recording system for line Bé-43 was an old system and employed only 24 channels. In order to record 48 channels per shot (and hence achieve 24-fold subsurface coverage) on Bé-43, each source point was energized twice, first with the 24 receiver channels placed at stations behind the source point, second with them placed ahead of the source point (e.g., with the source point at station 25, channels 1-24 were located at stations 1-24 for the first shot and at stations 26-49 for the second shot).

Line Bé-2 and the walk-away VSP were recorded concurrently. The surface line extended 4 km on either side of the Békés-2 well. Vibroseis techniques were used to

Table 1. Recording parameters for lines Gyu-29,-30, Bé-2 and Bé-43. NA: not applicable.

PARAMETERS	Gyu-29, Bé-43, Gyu-30	Bé-2
Number of Channels:	48, 48, 24	120
Sample interval:	2 ms	4 ms
Record length:	5 sec	14 sec
Energy source:	dynamite	Vibroseis
Sweep frequencies:	NA	8-64 Hz
Sweep length:	NA	9 sec
Source depth:	18 m	surface
Charge size:	4 kg	NA
Group interval:	80 m, 80 m, 70 m	50 m
Fold coverage:	2400 %	3000 %
Receiver pattern:	split-spread	variable

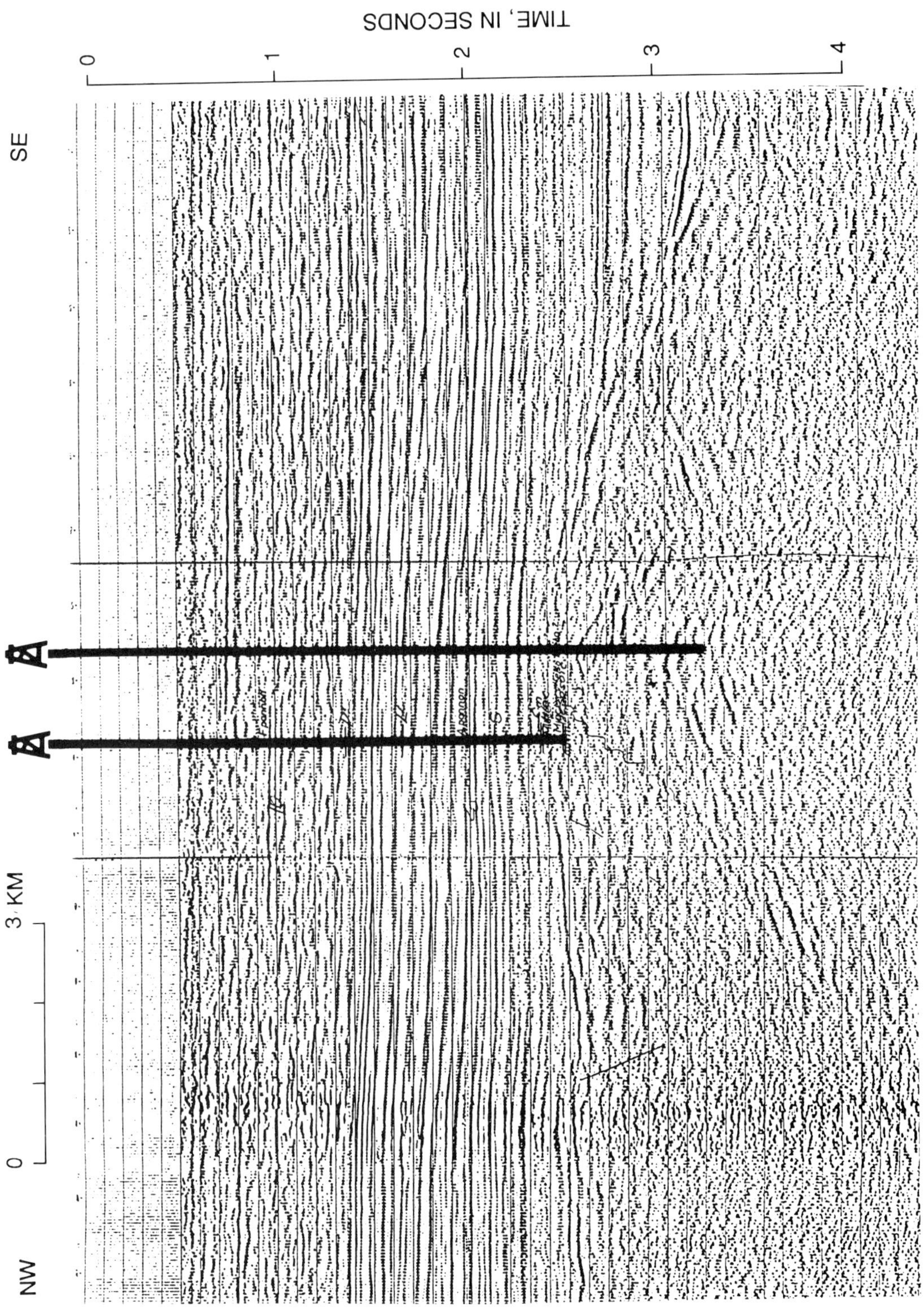

Figure 2. Line Gyu-30 (migrated stack) as originally processed.

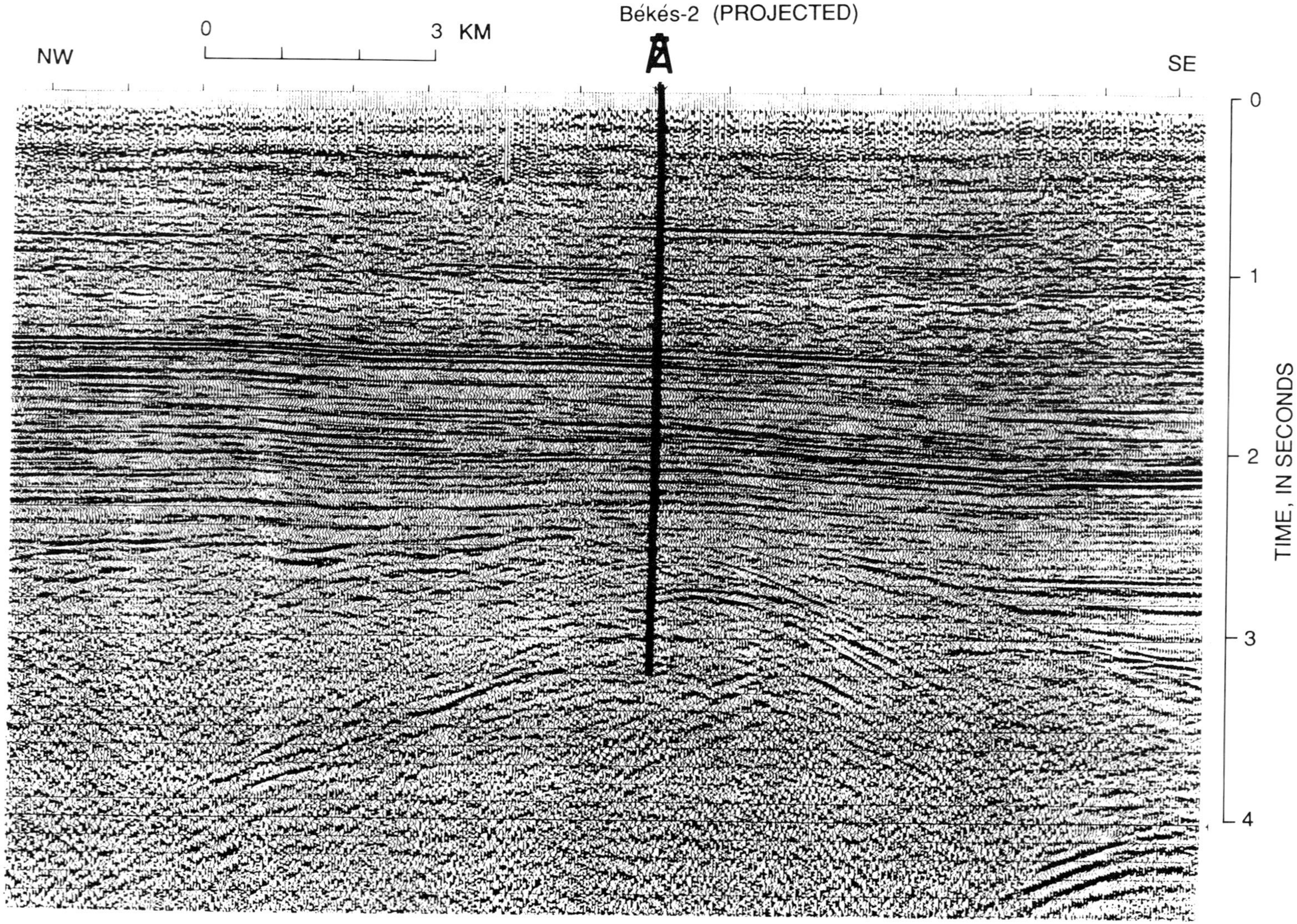

Figure 3. Line Gyu-30 reprocessed to the stage of unmigrated stack.

Table 2. Processing parameters for lines Gyu-29, Gyu-30, Bé-43 and Bé-2.

PROCESSING STEP	PARAMETER(S)
Demultiplex:	Reformat field tapes
Gain Recovery:	T^2 function[1]
Scale:	One window balance
Deconvolution: (Gyu-29, -30 & Bé-43)	Spiking, 3 windows, 300 ms filter
Spectral whitening: (Bé-2)	10-45 Hz
Datum Statics:	Datum = 50 m
Velocity analysis: (NMO[2], brute stack)	
Residual statics: (Surface consistent)	1 window max. shift = 24 ms
Velocity Anal., final NMO	
Mute, Stack, Scale:	750 ms AGC[3]
Wave equation migration:	V = stacking-10%

1) T: Two-way travel time in seconds
2) NMO: Normal moveout
3) AGC: Automatic gain control

generate energy and 120 channels were recorded per source location. One channel was reserved for recording signals from the downhole geophone of the VSP experiment. The recording parameters and geometric configuration of sources and receivers (Table 1) were designed to achieve maximum fold and a nominal value of 30-fold was obtained.

PROCESSING OF THE SURFACE DATA

Reprocessing

Lines Gyu-29, Gyu-30 and Bé-43 were reprocessed using a conventional processing sequence that included the following: (1) scaling to recover amplitude losses due to geometrical spreading, transmission and absorption, (2) deconvolution, (3) detailed velocity analyses, (4) residual statics determination, (5) stacking, and (6) migration. The processing sequence and parameters used are shown in Table 2.

Line Gyu-30, as originally processed (through migration), is shown in Figure 2. The reprocessed version of line Gyu-30 is shown in Figure 3 (unmigrated) and Figure 4 (migrated). The reprocessed versions (through migration) of line Gyu-29 and Bé-43 are shown in Figures 5 and 6, respectively.

Processing of line Bé-2

A data processing sequence, identical to that used in reprocessing lines Gyu-30, Gyu-29 and Bé-43, was used in processing line Bé-2. However, a spectral whitening algorithm (Lee, 1986) was used in place of spiking deconvolution because its application provided an increased signal bandwidth.

The results of the VSP experiment were used in developing a velocity model for migrating line Bé-2. Seismic velocities determined from the VSP experiment are assumed to be more accurate than those determined from the surface data, because in the VSP technique velocities are measured directly from receivers positioned at known depths and the resulting travel times can be converted directly to velocity. Lee and Göncz (this volume) discuss the VSP data and its geologic interpretation. The migrated version of line Bé-2 is shown in Figure 7.

MODELING

A unique geologic interpretation of seismic reflection data is sometimes difficult to obtain. Forward, or direct, modeling (i.e., calculation of the seismic response of an assumed geologic structure and lithology) can be helpful in resolving some of the non-uniqueness inherent in interpretations of seismic reflection data (Sheriff, 1980). The normal procedure is as follows:

1) A geologic interpretation is made from the seismic record section and all other available geologic information. This interpretation includes, at a minimum, the geometrical configuration of the rock sequences and faults (usually interpreted directly from the seismic record-section) and the physical properties of the rocks obtained from well logs or calculated from the seismic data.

2) The two-dimensional seismic response (synthetic response) of the interpreted structure is calculated.

3) The synthetic response is compared to the actual, observed seismic data. If there is little similarity between

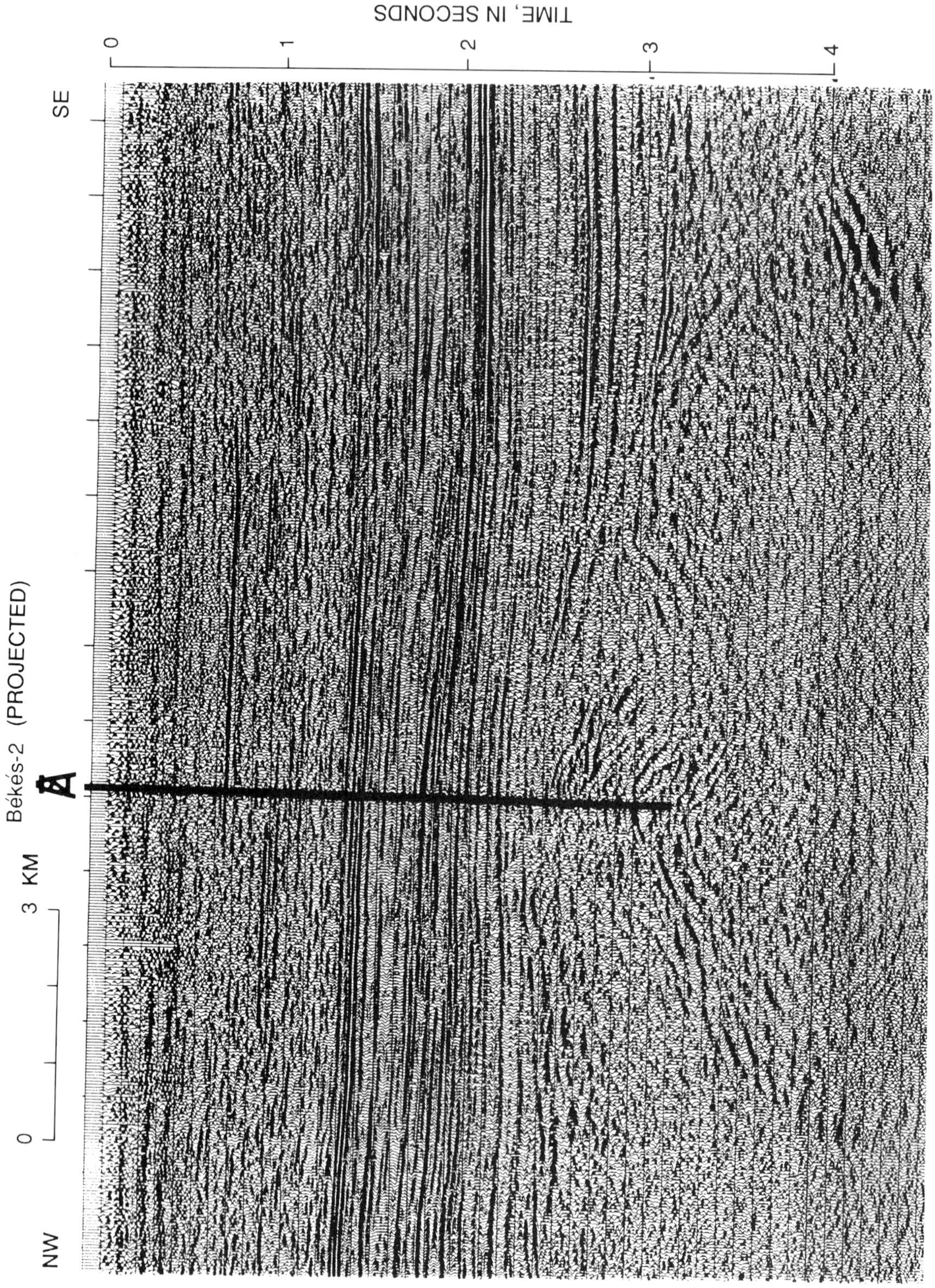

Figure 4. *Line Gyu-30 processed to the stage of migrated stack.*

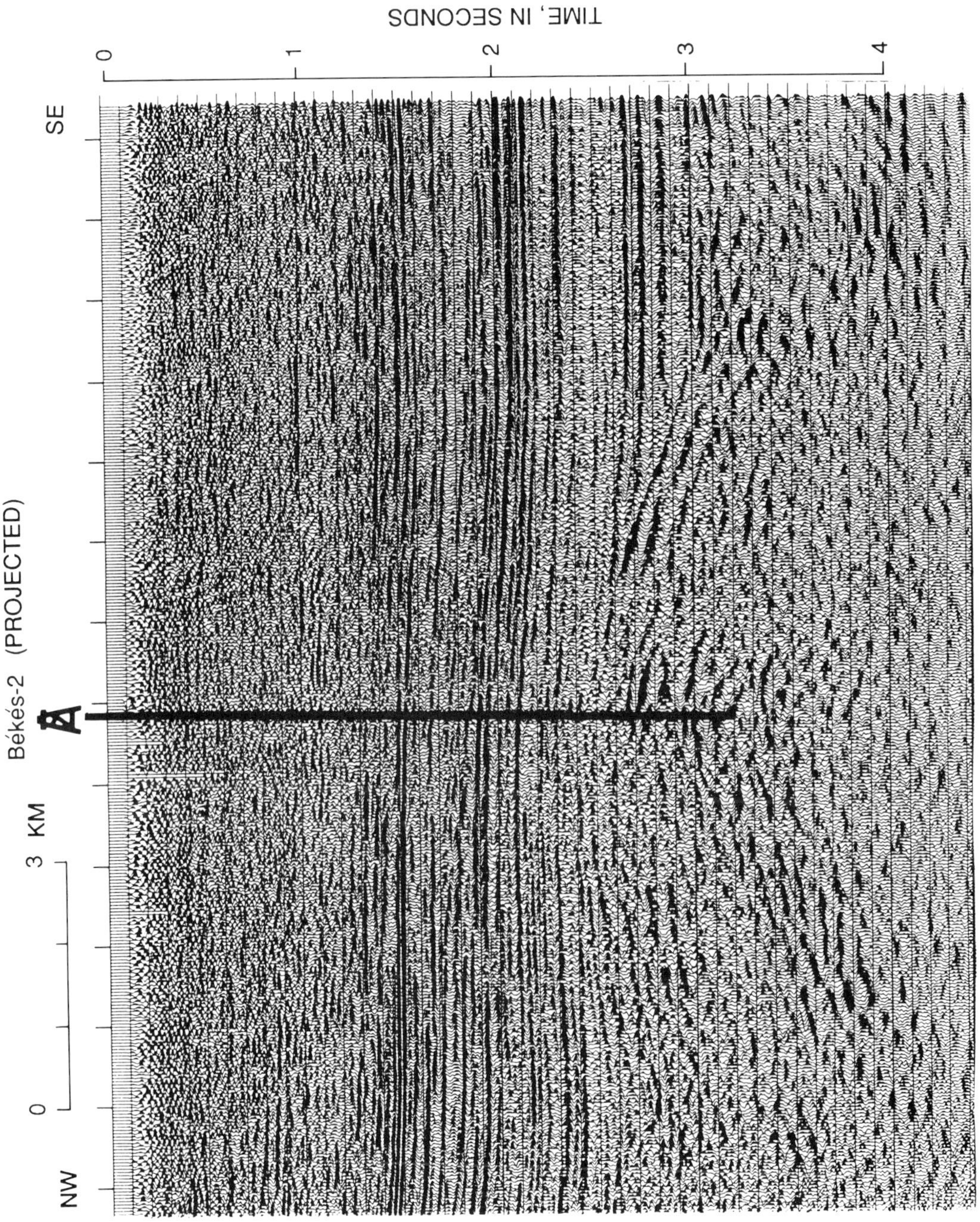

Figure 5. Line Gyu-29 reprocessed to the stage of migrated stack.

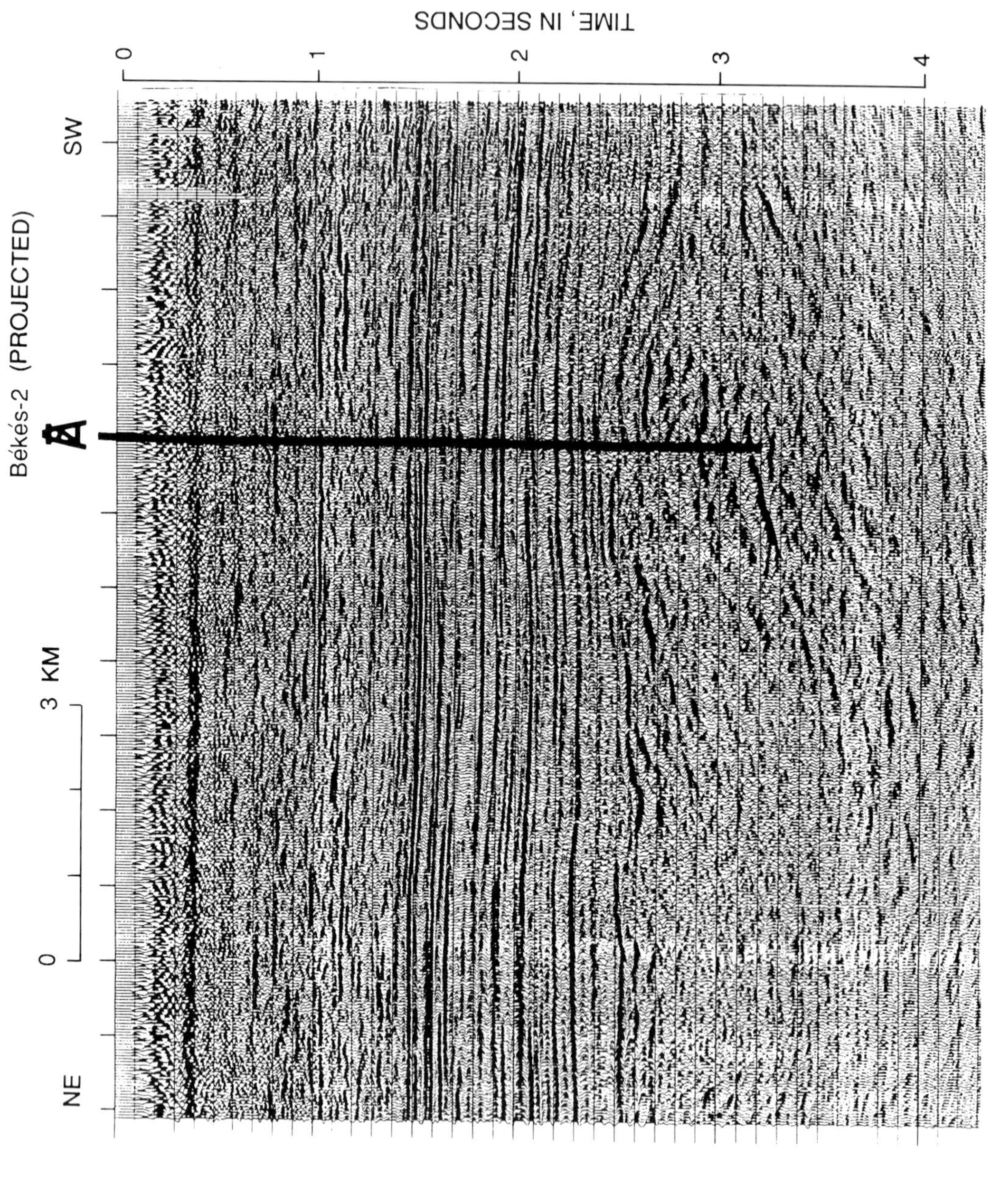

Figure 6. Line Bé-43 reprocessed to the stage of migrated stack.

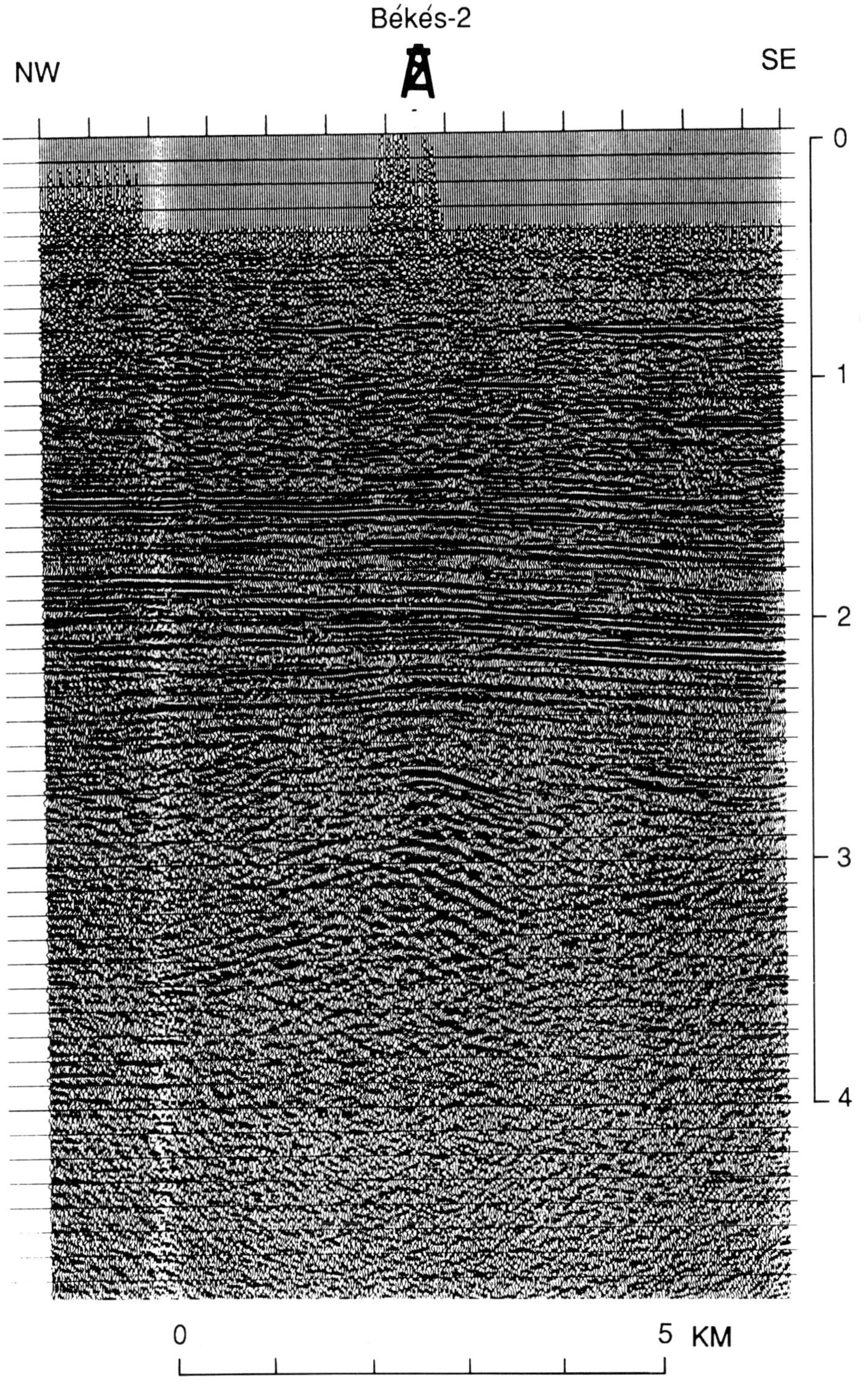

Figure 7. Line Bé-2 processed to the stage of migrated stack.

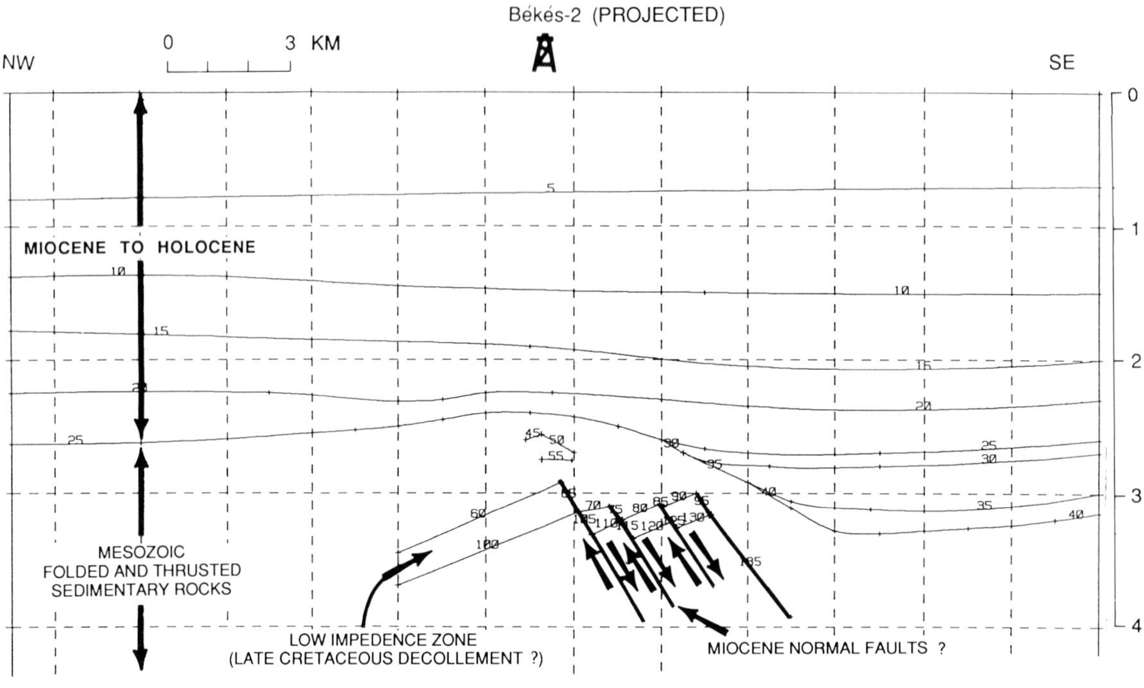

Figure 8. Initial structural model (in time) calculated from interpretation of seismic reflection line Gyu-30 (Fig. 2). This is a computer-generated display and the numbers on this and subsequent displays of modeled data are annotated for computer calculation purposes only.

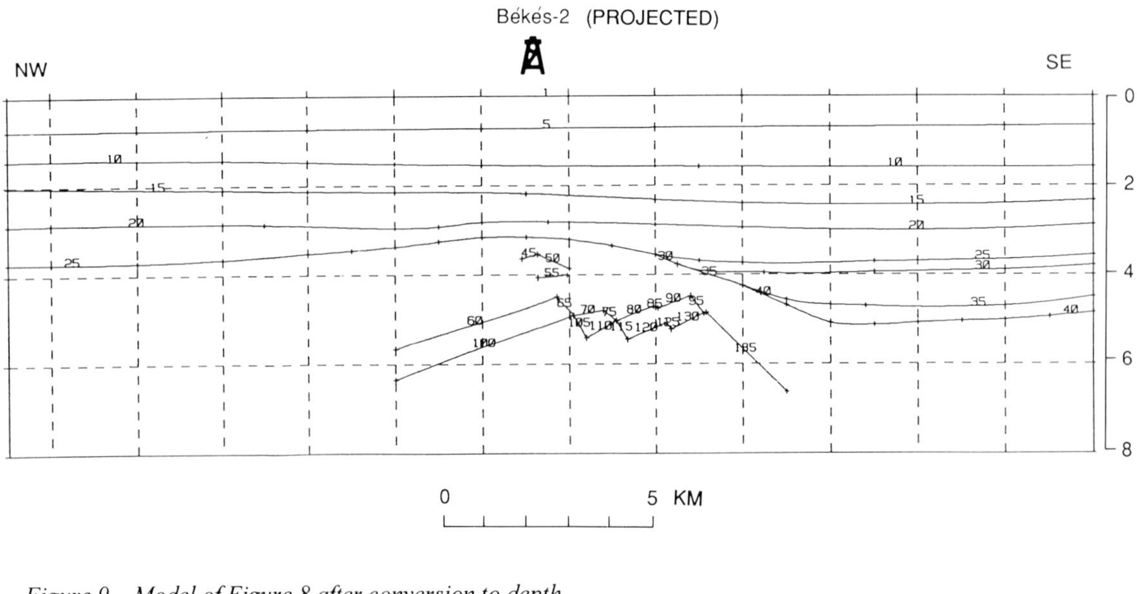

Figure 9. Model of Figure 8 after conversion to depth.

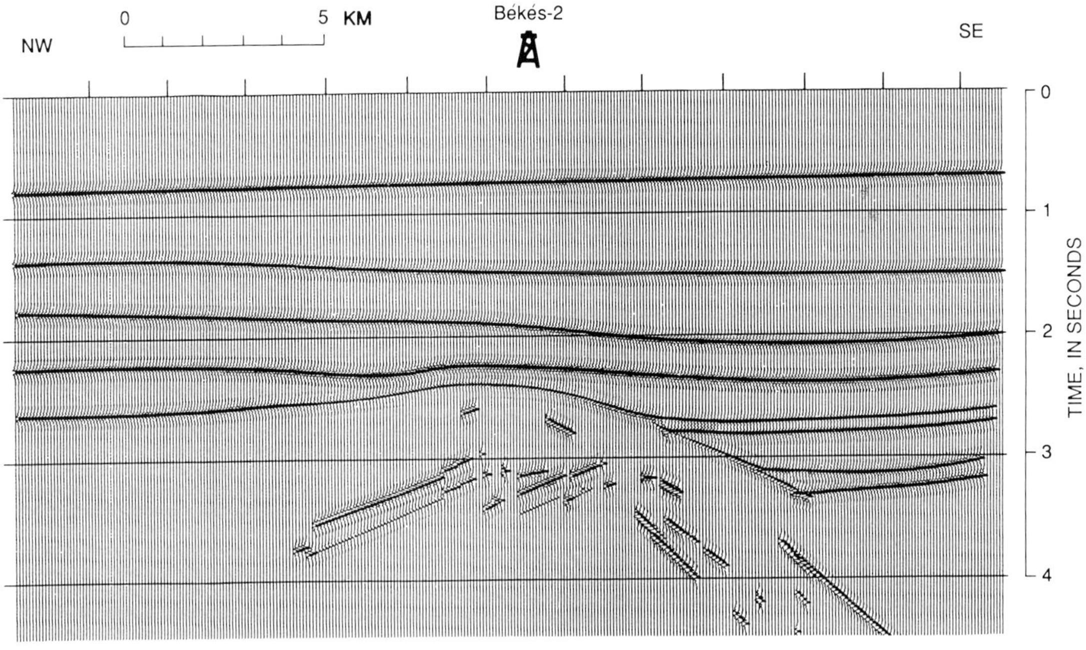

Figure 10. Calculated synthetic response from model shown in Figure 9. Normal-incidence ray tracing techniques were used in the calculation.

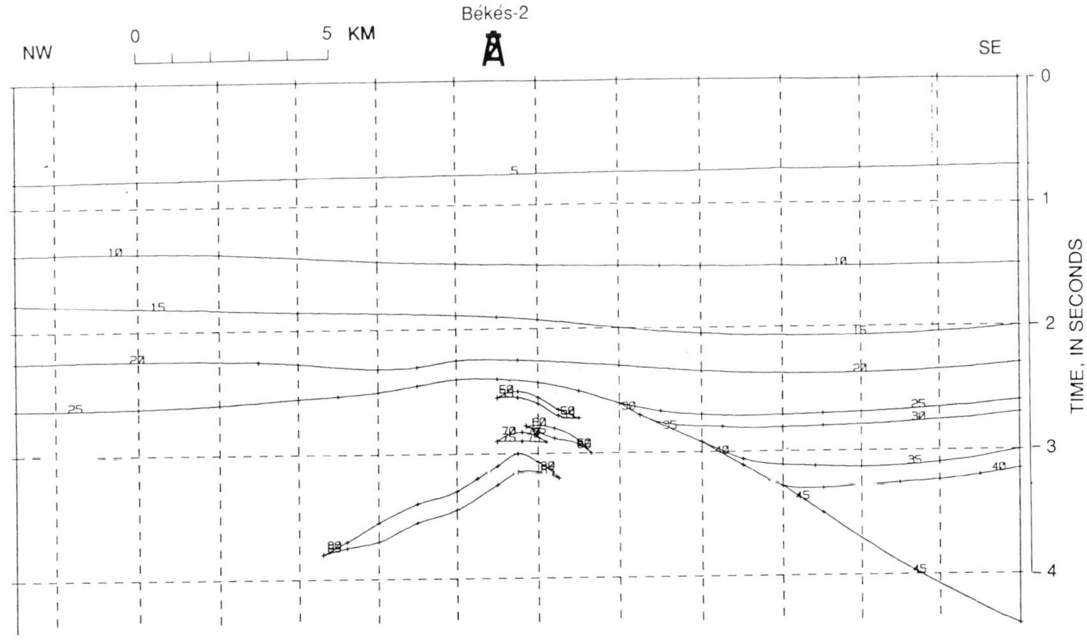

Figure 11. Second model (in time) which was constructed from a revised interpretation of seismic profile Gyu-30. Information from the Békés-2 surface line and Vertical Seismic Profiles (VSPs) were used to construct this model.

the synthetic and observed data, then the geologic interpretation is assumed to be incorrect. A high degree of similarity indicates that the calculated model probably represents the subsurface structure. Hence, modeling cannot prove that an interpretation is correct, but it can prove that an interpretation is incorrect. If the modeled results indicate an incorrect interpretation, then a new interpretation is made and the modeling procedure repeated. This iterative technique is continued until a reasonable match between modeled and observed data is obtained.

An attempt was made to calculate one possible geometric interpretation of the complex structure of the buried Mesozoic rocks at the Békés-2 well site. In the resulting interpretation, modeling the amplitude anomalies present in the seismic data was not successful, because density logs recorded in the Békés-2 well did not extend to the required depths. From VSP experiments, Lee and Göncz (this volume) interpreted these high amplitude reflections as low impedance strata surrounded by high impedance media.

In the modeling calculations, available commercial software (AIMS: Advanced Interpretive Modeling System, Geoquest International) was used. This software utilizes two-dimensional ray-tracing and wave-equation methods and is described by Miller and others (1985).

Line Gyu-30, as originally processed (Fig. 2), was interpreted and the result is shown in Figure 8. The discontinuous reflection segments below 2.5 sec. (Fig. 2) were interpreted to represent strata offset by Miocene normal faults (Fig. 8), and the continuous, high-amplitude reflection that dips northwest from a travel time of 3 sec. is interpreted to represent a late Cretaceous decollement (Fig. 8), as reported by Grow and others (this volume). The high-amplitude reflection which begins at 4 sec. and dips to the left (right side of Figs. 3 and 4) was not modeled. This seismic event does not occur on any of the other lines recorded in the area, and it is possible that it arises from wave focussing from a point outside of the plane of the seismic section, an effect that cannot be compensated for by the modeling software used.

The initial interpretation, in time (Fig. 8), was converted to depth using interval velocities (from a depth of 0 - 3400 m) derived from the near-offset VSP recorded at the Békés-2 well site and velocity estimates from nearby wells for depths below 3500 m (Fig. 9). The synthetic, ray-trace response of the data in Figure 9 is shown in Figure 10. This synthetic response can be compared directly to the unmigrated seismic data (Fig. 3). Below 2.5 sec., in the vicinity of the well, the similarity between the synthetic and observed data disappears. Noticeably absent are the curved reflection, with its apex at 2.55 sec., and the high-amplitude reflection at 2.7 sec. These features correspond, respectively, to dipping and horizontal reflections observable on the migrated section (Figs. 2 and 4). Because of the poor coherence between the modeled and observed data, this initial interpretation is assumed to be incorrect.

A new interpretation was made, therefore, from the reprocessed version of line Gyu-30 (Fig. 4), line Bé-2 (Fig. 7), and information from the VSP experiments. Lee and Göncz (this volume) had interpreted the reflecting boundaries deeper than 2.5 sec. as low-velocity, lenticular strata (referred to as lenses) surrounded by high velocity media. The revised interpretation (shown with a time scale in Figure 11) represents the geometric configuration of these lenses as interpreted from Figure 4. This interpretation was converted from time to depth using all available velocity information, including that derived from the VSP data, and is shown in Figure 12.

The ray-trace response (primaries only) of Figure 12 was calculated and the result is shown in Figure 13. The similarity between the synthetic response and the unmigrated recorded section (Fig. 3) is good, except for two features. First, the so-called bow-tie patterns, which occur between 3 and 4 seconds and are associated with the lowest two reflections at the left side of the synthetic data (A on Fig. 13), are not present on the unmigrated recorded section (Fig. 3). Second, the downward curving reflections at the right side of the synthetic section, between 3 and 4 sec. (B on Fig. 13), extend much further southeastward than in the unmigrated recorded section (Fig. 3). Had diffractions been included in the modeled response, these events would have extended even further southeastward and the bow-tie effects would be more evident. Selected rays, traced upward from the modeled lenses, show the crossing pattern which causes the bow-tie effects (A on Fig. 14). Also shown are those rays which originate at the right edges of the lenses and emerge at the ground surface far to the right of the lenses (B on Fig. 14). The model of Figure 12 was modified to smooth that part of the lenses which caused the rays to cross and to truncate the right edges of the lenses. These modifications produced the final interpretation (Fig. 15).

The ray-trace response (primary reflections only) of the final interpretation is shown in Figure 16. The bow-ties

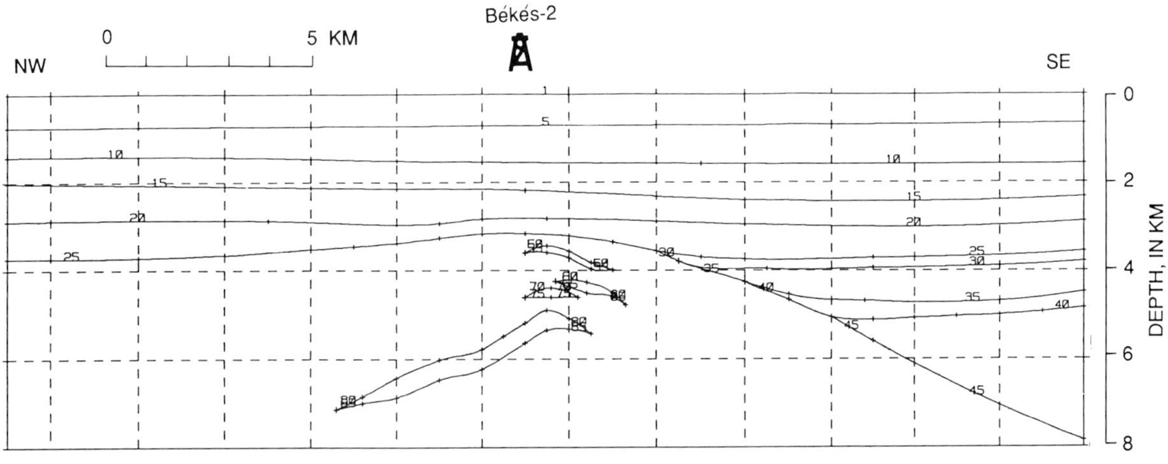

Figure 12. Model shown in Figure 11 after conversion to depth.

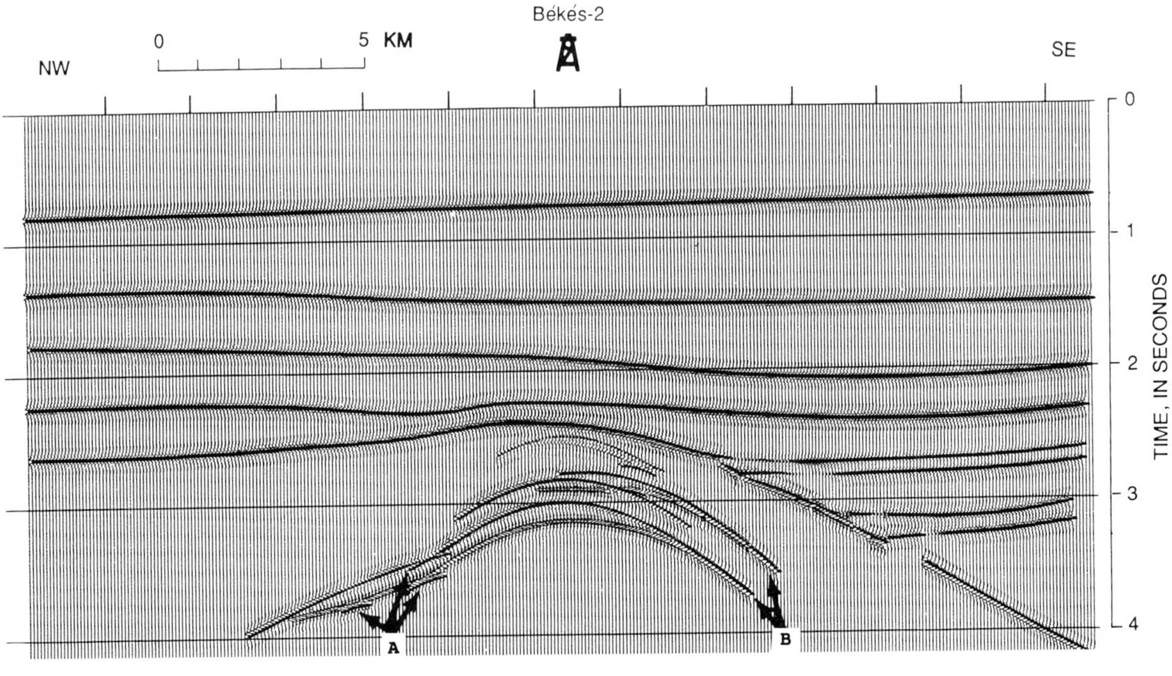

Figure 13. Synthetic response calculated from model shown in Figure 12. Normal-incidence ray tracing methods were used in the calculations. "Bow-tie" patterns (indicated by A) and reflections extending southeastward (indicated by B) cannot be seen on unmigrated stack of line Gyu-30 (Fig 3).

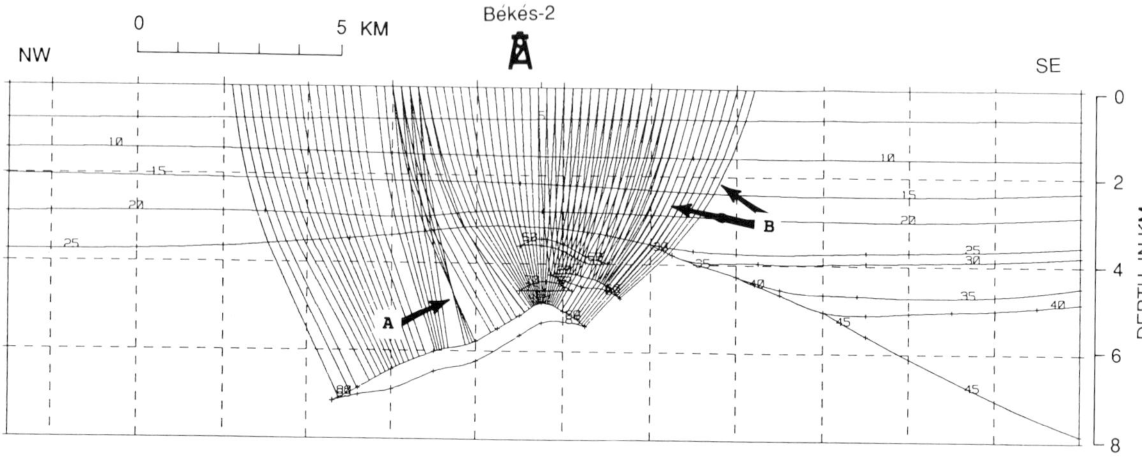

Figure 14. Figure shows normal-incidence rays that were traced from selected interfaces on Figure 12. Crossing rays are indicated by A, and rays emerging far to the right of their origin are indicated by B.

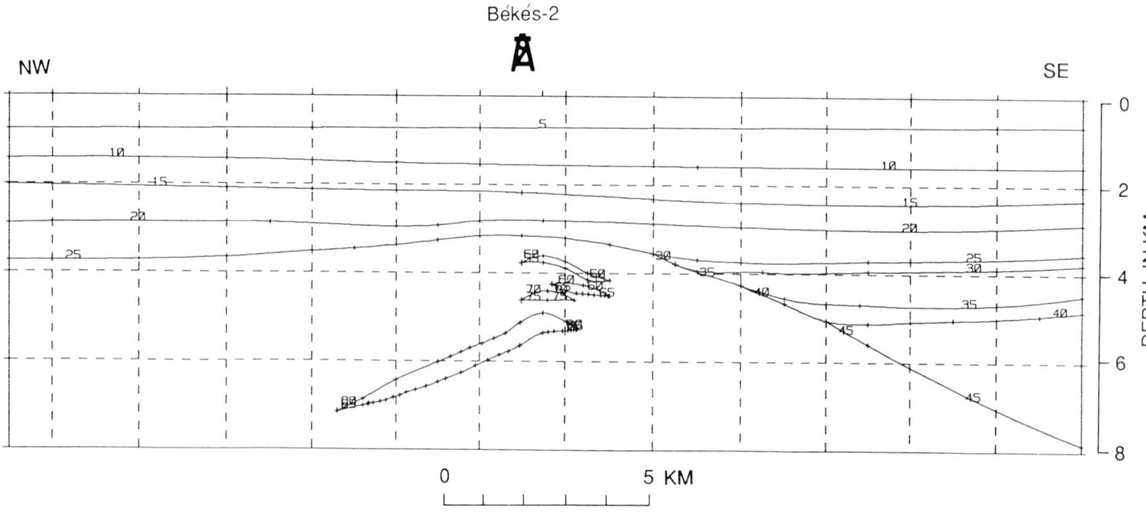

Figure 15. Final model modified from Figure 12. Crossing rays were eliminated, and the right edges of lenses were truncated.

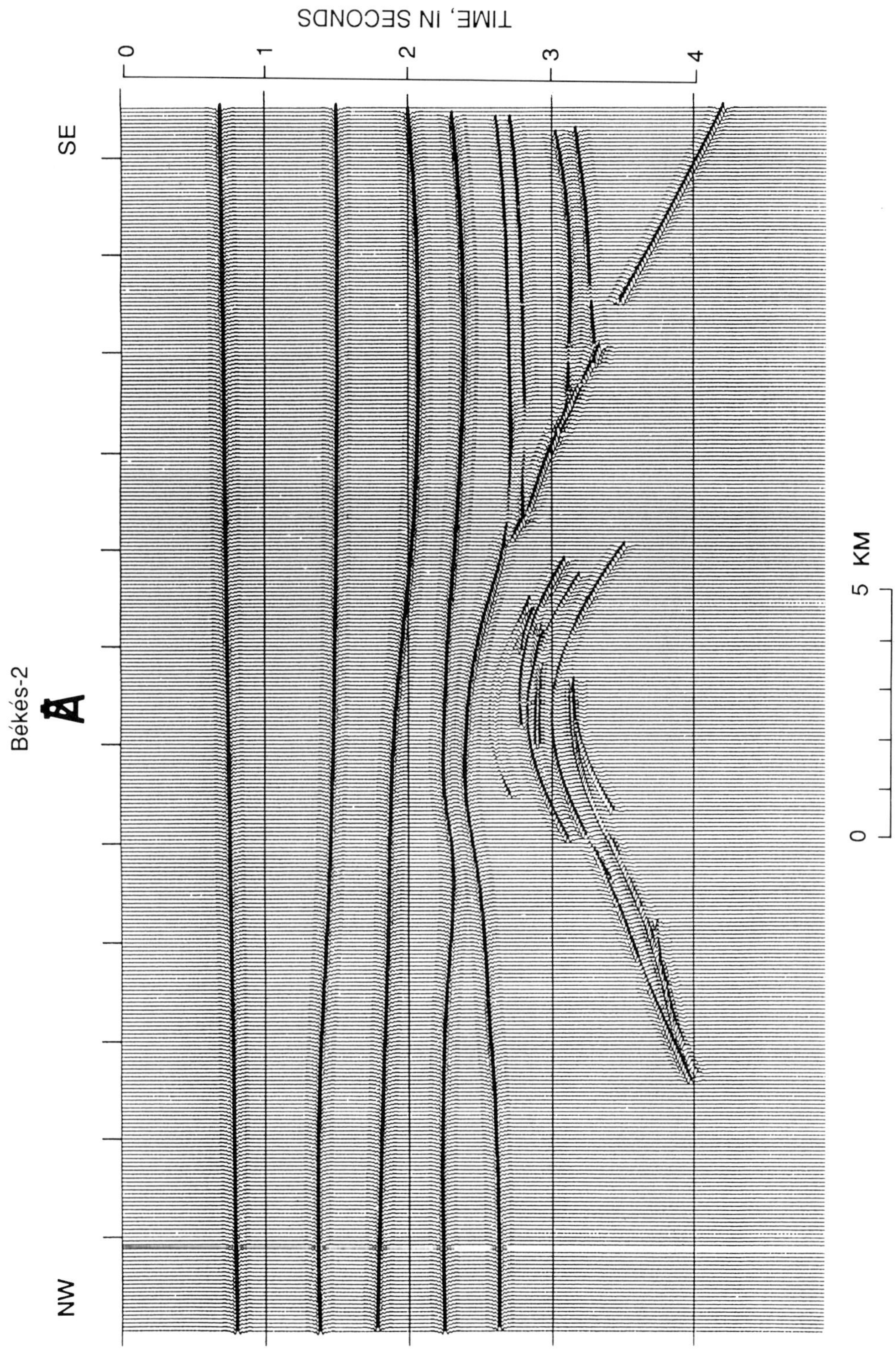

Figure 16. Synthetic response of final model (Figure 15) calculated using normal-incidence ray tracing methods.

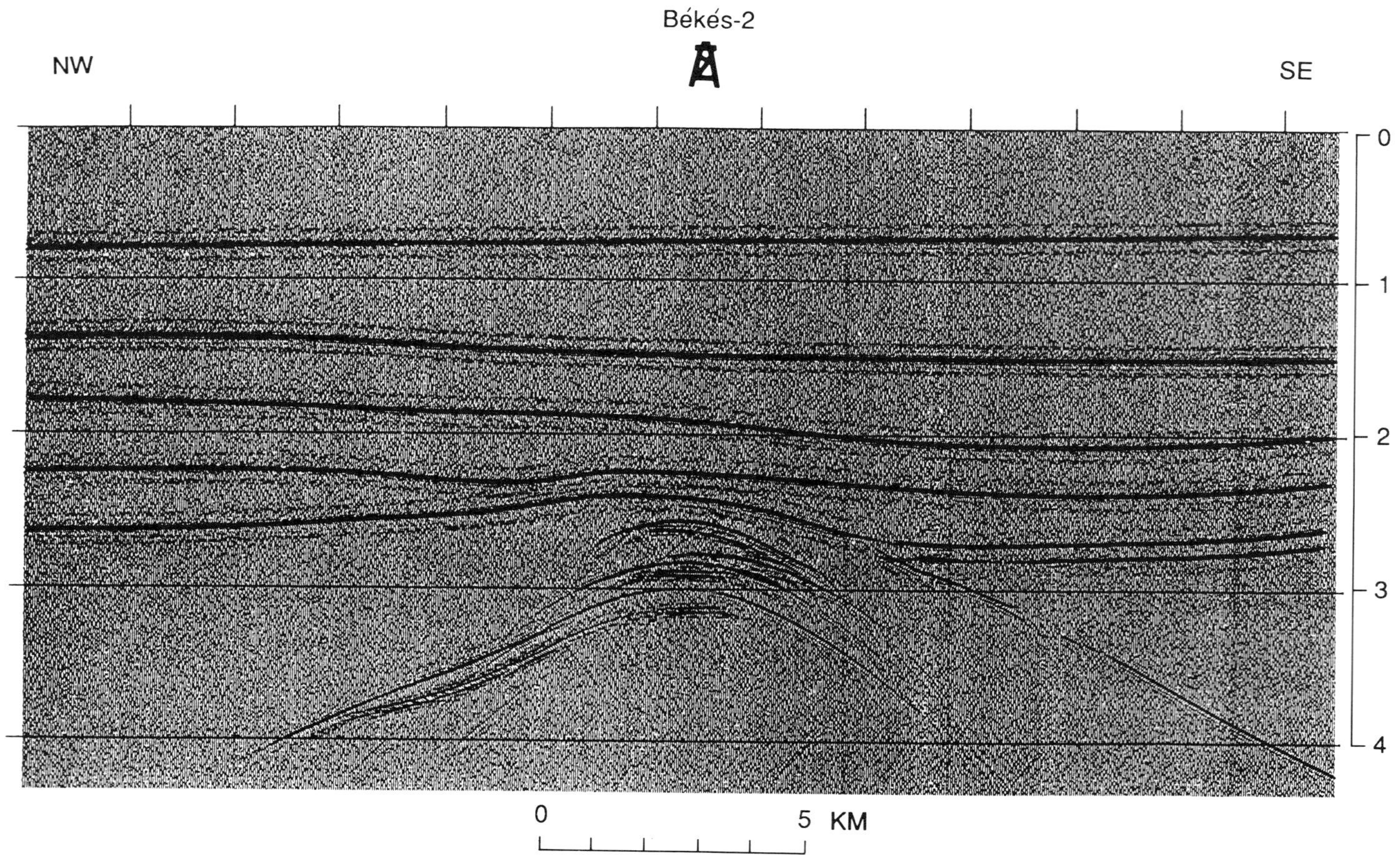

Figure 17. Synthetic response of final model (Figure 15) using a scalar wave-equation algorithm.

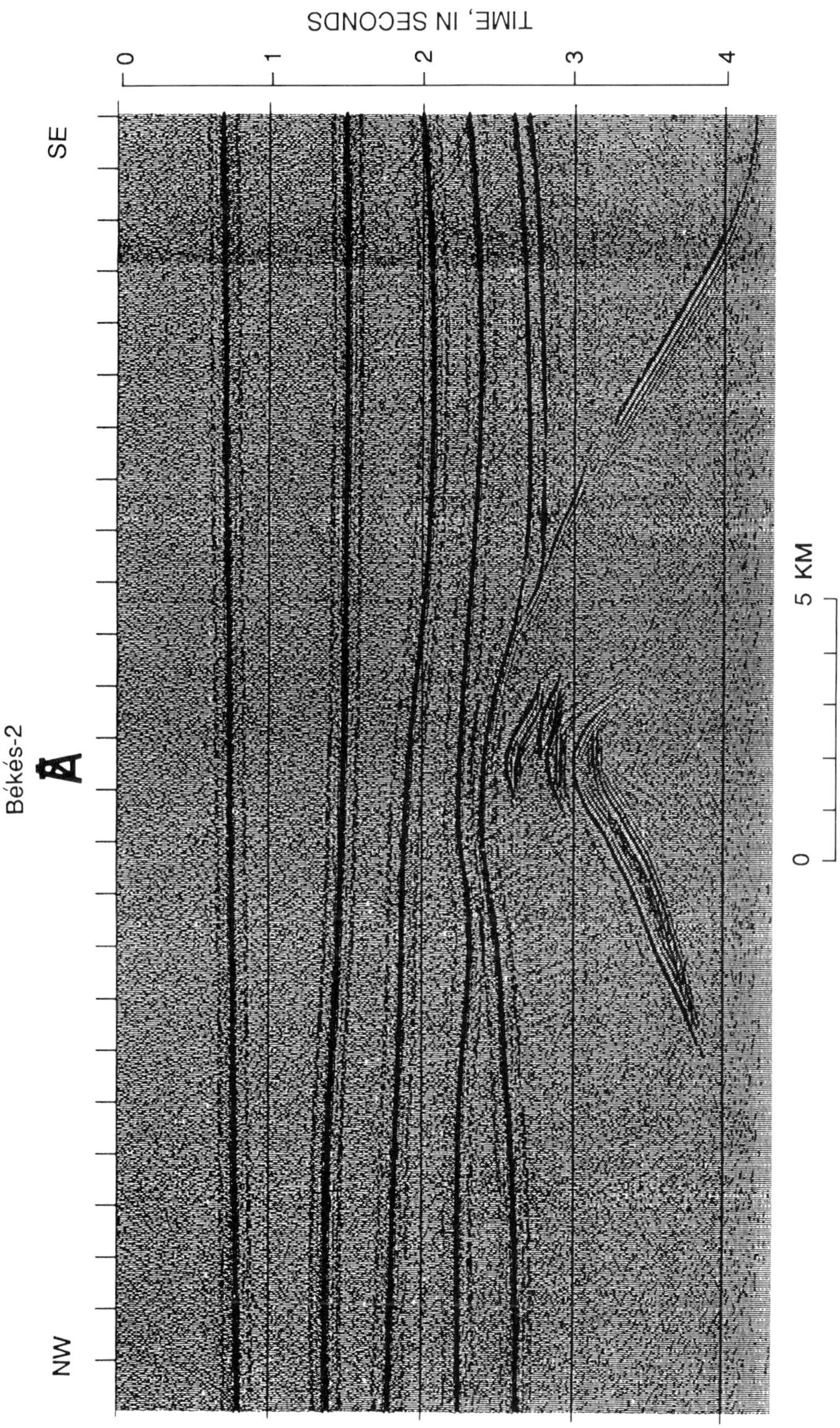

Figure 18. Migration of data shown in Figure 17. Velocities used in the migration are the same as those used in migrating the data on line Gyu-30.

patterns have, for the most part, disappeared and the downward curving reflections at the right of the figure closely resemble those of the unmigrated recorded section. The calculated synthetic response of Figure 15, using a scalar wave-equation which accounts for diffractions as well as primary reflections, is shown in Figure 17. In the constructed model (Fig. 17), a small amount of random noise has been added to the synthetic data. This figure shows a reasonable match to the unmigrated recorded section shown in Figure 3.

As a last check on the validity of our interpretation, the data displayed in Figure 17 were migrated and the result is shown in Figure 18. The same velocity model used in migrating line Gyu-30 was used for this migration. The match between the synthetic data (Fig. 18) and the migrated version of line Gyu-30 (Fig. 4) is quite good. Below the Békés-2 well, the curved, southeast dipping reflection with its apex at 2.55 sec., the horizontal reflection at 2.75 sec., the reflections between 2.8 and 2.9 sec., and the reflection that curves downward in two directions from a travel time of 3 sec. are all present on modeled data.

CONCLUSIONS

Seismic line Bé-2 and reprocessed versions of lines Gyu-29 and Bé-43 proved to be useful in interpreting the associated VSP data and in revising the interpretation of line Gyu-30. A geologic interpretation of the complex structure of the Mesozoic rocks in the area of the Békés-2 well site was developed by an iterative, forward-modeling procedure. The good match between the observed seismic events on line Gyu-30 and those generated by the synthetic model indicates that low-velocity lenses within the deformed Mesozoic sedimentary rocks are probably the sources of deep, seismic reflection events.

REFERENCES

Anderson, N.L., Brown, R.J., Hinds, R.C., and Hills, L.V., 1989, Seismic signature of a Swan Hills (Frasnian) reef reservoir, Snipe Lake, Alberta; Geophysics, 54, (2), 148-157.

Davis, T.L. and Jackson, G.M., 1988, Seismic study of algal mound reservoirs, Patterson and Nancy fields, Paradox basin, San Juan County, Utah; Geophysics, 53, (7), 875-880.

Grow, J. A., Mattick, R.E., Bérczi-Makk, A., Péró, C., Hajdú, D., Pogácsás, Gy., Várnai, P., and Varga, E., Structure of the Békés basin inferred from seismic reflection, well and gravity data; this volume, 1-38.

Lee, M., W., and Göncz, G., Vertical seismic profile experiments at the Békés-2 well, Békés basin; this volume, 257-275.

Lee, M.W., 1986, Spectral whitening in the frequency domain; U.S. Geological Survey Open-file Report, 86-108, 1-15.

Mattick, R., Rumpler, J., Ujfalusy, A., Szanyi, B. and Nagy, I., Sequence stratigraphy of the Békés basin; this volume, 39-65.

May, B.T., and Covey, J.D., 1983, Structural inversion of salt dome flanks Geophysics, 38, (8), 1039-1050.

Miller, J.J., Lee, M. W., Kilényi, E., Petrovics, I., Braun, L., and Korvin G., 1985, Seismic modeling in a complex tectonic environment; Eötvös Lóránd Geophys. Inst., Geophys. Trans., 31, (1-3), 213-255.

Molenaar, C.M., Révész, I., Bérczi, I., Kovács, A., Juhász, G.K., Gajdos I., and Szanyi, B., Stratigraphic framework and sandstone facies distribution of the Pannonian sequence in the Békés basin; this volume, 99-110.

Ryder, R. T., Lee, M. W., and Smith, G. N., 1981, Seismic models of sandstone stratigraphic traps in Rocky Mountain basins; Amer. Assoc. of Petrol. Geol., Methods in Exploration Series - 2, 1-77.

Sheriff, R. E., 1980, Seismic Stratigraphy; International Human Resources Development Corp., Boston, Massachusetts, 141-159.

15 Geologic Model, Probabilistic Methodology and Computer Programs for Petroleum Resource Assessment

Robert A. Crovelli[1] *and Richard H. Balay*[2]

ABSTRACT

The objective of this study was to develop a geostochastic system for estimation of undiscovered petroleum resources by play analysis of the Békés basin. A resource appraisal system was designed for play analysis using a reservoir-engineering geologic model and an analytic probabilistic methodology. The system resulted in a package of computer programs, called FASPUM, for play analysis and play aggregation. Resource estimates of crude oil, nonassociated gas, associated-dissolved gas, and total gas were calculated in terms of probability distributions. The FASPUM computer package is a complex system that is very flexible and efficient, and runs on IBM PC/XT/AT and compatible microcomputers.

INTRODUCTION

The geostochastic system for estimation of undiscovered petroleum resources by play analysis of the Békés basin is called FASPUM. FASPUM means Fast Appraisal System for Petroleum - Universal Metric version. The FASPUM system developed in this study is an extensive modification of the FASP system which was used by the U.S. Geological Survey (USGS) for a petroleum resource assessment of undiscovered oil and gas resources in the Arctic National Wildlife Refuge (ANWR) in Alaska (U.S. Department of the Interior, 1987). A description of the geologic model and probabilistic methodology applied in the 1987 ANWR assessment can be found in Crovelli (1988). An early version of the FASP computer programcomputer program for petroleum play analysis is given in Crovelli and Balay (1986).

The FASP play-analysis system possessed the following properties: (1) site-specific geologic model, (2) English units, (3) mainframe computer program, and (4) Fortran language. The geologic model in the FASP system was site-specific, that is, only applicable for a petroleum resource assessment in the North Slope of Alaska. It was necessary to modify the existing reservoir-engineering equations to make the geologic model universal in order to be applicable most anywhere in the world. Therefore, the geologic model used in this study is a generalization of the site-specific geologic model used in the 1987 ANWR assessment. Several other modifications of the FASP system were required.

[1] U.S. Geological Survey, Denver, Colorado 80225, USA
[2] Metropolitan State College of Denver, Denver, Colorado 80217, USA

P. G. Teleki et al. (eds.), Basin Analysis in Petroleum Exploration, 295–304.

Hence, the FASPUM play-analysis system was designed with the following properties: (1) universal-site geologic model, (2) metric units, (3) microcomputer program, and (4) Turbo Pascal language.

The basic geologic model used in the FASP system was developed by the U.S. Department of the Interior and applied by the USGS in petroleum assessments of the National Petroleum Reserve in Alaska (U.S. Department of the Interior, 1979; White, 1981) and the Arctic National Wildlife Refuge (Mast and others, 1980). The probabilistic methodology used in those two assessments was a Monte Carlo simulation method. Play analysis is a general term used for various geologic models and quantitative methods of analyzing a geologic play in petroleum assessments. The play-analysis approach in this study is described as a geostochastic system. The geostochastic system consists of (1) geologic model, (2) probabilistic methodology, and (3) computer program-computer programs. The geologic model is a particular type of probability model using reservoir-engineering equations. The probabilistic methodology is an analytic method derived from probability theory, in contrast to Monte Carlo simulation. The computer programs based upon the probabilistic methodology perform the arithmetic operations in Turbo Pascal language on IBM PC/XT/AT and compatible microcomputers.

GEOLOGIC MODEL

In play analysis, an area with petroleum potential is partitioned into geologic plays, and the individual plays are analyzed. A play consists of a collection of prospects having relatively homogeneous geologic characteristics. A prospect is a potential hydrocarbon accumulation. A hydrocarbon accumulation is a discrete oil or gas deposit, which may consist of one or more pools depending upon the specific play concept. A prospect is modeled by considering separately the uncertainty of the presence of a hydrocarbon accumulation, and its size, if it is present. A hydrocarbon accumulation is modeled as being either crude oil with associated-dissolved gas or solely as nonassociated gas. The amount of associated-dissolved gas present in an oil accumulation is calculated from a gas-oil ratio. Because gas refers to either nonassociated gas or associated-dissolved gas, the amount of gas in a play is the sum of the two types of gas from the prospects. There are three sets of geologic attributes or random variables involved in this play-analysis approach; these are for the play, the prospect, and the hydrocarbon volume. The play and prospect attributes are concerned with the presence or absence of certain geologic characteristics at the play and prospect levels, respectively. The hydrocarbon-volume attributes are concerned with the size of the hydrocarbon accumulation.

The play attributes are (1) existence of a hydrocarbon source, (2) favorable timing for migration of hydrocarbons from source to trap, (3) existence of potential migration paths, and (4) existence of potential reservoir facies. The presence of all four play attributes (in which case the play is said to be "favorable") is a necessary, but not sufficient, condition for the existence of oil and/or gas accumulations in the play. Thus, if one or more of these attributes is not present, all the prospects within the play are dry. Geologic experts make subjective estimates of the probability of the presence of each play attribute. The product of these four probabilities is the probability that the play is favorable for the existence of hydrocarbon accumulations and is called the marginal play probability.

The prospect attributes are the existence of (1) trapping mechanism, (2) effective porosity, and (3) hydrocarbon accumulation. Given a favorable play, the presence of all three prospect attributes is a necessary and sufficient condition for the existence of a hydrocarbon accumulation in the prospect. Geologic experts make subjective estimates of the probability of the presence of each prospect attribute. The product of these three probabilities is the probability that a prospect has a hydrocarbon accumulation, given the play is favorable, and is called the conditional deposit probability.

The hydrocarbon-volume attributes are (1) area of closure, (2) thickness of reservoir rock, (3) effective porosity, (4) trap fill, (5) depth to reservoir reservoir, and (6) hydrocarbon saturation. The hydrocarbon-volume attributes jointly determine the volume of the hydrocarbon accumulation within the prospect. The following reservoir engineering equations are used to calculate the in-place volumes of oil in metric tons and nonassociated gas in cubic meters, respectively:

$$\text{Oil in place} = 0.84*10^6*A*F*H*P*S_h/B_o$$

$$\text{Nonassociated gas in place} = 288.15*10^6*A*F*H*P*S_h*(P_e/T)*(1/Z)$$

where A = area of closure (square kilometers)
F = trap fill (decimal fraction)
H = reservoir thickness (meters)
P = effective porosity (decimal fraction)
S_h = hydrocarbon saturation (decimal fraction)
B_o = oil formation volume factor (no units)

P_e = original reservoir pressure (bars)
T = reservoir temperature (degrees Kelvin)
Z = gas compressibility factor (no units)
R_s = gas-oil ratio (cubic meters/ton)

To generalize the geologic model, a variety of mathematical functions needs to be available for modeling the five geologic variables P_e, T, R_s, B_o, and Z as functions of depth. Four types of mathematical functions were established: zoned linear, exponential, power, and logarithmic. Zoned linear means piece-wise linear with as many as four zones, or levels, and three transition depths. Each of the four types of mathematical functions has two parameters A and B, except zoned linear which has a set of A and B coefficients for each zone. The four types of mathematical functions are:

1. Zoned Linear Function: [A * Depth] + B
   Maximum of 4 zones with 3 transition depths (meters)

2. Exponential Function: A * [exp (B * Depth)]

3. Power Function: A * [Depth ** B]

4. Logarithmic Function: A * [Ln (B * Depth)]

When assessing a play, the following procedure is applied: for each of the five geologic variables, one type of function is selected and assigned values for the parameters A and B.

Both equations for oil in place and nonassociated gas in place consist of a product of factors that are functions of the hydrocarbon-volume attributes. The attributes are treated as continuous independent random variables, with the exception of effective porosity which possesses near perfect positive correlation with hydrocarbon saturation. The probability distribution for an attribute is determined from subjective judgments made by experts, usually geologists, based on actual geological and geophysical data, when available, and on the experience and knowledge of the experts using analog data and geologic extrapolations. The probability distribution for each attribute is described by a complementary cumulative distribution function determined from seven estimated fractiles (100th, 95th, 75th, 50th, 25th, 5th, 0th). The 5th fractile, for example, is an attribute value such that there is a 5% chance of at least that value. In each play analyzed, the seven fractiles are estimated for all six of the hydrocarbon-volume attributes. The experts also estimate the hydrocarbon-type probabilities which are the respective probabilities of a given accumulation being either oil or nonassociated gas. However, if the reservoir;depthreservoir depth is greater than a specified depth, called the oil floor depth, the accumulation is assumed to be nonassociated gas. The oil floor depth is assigned, along with recovery factors for oil and gas, in the case of recoverable estimates and is 100% for each in the case of in-place estimates.

The number of drillable prospects in the play is treated as a discrete random variable, and seven fractiles are estimated.

Probability judgments concerning each of the three sets of attributes are developed by experts familiar with the geology of the area of interest. The experts first review all existing data relevant to the appraisal, identify the major plays within the assessment area (e.g., basin or province), and then assess each identified play. All of the geologic data required by this model for a play are entered on a primary oil and gas appraisal data form (Fig. 1) and an addendum data form (Fig. 2). Information from the data forms is entered into computer data files as the input for a computer programcomputer program based upon an analytic method.

ANALYTICAL METHOD OF PLAY ANALYSIS

An analytical method using probability theory was developed as a more efficient alternative to the costly and time-consuming Monte Carlo simulation method for petroleum play analysis (Crovelli, 1987). The analytic method was developed by the application of laws of expectation and variance in probability theory. The analytical method systematically tracks through the geologic model, computes all of the means and variances of the appropriate random variables, and calculates all of the probabilities of occurrence. The lognormal distribution is used as a model for two unknown distributions in order to arrive at probability fractiles (Crovelli, 1984). Oil, nonassociated gas, dissolved gas, and gas resources are each assessed in turn. Separate methodologies have been developed for analyzing individual plays and for aggregating the plays.

The basic steps of the analytic method of play analysis are:

1. Select the play.

2. Oil is the first resource to be assessed.

Evaluator : ____________________ Play Name ____________________

Date Evaluated: ____________________

	Attribute		Probability of Favorable or Present	Comments
Play Attributes	Hydrocarbon Source			
	Timing			
	Migration			
	Potential Reservoir Facies			
	Marginal Play Probability			
Prospect Attributes	Trapping Mechanism			
	Effective Porosity (>3%)			
	Hydrocarbon Accumulation			
	Conditional Deposit Probability			
Hydrocarbon Volume Parameters	Reservoir Lithology	Sand		
		Carbonate		
	Hydrocarbon	Gas		
		Oil		

	Attribute \ Fractiles	Probability of equal to or greater than 100	95	75	50	25	5	0	
Hydrocarbon Volume Parameters	Area of Closure (Km^2)								
	Reservoir Thickness/vertical closure (meters)								
	Effective Porosity %								
	Trap Fill (%)								
	Reservoir Depth (m)								
	HC Saturation (%)								
	No. of drillable prospects (a play characteristic)								

Figure 1. Primary oil and gas appraisal data form used in play analysis of the Békés basin.

ADDENDUM DATA FORM FOR FASPUM

Geological Variables

Four Types of Mathematical Functions

1. Zones Linear Function: A * Depth + B

   Maximum of 4 zones with 3 transition depths (meters)

2. Exponential Function: A * exp (B * Depth)
3. Power Function: A * Depth ** B
4. Logarithmic Function: A * Ln (B * Depth)

For each of the five geological variables below, select one type of function and assign values for the parameters A and B.

Pe: Original Reservoir Pressure (Bars)

T: Reservoir Temperature (Deg K)

Rs: Gas-oil Ratio (m^3/Ton)

Bo: Oil Formation Volume Factor (no units)

Z: Gas Compressibility Factor (no units)

		Parameters										
Variable	Function	A	B	D	A	B	D	A	B	D	A	B
Pe												
T												
Rs												
Bo												
Z												

Oil Floor Depth (meters):__________

Oil Recovery Factor (percent):__________

Gas Recovery Factor (percent):__________

Figure 2. Addendum oil and gas appraisal data form used in play analysis of the Békés basin.

3. The following volume attributes are estimated: (1) area of closure, (2) thickness of reservoir;thicknessreservoir rock, (3) effective porosity, (4) trap fill, (5) depth to reservoir, and (6) hydrocarbon saturation. Determine the mean and variance from the estimated seven fractiles, assuming a uniform distribution between fractiles, that is, a piecewise uniform probability density function. Calculate the mean and variance of the product of effective porosity and hydrocarbon saturation, assuming they possess near perfect positive correlation. Also compute the mean and variance for the reciprocal of the oil formation volume factor, which is a function of reservoir depth.

4. Compute the mean and variance of the accumulation size of oil in place using a reservoir engineering equation. The equation involves the product of a constant, area of closure, reservoir thickness, trap fill, effective porosity, hydrocarbon saturation, and the reciprocal of the oil formation volume factor. Various laws of expectation and variance are involved in the calculations.

5. Model the accumulation-size distribution by the lognormal probability distribution with mean and variance from step 4. Calculate various lognormal fractiles of the accumulation size for oil.

6. Compute the probability that a prospect has an oil accumulation, given the play is favorable. This is called the conditional prospect probability of oil. This probability is the product of the conditional deposit probability, the probability that the reservoir depth is less than the oil floor depth, and the hydrocarbon-type probability of oil.

7. Compute the mean and variance of the conditional prospect potential for oil, which is the quantity of oil in a prospect, given the play is favorable. They are derived by applying the conditional prospect probability of oil to the mean and variance of the accumulation size of oil.

8. Compute various fractiles of the conditional prospect potential for oil by a transformation to appropriate lognormal fractiles of the accumulation size of oil using the conditional prospect probability of oil.

9. Compute the mean and variance of the number of prospects from the estimated seven fractiles, assuming a uniform distribution between fractiles.

10. Compute the mean and variance of the number of oil accumulations, given the play is favorable. They are derived by applying the conditional prospect probability of oil to the mean and variance of the number of prospects.

11. Compute the mean and variance of the conditional (A) play potential for oil, which is the quantity of oil in the play, given the play is favorable. They are determined from the probability theory of the expectation and variance of a random number (number of prospects) of random variables (conditional prospect potential).

12. Compute the conditional play probability of oil, which is the probability that a favorable play has at least one oil accumulation, and is a function of the conditional prospect probability of oil and the number-of-prospects distribution.

13. Compute the mean and variance of the conditional (B) play potential for oil, which is the quantity of oil in the play, given the play is favorable and there is at least one oil accumulation within the play. They are obtained by applying the conditional play probability of oil to the mean and variance of the conditional (A) play potential for oil.

14. Compute the unconditional play probability of oil, which is the probability that the play has at least one oil accumulation, and is the product of the conditional play probability of oil and the marginal play probability.

15. Compute the mean and variance of the unconditional play potential for oil, which is the quantity of oil in the play. They are derived by applying the unconditional play probability of oil to the mean and variance of the conditional (B) play potential for oil.

16. Model the probability distribution of the conditional (B) play potential for oil by the lognormal distribution with mean and variance from step 13. Calculate various lognormal fractiles.

17. Compute various fractiles of the conditional (A) play potential for oil by a transformation to appropriate lognormal fractiles of the conditional (B) play potential for oil using the conditional play probability of oil.

18. Compute various fractiles of the unconditional play potential for oil by a transformation to appropriate lognormal fractiles of the conditional (B) play potential for oil using the unconditional play probability of oil.

19. Nonassociated gas is the second resource to be assessed. Repeat steps 3 through 18, substituting nonassociated gas for oil, with two basic modifications as

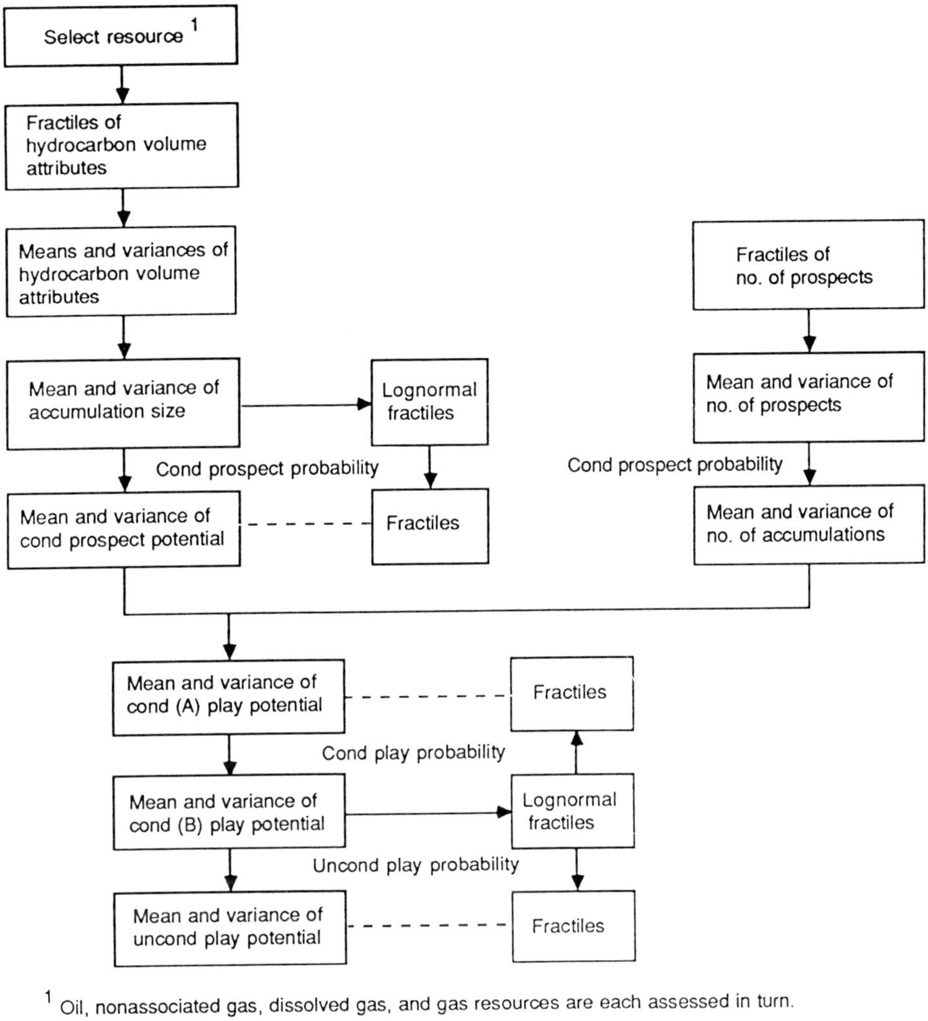

Figure 3. Flowchart of the analytic method of play analysis.

follows. A different reservoir engineering equation is used to calculate the accumulation size of nonassociated gas in place. The conditional prospect probability of nonassociated gas is equal to the conditional deposit probability minus the conditional prospect probability of oil.

20. Associated-dissolved gas is the third resource to be assessed. Repeat steps 3 through 18, substituting associated-dissolved gas for oil, with two basic modifications as follows. The reservoir engineering equation for the accumulation size of oil in-place is multiplied by a gas-oil ratio which is a function of reservoir depth. The conditional prospect probability of dissolved gas is the same as the conditional prospect probability of oil.

21. Gas is the fourth resource to be assessed. Repeat steps 4 through 18, substituting gas for oil, with two basic modifications as follows. Replace step 4 to compute the mean and variance of the accumulation size of gas in-place by using conditional probability theory and conditioning on the type of gas. The conditional prospect probability of gas is the same as the conditional deposit probability.

A simplified flowchart of the analytic methodology of play analysis is presented in Figure 3.

On the basis of the analytic method, a computer program was designed and called Fast Appraisal System for Petroleum - Universal Metric (FASPUM). Because both cost and running time are negligible, FASPUM allows for quick feedback evaluation of geologic input data.

ANALYTICAL METHOD OF PLAY AGGREGATION

A separate methodology was developed for aggregating a set of plays. In this method, the resource estimates of the individual plays from the analytic method of play analysis using the FASPUM program are aggregated by means of probability theory. Oil, non-associated gas, dissolved gas, and gas resources are each aggregated in turn.

The basic steps of the analytical method of play aggregation are:

1. Select plays to aggregate.

2. Oil is the first resource to be aggregated.

3. Compute the mean and variance of the unconditional aggregate potential for oil, which is the quantity of oil in the assessed area. They are determined by adding all of the individual play means and variances of the unconditional play potential for oil, respectively, assuming independence among the plays.

4. Compute the unconditional aggregate probability of oil, which is the probability that the assessed area has at least one play with oil, and is a function of the individual unconditional play probabilities of oil.

5. Compute the mean and variance of the conditional aggregate potential for oil, which is the quantity of oil in the assessment area, given the assessed area contains at least one play with oil. These are obtained by applying the unconditional aggregate probability of oil to the mean and variance of the unconditional aggregate potential for oil.

6. Model the probability distribution of the conditional aggregate potential for oil by the lognormal distribution with mean and variance from step 5. Calculate various lognormal fractiles.

7. Compute various fractiles of the unconditional aggregate potential for oil by a transformation to appropriate lognormal fractiles of the conditional aggregate potential for oil using the unconditional aggregate probability of oil.

8. Nonassociated gas is the second resource to be aggregated. Repeat steps 3 through 7 using play analysis estimates of nonassociated gas, namely the individual play means and variances of the unconditional play potential for nonassociated gas, along with the individual unconditional play probabilities of nonassociated gas.

9. Dissolved gas is the third resource to be aggregated. Repeat steps 3 through 7 using play analysis estimates of dissolved gas, namely the individual play means and variances of the unconditional play potential for dissolved gas, along with the individual unconditional play probabilities of dissolved gas.

10. Gas is the fourth resource to be aggregated. Repeat steps 3 through 7 using play-analysis estimates of gas, namely the individual play means and variances of the unconditional play potential for gas, along with the individual unconditional play probabilities of gas.

A simplified flowchart of the analytic methodology of play aggregation is presented in Figure 4.

A computer program was designed on the basis of the analytic method for the aggregation of plays and called the Fast Appraisal System for Petroleum Aggregation (FASPAG). FASPAG interfaces with FASPUM as follows. FASPUM not only generates a file of resource estimates for an individual play, but also generates a second file of results which consists of the mean and standard deviation of the unconditional play potential for each of the four resources, along with the corresponding unconditional play probabilities. The second file is needed for an aggregation of plays and forms an input file for FASPAG. Therefore, after FASPUM is applied to each play in a set of plays, any subset of plays can be aggregated by running FASPAG on the corresponding subset of aggregated input files. FASPAG not only generates a file of resource estimates for an aggregation of plays, but also generates a second file of results needed for an aggregation of aggregations, which forms yet another input file for FASPAG. Hence, after FASPAG has been applied to each aggregation in a set of aggregations, any subset of aggregations can be aggregated at once. FASPAG also possesses the capacity of aggregating a set of plays under a dependency assumption. Under the assumption of perfect positive correlation, all of the

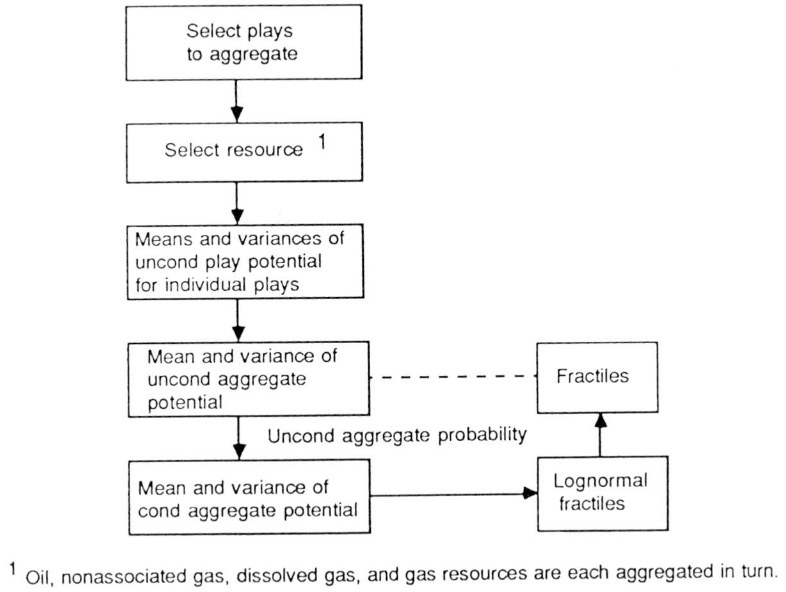

Figure 4. Flowchart of analytic methodology of play aggregation.

individual play standard deviations (instead of the variances) of the unconditional play potential for a resource are added together. In terms of the standard deviation of the unconditional aggregate potential, any degree of dependency is possible between 0 and 1, where 0 corresponds to independence and 1 denotes perfect positive correlation.

APPLICABLE COMPUTER PROGRAMS

A package of computer program;FASPUMcomputer programs was created to carry out (a) the assessment of hydrocarbon resources in a set of plays, and (b) the aggregation of the assessments in any subset of plays. The assessment module of the system is called FASPUM, and the aggregation module is called FASPAG. FASPUM is also the generic name of the entire package of software.

A user's manual was also written that included various examples along with user's guides for installation, operation, modification, maintenance, and documentation of the system. The operational details of the FASPUM system are contained in Crovelli and Balay (1987).

CONCLUSIONS

The FASPUM computational program was developed in support of making estimates of the undiscovered recoverable resources of crude oil, non-associated gas, associated-dissolved gas, and total gas in 12 plays in the Békés basin in terms of probability distributions (see Charpentier and others, this volume).

REFERENCES

Charpentier, R., Völgyi, L., Dolton, G., Mast, R., and Pályi, A., Undiscovered recoverable oil and gas resources: this volume, 305-319.

Crovelli, R.A., 1984, Procedures for petroleum resource assessment used by the U.S. Geological Survey--statistical and probabilistic methodology; *in* Masters, C.D., (ed.), Petroleum resource assessment; International Union of Geological Sciences, 17, 24-38.

Crovelli, R.A., 1987, Probability theory versus simulation of petroleum potential in play analysis; *in* Albin, S.L., and Harris, C.M., (eds.), Statistical and computational issues in probability modeling, Part 1; Annals of Operations Research, 8, 363-381.

Crovelli, R.A., 1988, U.S. Geological Survey assessment methodology for estimation of undiscovered petroleum resources in play analysis of the Arctic National Wildlife Refuge; *in* Chung, C.F., Fabbri, A.G., and Sinding-Larsen, R., (eds.), Quantitative analysis of mineral and energy resources; Dordrecht, Holland, D. Reidel Publishing, NATO ASI Series C: Mathematical and Physical Sciences, 223, 145-160.

Crovelli, R.A. and Balay, R.H., 1986, FASP, an analytic resource appraisal program for petroleum play analysis; Computers and Geosciences, 12, (4B), 423-475.

Crovelli, R.A. and Balay, R.H., 1987, FASPUM Metric Version: Analytic petroleum resource appraisal microcomputer programs for play analysis using a reservoir-engineering model; U.S. Geological

Survey Open-File Report 87-414A, 14 p. [U.S. Geological Survey Open-File Report 87-414B, Diskette].

Mast, R.F., McMullin, R.H., Bird, K.J., and Brosge, W.P., 1980, Resource appraisal of undiscovered oil and gas resources in the William O. Douglas Arctic Wildlife Range; U.S. Geological Survey Open-File Report 80-916, 62 p.

U.S. Department of the Interior, 1979, Final report of the 105(b) economic and policy analysis, alternative overall procedures for the exploration, development, production, transportation and distribution of the petroleum resources of the National Petroleum Reserve in Alaska (NPRA); U.S. Dept. Int., Office of Minerals Policy and Research Analysis, Washington, D.C., 145 p.

U.S. Department of the Interior, 1987, Arctic National Wildlife Refuge, Alaska, coastal plain resource assassment report and recommendation to the Congress of the United States and final legislative environmental impact statement; U.S. Fish and Wildlife Service, U.S. Geological Survey, and Bureau of Land Management, Washington, D.C., 208 p.

White, L.P., 1981, A play approach to hydrocarbon resource assessment and evaluation, *in* Ramsey, J.B., (ed.), The economics of exploration for energy resources; Contemporary studies in economic and financial analysis, Jai Press, Greenwich, Connecticut, 26, 51-67 (Proc., Energy exploration conference, May 1979, New York).

16 Undiscovered Recoverable Oil and Gas Resources

Ronald Charpentier[1], *László Völgyi*[2], *Gordon Dolton*[1], *Richard Mast*[1], *and András Pályi*[2]

ABSTRACT

An assessment was made of the undiscovered recoverable oil and gas resources of the Békés basin. The resulting mean estimates were 5.22 million metric tons* of undiscovered oil and 18.05x10^9 m^3 of undiscovered natural gas (recoverable). This compares with discovered amounts (through 1985) of 3.81 million metric tons of oil and 25.40x10^9 m^3 of natural gas. Among the twelve plays assessed, the delta front play has the highest mean potential for oil, 1.90 million metric tons, and the basal turbidite drape structure play has the highest mean potential for undiscovered gas, 5.22x10^9 m^3.

PLAY ANALYSIS APPROACH

Geologic Model

Play analysis is a quantitative approach for estimating undiscovered oil and gas resources at a play scale. A play is a set of pools or prospects which are conceived as having similar geologic characteristics and sharing common geologic elements. Most of the input variables used in the model are expressed in a probabilistic form, that is either as a probability of occurrence or as a probability distribution. This allows the uncertainty about the input variables to be expressed quantitatively. Likewise, the resulting resource estimates are expressed as probability distributions in order to show the uncertainty of the estimates. The present model has its antecedents in the play analysis technique developed by the Geological Survey of Canada (Canada Department of Energy, Mines and Resources, 1977).

Play analysis can be divided into two parts. In the first part, estimates are made of the favorability for resource in the play as a whole, as well as in a random prospect in the play. In the second part, estimates are made of the number of prospects and the sizes of the possible accumulations.

In the first part, the probability of favorability for resource in the play as a whole (the marginal play probability) is estimated by judging the probabilities of exist-

[1] U. S. Geological Survey, Denver, Colorado 80225, USA
[2] MOL Rt. (Hungarian Oil and Gas Co., Ltd.), H-1117, Budapest, Hungary
* Note = 1 metric ton of oil (at 0.84 metric tons/m^3) = 7.5 barrels; 1 m^3 = 35.31 ft^3

P. G. Teleki et al. (eds.), Basin Analysis in Petroleum Exploration, 305–319.

ence of the subsidiary attributes of hydrocarbon source, timing, migration, and potential reservoir facies. The play can contain resources only if all four of these attributes exist. The play risk, the probability that the play contains no resource, is 1.0 minus the marginal play probability. Next, the conditional deposit probability, the probability of occurrence of an accumulation in a random prospect (conditioned on the play attributes being favorable), is estimated by judging the probabilities of existence of the subsidiary attributes of trapping mechanism, effective porosity, and hydrocarbon accumulation. Prospect risk, the probability that a prospect contains no resource (again conditioned on the play attributes being favorable), is 1.0 minus the conditional deposit probability.

In the second part of play analysis, estimates are made of the number of prospects and also of various parameters which deal with the size of accumulations. Many of these variables are input as seven fractiles describing a probability distribution. Thus, the uncertainty about number of prospects is expressed by the seven fractiles for number of drillable prospects. The accumulation volumes are analytically calculated from the variables: area of closure, reservoir thickness/vertical closure, effective porosity, trap fill, reservoir depth, and hydrocarbon saturation. Also used in this calculation are such engineering variables as original reservoir pressure, reservoir temperature, gas-oil ratio, oil formation volume factor, gas compressibility factor, oil floor depth, oil recovery factor, and gas recovery factor.

The probabilities determined in the first part are used with the distributions of prospect number and accumulation size to estimate resource amounts. These resource appraisal estimates are calculated using an analytic probabilistic methodology. The details of how the calculations are performed can be found in Crovelli and Balay (this volume).

Computer Programs

The program for calculating estimates in this study was a modification of the FASP program (Crovelli and Balay, 1986) which, in turn, was an analytic version of the RASP program (White, 1981). RASP was a Monte-Carlo simulation program, but FASP greatly improved performance by using a probability-theory approach.

The specific program used was called FASPUM, for FASP-Universal-Metric (Crovelli and Balay, 1987). A detailed discussion of the FASPUM computer program is given by Crovelli and Balay (this volume).

APPRAISAL OF THE BÉKÉS BASIN

The undiscovered recoverable petroleum resources of the Békés basin were appraised in September 1987 in Budapest. In these meetings, twelve plays were defined. These included both producing plays and hypothetical plays. No significant resource potential in the Békés basin is expected in addition to these twelve plays.

In assessing by the play analysis method, it is necessary to be explicit about the size of the smallest occurrence considered to be exploitable. This is to guard against considering very small accumulations which could never be exploited practically. Also, it is important to have a consistent view of the minimum exploitable size when using different variables, each of which are affected by the minimum size considered.

The minimum area of closure considered was 0.75 km^2. The trap fill was in every case expressed as 100 percent, and, thus, the area of closure figure actually reflects an estimate of the area of accumulation. The minimum reservoir thickness considered was generally 5 m, but in a few cases thicknesses as small as 3 m, sufficient to contain exploitable resources, were taken into account. Reservoir porosity of 5 percent was the usual minimum value, but this minimum was lowered for a few plays involving fracture porosity. Hydrocarbon saturation of 35 percent was generally used as the minimum value, except in two plays, where a minimum of 30 percent was used based on analog data from discovered pools. Minimum depth and the number of prospects were not specified.

Several standard engineering factors were used for the entire basin. The pressure/depth gradient above 2400 m was 0.1 bars/m with an intercept of 1 bar at 0 m depth. Below 2400 m, the gradient was 0.15 bars/m with the same intercept. This pressure/depth relationship reflects overpressure below a depth of 2400 m. Temperature varied linearly with depth, with a gradient of 0.05 degrees K/m and a temperature of 284.15 degrees K at the ground surface. The gas-oil ratio was generally 200 m^3 per ton, the oil formation volume factor was 1.5, and the gas compressibility factor was 0.9. The oil floor was 3100 m, and standard oil and gas recoveries were 30 and 70 percent respectively. The input values used for each play are presented in Tables 1 to 12. Some of the values presented in the tables are fractiles from probability

Table 1. Play # 1: Fractured crystalline basement.

INPUT							
Play attributes		Hydrocarbon source			1.0		
		Timing			1.0		
		Migration			1.0		
		Potential reservoir facies			1.0		
		Marginal play probability			1.0		
Prospect attributes		Trapping mechanism			0.8		
		Effective porosity (>1.5)			0.9		
		Hydrocarbon accumulation			0.4		
		Conditional deposit probability			0.288		
Hydrocarbon mix		Gas			0.8		
		Oil			0.2		
Geologic variables	F_{100}	F_{95}	F_{75}	F_{50}	F_{25}	F_{05}	F_{0}
Closure (km^2)	0.75	1.5	1.9	2.3	3.0	3.5	5.0
Thickness (m)	5.0	12.0	37.0	40.0	50.0	70.0	150.0
Porosity (%)	1.5	2.5	3.5	4.0	5.0	7.5	10.0
Depth (m)	1100	1500	2400	3300	3800	4400	5000
HC saturation (%)	40	45	53	60	66	70	80
Number of prospects	4	4	5	6	7	8	10
Oil floor		3100 meters					
Recovery factors		Oil			30 %		
		Free gas			35%		
Reservoirs pressure (bars)	= 0.10 (depth) + 1.0			(from 0 to 2400 m)			
Reservoirs pressure (bars)	= 0.15 (depth) + 1.0			(below 2400 m)			
Reservoir temperature (^0K)	= 0.05 (depth) + 284.15						
Gas-oil ratio (m^3/ton)	= 200						
Oil formation volume factor	= 1.7						
Gas compressibility factor	= 0.95						
RESULTS							
		Low (F_{95})		Mean		High (F_{05})	
Oil (millions of metric tons)		0.0		0.0627		0.4532	
Non-associated gas (10^9 m^3)		0.0		0.4330		1.2854	
Dissolved gas (10^9 m^3)		0.0		0.0125		0.0906	
Total gas (10^9 m^3)		0.0		0.4455		1.2978	

distributions. F_{95}, for example, is the 95th fractile and is interpreted to mean that there is a 95 percent chance of more than that specific value.

PLAY DESCRIPTIONS

Play #1: Fractured Crystalline Basement

The fractured crystalline basement play (Table 1) consists of prospects with reservoirs in metamorphic and igneous rocks (granite, micaceous shale, and gneiss) of Precambrian and Paleozoic age. The prospects are related to fracturing on basement structural highs. Source rocks are within the overlying Neogene sedimentary rocks. Only one pool has been discovered in the play.

Play attributes are shown as favorable (Table 1) because a discovery has been made. Prospect risk, however, is high due to the absence of source rocks within the basement.

Hydrocarbon volume parameters were estimated partially based on data from the discovered pool and from analogous pools in similar plays. The high uncertainty reflects concern about sufficient porosity and hydrocarbon saturation in fractured rocks. Seismic data gave information on identified, but as yet undrilled, structures. The gas recovery factor was adjusted to reflect an expected 50% CO_2 content in the gas.

Play #2: Mesozoic Fractured Basement (Pre-Upper Cretaceous)

The Mesozoic fractured basement play (Table 2) consists mostly of traps involving unmetamorphosed Mesozoic carbonates (dolomite and limestone), and siltstones and sandstones. The prospects are related to paleogeomorphologic or faulted highs of the Mesozoic basement sealed by overlying Neogene rocks. Although minor amounts of hydrocarbons may have been generated by Jurassic carbonates, the major source rocks are likely to be in the overlying Neogene sequence. No discovery has been made in this play in the Békés basin.

In spite of the lack of current production, the play was considered favorable. The individual prospect risk was considered fairly high, because clearly definable migration paths are absent and lateral seals on the faulted structures are likely discontinuous.

Hydrocarbon volume factors and number of drillable prospects were estimated by using seismic data to identify possible structures and by analogy of similar accumulations in nearby basins. Standard engineering values were used for most of the variables, but the gas recovery factor was adjusted to reflect an expected CO_2 content of more than 70%.

Play #3: Upper Cretaceous of North Békés

The Upper Cretaceous of North Békés play (Table 3) consists mainly of prospects in relatively unmetamorphosed Upper Cretaceous sandstones at depths between 2500 and 4200 m. The prospects include both structural and stratigraphic traps. Possible source rocks include the Upper Cretaceous clayey marls. No discovery has been made in this play in the Békés basin, although hydrocarbon shows were reported from several wells.

Although hydrocarbons are not produced currently from this type of play, the play was considered favorable. Namely, some chance exists for hydrocarbon occurrence in the play. The individual prospect risk was considered high because the Upper Cretaceous is often overlain by fractured pre-Pannonian rocks which do not provide an adequate seal. Also, some of the prospects are not favorably located in relation to known source rocks in the Neogene section.

Table 2. Play # 2: Mesozoic fractured basement (pre-Upper Cretaceous).

INPUT							
Play attributes		Hydrocarbon source			1.0		
		Timing			1.0		
		Migration			0.75		
		Potential reservoir facies			1.0		
		Marginal play probability			0.75		
Prospect attributes		Trapping mechanism			0.7		
		Effective porosity (>1.5)			0.8		
		Hydrocarbon accumulation			0.5		
		Conditional deposit probability			0.28		
Hydrocarbon mix		Gas			0.9		
		Oil			0.1		
Geologic variables	F_{100}	F_{95}	F_{75}	F_{50}	F_{25}	F_{05}	F_0
Closure (km^2)	0.75	1.3	1.7	2.0	3.0	5.0	7.0
Thickness (m)	5.0	15.0	40.0	60.0	70.0	90.0	180.0
Porosity (%)	2.0	2.5	4.0	5.0	6.0	8.0	10.0
Depth (m)	1500	1900	2700	3500	3900	4300	4700
HC saturation (%)	40	45	53	60	66	70	80
Number of prospects	10	10	11	12	13	15	16
Oil floor		3100 meters					
Recovery factors		Oil			30 %		
		Free gas			20%		
Reservoirs pressure (bars)		= 0.10 (depth) + 1.0		(from 0 to 2400 m)			
Reservoirs pressure (bars)		= 0.15 (depth) + 1.0		(below 2400 m)			
Reservoir temperature (^0K)		= 0.05 (depth) + 284.15					
Gas-oil ratio (m^3/ton)		= 200					
Oil formation volume factor		= 1.7					
Gas compressibility factor		= 0.95					

RESULTS	Low (F_{95})	Mean	High (F_{05})
Oil (millions of metric tons)	0.0	0.0634	0.4587
Non-associated gas (10^9 m^3)	0.0	0.6981	2.0254
Dissolved gas (10^9 m^3)	0.0	0.0127	0.0917
Total gas (10^9 m^3)	0.0	0.7108	2.0467

Table 3. Play # 3: Upper Cretaceous of North Békés.

INPUT							
Play attributes		Hydrocarbon source			1.0		
		Timing			1.0		
		Migration			0.8		
		Potential reservoir facies			1.0		
		Marginal play probability			0.8		
Prospect attributes		Trapping mechanism			0.6		
		Effective porosity (>1.5)			0.8		
		Hydrocarbon accumulation			0.6		
		Conditional deposit probability			0.288		
Hydrocarbon mix		Gas			0.95		
		Oil			0.05		
Geologic variables	F_{100}	F_{95}	F_{75}	F_{50}	F_{25}	F_{05}	F_{0}
Closure (km^2)	0.75	1.7	2.5	3.5	4.5	6.0	8.0
Thickness (m)	3.0	4.0	5.5	7.0	9.0	12.0	15.0
Porosity (%)	3.0	3.1	3.5	4.0	5.0	7.5	10.0
Depth (m)	2500	2650	2900	3300	3600	4000	4200
HC saturation (%)	35	38	45	50	55	62	65
Number of prospects	3	3	4	5	6	7	8
Oil floor		3100 meters					
Recovery factors		Oil			30 %		
		Free gas			46%		
Reservoirs pressure (bars)	= 0.10 (depth) + 1.0			(from 0 to 2400 m)			
Reservoirs pressure (bars)	= 0.15 (depth) + 1.0			(below 2400 m)			
Reservoir temperature (^{o}K)	= 0.05 (depth) + 284.15						
Gas-oil ratio (m^3/ton)	= 200						
Oil formation volume factor	= 1.5						
Gas compressibility factor	= 1.1						

RESULTS	Low (F_{95})	Mean	High (F_{05})
Oil (millions of metric tons)	0.0	0.0021	0.0
Non-associated gas (10^9 m^3)	0.0	0.0880	0.2968
Dissolved gas (10^9 m^3)	0.0	0.0004	0.0
Total gas (10^9 m^3)	0.0	0.0885	0.2971

Hydrocarbon volume factors and number of drillable prospects were estimated using geophysical information and analogy. Most of the engineering values used were standard, but the gas recovery factor was adjusted to reflect an expected 35% CO_2 content.

Play #4: Overpressured Miocene

The overpressured Miocene play (Table 4) consists mainly of sandstone reservoirs of pre-Pannonian Miocene age in structural traps located, mostly deeper than 2400 m. Quantities of oil and gas are also produced from algal carbonates of the same age. Seals are marls. Middle Miocene (Badenian) age rocks overlie the basement, with the exception of a few basement highs (Szentgyörgyi and Teleki, this volume). The Badenian sequence contains some of the best source rocks in the basin and both oil and gas are expected to be present; however, gas is expected to be dominant at depths greater than 3100 m. Only 3 pools within 1 field have been discovered from this play in the Békés basin.

Play attributes are shown to be favorable by previous discoveries. Prospect risk, however, is moderately high, due to the uncertainty of basal seals, where thin, fractured Miocene rocks rest on fractured basement rocks. Low porosity is expected at great depths. The hydrocarbon accumulation factor has little to no risk involved, because the Miocene sequence contains excellent source rocks.

Hydrocarbon volume parameters were estimated primarily from seismic data for identified, but as yet undrilled structures and in part using data on discovered pools. Standard engineering values were used, except for the gas compressibility factor (based on data from the discovered pools) and the gas recovery factor (adjusted to reflect expected 20% CO_2).

Play #5: Lower Pannonian Basal Conglomerates

The lower Pannonian basal conglomerate play (Table 5) consists of conglomerate and sandstone reservoirs deposited on the top of basement highs in the area of the Battonya-Pusztaföldvár high. Well data from 7 discovered oil and gas pools indicates that these rocks are overlain by basal clay marls of Pannonian age. The oil and gas contained in the pools was probably generated either from Miocene marls and shales or from Pannonian shales located downdip. Migration was through fractures in the basement rocks.

The play attributes are all shown to be favorable based on previous discoveries. Prospect attributes are also all considered to be very favorable.

Data from the discovered pools, as well as seismic data from identified prospects, were used to estimate the hydrocarbon volume parameters. Some modification was necessary due to the generally greater than expected depth of undiscovered pools. The gas-oil ratio, oil formation volume factor, and gas compressibility factor were based on data from the discovered pools. Also, the gas recovery factor was reduced to reflect an expected 50% CO_2 content.

Play #6: Basal Lower Pannonian Fractured and Silty Marls

The basal lower Pannonian fractured clayey marl and silty marl play (Table 6) consists of hydrocarbon accu-

Table 4. Play # 4: Overpressured Miocene. (Badenian-Sarmatian)

INPUT							
Play attributes		Hydrocarbon source			1.0		
		Timing			1.0		
		Migration			1.0		
		Potential reservoir facies			1.0		
		Marginal play probability			1.0		
Prospect attributes		Trapping mechanism			0.6		
		Effective porosity (>1.5)			0.6		
		Hydrocarbon accumulation			0.9		
		Conditional deposit probability			0.324		
Hydrocarbon mix		Gas			0.75		
		Oil			0.25		
Geologic variables	F_{100}	F_{95}	F_{75}	F_{50}	F_{25}	F_{05}	F_0
Closure (km^2)	0.75	1.5	2.25	3.0	4.0	5.5	7.0
Thickness (m)	3.0	4.0	5.5	7.0	9.0	12.0	15.0
Porosity (%)	5.0	5.5	6.5	7.0	9.0	15.0	25.0
Depth (m)	1500	1800	2400	3200	3700	4200	4500
HC saturation (%)	35	38	45	50	55	62	65
Number of prospects	12	13	15	16	17	19	20
Oil floor		3100 meters					
Recovery factors		Oil			30 %		
		Free gas			56%		
Reservoirs pressure (bars)	= 0.10 (depth) + 1.0			(from 0 to 2400 m)			
Reservoirs pressure (bars)	= 0.15 (depth) + 1.0			(below 2400 m)			
Reservoir temperature (°K)	= 0.05 (depth) + 284.15						
Gas-oil ratio (m^3/ton)	= 200						
Oil formation volume factor	= 1.5						
Gas compressibility factor	= 1.1						

RESULTS			
	Low (F_{95})	Mean	High (F_{05})
Oil (millions of metric tons)	0.0	0.1076	0.4520
Non-associated gas (10^9 m^3)	0.2177	0.6618	1.4338
Dissolved gas (10^9 m^3)	0.0	0.0215	0.0904
Total gas (10^9 m^3)	0.2348	0.6833	1.4566

Table 5. Play #5 Lower Pannonian basal conglomerates.

INPUT							
Play attributes		Hydrocarbon source			1.0		
		Timing			1.0		
		Migration			1.0		
		Potential reservoir facies			1.0		
		Marginal play probability			1.0		
Prospect attributes		Trapping mechanism			0.9		
		Effective porosity (>1.5)			0.9		
		Hydrocarbon accumulation			0.9		
		Conditional deposit probability			0.729		
Hydrocarbon mix		Gas			0.7		
		Oil			0.3		
Geologic variables	F_{100}	F_{95}	F_{75}	F_{50}	F_{25}	F_{05}	F_{0}
Closure (km^2)	0.75	1.0	1.5	2.0	2.5	3.0	3.5
Thickness (m)	4.0	5.0	7.0	9.0	12.0	17.0	20.0
Porosity (%)	9.0	11.0	15.0	17.0	18.0	19.0	20.0
Depth (m)	1000	1200	1500	1800	2200	2700	3000
HC saturation (%)	50	51	53	55	60	74	85
Number of prospects	3	3	4	4	5	6	6
Oil floor		3100 meters					
Recovery factors		Oil			30 %		
		Free gas			35%		

Reservoirs pressure (bars)	= 0.10 (depth) + 1.0 (from 0 to 2400 m)
Reservoir temperature (^oK)	= 0.05 (depth) + 284.15 (below 2400 m)
Gas-oil ratio (m^3/ton)	= 80
Oil formation volume factor	= 1.2
Gas compressibility factor	= 0.87

RESULTS	Low (F_{95})	Mean	High (F_{05})
Oil (millions of metric tons)	0.0	0.3520	1.1034
Non-associated gas (10^9 m^3)	0.0	0.2228	0.5058
Dissolved gas (10^9 m^3)	0.0	0.0282	0.0883
Total gas (10^9 m^3)	0.0863	0.2509	0.5307

mulations in the lower Pannonian section associated with two types of traps. In the area of the Battonya-Pusztaföldvár high, the traps have lateral variations in the porous siltstone content of marl beds. In other parts of the basin, the traps are formed by differential fracturing. Both trap types are enhanced by structure. The marls are good source rocks and overlie high-quality, older Miocene source rocks. Oil is the more likely hydrocarbon to occur at depths shallower than 3000 m. To date, 6 pools have been discovered in this play.

Previous discoveries indicate that the play attributes are favorable, but the prospect risk is high, mainly due to expectations of poor porosity. Access to hydrocarbon sources is considered very good and trapping mechanism is considered moderately good because of the structural component.

Hydrocarbon volume parameters have been estimated using analog data from the discovered pools. These estimates were modified to reflect some chance of deeper structures being present, but the deeper structures were considered less prospective in light of poor drilling results to date. Numbers and depths of drillable structures were based on available seismic data .

All of the engineering factors were set to the standards except the recovery factors. The oil recovery factor was reduced because of the expected decrease in porosity with increasing depth. Gas recovery was reduced to reflect an expected 50% CO_2 content.

Play #7: Basal Turbidite Onlap

The basal turbidite onlap play (Table 7) consists of stratigraphic pinchouts of multiple turbidite sandstones

Table 6. Play #6 Basal lower Pannonian fractured and silty marls.

INPUT						
Play attributes		Hydrocarbon source			1.0	
		Timing			1.0	
		Migration			1.0	
		Potential reservoir facies			1.0	
		Marginal play probability			1.0	
Prospect attributes		Trapping mechanism			0.8	
		Effective porosity (>1.5)			0.3	
		Hydrocarbon accumulation			0.9	
		Conditional deposit probability			0.216	
Hydrocarbon mix		Gas			0.2	
		Oil			0.8	
Geologic variables	F_{100}	F_{95}	F_{75}	F_{50}	F_{25}	F_{05}	F_0
Closure (km^2)	0.75	1.0	1.5	2.0	2.7	3.5	4.0
Thickness (m)	5.0	6.0	7.0	8.0	10.0	13.0	15.0
Porosity (%)	2.0	4.0	7.0	10.0	14.0	19.0	22.0
Depth (m)	1000	1200	1600	2000	2500	3300	4000
HC saturation (%)	30	34	40	45	50	58	60
Number of prospects	13	15	16	17	19	21	22
Oil floor		3100 meters					
Recovery factors		Oil			22 %		
		Free gas			35%		
Reservoirs pressure (bars)	= 0.10 (depth) + 1.0			(from 0 to 2400 m)			
Reservoirs pressure (bars)	= 0.15 (depth) + 1.0			(below 2400 m)			
Reservoir temperature (^{0}K)	= 0.05 (depth) + 284.15						
Gas-oil ratio (m^3/ton)	= 200						
Oil formation volume factor	= 1.5						
Gas compressibility factor	= 0.90						

RESULTS			
	Low (F_{95})	Mean	High (F_{05})
Oil (millions of metric tons)	0.0	0.3147	0.7562
Non-associated gas (10^9 m^3)	0.0	0.0732	0.2521
Dissolved gas (10^9 m^3)	0.0	0.0629	0.1512
Total gas (10^9 m^3)	0.0313	0.1361	0.3358

within the Szolnok Formation where they onlap basement highs. Seismic data indicate that these pinchouts occur over a wide range of depths. The expected trapping mechanism is the updip onlap of reservoir sandstones against older rocks, but updip facies changes may also be traps. Seals are primarily the interbedded shales and marls within the Szolnok Formation. The underlying basal clayey marls are expected to restrict lateral migration into fractured basement rocks. Potential hydrocarbon source beds include the interbedded shales as well as the basal clay marls and Miocene rocks. Both oil and gas have been produced from this play with 38 discovered pools in 4 fields.

Previous production from the play indicates that all play attributes are favorable. However, exploration risks of individual prospects are high. Trapping risk is high because of the stratigraphic nature of the trap and the possibility of inadequate lateral seals if the basal clayey marl is fractured. Porosity risk is moderate because of the substantial depth of many of the prospects. Hydrocarbon accumulation risk is relatively low because the interbedded shales could be both source rocks and seals.

Hydrocarbon volume parameters were estimated by analogy with accumulations already discovered within the play. Some modification of the distributions was necessary, however, in order to accommodate the generally greater depth of the unexplored prospects.

Standard pressure/depth and temperature/depth relations for the basin were used, as were the standard oil floor value and oil recovery factor. The standard gas recovery factor was decreased, assuming an average CO_2 content of 15%. The gas-oil ratio, oil formation volume factor, and gas compressibility factor were estimated

Table 7. Play #7 Basal turbidite onlap.

INPUT							
Play attributes		Hydrocarbon source			1.0		
		Timing			1.0		
		Migration			1.0		
		Potential reservoir facies			1.0		
		Marginal play probability			1.0		
Prospect attributes		Trapping mechanism			0.3		
		Effective porosity (>1.5)			0.75		
		Hydrocarbon accumulation			0.8		
		Conditional deposit probability			0.18		
Hydrocarbon mix		Gas			0.8		
		Oil			0.2		
Geologic variables	F_{100}	F_{95}	F_{75}	F_{50}	F_{25}	F_{05}	F_0
Closure (km^2)	0.75	1.5	2.0	3.0	5.0	10.0	20.0
Thickness (m)	5.0	6.0	8.0	10.0	13.0	18.0	25.0
Porosity (%)	5.0	6.0	7.0	8.0	10.0	14.0	18.0
Depth (m)	1900	2300	2700	3000	3300	3700	4000
HC saturation (%)	35	38	42	45	50	58	65
Number of prospects	9	16	26	35	44	52	60
Oil floor		3100 meters					
Recovery factors		Oil			30 %		
		Free gas			60%		
Reservoirs pressure (bars)	= 0.10 (depth) + 1.0			(from 0 to 2400 m)			
Reservoirs pressure (bars)	= 0.15 (depth) + 1.0			(below 2400 m)			
Reservoir temperature (°K)	= 0.05 (depth) + 284.15						
Gas-oil ratio (m^3/ton)	= 200						
Oil formation volume factor	= 1.5						
Gas compressibility factor	= 0.92						
RESULTS							
			Low (F_{95})		Mean	High (F_{05})	
Oil (millions of metric tons)			0.0		0.2452	1.0208	
Non-associated gas (10^9 m^3)			0.5488		2.0568	4.8137	
Dissolved gas (10^9 m^3)			0.0		0.0490	0.2042	
Total gas (10^9 m^3)			0.5875		2.1058	4.8846	

based on data from producing pools, taking into account expected depths.

Play #8: Basal Turbidite Drape Structures

The basal turbidite drape structure play (Table 8) consists of structural traps involving turbidite sands of the Szolnok Formation. Such traps are usually interpreted as compaction structures and can occur throughout the basin over basement highs. The reservoir turbidite sands are interbedded with shales which form the seals. Likely hydrocarbon sources are the underlying basal clayey marls, the pre-Pannonian Miocene rocks, and also the interbedded turbidite shales themselves. Discoveries in this play include 32 oil and gas pools in 2 fields.

All play attributes are inferred to be favorable because of previous production in the play. Individual prospect risk is considered moderate, with a fairly even distribution of risk among the trapping mechanism, effective porosity, and hydrocarbon accumulation factors.

Hydrocarbon volume parameters were estimated primarily by analogy with those of discovered accumulations, although with modifications, because the undiscovered accumulations are expected to occur at a greater average depth than the discovered accumulations. Some information pertaining to numbers, areas, and depths of prospects was obtained from seismic data for identified, but as yet undrilled, structures. Structures deeper than 3500 m were not assessed because of the likelihood of insufficient porosity below that depth.

Standard values for oil floor depth and percent oil recovery were used. Standard pressure and temperature relations were applied. The gas-oil ratio, oil formation volume factor, and gas compressibility factor were estimated based on data from the discovered accumulations.

Table 8. Play #8 Basal turbidite drape structures.

INPUT							
Play attributes		Hydrocarbon source			1.0		
		Timing			1.0		
		Migration			1.0		
		Potential reservoir facies			1.0		
		Marginal play probability			1.0		
Prospect attributes		Trapping mechanism			0.75		
		Effective porosity (>1.5)			0.75		
		Hydrocarbon accumulation			0.8		
		Conditional deposit probability			0.45		
Hydrocarbon mix		Gas			0.9		
		Oil			0.1		
Geologic variables	F_{100}	F_{95}	F_{75}	F_{50}	F_{25}	F_{05}	F_{0}
Closure (km^2)	0.75	1.25	2.25	3.0	4.0	5.25	6.0
Thickness (m)	5.0	8.0	14.0	20.0	30.0	50.0	150.0
Porosity (%)	5.0	6.0	8.0	10.0	13.0	17.0	20.0
Depth (m)	1800	2000	2400	2700	3000	3300	3500
HC saturation (%)	40	42	46	50	56	64	70
Number of prospects	13	14	16	17	18	19	20
Oil floor		3100 meters					
Recovery factors		Oil			30 %		
		Free gas			60%		
Reservoirs pressure (bars)		= 0.10 (depth) + 1.0		(from 0 to 2400 m)			
Reservoirs pressure (bars)		= 0.15 (depth) + 1.0		(below 2400 m)			
Reservoir temperature (°K)		= 0.05 (depth) + 284.15					
Gas-oil ratio (m^3/ton)		= 200					
Oil formation volume factor		= 1.5					
Gas compressibility factor		= 0.92					
RESULTS							
		Low (F_{95})		Mean		High (F_{05})	
Oil (millions of metric tons)		0.0		0.4773		2.1067	
Non-associated gas (10^9 m^3)		1.8811		5.1288		10.615	
Dissolved gas (10^9 m^3)		0.0		0.0955		0.4213	
Total gas (10^9 m^3)		1.9538		5.2243		10.705	

The gas recovery factor was adjusted because the CO_2 content of the gas is estimated to be 15%.

Play #9: Fans and Slumps

The fan and slump play (Table 9) consists of stratigraphic prospects at and near the base of the delta slope and overlying the basal turbidite sequences. These prospects are located throughout the basin and consist of sandstone deposits sealed by clayey marls deposited in a slope environment. Seismic records indicate that traps in this play may be associated with mound-shaped structures which represent either submarine fan or slump deposits. Source rocks are underlying organic-rich rocks deposited in a slope environment. No discoveries in this play have been made in the Békés basin.

In spite of the lack of current production, it was estimated that the play has at least one accumulation somewhere in the basin. However, individual prospect risk was considered very high. Trapping risk is relatively high because of the uncertainty concerning closure and seals. Also, the likelihood of poor reservoir conditions in the slump deposits contributed to the high risk estimation.

Hydrocarbon volume factors and number of drillable prospects were estimated using seismic data to identify possible fans and slumps, and analogy with similar accumulations outside the basin. Standard engineering factors were used.

Play #10: Delta Slope

The delta slope play (Table 10) consists of stratigraphic and combination structural-stratigraphic traps within the delta slope sequence (Algyö Formation). This sequence consists of an obliquely-bedded succession of submarine channel sandstones, shales, and marls. The reservoirs are believed to be the channel sandstones sealed by slope shales. Trapping can be enhanced where the channel sands drape over a structural high. Possible delta-slope traps exist throughout the basin. Likely source rocks for oil and gas are the surrounding shales, supplemented by vertical migration from deeper overpressured basal clayey-marl source beds. The Békés basin contains 32 known delta slope pools in 7 fields.

The play attributes are considered favorable because previous discoveries exist. Prospect risk is high, however, because of the low probability that updip seals are present. The relatively shallow depth of prospects, however, suggests that reservoir quality will likely be good.

Analog data from discovered accumulations were used to estimate hydrocarbon volume factors. Also, seismic data were used to estimate depths and number of drillable prospects. Standard engineering values were used except for the gas-oil ratio and the oil formation volume factor, both of which were reduced as a function of differences in gas content of oil in the discovered shallow reservoirs.

Play #11: Delta Front

The delta front play (Table 11) consists of sandstone bodies which accumulated along the shelf break between the delta plain and delta slope environments. traps are either stratigraphic or combination structural-stratigraphic, with most of the undrilled prospects probably stratigraphic. Seals are the flanking and overlying clayey marls of the delta plain. Delta-front prospects exist throughout the basin. Source rocks are most likely interbedded lignites, and biogenic gas is likely to be present. However, vertical migration from deep, overpressured source beds may have contributed thermally generated

Table 9. Play #9 Fans and slumps.

INPUT							
Play attributes		Hydrocarbon source			1.0		
		Timing			1.0		
		Migration			1.0		
		Potential reservoir facies			1.0		
		Marginal play probability			1.0		
Prospect attributes		Trapping mechanism			0.2		
		Effective porosity (>1.5)			0.5		
		Hydrocarbon accumulation			0.8		
		Conditional deposit probability			0.08		
Hydrocarbon mix		Gas			0.8		
		Oil			0.2		
Geologic variables	F_{100}	F_{95}	F_{75}	F_{50}	F_{25}	F_{05}	F_0
Closure (km^2)	0.75	1.5	2.5	3.5	5.0	7.0	9.0
Thickness (m)	5.0	10.0	12.5	25.0	40.0	75.0	150.0
Porosity (%)	5.0	6.0	8.0	10.0	13.0	17.0	20.0
Depth (m)	1500	1600	1800	2000	2300	2700	3000
HC saturation (%)	40	42	46	50	56	64	70
Number of prospects	18	20	25	30	36	44	50
Oil floor		3100 meters					
Recovery factors		Oil			30 %		
		Free gas			70%		
Reservoirs pressure (bars)	= 0.10 (depth) + 1.0			(from 0 to 2400 m)			
Reservoirs pressure (bars)	= 0.15 (depth) + 1.0			(below 2400 m)			
Reservoir temperature (^{0}K)	= 0.05 (depth) + 284.15						
Gas-oil ratio (m^3/ton)	= 200						
Oil formation volume factor	= 1.5						
Gas compressibility factor	= 0.90						

RESULTS	Low (F_{95})	Mean	High (F_{05})
Oil (millions of metric tons)	0.0	0.5966	2.8688
Non-associated gas (10^9 m^3)	0.0	1.8551	5.6997
Dissolved gas (10^9 m^3)	0.0	0.1193	0.5738
Total gas (10^9 m^3)	0.0	1.9744	5.8434

Table 10. Play #10 Delta slopes.

INPUT							
Play attributes		Hydrocarbon source			1.0		
		Timing			1.0		
		Migration			1.0		
		Potential reservoir facies			1.0		
		Marginal play probability			1.0		
Prospect attributes		Trapping mechanism			0.3		
		Effective porosity (>1.5)			0.8		
		Hydrocarbon accumulation			0.5		
		Conditional deposit probability			0.12		
Hydrocarbon mix		Gas			0.7		
		Oil			0.3		
Geologic variables	F_{100}	F_{95}	F_{75}	F_{50}	F_{25}	F_{05}	F_{0}
Closure (km^2)	0.75	1.0	1.5	2.0	3.0	8.0	18.0
Thickness (m)	5.0	6.0	8.0	10.0	14.0	30.0	50.0
Porosity (%)	5.0	8.0	12.0	15.0	20.0	26.0	30.0
Depth (m)	700	1100	1500	1800	2000	2200	2400
HC saturation (%)	30	34	40	45	54	68	75
Number of prospects	22	26	34	40	48	64	80
Oil floor		3100 meters					
Recovery factors		Oil			30 %		
		Free gas			70%		
Reservoirs pressure (bars)	= 0.10 (depth) + 1.0 (from 0 to 200 m)						
Reservoir temperature (^0K)	= 0.05 (depth) + 284.15 (below 2400 m)						
Gas-oil ratio (m^3/ton)	= 110						
Oil formation volume factor	= 1.2						
Gas compressibility factor	= 0.90						

RESULTS	Low (F_{95})	Mean	High (F_{05})
Oil (millions of metric tons)	0.0	1.0905	3.8787
Non-associated gas (10^9 m^3)	0.1372	1.2620	3.6087
Dissolved gas (10^9 m^3)	0.0	0.1200	0.4267
Total gas (10^9 m^3)	0.2449	1.3820	3.7951

gas and oil. In the Békés basin, 32 pools have been discovered in 4 fields.

Previous discoveries show that the play attributes are favorable. Prospect risk is moderately high because the remaining prospects are mainly stratigraphic and the amount of hydrocarbon fill in the traps is uncertain because of the lack of information on rates of biogenic gas generation. However, reservoir quality is expected to be excellent.

The analog data from the discovered pools were modified to take into account likely greater depth of the prospects. These analogues and also seismic data were used to estimate the hydrocarbon volume parameters and the number of drillable prospects. Standard engineering factors were used except for the gas-oil ratio, the oil formation volume factor, and the gas compressibility factor, which were estimated using data from discovered pools.

Play #12: Upper Pannonian Fluvial and Low-Relief Structures

The upper Pannonian fluvial and low-relief structure play (Table 12) consists of low relief structures and stratigraphic traps in the delta-plain sequence. Stratigraphic traps in fluvial sandstone pinchouts and combination traps containing fluvial sandstones on structures are both included. The fluvial sandstones may be channel or crevasse-splay deposits. Seals are the floodplain shales. Possible traps exist throughout the basin. Source rocks are the associated coals and lignites, and only gas (probably biogenic) is likely to be present. A total of 80 pools have already been discovered within 8 fields.

Table 11. Play #11 Delta front.

INPUT							
Play attributes		Hydrocarbon source			1.0		
		Timing			1.0		
		Migration			1.0		
		Potential reservoir facies			1.0		
		Marginal play probability			1.0		
Prospect attributes		Trapping mechanism			0.5		
		Effective porosity (>1.5)			1.0		
		Hydrocarbon accumulation			0.5		
		Conditional deposit probability			0.25		
Hydrocarbon mix		Gas			0.8		
		Oil			0.2		
Geologic variables	F_{100}	F_{95}	F_{75}	F_{50}	F_{25}	F_{05}	F_0
Closure (km^2)	0.75	1.0	1.5	2.0	2.75	4.0	5.0
Thickness (m)	5.0	7.0	10.0	13.0	18.0	30.0	40.0
Porosity (%)	15.0	17.0	20.0	22.0	25.0	30.0	35.0
Depth (m)	600	800	1200	1500	1700	1900	2100
HC saturation (%)	50	56	65	70	74	78	80
Number of prospects	17	20	26	30	36	44	50
Oil floor		3100 meters					
Recovery factors		Oil			30 %		
		Free gas			70%		
Reservoirs pressure (bars)	= 0.10 (depth) + 1.0						
Reservoir temperature (^{o}K)	= 0.05 (depth) + 284.15						
Gas-oil ratio (m^3/ton)	= 80						
Oil formation volume factor	= 1.1						
Gas compressibility factor	= 0.93						

RESULTS	Low (F_{95})	Mean	High (F_{05})
Oil (millions of metric tons)	0.0	1.9035	5.4889
Non-associated gas (10^9 m^3)	1.0954	2.8911	5.8618
Dissolved gas (10^9 m^3)	0.0	0.1523	0.4391
Total gas (10^9 m^3)	1.2043	3.0433	6.0466

Play attributes are favorable, as shown by the existing production. Prospect risk is moderate, mainly due to the hydrocarbon accumulation factor, which is moderate because of the uncertainties about the rates of biogenic gas generation. Good porosity in prospects is expected because depths are shallow. Trapping risk is considered to be moderately low because most of the remaining traps are likely to be combination traps, rather than stratigraphic traps.

Analog data from the discovered pools were slightly modified due to an expected greater average depth of undiscovered accumulations. Area of closure was also increased to take into account the larger possible size of some purely stratigraphic traps. The gas compressibility factor was estimated from data pertaining to discovered pools in the play, otherwise the standard engineering factors were used.

RESULTS

Estimates of the total undiscovered recoverable quantities of oil and gas in the Békés basin are summarized in Table 13. The mean estimate for undiscovered oil is 5.22 million metric tons and the mean estimate for undiscovered gas is 18.05x10^9 m^3. These, as well as all other resource estimates in the tables, are of recoverable quantities.

Through 1985, a total of 3.81 million metric tons of oil and 25.40x10^9 m^3 of gas were discovered in the Békés basin. Thus, the mean estimates suggest that approximately 42 percent of the recoverable oil and 58 percent of the recoverable natural gas have been discovered.

Fractiles of the probability distributions are also given in the tables. These are "more than" fractiles; hence, when the F_{05} (5% fractile) for oil is cited as 17.419

Table 12 Play #12 Upper Pannonian fluvial and low-relief structures.

INPUT							
Play attributes		Hydrocarbon source			1.0		
		Timing			1.0		
		Migration			1.0		
		Potential reservoir facies			1.0		
		Marginal play probability			1.0		
Prospect attributes		Trapping mechanism			0.8		
		Effective porosity (>1.5)			1.0		
		Hydrocarbon accumulation			0.5		
		Conditional deposit probability			0.4		
Hydrocarbon mix		Gas			1.0		
		Oil			0.0		
Geologic variables	F_{100}	F_{95}	F_{75}	F_{50}	F_{25}	F_{05}	F_0
Closure (km^2)	0.75	1.0	1.25	1.5	2.0	3.0	4.0
Thickness (m)	5.0	6.0	8.0	10.0	14.0	22.0	30.0
Porosity (%)	13.0	18.0	22.0	25.0	27.0	30.0	35.0
Depth (m)	400	550	800	1000	1200	1500	1800
HC saturation (%)	50	56	65	70	74	78	80
Number of prospects	10	12	16	20	26	38	50
Oil floor		3100 meters					
Recovery factors		Oil			30 %		
		Free gas			70%		

Reservoirs pressure (bars)	= 0.10 (depth) + 1.0	(from 0 to 2400 m)
Reservoirs pressure (bar)	= 0.15 (depth) + 1.0	(below 2400 m)
Reservoir temperature (^oK)	= 0.05 (depth) + 284.15	
Gas-oil ratio (m^3/ton)	= 200	
Oil formation volume factor	= 1.5	
Gas compressibility factor	= 0.92	

RESULTS			
	Low (F_{95})	Mean	High (F_{05})
Oil (millions of metric tons)	0.0	0.0	0.0
Non-associated gas (10^9 m^3)	0.7739	2.0047	4.0375
Dissolved gas (10^9 m^3)	0.0	0.0	0.0
Total gas (10^9 m^3)	0.7739	2.0047	4.0375

million metric tons, that means that there is only a 5% chance of more than 17.419 million metric tons.

Note that there are two aggregations in Table 13. The first assumes a degree of dependency of 0.0, in other words the plays were considered independent of one another. The second aggregation assumes a degree of dependency of 1.0, in other words the plays were considered completely dependent on one another. The assumption of dependency leads to a larger range in uncertainty for the aggregate distribution. In actuality the plays are partially dependent, therefore, the range should be between these two extreme cases.

The individual plays vary greatly in their contribution to the total amount of undiscovered resource. For oil, the delta front play has the highest mean potential, 1.90 million metric tons or 36 percent of the total mean oil potential. For natural gas, the basal turbidite drape structure play has the highest mean potential, 5.22×10^9 m^3 or 29 percent of the total mean natural gas potential.

ACKNOWLEDGMENTS

Results of the play analyses are based on the geological, geophysical and geochemical analyses and interpretations of all authors this volume, who have participated in defining the plays and their principal attributes.

REFERENCES

Canada Department of Energy, Mines and Resources, 1977, Oil and natural gas resources of Canada, 1976; Canada Department of Energy, Mines and Resources Report EP 77-1, 76 p.

Table 13. Estimates of the total undiscovered recoverable quantities of oil and gas in the Békés basin.

TOTAL BÉKÉS BASIN			
	Low (F_{95})	Mean	High (F_{05})
Degree of dependency = 0.0			
Oil (millions of metric tons)	1.784	5.216	11.187
Non-associated gas (10^9 m^3)	10.905	17.376	25.844
Dissolved gas (10^9 m^3)	0.219	0.674	1.484
Total gas (10^9 m^3)	11.510	18.050	26.539
Degree of dependency = 1.0			
Oil (millions of metric tons)	0.0	5.216	17.419
Non-associsted gas (10^9 m^3)	4.894	17.376	40.788
Dissolved gas (10^9 m^3)	0.0	0.674	2.336
Total gas (10^9 m^3)	5.287	18.050	41.648

Crovelli, R.A., and Balay, R.H., 1986, FASP, an analytic resource appraisal program for petroleum play analysis; Computers and Geosciences, 12, (4B), 423-475.

Crovelli, R.A., and Balay, R.H. 1987, FASPUM metric version - Analytic petroleum resource appraisal microcomputer programs for play analysis using a reservoir-engineering model; U.S. Geological Survey Open-File Report 87-414A, 14 p. [U.S. Geological Survey Open-File Report 87-414B, diskette].

Crovelli, R.A., and Balay, R.H., Geologic model, probabilistic methodology and computer programs for petroleum resource assessment: this volume, 295-304.

Szentgyörgyi, K. and Teleki, P.G., Facies and depositional environments of Miocene sedimentary rocks; this volume, 83-97.

White, L. P., 1981, A play approach to hydrocarbon resource assessment and evaluation; in Ramsey, J. B., (ed.), The economics of exploration for energy resources; JAI Press, Greenwich, Connecticut, 51-67.

Index

E

F

G

N

O

R

S